An Introduction to Biomedical Optics

Series in Optics and Optoelectronics

Recent titles in the series

High-Speed Photonic Devices
Nadir Dagli

Lasers in the Preservation of Cultural Heritage: Principles and Applications
C Fotakis, D Anglos, V Zafiropulos, S Georgiou, V Tornari

Modeling Fluctuations in Scattered Waves
E Jakeman, K D Ridley

Fast Light, Slow Light and Left-Handed Light
P W Milonni

Diode Lasers
D Sands

Diffractional Optics of Millimetre Waves
I V Minin

Handbook of Electroluminescent Materials
D R Vij

Handbook of Moire Measurement
C A Walker

Next Generation Photovoltaics
A Martí

Stimulated Brillouin Scattering
M J Damzen

Laser Induced Damage of Optical Materials
R M Wood

Optical Applications of Liquid Crystals
L Vicari

Optical Fibre Devices
J P Goure

Applications of Silicon-Gremanium Heterostructure Devices
C K Maiti

Optical Transfer Function of Imaging Systems
T L Williams

Series in Optics and Optoelectronics

An Introduction to Biomedical Optics

R Splinter
Analytica Sciences, Inc.
Concord, North Carolina, USA

B A Hooper
Areté Associates
Arlington, Virginia, USA

Taylor & Francis
Taylor & Francis Group
New York London

Taylor & Francis is an imprint of the
Taylor & Francis Group, an informa business

CRC Press
Taylor & Francis Group
6000 Broken Sound Parkway NW, Suite 300
Boca Raton, FL 33487-2742

Printed in the United States of America on acid-free paper
10 9 8 7 6 5 4 3 2 1

International Standard Book Number-10: 0-7503-0938-5 (Hardcover)
International Standard Book Number-13: 978-0-7503-0938-7 (Hardcover)

Library of Congress Cataloging-in-Publication Data

Splinter, Robert.
An introduction to biomedical optics / Robert Splinter and Brett A. Hooper.
p. ; cm. -- (Series in optics and optoelectronics ; 3)
Includes bibliographical references and index.
ISBN-13: 978-0-7503-0938-7 (hardcover : alk. paper)
ISBN-10: 0-7503-0938-5 (hardcover : alk. paper)
1. Optoelectronic devices. 2. Biomedicine. 3. Biotechnology. 4. Optical fibers in medicine. I. Hooper, Brett A. II. Title. III. Series: Series in optics and optoelectronics (CRC Press) ; 3. [DNLM: 1. Biomedical Technology. 2. Light. 3. Biophysics--methods. 4. Optics. 5. Photobiology--methods. WB 117 S761i 2007]

R857.B54I584 2007
610.28--dc22 2006025589

Visit the Taylor & Francis Web site at
http://www.taylorandfrancis.com

and the CRC Press Web site at
http://www.crcpress.com

To: Jackie and Anne, Lauren, Christine, and Catherine

Preface

Biomedical Use of Light

The continuing expansion of the medical applications of light demands a solid foundation of all the theoretical concepts involved. The primary requirements for a good understanding of the interaction of light with biological media is the knowledge of all the optical parameters that determine the interaction process and when and where approximations are permitted.

The field of biomedical optics is also called biophotonics and medical optics depending on the background of the reporting organization or the focus of the target group. The number of publications concerning biomedical optics is growing at a steady pace and many institutions, both at the educational as well as at the professional level, are initiating biophotonics core groups and organizing dedicated facilities.

The key interests in the development of biomedical optics are the persistent search for faster, better, and cheaper approaches to medical procedures in addition to the potential for noninvasive, nonionizing diagnostic modalities that are highly accurate and reproducible.

Especially with the current developments in nanotechnology, the combination of size and speed offered by biophotonic technology provides a unique window of opportunities. The most attractive opening for biomedical optics is in early warning systems based on the development of various types of diagnostic devices.

The use of optics in medicine has a long history, but only recently has started to take full advantage of all the potential capabilities of the interaction that electromagnetic radiation can establish with biological entities.

Some examples of developments in biomedical optics are nano surgery, miniaturization, and advanced imaging devices. In nano surgery, the combination of small-scale devices and light offers the opportunity to manipulate biological media on a cellular and even molecular level. Additionally, the miniaturization of optical devices has resulted in miniature endoscopes for fetal and minimally invasive surgical procedures. The advanced optical devices developed over the past decade have moved imaging beyond the theoretically established diffraction limits of classical optics theory. There are even nano-scale lasers that can be incoporated with biological systems to perform continuous real-time physiologic feedback. Another level of the impact of biomedical optics is the coming of age of light-induced cancer treatment as well.

Since the invention of the laser in 1960, several biophotonics applications have been dramatically improved while a wide range of new uses of light were discovered and developed.

All properties of electromagnetic radiation are capable of delivering certain specific diagnostic applications, while the therapeutic applications are less stringent in their demand for basic theoretical support. Nonetheless all electromagnetic characteristics need to be taken into consideration to optimize all the particulars of the light interaction course of action. Several aspects of biophotonics development were "borrowed" from other disciplines, frequently not directly medical related. The close collaboration between different medical disciplines is found in the development of optical coherence tomography, which adopted the process from ultrasound. Other imaging techniques that derived their inspiration from ultrasound are diffuse light tomography and photoacoustic imaging. However, diffuse optical tomography can also be compared with an imaging technique that uses a completely different part of the electromagnetic spectrum: x-ray radiography. One example of the crossover between medicine and technology is the imaging of the index of refraction dispersion, which was a common method for decades to perform quality control on predominantly transparent mechanical structures, and was transformed in (so-far limited) medical use in the Schlieren imaging technique.

The main difference between ultrasound and the newly developed optical imaging techniques is that the majority of the theoretical approximations in the technical description of the mechanical ultrasound interaction process are not allowed for the electromagnetic fields and actually provide a practical advantage in resolution and contrast obtained by the optical techniques.

About the Book

With many universities starting new biophotonics courses and even founding entire biomedical optics centers, this book will provide the fundamental theoretical background for the aspiring physics, biology, electrical engineering, and medical student to learn about several biomedical optics issues. Additionally, the corporate biophotonics divisions being embarked upon by various large and mid-sized companies as well as many new biomedical optics-based start-up companies will benefit from the comprehensive coverage of a wide range of theoretical and practical issues involved with the use of light for diagnostic and therapeutic applications. This book offers a solid reference manual for the inquisitive novice in the field of biomedical optics.

The book is generally divided into three categories: biophysical, biochemical, and biological applications of optical technology for therapeutic as well as diagnostic intentions. The physical fundamental optics theory is subdivided into fundamental electromagnetic wave philosophy, classical optics culminating in the definition of theory, and parameters involved in the actual light–tissue interaction process.

Several chapters can be approached without in-depth knowledge of the preceding chapters, but will rely heavily on the various concepts and parameters that are introduced in the respective chapters.

Brief Description of the Chapters

The book is subdivided into three main sections, which are general optics theory, therapeutic applications of light, and diagnostic optical methods. Chapter 1 gives a brief historical overview of the optical developments over the centuries and how they have influenced the design of present-day biomedical optics use. Chapters 2 through 4 give an in-depth review of optical theory. Chapter 2 discusses electromagnetic wave theory and laser operation, and establishes the basis of optics terminology. Chapter 3 reviews the classical optics theory before discussing the core area in absorption and scattering applied in biomedical optics in Chapter 4. The light and tissue variables applicable to biomedical optics are outlined in Chapter 5. A more detailed discussion of the theoretical approximations pertaining to light–tissue interaction concepts are covered in Chapter 6. Additional theoretical review of the practical approach on how to determine the actual light–tissue interaction is described in Chapter 7. In Chapters 8 through 10, the three distinct interaction principles of photophysical, photochemical, and photobiological mechanisms are discussed, respectively. In Chapters 11 through 13, the same differentiation is applied to the description of selected therapeutic applications of light, while this subdivision is repeated for the diagnostic medical use of light in Chapters 14 through 16, respectively.

Acknowledgments

We both would like to thank our parents for their inspiration and loving support.

The authors acknowledge the help of various students and colleagues who made this book possible by their discussions and suggestions. The authors also thank all the hospitals, colleagues, corporations, clinics, and all other individuals who contributed their biomedical optics work for publication in this book. All the images that are obtained from outside sources are referenced in the text, and we also thank all the people who inspired us in our work and illustrations but are not mentioned.

Lastly, we would like to mention the influence and inspiration of several mentors who helped shaped our view of photomedicine: Martin van Gemert, Willem Star, A.J. Welch, R. Rox Anderson, Richard Straight, John M. J. Madey, Robert H. Svenson and Olaf von Ramm. It has been an honor to work with these men.

Authors

Robert Splinter received his master's degree from the University of Technology in Eindhoven, The Netherlands and his doctoral degree from the University of Amsterdam in The Netherlands. Dr. Splinter has worked in biomedical imaging and optical diagnostics for over 20 years. He has been affiliated with the University of North Carolina in Charlotte for more than 15 years as an adjunct professor with the Department of Physics and has held a research scientist position with the local Carolinas HealthCare System hospital affiliate and research facility for over 16 years.

Dr. Splinter is vice-president and a founding partner of a radiation technology company. He has cofounded various other biomedical start-up companies and has served as Chief Technology Officer.

Dr. Splinter is a consultant for the medical device industry and has served on the NIH grant review board. He has published over 90 papers and book chapters and is the coauthor of a book on biomedical signal and image processing. He has served on the graduate committees of the University of Alabama and the University of Tennessee. He also is the co-owner of several U.S. and international patents pertaining to biomedical diagnostic methodologies.

Brett A. Hooper received his bachelor of science degree from the University of Utah and his Ph.D. in physics from Duke University. Dr. Hooper has worked in the field of biomedical optics for over 15 years and has served on the faculty of both Wellman Laboratories of Photomedicine at Harvard Medical School and the Department of Biomedical Engineering at Duke University. While at Wellman Laboratories, he demonstrated the first use of evanescent optical waves for use in surgery and medicine. At Duke University, he led a multidisciplinary research group investigating the interaction of midinfrared light with biological systems, created and taught the department's first biomedical optics course, and served on several graduate committees.

Currently, Dr. Hooper works for a firm that specializes in remote sensing of the near ocean. He is leading a team to develop a new remote sensing capability utilizing a multispectral, polarimetric sensor system. Dr. Hooper continues to be a consultant and peer reviewer for the biomedical optics field.

Contents

Section I General Biomedical Optics Theory

1 Introduction to the Use of Light for Diagnostic and Therapeutic Modalities 3

1.1 What Is Biomedical Optics? 3
1.2 Biomedical Optics Timeline 4
1.2.1 Elementary Optical Discoveries 4
1.2.2 Development of Optical Devices 6
1.2.3 Scientific Advancements in Optics Theory 9
1.3 Historical Events in Therapeutic and Diagnostic Use of Light 14
1.3.1 Development of Therapeutic Applications of Light in Medicine 14
1.3.1.1 Development of Diagnostic Optical Applications 17
1.4 Light Sources 18
1.5 Current State of the Art 19
1.6 Summary 20

2 Review of Optical Principles: Fundamental Electromagnetic Theory and Description of Light Sources 23

2.1 Definitions in Optics 23
2.2 Kirchhoff's Laws of Radiation 27
2.2.1 Planck Function for BlackBody Radiation 28
2.3 Electromagnetic Wave Theory 30
2.3.1 Gauss's Law 30
2.3.2 Faraday's Law 32
2.3.3 Maxwell Equations 33
2.3.4 Energy and Momentum of Electromagnetic Waves 35
2.3.5 Coherence of Electromagnetic Waves 36
2.3.5.1 Temporal Coherence 36
2.3.5.2 Spatial Coherence 37
2.3.6 Interference of Electromagnetic Waves 37
2.3.7 Phase Velocity 38
2.3.8 Group Velocity 38
2.3.9 Cauchy Theorem 39

2.3.10 Electromagnetic Wave Spectrum 42
2.3.10.1 Television and Radio Waves 42
2.3.10.2 Microwaves Waves 42
2.3.10.3 Infrared Waves 42
2.3.10.4 Visible Light Waves 42
2.3.10.5 Ultraviolet Waves 43
2.3.10.6 X-Ray Waves 43
2.3.10.7 Gamma Radiation Waves and Cosmic Rays 43
2.4 Light Sources 44
2.4.1 Broadband Light Sources 45
2.4.2 Laser Operation 46
2.4.2.1 Einstein Population Inversion 47
2.4.2.2 Achieving a Metastable State 50
2.4.2.3 Harnessing the Photons in the Laser Medium 51
2.4.2.4 Amplification 52
2.4.3 Laser Light Sources 53
2.4.3.1 Chemical Laser 53
2.4.3.2 Diode Laser or Semiconductor Laser 54
2.4.3.3 Dye Laser 55
2.4.3.4 Gas Laser 56
2.4.3.5 Solid-State Lasers 57
2.4.3.6 Free-Electron Laser (FEL) 58
2.5 Applications of Various Lasers 61
2.6 Summary 63

3 Review of Optical Principles: Classical Optics 69

3.1 Geometrical Optics 69
3.1.1 Huygens' Principle 70
3.1.2 Laws of Geometrical Optics 70
3.1.2.1 Law of Rectilinear Propagation 71
3.1.2.2 Law of Reflection 71
3.1.2.3 Law of Refraction (Snell's Law) 72
3.1.2.4 Law of Diffraction 73
3.1.2.5 Law of Conservation of Energy 78
3.1.3 Fermat's Principle 78
3.1.4 Ray Optics 78
3.1.4.1 Critical Angle 80
3.1.4.2 Brewster Angle of Polarization 80
3.1.4.3 Lens Makers' Equation 82
3.1.4.4 Optical Instruments 84
3.2 Other Optical Principles 85
3.2.1 Dispersion 85
3.3 Quantum Physics 85
3.3.1 Schrödinger Equation 86
3.4 Gaussian Optics 87

3.4.1 Matrix Methods in Gaussian Optics 88
3.4.2 Fourier Optics 90
3.5 Summary 91

4 Review of Optical Interaction Properties 95

4.1 Absorption and Scattering 95
4.1.1 Light and Atom Interaction Overview 95
4.1.2 Absorption 97
4.1.2.1 Classical Theory of Absorption 97
4.1.2.2 Lorentz Model 98
4.1.2.3 Beer–Lambert–Bouguer Law of Absorption 99
4.1.3 Scattering 99
4.1.3.1 Rayleigh Scattering 99
4.1.4 Mie Scattering 101
4.1.4.1 Scattering Phase Function 101
4.1.5 Raman Scattering 101
4.1.5.1 Simple Atom and Molecule Models 102
4.1.5.2 Material Polarization 109
4.1.5.3 Biochemistry Applications 114
4.2 Doppler Effect 115
4.3 Summary 118

5 Light–Tissue Interaction Variables 121

5.1 Laser Variables 121
5.1.1 Laser Power 121
5.1.2 Light Delivery Protocol 121
5.1.3 Power Density Profile of the Beam 125
5.1.4 Gaussian Beam Profile 126
5.1.5 Top-Hat Beam Profile 127
5.1.6 Irradiation Spot Size 128
5.1.7 Power Density and Energy Density of a Light Source 129
5.1.8 Local Beam Angle of Incidence with the Tissue 130
5.1.9 Collimated or Diffuse Irradiation 130
5.1.10 Light Wavelength 132
5.1.11 Repetition Rate 132
5.1.12 Pulse Length 132
5.1.13 Light Delivery Pulse Modulation 132
5.2 Tissue Variables 133
5.2.1 Optical Properties of the Tissue 133
5.2.1.1 Tissue Index of Refraction 134
5.2.2 Surface Contour 135
5.2.3 Tissue Temperature 135
5.2.4 Thermodynamic Tissue Properties 135
5.2.5 Tissue Blood Flow and Blood Content 136
5.3 Light Transportation Theory 137

5.3.1 The Time-Dependent Angular and Spatial Photon Energy Rate Distribution 138
5.3.2 Steady-State Angular and Spatial Photon Energy Rate Distribution 140
5.3.3 Boundary Conditions 140
5.3.4 Radiation Definitions for Turbid Media 140
5.3.4.1 Photon Power Related to Radiance 140
5.3.4.2 Radiance Incident on a Volume (Fluence Rate) 141
5.3.4.3 Photon Power Density in Medium 141
5.3.4.4 Photon Radiative Flux 141
5.3.4.5 Photon Radiation Pressure 141
5.3.4.6 Penetration Depth 142
5.4 Light Propagation under Dominant Absorption 142
5.4.1 Description of Angular Distribution of Scatter 142
5.5 Summary 146
5.6 Nomenclature 146

6 Light–Tissue Interaction Theory 155

6.1 Approximations of the Equation of Radiative Transport 155
6.1.1 Spherical-Harmonics Substitution Method to Solve the Equation of Radiative Transport 155
6.1.1.1 Taylor Expansion of Radiance 156
6.1.2 Diffusion Approximation of the Equation of Radiative Transfer 158
6.1.2.1 Isotropic Source Solution 166
6.1.3 Discrete-Ordinate Method for Solving the Equation of Radiative Transport 166
6.1.3.1 Two-Flux Theory 166
6.1.3.2 Three-Flux Theory 173
6.1.3.3 Special Situation for Infinite Slab 176
6.1.4 Light Distribution under Infinite Wide Beam Irradiation 177
6.1.5 Numerical Method to Solve the Equation of Radiative Transport 177
6.2 Summary 177

7 Numerical and Deterministic Methods in Light–Tissue Interaction Theory 181

7.1 Numerical Method to Solve the Equation of Radiative Transport 181
7.1.1 Monte Carlo Simulation 182
7.1.1.1 Simulation Process 182
7.1.1.2 Light–Tissue Interaction Model 186
7.1.1.3 Random Walk Model 187
7.1.1.4 Light Propagation Model Components 190
7.2 Measurement of Optical Parameters 195
7.2.1 Invasive Techniques 197

7.2.1.1 Integrating Sphere Measurements 197
7.2.1.2 Fiberoptic Probing to Determine the Optical Properties 201
7.2.2 Noninvasive Techniques 203
7.2.2.1 Specular Reflectance Measurement 203
7.2.2.2 Diffuse Backscatter Measurement by CCD Camera Imaging 203
7.2.2.3 Diffuse Backscatter by Fiberoptic Area Measurement 206
7.2.2.4 Diffuse Backscatter Measurement by Interferometric Probing 207
7.2.2.5 Diffuse Forward Remittance Measurement 207
7.2.2.6 Diffuse Forward Scatter Measurement by CCD Camera Imaging 207
7.2.2.7 Time-Resolved Reflectance and Transmittance Measurement 208
7.3 Temperature Effects of Light–Tissue Interaction 211
7.3.1 The Bioheat Equation 212
7.3.2 Computer Modeling of Absorption, Heat Dissipation, and Photocoagulation 213
7.3.3 Damage Integral 213
7.3.4 The Damage State 214
7.4 Summary 214

8 Light–Tissue Interaction Mechanisms and Applications: Photophysical 219

8.1 Range of Photophysical Mechanisms 219
8.2 Photoablation 220
8.2.1 Ablation Threshold 221
8.2.2 Pulsed Laser–Tissue Interaction 222
8.2.3 Pulsed Vaporization 223
8.2.4 Definition of a Pulsed Laser 227
8.2.5 Ultraviolet Laser Ablation 228
8.2.5.1 Photochemical Breakdown 228
8.2.5.2 Rate of Bond-Breaking 230
8.2.5.3 Bond-Breaking Depth 231
8.2.5.4 Gas Production during Photoablation 232
8.3 Photoacoustics 233
8.3.1 History of the Photoacoustic Effect 233
8.3.2 The Photoacoustic Effect 234
8.3.2.1 Displacement 236
8.3.2.2 Sound Propagation 238
8.3.3 Detection of the Photoacoustic Signal 241
8.3.4 Theory of Photoacoustic Wave Propagation 241
8.3.5 Electromechanical Effects during Photoacoustic Interaction 242

8.4 Birefringence Effects 243
8.4.1 Schlieren Imaging 244
8.4.2 Conventional Töepler Schlieren Configuration 248
8.4.3 Ronchi Ruling Focusing Schlieren Configuration 249
8.5 Polarization Effects 250
8.5.1 Polarization in Nature 251
8.5.2 Polarization in Medical Imaging 252
8.6 Optical Activity 253
8.6.1 Glucose Concentration Determination 256
8.7 Evanescent Wave Interaction in Biomedical Optics 257
8.7.1 Evanescent Optical Waves 257
8.7.2 Precise, Controlled Light Delivery with Evanescent Optical Waves 261
8.7.3 Tissue Ablation with FEL-Generated Evanescent Optical Waves 264
8.8 Phase Interference Effects 268
8.8.1 Interferometry in Medical Imaging 268
8.9 Spectroscopy 269
8.9.1 Light Scattering Spectroscopy (LSS) 269
8.9.2 Fourier Transform Infrared (FTIR) Spectroscopy 270
8.9.3 Ultrafast Spectroscopy 271
8.9.4 Time-Resolved Spectroscopy 271
8.9.5 Raman Scattering Spectroscopy 271
8.9.6 Coherent Anti-Stokes Raman Spectroscopy 279
8.9.7 Time-Resolved Raman Spectroscopy 279
8.9.8 Raman Spectroscopy Advantages and Disadvantages 280
8.10 Endoscopy 281
8.10.1 Light Delivery with Optical Fibers 282
8.10.2 Medical Applications Using Endoscopy 282
8.10.2.1 Arthroscopy 282
8.10.2.2 Bronchoscopy 283
8.10.2.3 Cardiology 283
8.10.2.4 Cystoscopy 283
8.10.2.5 Fetoscopy 283
8.10.2.6 Gastrointestinal Endoscopy 283
8.10.2.7 Laparoscopy 284
8.10.2.8 Neuroendoscopy 284
8.10.2.9 Otolaryngology 284
8.11 Summary 284

9 Light–Tissue Interaction Mechanisms and Applications: Photochemical 289

9.1 Basic Photochemical Principles 289
9.1.1 Photosynthesis 290
9.1.2 Sun Tanning 291
9.1.3 Light-Reactive Biological Chromophores 291

9.2 Photochemical Effects 291
9.2.1 Photodynamic Therapy 292
9.2.1.1 The Photosensitizer Excitation Process 292
9.2.1.2 The Role of Oxygen 294
9.2.1.3 Mechanisms in PDT to Induce Cell Death 295
9.2.1.4 Light Delivery 296
9.2.1.5 Photosensitizers 297
9.2.1.6 Dosimetry 300
9.2.1.7 Clinical Applications 305
9.2.1.8 Light Sources 310
9.2.1.9 Advantages of PDT 311
9.2.1.10 Disadvantages and Limitations 312
9.2.2 Wound Healing 312
9.3 Summary 314

10 Light–Tissue Interaction Mechanisms and Applications: Photobiological 317

10.1 Photobiological Biostimulation 317
10.2 Photobiological Effects 320
10.2.1 Photothermal 321
10.2.1.1 Photocoagulation 321
10.2.1.2 Reversible Photocoagulation 322
10.2.1.3 Irreversible Photocoagulation 322
10.2.1.4 Damage Integral 325
10.2.2 Medical Applications of the Photothermal Effect 327
10.2.2.1 Thermal Effects in Treatment of Port-Wine Stains 327
10.2.2.2 Tissue Closure and Welding 333
10.2.2.3 Current Tissue Closure Methods 335
10.2.3 Photobiological Nonthermal Interaction 335
10.3 Excitation of Chromophores 335
10.4 Optic Nerve-Cell Depolarization under the Influence of Light (Vision) 336
10.4.1 Quantum Photon Dots as Biological Fluorescent Markers 337
10.4.2 Optical Properties of Quantum Dots 337
10.5 Summary 339

Section II Therapeutic Applications of Light

11 Therapeutic Applications of Light: Photophysical 345

11.1 Delivery Considerations 346
11.2 Pulsed Laser Use in Cardiology 349
11.2.1 Pulsed Laser–Tissue Interaction 349
11.2.1.1 Ablation Rate 351
11.2.2 Laser Plaque Molding (Angioplasty) 354

11.2.3 Laser Thrombolysis 354
11.2.4 Laser Valvulotomy and Valve Debridement 354
11.2.5 Transmyocardial Revascularization 356
11.3 Dentistry and Oral Surgery 359
11.3.1 Photo Curing 359
11.3.2 Dental Drill 359
11.3.3 Etching 361
11.3.4 Tooth Hardening 362
11.3.5 Scaling 362
11.4 Ophthalmology 362
11.5 Optical Tweezers 364
11.5.1 Rayleigh Regime Particle Force 367
11.5.2 Mie Regime Particle Force 368
11.5.3 Size Region between the Rayleigh and Mie Regime 371
11.5.4 Applications of Optical Tweezers 373
11.6 Summary 373

12 Therapeutic Applications of Light: Photochemical 377

12.1 Vascular Welding 377
12.2 Cosmetic Surgery 378
12.2.1 Inflammatory Disease Lesion Treatment 380
12.2.2 Pigmented Lesion Treatment 382
12.3 Oncology 384
12.3.1 Photodynamic Therapy 384
12.4 Summary 388

13 Therapeutic Applications of Light: Photobiological 389

13.1 Cardiology and Cardiovascular Surgery 389
13.1.1 Arrhythmogenic Laser Applications 389
13.1.2 Laser Photocoagulation 391
13.1.3 Arrhythmic Node Ablation 401
13.1.4 Atrial Ablation 401
13.2 Soft Tissue Treatment 402
13.3 Dermatology 402
13.3.1 Vascular Lesion Treatment 403
13.4 Fetal Surgery 407
13.5 Gastroenterology 409
13.6 General Surgery 410
13.7 Gynecology 411
13.8 Neurosurgery 411
13.9 Ophthalmology 412
13.10 Pulmonology and Otorhinolaryngology 413
13.11 Otolaryngology, Ear, Nose and Throat (ENT), and Maxillofacial Surgery 414
13.12 Podiatry 416

13.13 Urology 416
13.13.1 Lasers in the Treatment of Benign Prostatic Hyperplasia 417
13.13.1.1 Nd:YAG for Prostate Tissue Ablation 419
13.13.1.2 Contact Tip Technology for Prostate Ablation 419
13.13.1.3 Free Beam KTP and KTP/Nd:YAG for Prostatectomy 419
13.13.1.4 Holmium:YAG for Prostate Tissue Ablation 419
13.13.1.5 Holmium Laser Enucleation of the Prostate—HoLEP 420
13.14 Summary 420

Section III Diagnostic Applications of Light

14 Diagnostic Methods Using Light: Photophysical 425
14.1 Optical Microscopy 425
14.1.1 Diffraction in the Far-Field 429
14.2 Various Microscopic Techniques 439
14.2.1 Confocal Microscopy 439
14.2.2 Multiphoton Imaging 442
14.2.2.1 Two-Photon Microscopy 442
14.2.2.2 Advantages of Two-Photon Microscopy 446
14.2.2.3 Limitations of Two-Photon Microscopy 447
14.3 Near-Field Scanning Optical Microscope 448
14.3.1 The Concept of the Near-Field Scanning Optical Microscope 449
14.3.2 General Design of the Near-Field Scanning Optical Microscope 449
14.3.3 Near-Field Scanning Optical Microscope Tip 450
14.3.3.1 Feedback Mechanisms Employed to Maintain a Constant Tip and Sample Separation 452
14.3.3.2 Shear-Force-Mode Tip Feedback 453
14.3.3.3 Tapping-Mode Tip Feedback 454
14.3.3.4 Intensity Imaging 455
14.3.3.5 Phase Imaging 456
14.4 Spectral Range Diagnostics 457
14.4.1 Lab-on-a-CHIP 459
14.4.2 Coherent X-Ray Imaging 460
14.4.2.1 Ultrafast X-Ray Pulses Reveal Atoms in Motion..460
14.4.2.2 Free Electron Laser Protein X-Ray Holography 461
14.5 Holographic Imaging 462
14.6 Polarization Imaging 463

14.7 Transillumination Imaging 464
14.7.1 Examination of the Male Genitalia of Infants 465
14.7.2 Transillumination for Detection of Pneumothorax in Premature Infants 465
14.7.2.1 Continuous Monitoring for Pneumothorax Detection 466
14.7.3 Transillumination of Infant Brain 467
14.7.4 Diffuse Optical Tomography 468
14.7.4.1 Theoretical Background of Diffuse Optical Tomography 469
14.7.4.2 Optical Heterodyning 472
14.8 Optical Coherence Tomography 476
14.8.1 Conventional Optical Coherence Tomography Systems 477
14.8.2 Light Sources and Coherence Length 479
14.8.3 Theory of Optical Coherence Tomography 482
14.8.4 Operation of the Fiberoptic Michelson Interferometer 483
14.8.5 Correlation Theory 487
14.8.6 The Effect of Scattering on the Visibility Function 492
14.8.7 Image Acquisition Process 494
14.8.8 Applications of Optical Coherence Tomography to Physical Problems 496
14.8.8.1 Optical Coherence Tomography in Dentistry 497
14.8.8.2 Polarization-Sensitive Optical Coherence Tomography 498
14.8.8.3 Phase-Resolved Optical Coherence Tomography 498
14.8.8.4 Spectroscopic Optical Coherence Tomography 499
14.8.8.5 Time-Domain Optical Coherence Tomography 500
14.8.8.6 Color-Doppler Time-Domain Optical Coherence Tomography 500
14.8.8.7 Spectral-Domain Optical Coherence Tomography or Fourier-Domain Optical Coherence Tomography 501
14.9 Ballistic Photon Imaging 503
14.10 Reflectometry 506
14.11 Evanescent Wave Imaging Applications 506
14.11.1 Evanescent Optical Wave Device Designs 507
14.12 Medical Thermography 509
14.13 Photoacoustic Imaging 512
14.13.1 Acoustic Wave 519
14.13.2 Medical Imaging Applications 520
14.13.2.1 Acoustooptical Imaging of Teeth 521

14.13.2.2 Photoacoustic Excitation and Wave Propagation 524
14.14 Terahertz Imaging 526
14.15 Summary 528

15 Diagnostic Methods Using Light: Photochemical 531

15.1 Fluorescence Imaging 531
15.1.1 Fluorescence Molecular Explanation 532
15.1.2 Fluorescent Molecules 535
15.1.3 Fading 536
15.2 Ratio Fluorescence Microscopy 537
15.2.1 Fluorescence Resonance Energy Transfer (FRET) Microscopy 537
15.2.2 Mechanism of Fluorescent Resonant Energy Transfer Imaging 538
15.2.3 Fluorescence Resonance Energy Transfer Pair 540
15.2.4 Problems with Fluorescence Resonance Energy Transfer Microscopy Imaging 541
15.3 Raman Spectroscopy with Near-Field Scanning Optical Microscopy (NSOM) Employed 541
15.3.1 Fluorescence Resonance Emission Transfer 542
15.3.2 Applications in Biology 543
15.4 Optical "Tongue" 543
15.4.1 Mechanism of Operation 544
15.4.2 Taste Stimuli Transduction 544
15.4.3 Taste Transduction Mechanisms 544
15.4.4 Taste Processes 545
15.4.5 Combinatorial Libraries 545
15.4.6 Charge-Coupled Device Detection 548
15.5 Summary 549

16 Diagnostic Methods Using Light: Photobiological 551

16.1 Immunostaining ("Functional Imaging") 551
16.2 Immunofluorescence 552
16.3 Diagnostic Applications of Spectroscopy 554
16.3.1 Detection of Dental Cavities and Caries 554
16.3.2 Optical Detection of Erythema 555
16.4 Fiberoptic Sensors 557
16.4.1 Biosensors 558
16.4.2 Fiberoptic Biosensor Design 559
16.4.2.1 Fiberoptic Fluorescence Sensors 559
16.4.2.2 Fiberoptic Sensors in Gastrointestinal Applications 559
16.4.2.3 Fiberoptic Immunosensors 560

16.4.2.4 Medical Immunosensor 560
16.4.2.5 Environmental Biosensor 561
16.4.2.6 Public Health 561
16.4.3 Distributed Fiberoptic Sensors 561
16.4.3.1 Time-of-Flight Measurement Using Laser Light 562
16.4.3.2 Time-of-Flight Measurement Using LED Light 562
16.4.4 Plastic-Clad Fiber 562
16.4.5 Limitations of Fiberoptic Sensors 563
16.5 Optical Coherence Tomography in Dentistry 563
16.6 Optical Biopsy 563
16.7 Determination of Blood Oxygenation 567
16.7.1 Pulse Oximetry 568
16.8 Electroluminescent Electrophysiologic Mapping 570
16.9 Quantum Dots as Biological Fluorescent Markers 572
16.9.1 Future Considerations of Quantum Dot Imaging 573
16.10 Compilation of the Optical Requirements for Wavelength Selection Based on the Desired Effects 573
16.11 Summary....574

Index 577

Section I

General Biomedical Optics Theory

1

Introduction to the Use of Light for Diagnostic and Therapeutic Modalities

Mankind has always had a fascination with light. In this introduction we present the historical development of the use of optical methods in medical diagnosis and therapy. The elementary discoveries made over the centuries in optics are described and the culmination of these findings appears in the sophisticated optical diagnostic and therapeutic modalities available to us today. Even though the diagnostic applications may seem obvious, in many cases the theoretical formulation of the phenomenon of electromagnetic radiation and the practical application thereof took a considerable time. However, history reveals that the therapeutic applications of light were on the vanguard. In some cases an application of light in medicine may have gone unnoticed for centuries before being revisited and developed for the benefit of the masses.

A brief description of the first use of light in the treatment of smallpox in the thirteenth century provides an example of the early use of light in medical applications, and the treatment of certain diseases of the eye using photons in the late nineteenth and early twentieth centuries is an example of the pioneering efforts in ophthalmology. The inspiration derived from this fueled other optical techniques in medicine.

1.1 What Is Biomedical Optics?

The interaction of light with biological media covers the general field of biomedical optics. However, it is difficult to distinguish between light applied to a cellular grouping in a living organism or on a Petri dish. The distinction becomes even more blurred when you consider that diagnostic methods develop from the in vitro stage to the in vivo stage and when the barriers have been removed from the design scaling. Miniaturization is often the pivotal step to go from the bench directly to the patient.

The theoretical description of the interaction of light with tissue is based on research on the development of the galaxy and the mathematical concepts that were derived to formulate the stellar light propagation through galactic dust clouds on its route to earth.

1.2 Biomedical Optics Timeline

Starting with the discoveries of the ancient Greek philosophers in the fourth century before Christ, the early concepts of present-day optics were originally proposed but were discarded at the time, to be revived almost two millennia later. Throughout history there has been a significant connection between astronomy and the elementary discoveries in optics, and more specifically biomedical optics. In the twentieth century we saw that discoveries and theoretical analysis in astronomy, dye industry, and offset printing provided the crucial benefits for the development of biomedical optics.

1.2.1 Elementary Optical Discoveries

The concept of vision seems to have presented a challenge for many scientists over the centuries. The ancient Greek philosophers believed that the eyes emitted a "fire" that provided man with the capability of vision. Plato (original name: Aristocle: 427 to 347 BC) wrote that light emanated from the eye, seizing objects with its rays. It was not until Aristotle (384–322 BC) that the first hypothesis was proposed that light emitted by a source, such as the sun or a candle, was captured by the eye when reflected from objects, providing the fundamental principle of sight. This idea was mainly based on the fact that no one could explain the inability to see at night without an external light source. In addition, Aristotle observed that damage was done to the eye after exposure to strong light, which could only be explained if the eye would respond to external rather than internal sources. These findings may have been prompted by complaints from observers of a solar eclipse without protecting their eyes, which we now know one should never attempt. There are 225 documented solar eclipses in the period 399 through 290 BC, with 57 total eclipses. The following translation from Titus Livius, *The History of Rome*, XXII, 1, 4; 8 from 339 BC gives the poetic description of the apparent conditions observed at the time of one of those solar eclipses:

> About the same time Consul Servilius entered upon his consulship at Rome, on the 15th of March.... To add to the general feeling of apprehension, information was received of portents having occurred simultaneously in several places. In Sicily several of the soldiers' darts were covered with flames; in Sardinia the same thing happened to the staff in the hand of an officer who was going his rounds to inspect the sentinels

> on the wall; the shores had been lit up by numerous fires; a couple of shields had sweated blood; some soldiers had been struck by lightning; an eclipse of the sun had been observed; at Praeneste there had been a shower of red-hot stones; at Arpi shields had been seen in the sky and the sun had appeared to be fighting with the moon; at Capena two moons were visible in the daytime; ....

Even though this narrative of the phenomena appears highly anecdotal and imaginative, the unfamiliarity with the solar eclipse and the lack of theoretical explanation are partly to blame for that. This could be the first documented correlation between biomedical optics (e.g., vision) and astronomical events, which will continue to evolve as time progresses.

Aristotle also proposed the first laws of reflection. However, the thought of external stimulation to obtain vision was abandoned by Aristotle's protégée, Theophrastus, and remained unsupported for centuries to come. Investigation into the nature of light and the use of optics has been going on for thousands of years beginning with Euclid, who, in 300 BC, postulated many laws of geometric optics, which were not taken seriously and disappeared into oblivion.

In the second century AD, the Greek anatomist, physiologist, and physician Claudius Galenus (130–201) (often referred to as Galen) also supported the extramission theory of vision. He provided great anatomical detail of the structure of the eye, and recognized the link between the brain and the eye through the optic nerve, identified by Galen as the *pneuma*. However, he saw the optic nerve as a duct providing a path for the emission of rays from the brain. His work was largely based on the findings of anatomists in Alexandria, such as Quintus Curtius Rufus of Ephesus. Galen described the retina, cornea, iris, uvea, tear ducts, eyelids, and the two fluids, which he called the vitreous and aqueous humors. He noted some of the peculiar features of sight such as binocular vision. Galen accurately described the purpose and composition of the lens, explained the crystalline lens as the principal instrument of vision, and identified cataracts.

Even the mideastern philosophers were fascinated by the eye; but they were still convinced that the eye emitted the power of vision instead of receiving images. Records of publications dating back to the ninth century by Yaqub ibn Ishaq al-Kindi (801–873) and Abu Zayd Hunayn ibn Ishaq al-Ibadi (808–873) (Johannitius) in *Ten Treatises on the Eye* and the *Book of the Questions on the Eye* elaborated on this concept on the basis of Galen's notes. In the tenth century the clinician Mohammad ibn Zakariya al-Razi (864–930) (Rhazes) in Baghdad realized that the pupil contracts in response to light. It was not until the eleventh century that Abu Ali al-Hasan ibn al-Haytham (965–1040) (Alhazen) provided records in his *Kitab al-Manazir* (*Book of Optics*) that the eye was injured by exposure to strong light. These conclusions were most likely again due to increased occurrences of blindness following the observation of a solar eclipse in his time. Details on solar eclipses of that time are documented. There are 227 reported solar eclipses in various parts of the world in the period from 900 to 1000 AD, with 66 total eclipses. Both researchers concluded that the eye is affected by external light and not by emitted light. Most of the mideastern achievements in optics were translated into Latin in the twelfth and thirteenth centuries, along with the Greek texts upon which the mideastern thinking on optics had been built.

The Italian artist and inventor Leonardo da Vinci (1452–1519) at first supported the extramission theory of vision in the 1480s; however, after more careful anatomical investigation and experimental observations he changed his viewpoint in support of external sources providing the opportunity to perceive one's direct environment. One of the triggers to this sway of conviction may have been his experiment with ox eyes, in which he scraped the rear surface off to make it transparent, and thus observed the image formation projected on a screen placed behind the eye.

It was not until 1604 that Johannes Kepler (1571–1630) provided a theory of retinal image formation, largely based on his development of the telescope, also working with ox eyes with the retina scraped off, to allow the image formed by the lens of the eye to be projected. Since Kepler was looking at distant objects, he had only to deal with small angles of incidence of the light on the optical lenses. It was not until 1621 that Willibrord Snell (1591–1626) made his observations on the refraction of light beams at larger angles and proposed his Law of Refraction. The same conclusions can be derived from the particle theory of light described by French philosopher, scientist, and mathematician René Descartes (1596–1650) in 1637, and hence the Law of Refraction is known as Descartes' Law in France.

In the next two sections, we will look at the development of early optical devices and how they helped shape early optical theories, respectively.

1.2.2 Development of Optical Devices

The use of devices for better observation took a long time to get established. Records of naturally occurring glasses date back as far as archeological evidence is available. The fabrication of glasses is known to have existed in about 12,000 BC (Morey, 1954). However, the use of glass as an optical instrument can only be originated from the ancient Assyrians who reliably used rock crystal in the ninth century BC for primitive magnifying purposes.

Around 1000 AD the reading stone made its entry into society; we recognize this as a magnifying lens. It enabled farsighted monks to read and was most likely the first recorded use of a reading aid. The origin of manufactured reading aids can be traced to the Venetians who learned how to produce glass for reading stones. They went on to construct lenses that could be held in a frame in front of the eyes instead of lying directly on the paper.

In 1268 the English philosopher Roger Bacon (1214–1294) described in his *Opus Majus* the structure of a crystal/glass instrument placed in front of his eyes that allowed him to observe the written text more accurately. In a later publication by di Popozo in 1289 the term spectacles is mentioned in the context of improved reading capabilities. We may conclude from this publication that the use of spectacles for reading probably originated in the twelfth century. Believed to have been invented in China, it later turned out that the Chinese used colored glasses only to ward off evil spirits, and were not intended to improve vision. Magnifying glasses (or spectacles) thus made its introduction in Europe by the twelfth century. The name of the true inventor of eyeglasses is lost in time. The first known artistic representation of the use of spectacles was by Tommaso da Modena in 1352, showing members of a monastery reading and copying manuscripts wearing eyeglasses.

Leonardo da Vinci described the altered vision he experienced when having his head submerged under water. On the basis of these observations, in 1508, he designed a glass bowl filled with water to act as an aid to improve vision. This step can be considered the first attempt in contact lens design. Descartes also tried to improve on the idea of contact lenses by proposing the use of a container filled with water pressed against the eye as a visual aid in 1636. However, this still did not provide a real solution in the form of portable contact lenses.

The roots of optical diagnostics date back to a time frame between 1590 and 1608, when the first compound microscope was developed by spectacle-maker father-and-son enterprise Hans and Zacharias Jansen (1580–1640) and later Hans Lippershey (1570–1619). The compound microscope was later perfected by Antonie van Leeuwenhoek (1632–1723). The microscope was constructed after the example of imaging instruments made from wooden slats with pinholes having water droplets held in place by surface tension, acting as lenses. In 1610 the astronomer Galileo Galilei (1564 –1642) improved the microscope by introducing double tubes to obtain better focusing capabilities. Galileo was also one of the first scientists who made a serious attempt to measure the speed of light by measuring the time light takes to travel between two mountaintops. He was unsuccessful due to the limited instruments available to him to accurately keep track of elapsed time. Later on, the Danish astronomer Ole Roemer (1644–1710) established that the speed of light was finite, however extremely large. Roemer based his opinion on his observations of the times of eclipses of Jupiter's moons with respect to the rotation of the Earth around the sun, and the generally accepted radius of the earth's orbit during his time. His observations were that the orbits of the moons seemed to speed up when the proximity of the Earth was closer to Jupiter. On the basis of these observations he concluded that the speed of light was involved, and that the speed of light in vacuum had to be approximately 200,000 km/s. In 1728 the English physicist James Bradley estimated that the speed of light in vacuum was approximately 301,000 km/s (or 3.01×10^8 m/s) on the basis of his observations of stellar aberrations. Stellar aberrations cause the apparent position of stars to be altered due to the elliptical orbital motion of the earth around the sun. At this time astrophysics started to take a more involved role in the development of general optical theory and as such affected the progress of biomedical optics indirectly.

Earth-bound measurements of the speed of light were continued by Armand Hippolyte Louis Fizeau (1819–1896) in 1849. Fizeau used a rotating sprocket at several hundred revolutions per second, with over 100 teeth to periodically interrupt the path of a light beam, and measured the time the light took to travel to and from a fixed mirror at a preset distance of over 8.5 km away to return through the same gap between the sprocket teeth. A schematic representation of the experimental setup used by Fizeau is illustrated in Figure 1.1. He established that the speed of light was 313,300 km/s (3.133×10^8 m/s) in air. In 1850 Jean Bernard Leon Foucault (1819–1868) derived the speed of light in water and compared it to the established speed of light in air derived by Fizeau. Foucault made his assesment of the speed of light with a rotating mirror. He measured the returning angle of the reflected light over a distance of 18 m to the mirror and back, due to the rotation of the mirror, and concluded that the speed of light had to be 2.99796×10^8 m/s in air. In 1879 the American scientist of Polish origin Albert Abraham Michelson (1852–1931) improved on Foucault's measurements. Michelson established the speed of light to be 2.9991×10^8 m/s. He repeated his measurements in 1926 between Mount Wilson and Mount San Antonio by increasing the distance from the sprocket to the mirror to 600 m, and using high-quality focusing lenses to target the mirror over this distance to obtain the speed of light equivalent to 2.99796×10^8 m/s. Other measurements were done in both the radio and light waves in the past decades. The currently accepted speed of light in vacuum is 2.99792458×10^8 m/s.

On another front there were developments in the optics of the eye. In 1850 the German scientist Hermann Ludwig Ferdinand von Helmholtz (1821–1894) presented his ideas for an ophthalmoscope during a presentation to the Physical Society of Berlin, and in 1856 he published his findings on eye accommodation by reshaping the lens in his *Handbuch der Physiologischen Optik.*

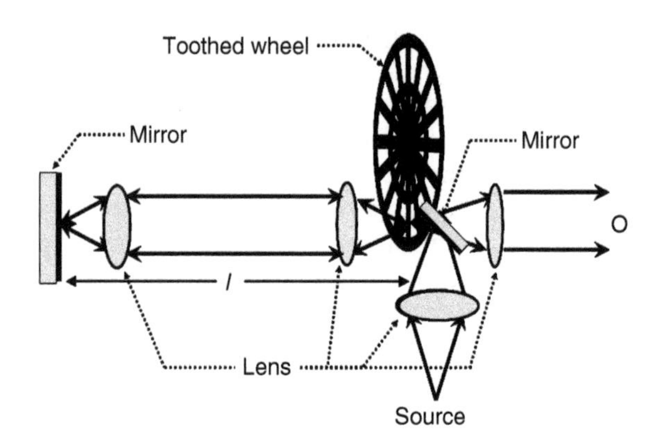

FIGURE 1.1
Fizeau's measurements of the speed of light as performed by Cornu, using Fizeau's apparatus with a distance of 22.9 km, between the two mirrors. The toothed wheel had 180 teeth and a diameter of 40 mm. The observation of the reflected light passing through the toothed wheel twice in opposite directions at consecutive gaps in the wheel is made at point O.

A milestone in the history of biomedical optics was the invention of the maser (microwave amplification by stimulated emission of radiation) in 1954, by Charles Townes (1905–) and Arthur Schawlow (1921–1999), using ammonia gas to produce microwave radiation. This was the predecessor to the laser (light amplification by stimulated emission of radiation). The technology of the maser is very similar to that of the visible light laser, both based on the concept of stimulated emission. The concept of stimulated emission was introduced by Albert Einstein in 1917 in his article *On the Quantum Theory of Radiation, Physikalische Zeitschrift,* translated into English by Van der Waerden, in his article *The Old Quantum Theory,* in the derivation of Planck's Law of Radiation. In 1958, Charles Townes and Arthur Schawlow theorized about a visible laser, a device that would produce infrared and/or visible spectrum light. However, the first practical construction of the laser was not until 1960 when Theodore H. Maiman (1927–), working for Hughes Aircraft, introduced his solid-state ruby laser (Cr_2O_3) to the world, operating in the visible red. There are many indications that Maiman invented the first optical laser; however, at the time there was some controversy that Gordon Gould (1920–2005) could have been the first to have conceived the laser concept.

Gordon Gould is reportedly the first person to use the word "laser," as noted in his laboratory notebook dated November 13, 1957. Gould was a doctoral student at Columbia University under Professor Polykarp Kusch. Kusch had worked closely with Isidor Isaac Rabi, chairman of the physics department, in molecular beam spectroscopy before World War II, and had an office next to Charles Townes, the inventor of the maser. There is documented evidence (his laboratory notebook) that Gould was inspired to build his optical laser starting in 1957. As he failed to file for a patent on his invention until 1959, the patent for his gas laser was refused. It was not until 1977 that Gould was finally recognized for his laser, and was awarded the patent.

1.2.3 Scientific Advancements in Optics Theory

With the diagnostic use of optics under development, we now turn our attention to the accepted theories of light propagation in the early seventeenth century. The corpuscular theory of light— light propagates as discrete particles—was the accepted theory and would be so for several hundred years. In 1637 René Descartes published his *La Dioptrique,* confirming the corpuscular theory of light once again. In the late seventeenth century, Sir Isaac Newton (1642–1727) supported this theory. Meanwhile, Christiaan Huygens (1629–1695), a Dutch physicist and astronomer who had experimental proof of the wave nature of light, was feverishly expanding the wave theory of light. Huygens made his experimental diffraction observations in 1678, thus clearing the way for new applications of optical imaging. The wave concept of light experienced severe difficulties in acceptance, since all other known waves required a medium to propagate, and light was known to traverse the vacuum of space. The theory of reflection proposed by Huygens however still supported the particle theory of light. Newton's support of the corpuscular theory would deter the wave theory from acceptance for the better part of 100 years after his death. It was not until the early nineteenth century and the work of Thomas Young that wave theory started to get a foothold.

The concept of diffraction was generally understood for mechanical waves, and so was the principle of interference. Even though Francesco Grimaldi (1618–1663) had proved that the diffraction principle also applied to light in an experiment performed in 1660, it was not until 1801 that the first conclusive proof was provided by Thomas Young (1773–1829) that made the wave theory receive general acceptance. The proof was provided by an interference pattern generated by light from a single source passing through two closely spaced thin slits projected on a screen at a distance away from the plate with the slits. The bands of light and dark projected on the screen resembled the knots and crests of mechanical waves in water in a similar experiment. The bright bands of Young's experiment were the product of constructive interference and the dark bands represented destructive interference. This behavior could only be explained by wave theory and not by the ruling particle theory for light.

The French physicist Augustin Fresnel (1788–1828) provided further proof of the wave nature of light using interference and diffraction. Fresnel's wave theory description led to an interesting development in 1818. When Fresnel entered his theory of diffraction in a competition sponsored by the French Academy, it created great discord. The judging committee was assembled of several prominent names such as Jean Biot, Siméon Poisson, Dominique Arago, and Pierre Laplace. Poisson, who dismissed the wave theory of light, asserted a seemingly untenable conclusion from Fresnel's theory. He claimed that if the wave theory was correct, a bright spot would be visible at the center of the shadow of an opaque circular object. This theoretical result, he felt, showed the absurdity of Fresnel's theory and considered it a deathblow to the wave theory. Arago, as a matter of supporting the Fresnel wave theory, almost immediately experimentally verified Poisson's prediction. It turned out that the Italian-French scientist Giacomo Filippo (Jacques Philippe) Maraldi (1665–1729), whose work had gone unnoticed, described the observation of a bright spot in the center of a solid shadow in 1723, as part of his work in astronomy observing the moon in the solar corona during a total eclipse.

In 1808 the French physicist Etienne Louis Malus (1775–1812), an engineer in Napoleon's army, discovered the polarization of light waves. Malus was one of Jean Baptiste Joseph Fourier's (1768–1830) pupils at the Ecole Polytechnique. After joining the Egyptian expedition of Napoleon in 1798, he remained in the East. He returned in 1801 and devoted himself to optical research. Malus began experiments on double refraction in 1807 inspired by the work of the Dutch physicist Huygens, and his assumptions of the wave-like character of light. The double refraction phenomenon causes a light beam to split into two on passing through Iceland spar and certain other crystals. Malus's results thus confirmed Huygens's laws. He stated his discovery of the polarization of light on the basis of his observations when holding an Iceland spar up to some light reflecting off a window. The light beam emanating from the crystal, if the crystal was held at a certain angle, was single, not double, supposedly caused by the reflection the light encountered in its interactions with the various surfaces. He published a memoir of his theory of double refraction and polarization. The phenomenon of polarization is specially associated with his name. In the following year he won a prize from the Institute with his memoir, *Theorie de la double refraction de la lumière dans les substances cristallisees*. The book illustrates that polarization plays an important role in certain biomedical optics diagnostic applications.

The interest in electricity rose in the early nineteenth century following the ground-breaking discoveries by the Danish scientist Hans Christian Oersted (1777–1851) and Andre-Marie Ampere (1775–1836). Their work led to the discovery of a link between magnetism and electricity in 1819. The existence of magnetism and magnetic particles was known as early as the thirteenth century BC. Almost 24 centuries later records show that Pierre de Maricourt a.k.a. Petri Pergrinus (1220–1290) used a needle made of magnetic material to map out the patterns of magnetic field lines of naturally occurring magnetic materials. In the 1820s, Joseph Henry (1797–1878) in the United States of America and Michael Faraday (1791–1867) in England, independently from each other, discovered far-reaching relations between electric and magnetic fields. Building on these earlier discoveries James Clerk Maxwell (1831–1879) developed the theories of electromagnetic wave propagation and proved that electromagnetic waves travel with the speed of light.

Another early contribution to biomedical optics came from Gustav Kirchhoff (1824–1887), during his collaboration with Robert Bunsen (1811–1899). They discovered the absorption spectra of molecules in the atmosphere, which was obtained while viewing the sun, and explained the Fraunhofer lines in the solar spectrum. The Fraunhofer lines were discovered by Josef von Fraunhofer (1787–1826) in 1814, who was also the first scientist to measure the wavelength of spectral lines. The work of Kirchhoff and Bunsen was instrumental in the development of what we now know as spectroscopy, used in daily analyses of the composition of substances. Their collaboration involved the early use of the spectroscope and the Bunsen burner.

The observations of Kirchhoff, Bunsen, and Fraunhofer led to optical theories that spawned the field of spectroscopy. Kirchhoff and Bunsen derived from their experimental observations that each substance emitted light with its own unique series of spectral lines. In addition, they concluded that substances absorb spectral lines from a continuous spectrum light source such as the sun. In 1859 Fraunhofer published his explanation of the solar spectral lines, and in 1862 he postulated his third law of radiation: "The rate of emission of energy by a substance is equal to the rate of absorption of energy by the same substance," as part of his philosophy on black-body radiation (a term he introduced) and the experimental verification of his ideas in 1861. This was a significant step in the development of spectroscopy as well as an introduction to thermography. In thermography the radiation emitted from a black body reveals the temperature of that body. Thermography proves invaluable in the detection of inflammatory responses either after surgery or under general substandard health conditions. The inflammatory process is accompanied by an increased metabolism, which explains the elevated temperature profile.

Kirchhoff's laws also form the theoretical basis for the concept of color vision. When a continuous spectrum of light hits an object that absorbs several spectral lines in varying degrees, as a result of the chemical composition of the components, the object will reflect the remaining part of the spectrum, forming a colored image, instead of white. Kirchhoff's first and second laws are as follows, respectively: a liquid, solid, or gas at high temperature will emit a continuum spectrum when heated to incandescence; and a gas, when heated under low pressure, will emit only bright spectral lines at certain characteristic wavelengths rather than a continuum.

When Fizeau performed his speed of light measurements in 1849, he also concluded that the light waves did not experience acceleration over this path length, and thus the velocity should equal the distance traveled divided by the time it takes to traverse this distance. More accurate measurements of the speed of light were performed by Albert A. Michelson (1852–1931), who meanwhile perfected the concept of interferometry now widely used in biomedical optical imaging. The measurements of Fizeau and Michelson honed in on the value of the speed of light propagating near the currently accepted value of 2.998×10^8 m/s.

The interaction of light with particles of different sizes and the resulting redirection of the rays of light were described by John William Strutt Lord Rayleigh (1842–1919) in 1871 for particles smaller than the wavelength of the incident light. His work was based on the observations reported by John Tyndall (1820–1893) describing the wavelength dependence in the process of light interaction with particles that are small compared to the wavelength of the light. Lord Rayleigh's work explained the concepts of polarization and wavelength-dependent scattering, mostly of sunlight interacting with the upper atmosphere. The interaction of light with large particles was described by Gustav Mie (1868–1957) in 1908. These two theoretical descriptions of interaction of light with objects of various sizes form the basis of the present biomedical optics theories. In general the particle size distribution of the biological medium is not uniform, requiring the combination of both Rayleigh and Mie scattering theory.

The quantum theory of light was introduced in 1900 by Maxwell Karl Ernst Ludwig Planck (1857–1947). This fundamental concept in the theory of light was followed by Albert Einstein's (1879–1955) photon particle-wave duality theory in 1905. Einstein's publications on the theory of the photoelectric effect earned him his Ph.D. from the University of Zurich and resulted in a Nobel Prize in 1921. There was however considerable opposition against the reintroduction of the particle concept for light, since it would revert the optical sciences several hundred years. The theoretical postulation of the quantum concept by Planck in 1900 also paved the way for the introduction of the photoelectric effect, used to measure light quantities, thus allowing the research in the use of biomedical optics to build a broader experimental and theoretical foundation. Einstein soon followed Planck in the rigorous theoretical explanation of the photoelectric concept, and in 1913–1914 this concept was experimentally ratified by the research of R.A. Millikan.

With the identification of electromagnetic forms of energy, radiative transfer came into its own as a field of study. A significantly improved contribution in the theoretical description of light in biological media came from the theoretical developments in astronomy. Some of those contributions can be accredited to the work of Arthur Schuster in 1905 and Karl Schwarzschild in 1906 when they formulated the propagation of stellar light through cosmic dust clouds. In 1906, the German astronomer and physicist Karl Schwarzschild (1873–1916), who developed the use of photography for measuring variable stars, discovered the principle of radiative equilibrium and used radiative processes to describe the transport of heat through stars. The theory of radiative transfer introduced by the works of Schuster and Schwarzschild formed the foundation for general and thorough biomedical optics theory.

The aforementioned developments form the major connection between biomedical optics and astronomy that marks the breakthrough of our understanding of the interaction of electromagnetic radiation with turbid media. Biological media are certainly turbid for a wide range of the electromagnetic spectrum. The scale of interaction is on a slightly different level than the interaction of stellar radiation with cosmic dust; however, it has significant similarities. In 1918 Friedrich Kottler (1886–1965) finds a solution to Schwarzschild's radiative transfer equation, without the need for Einstein's vacuum field equations related to their work on the light transport in black holes. The solution proposed by Kottler advanced the development of printing and dye industry processes, fueling the theoretical development of biomedical optics. Another contribution to the understanding of the interaction of electromagnetic waves with biological media came from a postulate by the French mathematician Louis Victor Pierre Raymond duc de Broglie (1892–1987) in 1923, stating the wave properties of particles, of electrons in particular. This hypothesis was experimentally confirmed by the experiments of the American physicists Clinton Joseph Davisson (1881–1958) and Lester Halbert Germer (1896–1971) in 1927, ultimately leading to a Nobel Prize for de Broglie in 1929. The discovery of the wave phenomenon of particles resulted in the development of the electron microscope.

Approximations to the Maxwell equations governing the transport of electromagnetic radiation in all media in a two-dimensional geometry assuming predominantly isotropic scattering (scattering in all 4π solid angle with equal probability irrespective of the angle) were derived with respect to an entirely different field of science. Paul Kubelka and Franz Munk made their mark in 1931 with the derivation of two- and three-flux models as part of their involvement in the manufacturing of pigments for industrial applications, glass manufacturing, and the offset quality for the printing industry. Later on they expanded their theoretical derivation to generalized interaction of light with turbid media in 1948. This was the first big step towards providing a platform for the development of a concise biomedical optics theory to start research in the specialized use of light for clinical and diagnostic purposes.

In 1947, the Indian mathematician Subrahmanyan Chandrasekhar (1910–1995) studied radiative transfer of energy within stellar atmospheres. Developments in radiative transfer proved useful in describing energy transport in environments such as furnaces. In 1986, James Kajiya realized that the methods of radiative heat transfer could be used to develop a single equation to describe light scattering off a surface.

1.3 Historical Events in Therapeutic and Diagnostic Use of Light

A brief overview of the light–tissue interaction mechanisms that determine the boundaries of selected therapeutic and diagnostic applications is presented in this section. The mechanisms described are: photophysical, photochemical damage (reversible and irreversible), photocoagulation, photodynamic therapy (PDT), and photobiological irradiation principles. The next two sections show a chronology of some of the therapeutic and diagnostic methods that use light, and review some of the technological developments resulting from the increasing demands for resolution and physiological details.

1.3.1 Development of Therapeutic Applications of Light in Medicine

The first recorded therapeutic use of light in a biomedical application was from the physician Henri de Mondeville (1260–1320) who used red light in the treatment of smallpox. Building on this initial endeavor, in the early twentieth century the Danish physician Niels Ryberg Finsen (1860–1904) received a Nobel Prize for his work on the treatment of smallpox with light. Accurate descriptions of optical eye examinations and treatment options with optical means date back to at least 1830 in the book *Treatise on Diseases of the Eye*, from the work of the English ophthalmologist Dr. William Mackenzie (1791–1868). Most physicians had been aware of eye tumors for centuries and used scalpels to remove the cancerous growth, often damaging the eye.

The first photochemical evidence of biomedical optics was described in 1888 by Marcacci, when he discovered the toxicity of quinine and cinchonamine in plant and animal cells when exposed to light. The next crucial step was made in 1898, when a German medical student Oscar Raab described the principle of phototoxicity and named it "photodynamic action" to describe the light-mediated enhanced cytotoxicity of eosin on skin. This term later evolved in what is now known as Photo Dynamic Therapy (PDT). He noticed that paramecia cells injected with the dye acridine orange when exposed to light resulted in cell death. The discovery of phototoxicity eventually led to the first medical applications of PDT. The pioneering efforts reported in the period from 1903 through 1907 used the phototoxic effect of the combination of certain dyes and the exposure to light to initially treat skin cancer, and this was rapidly followed by the treatment of eye tumors. An illustration of the use of light in PDT applications is shown in Figure 1.2. Other applications of light in the treatment of tumors were the use of highly focused light to denature the malignant cell formations on the skin and on the eye reported initially in 1912. Figure 1.3 shows the location of an eye tumor. These cases are the first reported application of what was later called photocoagulation.

Photocoagulation is the thermally induced denaturization of proteins resulting from the conversion of photon energy into mechanical energy. The mechanical energy expresses itself as vibration and/or rotation of molecules and atoms. The kinetic energy of a vibrating/rotating atom and molecule is directly related to the thermal energy of the particle and thus the average temperature of the system of particles. The initial photocoagulation was achieved by reflecting sunlight from the roof of the hospital to the operating room with the use of mirrors and lenses.

The next application of light in medicine was the use of sunlight for correction of retinal detachment and destruction of tumors in the eye by means of photocoagulation, influenced by diagnosed macular burns suffered while observing a solar eclipse in the late eighteenth century. In the 1950s the use of sunlight was substituted by high-power arc light sources. In a short report, Daniel Vorosmarthy describes the history of light application as a therapeutic method in medicine and the xenon- and sunlight-photocoagulators first used. He commemorates professor Gerd Meyer-Schwickerath, who was not only the inventor of this new method but also a leading person in ophthalmology in the postwar years.

The pinnacle of the therapeutic use of light is photocoagulation with laser light, where the laser is used selectively to denature specific regions inside an organ or tissue. A highly specialized area of laser–tissue interaction is PDT, in which a laser is used to initiate a chemical reaction ultimately resulting in the cell death of a specific region or cell structure that is selective to a particular wavelength.

There are numerous applications of light in the aesthetic feature of medicine or biomedical optics. Examples are port-wine stain (Naevus Flammaeus) and rosacea treatment as well as laser hair removal. Port-wine stain is a congenital condition from birth, caused by excessive capillary vascularity in the dermis. Rosacea is a condition involving broken blood vessels under the skin, which has a light treatment similar to port-wine stain removal. Hair removal is performed by destroying the hair follicles through the selective absorption of light in the darker follicles compared to the

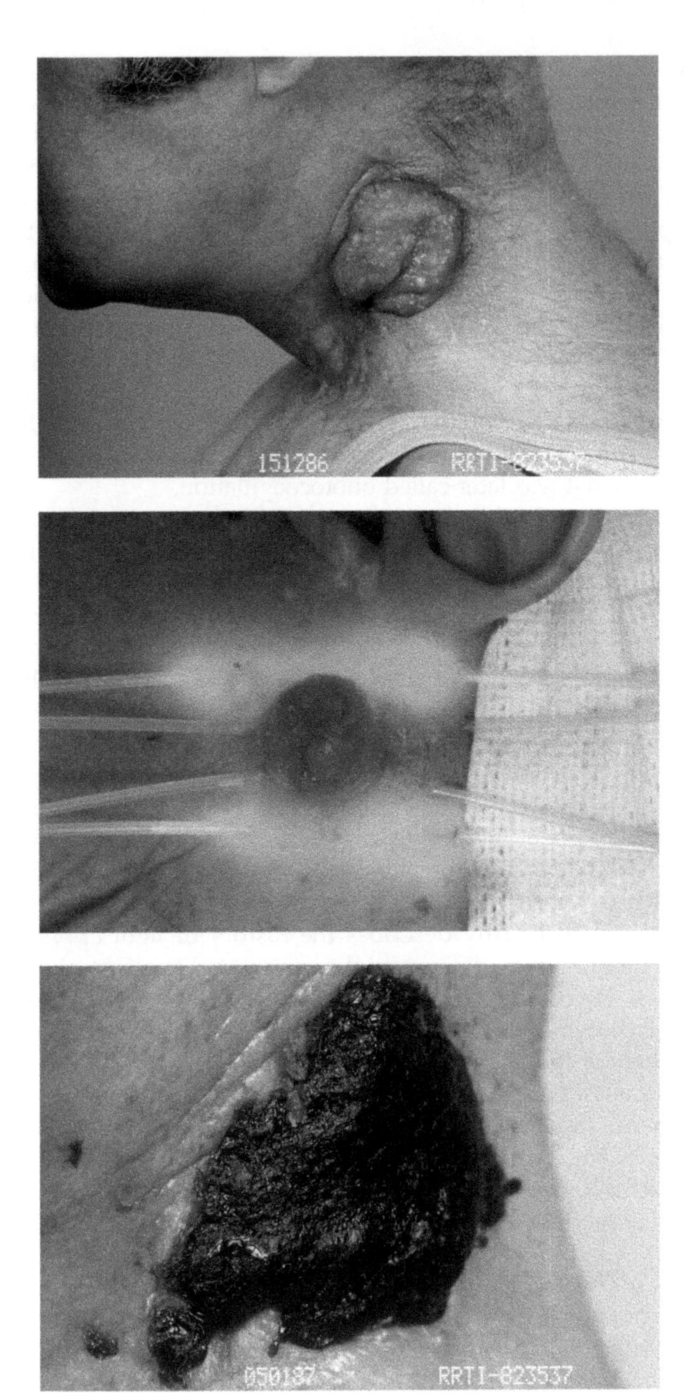

FIGURE 1.2
Illustration of the use of light in photodynamic therapy for treatment of a tumor located on the neck: (a) tumor location, (b) fiberoptic illumination, (c) necrotized tumor tissue. *Source*: Hans Marijnissen, Daniel Den Hoed Kliniek/Erasmus Medical Center.

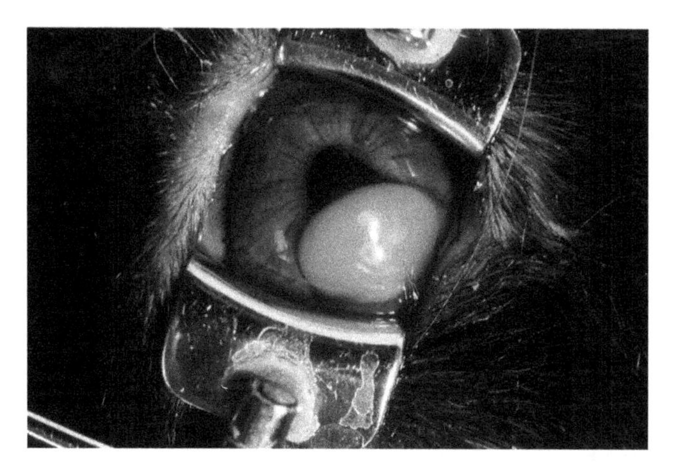

FIGURE 1.3
Illustration of an eye tumor that can be treated by laser irradiation. *Source*: Dr. J.P.A. Marijnissen Daniel Den Hoed Kliniek/Erasmus University Medical Center, Rotterdam, The Netherlands.

surrounding skin, and the subsequent photocoagulation of the root of the hair. Another application is the removal of in-grown hairs, which is a genuine medical condition and not purely cosmetic.

1.3.1.1 Development of Diagnostic Optical Applications

After the discovery of the microscope in the late seventeenth century it was not until the end of the nineteenth century that other biomedical optics applications were being developed for imaging and diagnostic purposes. August Toepler recognized the birefringence induced by heat in various inorganic materials in 1894. Similar conditions apply when heating biological materials. He referred to this type of imaging as Schlieren imaging, which has so far had limited exposure in medical imaging, and was actually not used in medical diagnostic research until the 1980s.

In 1895 a more significant discovery was made by the German physicist Wilhelm Conrad Röntgen (1845–1923). When he experimented with the flow of current through a partially evacuated glass tube, he observed that a piece of barium platinocyanide would emit light. Röntgen theorized that the cathode rays from the tube (electrons) induced another ray resulting from the cathode rays hitting the glass of the tube. This radiation in turn induced the fluorescence of the barium platinocyanide. He discovered that the radiation formed by the electrons hitting the enclosure was capable of penetrating paper, wood, aluminum, and other materials, but did not relate the newly found radiation to electromagnetic radiation, since he could not recognize the attributes one associates with electromagnetic radiation, such as reflection and refraction. He also discovered that the unknown radiation (x-ray) was capable of exposing photographic plates. He subsequently made images of the insides of metallic objects and the first biomedical application was a picture of his wife's hand. He received a Nobel Prize in Physics in 1901 for his publications on this phenomenon.

Additional imaging techniques evolved over the course of the twentieth century such as phase contrast microscopy, medical infrared imaging, fluorescent imaging, interference imaging, ballistic photon imaging, multiphoton imaging, optical coherence tomography, nearfield scanning optical microscopy, scanning confocal microscopy, spectroscopy, optoacoustic imaging, terahertz imaging, spectral imaging (hyperspectral imaging), and evanescent wave diagnostics. More biomedical optics resources were used in blood gas analysis, blood oxygenation measurement, blood glucose level sensing, blood alcohol level assessment, and in the fiberoptic application of fiber Bragg grating temperature measurements.

One biomedical optics imaging technique based on transillumination for detection of breast cancer dates back to the beginning of the nineteenth century. More recently a time-tested method of determining if a premature baby on a ventilator has been damaged by the use of the ventilator, and suffered a collapsed lung, is to hold the baby up to a bright light and observe the shadow fields created by the lungs in the chest. This is a quick but inaccurate method; in addition, one must be careful not to harm the baby due to the intense heat generated by the high-powered light source.

1.4 Light Sources

The initial light sources used in biomedical optics applications were incoherent, broadband sources such as sunlight, incandescent lamps, and noble gas arc lamps. With the advent of the laser, the selection of treatment or diagnostic wavelength became much more specific and selective.

In 1960, the use of light in biomedical optical applications has become more selective, resulting in less peripheral damage. This is mostly because a single wavelength can be selected to effectively produce the desired effect obtained with the light energy. This light source can be geared to the criteria involved for the treatment or diagnostics of a particular procedure. The omission of the least relevant portions of the spectrum results in minimized extravagant damage adjacent to the treatment area or diagnostic distortion from photons irrelevant to the process which are usually absorbed and converted into heat, or scattered resulting in image distortion and redirection of the light energy, respectively. An illustration of a helium–neon laser operating at 632.8 nm in the visible red in a bare beam is shown in Figure 1.4. Laser light–tissue interaction and distribution of light energy in tissue depend on both laser characteristics and tissue properties. Lasers are available in the ultraviolet, visible, and near- and mid-infrared regions of the electromagnetic spectrum.

Light is scattered and absorbed in the tissue. When light is absorbed it is converted into another form of energy, mostly into heat, and to a smaller extent into light of another wavelength. Since conversion of light into heat

FIGURE 1.4
Illustration of a helium-neon laser operating in bare beam mode.

occurs within tissue areas that the photons have managed to reach, heat generation by laser follows a pattern of the light distribution.

In many cases the incoherent nature and broad spectrum of conventional light sources were not necessarily a disadvantage. The beneficial part of the spectrum could be utilized since either the photochemical or photothermal effects were desired, and the wide bandwidth was acceptable at the time. The fact that lasers are coherent light sources is still not necessarily used to its full potential, since the user is often generally interested in the photothermal capacity of the irradiation. In brief, the coherence property of light refers to the fact that all light travels in unison, allowing the phase properties to be used for location information of the source of the (secondary) light source. The secondary light source refers to the light scattered from within the tissue.

Only in certain diagnostic applications, the coherence properties of a laser are capitalized on to deliver the unique performance of the instrumentation.

1.5 Current State of the Art

Since 1960, the laser has found application in many medical applications. From the very beginning lasers have been used in ophthalmology. Later, the laser found applications in dermatology, oncology, cardiology, and in many other fields of medicine.

Laser can be used in minimally invasive surgery to cut or weld tissues by photoablation or photowelding. It is also used to evaporate or shatter hard materials (e.g., calcium deposits in kidney or gall bladder) that are formed in organs and interfere with the normal functioning; this treatment is called photodisruption, examples of this method are laser angioplasty and laser lithotripsy. Laser

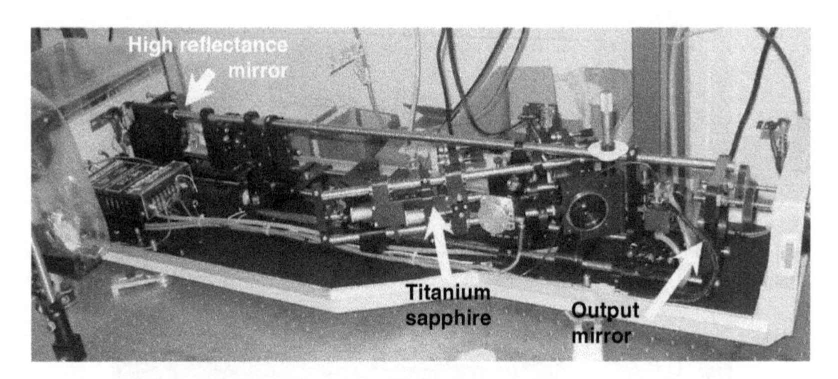

FIGURE 1.5
Laser cavity of a titanium:sapphire laser Fizeau Cornu tooth wheel mirror.

light can be used to denature proteins (cook tissue) and kill harmful tissues without dramatically affecting the structural integrity of the organ; it is called photocoagulation.

The use of laser light is playing a greater role in more disciplines of medicine every year. The recovery time from a laser procedure is often shorter than when conventional methods are employed. The faster recovery time is also in part due to the minimally invasive nature of the surgery, and hence the hospital stay is considerably reduced with substantial cost benefits.

The interest in the use of high-intensity light in treatment and diagnosis is still growing. So far the basics of light–tissue interaction are not completely understood, particularly for a large range of unexplored wavelengths in the ultraviolet and infrared. This is largely due to the diverse chemical nature and particular structure of biological cells and tissues that form organs. In the various disciplines of medicine different lasers at different wavelengths and modes of operation (continuous, pulsed, and super pulsed) are used. Laser–tissue interaction is still insufficiently developed concerning the assessment of optimal wavelengths, and laser mode for each individual therapeutic technique and tissue involved. As far as the availability of laser sources is concerned there are six distinct types: chemical lasers, diode or semiconductor lasers, dye lasers, free-electron lasers, gas lasers, and solid-state lasers. Figure 1.5 shows the laser cavity of a tunable titanium:sapphire laser.

1.6 Summary

A rough outline of the biomedical optics developmental timeline has been presented with the main connections to astrophysics illustrated. Additionally a brief overview of therapeutic and diagnostic methods using light as the main source of energy is presented.

Further Reading

Abramson, D.H., Retinoblastoma in the 20th century, past success and future challenges. The Weisenfeld Lecture, *Invest. Ophthalmol. Vis. Sci.*, 46, 2684–2691, 2005.
Allen, R.M., *The Microscope*, D. Van Norstrand Company, New York, 1940.
Boorstin, D.J., The Discoverers: A History of Man's Search to Know his World and Himself, Random House, New York, 1983.
Bradbury, S., *The Evolution of the Microscope*, Pergamon Press, Oxford, 1967.
Descartes, R., *La Dioptrique*, Jean Maire, Leiden, 1673.
Ditchburg, R.W., *Light*, Dover Publications, New York, 1991.
Halliday, D., Resnick, R., and Walker, J., *Fundamentals of Physics*, 7th ed., Wiley, Hoboken, New Jersey, 2005.
Fizeau, H.L., Sur une experience relative a la vitesse de propogation de la lumiere, *Comptes Rendus*, 29, 90–92, 132, 1849.
Inc. Magazine, March 1989; http://www.inc.com/magazine/19890301/5568.html
Johnson, B.K., *Optics and Optical Instruments*, Dover Publications, New York, 1960.
Jones, T.E., *History of the Light Microscope*, University of Tennessee, Memphis, 1997.
Jones, R.McC., *Basic Microscopic Techniques*, University of Chicago Press, Chicago and London, 1966.
Malus, E.L., Theorie de la double refraction de la lumière dans les substances cristallisees, *Mém. Sav. Étr.*, 2, 303–508, 1810.
Mandel, L., and Wolf, E., *Optical Coherence and Quantum Optics*, Cambridge University Press, Cambridge, 1995.
Morey, G.W., *The Properties of Glass*, Reinhold, New York, 1954.
von Helmholtz, H.L.F., *Handbuch der Physiologischen Optik*, L. Voss, Leipzig, 1856.
Vorosmarthy, D., Solar coagulation, *Bibl. Ophthalmol.* 60, 6–8, 1962

Problems

1. How long did it take for light to travel between Galileo's hilltops, which were 1.5 km apart? (Use the accepted speed of light as 2.99792458×10^8 m/s.)
2. The distance from the earth to the moon varies during its annual orbit, but suppose an eclipse of the sun by the moon takes place when the distance from the moon to the earth is 363.700 km, and the distance between the sun and the earth equals 1.496×10^8 km. The radii of the earth, moon, and sun are 6.38×10^3, 1.74×10^3, and 6.96×10^5 km on average, respectively. Compute the diameter of the conical umbra of the moon's shadow and compare this distance with the distance from the moon to the earth's surface.
3. The average radius of the earth's orbit around the sun is 1.496×10^5 km; use the accepted speed of light as 2.99792458×10^8 m/s.
   a. Compute the time required for light to travel a distance equal to the diameter of the earth's orbit.
   b. Compare this value with the time of 22 min obtained by Roemer.

4. Fizeau's measurements of the speed of light were repeated by Cornu, using Fizeau's apparatus but with a distance of 22.9 km between the two mirrors in Figure 1.1. The toothed wheel had 180 teeth and a diameter of 40 mm. Calculate the angular velocity, when one turn equals radians for each complete rotation, to create the conditions that light exits through one gap and returns through the next gap.
5. What was the famous experiment performed by Thomas Young in 1801?
6. State Malus's law.
7. What experiment in 1887 led to the collapse of the concept of the ether?
8. What part of the eye might be seriously damaged by looking directly at the sun?
9. Are there other units in which the wavelengths of spectral lines could be measured, in addition to nanometers and Ångstroms?
10. Considering the accepted speed of light as 2.99792458×10^8 m/s:
    a. How far does light propagate in a nanosecond?
    b. How far in a light year?
11. How does the theory of ray optics described by Aristotle compare to the postulates made by Willebrord Snell in 1621?
    a. Who else proposed similar concepts of light propagation in the period between Aristotle and Snell?
12. When was the idea of vision by emission from the eyes first discredited?
    a. When was the concept that the eyes capture light as it is reflected from objects to produce an image in the eye ultimately accepted as the explanation of vision?
13. How much time elapsed between the first documented use of optical instrumentation and the first use of optical instruments in biological experimentation?
14. Who published the first records of the wave nature of light?
    a. What experimental observations were these conclusions based upon?
15. When was the wave theory of light evocatively introduced and by whom?
16. When was the wave-particle duality proposed, and by whom?
    a. When was the wave-particle duality generally accepted by the scientific community?
17. Describe the first documented therapeutic use of biomedical optics.
18. Name four commonly used light sources in therapeutic and diagnostic biomedical applications.
19. Give at least one example of the respective usage of optics in the diagnostics or therapeutic application.
20. Which famous natural scientist was the first to postulate many of the properties of geometrical optics, including the first law of reflection?
21. How do the maser and the laser differ?
    a. Are both these devices considered to be generators of electromagnetic radiation?
22. What was the first recorded historical use of light in a biomedical application? Who performed this first use and when was it performed?
23. Place the infrared, radio, ultraviolet, visible, and x-ray regions of the electromagnetic spectrum in order of increasing photon energy (decreasing wavelength).

2

Review of Optical Principles: Fundamental Electromagnetic Theory and Description of Light Sources

This chapter provides an in-depth theoretical description of how electromagnetic waves are physically generated and what the theoretical description involves. It also elaborates on what laws electromagnetic waves obey, and the different classifications of electromagnetic waves ranging from television waves to gamma radiation. It also briefly illustrates the physical application of electromagnetic radiation in medical applications, which will be described in detail in the later chapters.

In addition, various sources of electromagnetic radiation are discussed. Moreover, the rules that were developed over the centuries regarding the modeling of light and image formation are presented and explained. From this view the quantum theory of light is also outlined. At the end of this chapter follows a compendium of the common terms and definitions in the arena of optics principles as they apply to biological interaction.

2.1 Definitions in Optics

Even though we consider our knowledge of light to be sufficient to explain everyday observations, there is more to light than just vision. In some cases light may be described as the motion of particles, but light may also be considered as a wave phenomenon. However, the true description of light is a combination of the wave and particle concept, and light is often referred to as a wavicle, which is a particle with zero rest mass. The particle-wave duality as recognized by Albert Einstein in 1905, resulting in a Nobel Prize in 1921, is still adhered to as the most accurate model for light. This particle-wave duality assigns light as packages of energy called photons. Light in general is considered as the visible part of the electromagnetic spectrum, ranging from blue to red. The electromagnetic spectrum describes the changing combined electric and magnetic fields emanating from a positive or negative electrical charge in an oscillatory motion. To begin with, the electromagnetic spectrum spans far beyond our daily perception, and ranges from very high-energy cosmic rays to x-rays, ultraviolet light, visible light, infrared radiation (near-, mid-, and far-infrared), microwave radiation, television waves, and at the other extreme, low-energetic radio waves.

The mathematical description of the propagation of waves relies on the fact that a description of amplitude as a function of position will repeat itself over a fixed distance. If we consider the amplitude A of a wave phenomenon to be represented by a function of position, the following generic description of a function can be posed: $A = f(x)$, when the position x is replaced by $(x-a)$ the same function has moved to a distance "a" in the positive x-direction, even so for $(x+a)$, the function has moved a distance a, but in the negative x-direction. Thus, $A = f(x \pm a)$ describes the amplitude as a function of position.

Consider now that distance traveled is velocity times the elapsed time; i.e., $a=Vt$; the equation can be seen in the form $A = f(x \pm Vt)$, where $V>0$, and the negative sign indicates a wave traveling in the positive direction and a positive sign for propagation in the negative x-direction. The velocity v is the phase velocity of the wave. The phase velocity of a wave is defined as the speed at which a phase of the periodicity of a wave propagates through space. Since the amplitude is a function of time and space, the equation for a wave traveling in one dimension can be written as $A(x,t) = f(x \pm vt)$, or a combination of the sum of a wave moving in the positive and negative x-direction as $A(x,t) = f(x-vt)+g(x+vt)$.

To find the general wave-equation function $A(x,t)$ will be differentiated twice with respect to both time and position; however, first the substitutions $x' = x-vt$ and $x'' = x+vt$ are introduced, giving $A(x,t) = f(x') + g(x'')$.

The first differentiation with respect to time yields the following equation:

$$\frac{\partial A}{\partial t} = \frac{\partial f}{\partial x'}\frac{\partial x'}{\partial t} + \frac{\partial g}{\partial x''}\frac{\partial x''}{\partial t} = -v\frac{\partial f}{\partial x'} + v\frac{\partial g}{\partial x''} \tag{2.1}$$

The first differentiation with respect to position yields

$$\frac{\partial A}{\partial x} = \frac{\partial f}{\partial x'}\frac{\partial x'}{\partial x} + \frac{\partial g}{\partial x''}\frac{\partial x''}{\partial x} = \frac{\partial f}{\partial x'} + \frac{\partial g}{\partial x''} \tag{2.2}$$

Next, Equation (2.1) is differentiated with respect to time again to give

$$\frac{\partial^2 A}{\partial t^2} = v\left[\frac{\partial^2 f}{\partial x'^2} + \frac{\partial^2 g}{\partial x''^2}\right] \tag{2.3}$$

And similarly for the second differentiation with respect to position for Equation (2.2):

$$\frac{\partial^2 A}{\partial x^2} = \frac{\partial^2 f}{\partial x'^2} + \frac{\partial^2 g}{\partial x''^2} \tag{2.4}$$

After equating Equations (2.3) and (2.4) the general wave equation follows:

$$\frac{\partial^2 A(x,t)}{\partial x^2} = \frac{1}{v^2}\frac{\partial^2 A(x,t)}{\partial t^2} \tag{2.5}$$

A special case is defined when the amplitude is sinusoidal or oscillatory; the following expression for the amplitude as a function of time and position can be derived:

$$A(x,t) = A_0 \sin k(x - vt) \tag{2.6}$$

Equation (2.6) is for a sinusoidal wave traveling in the positive x-direction, where k is a scaling factor and will be defined later. For a sinusoidal function the periodicity is also 2π, and x can be replaced by $x+(2\pi/k)$; the equation is rewritten as

$$\begin{aligned} A\left(x+\frac{2\pi}{k},t\right) &= A_0 \sin k\left(x+\frac{2\pi}{k}-vt\right) \\ &= A_0 \sin(kx - vt + 2\pi) \end{aligned} \tag{2.7}$$

The periodicity in the x-direction $(2\pi/k)$ is referred to as λ, the wavelength. The waveform repeats itself after every distance λ. The variable k indicates the number of wavelengths over a phase difference of 2π and is called the wave number.

Rewriting Equation (2.7) with a substitution for the linear velocity of the moving particle we obtain

$$\begin{aligned} A(x,t) &= A_0 \sin k(x - vt) = A_0 \sin\frac{2\pi}{\lambda}(x - vt) \\ &= A_0 \sin(kx - \omega t) \end{aligned} \tag{2.8}$$

where $\omega = kv = (2\pi v/\lambda)$ is the angular velocity of a circular motion, and $\omega=2\pi\nu$, with ν being the frequency of the oscillation. This provides the equality $\lambda\nu = v$ among the wavelength, frequency, and phase velocity. The term of the goniometric function describes the phase angle of the periodically repetitive single frequency wave.

Every traveling wave can be described by a sine or cosine equation where the amplitude depends on the phase of the position, either in time or in space. Both the time and space information provide the phase of the wave, and thus tie position and time together.

$A = A_0 \sin(\omega t - kx)$ thus can also be written in complex form as

$$A = A_0 \exp[i(\omega t - kx)] \tag{2.9}$$

where k designates the wave number of the medium. Then wave number k and the angular wave frequency ω are related by the expression $(\omega/k) = c$. The phase of the wave $\phi(x,t)$ is the entire argument of the sine or cosine term, or, as shown in the following equation:

$$A(x,t) = A_0 \exp[i(\omega t - kx)] = A_0 \exp[i\phi(x,t)] \tag{2.10}$$

The distance over which the sine (or cosine) function repeats itself is the wavelength λ.

The angular frequency ω relates to the oscillatory frequency (the number of repetitions of the oscillatory motion per second) v by the circumference of one revolution 2π, as $\omega = 2\pi v$.

Both descriptions are a solution to the wave equation given by

$$\frac{\partial^2 \Psi}{\partial x^2} = \frac{1}{v^2} \frac{\partial^2 \Psi}{\partial t^2} \tag{2.11}$$

where v equals the speed of propagation of the wave, and for light Equation (2.11) is rewritten as

$$\frac{\partial^2 \Psi}{\partial x^2} = \frac{1}{c^2} \frac{\partial^2 \Psi}{\partial t^2} \tag{2.12}$$

with c being the speed of light in vacuum. In general, this wave equation will have a solution satisfying $\Psi = \Psi(ct - x)$.

From Equation (2.12) we can derive the solutions for the electric field traveling in the x-direction as expressed by

$$E = E_0 \sin\left(\omega t - \frac{x}{\lambda} 2\pi\right) = E_0 \sin(\omega t - kx) = E_0 \sin[k(ct - x)] \tag{2.13}$$

Another property of electromagnetic waves is the fact that they have all the characteristics found in other waves. As a result an electromagnetic wave carries both energy and momentum.

For any ordinary wave function, the square of the amplitude is called the intensity of the wave; however, since electromagnetic waves are a combination of two waveforms, the electric and the magnetic field, this requires an entirely different approach.

For electromagnetic waves the "intensity" is regarded as how much energy is flowing in or out of a surface. This is somewhat in contrast to the general definition of intensity, which is the square of the amplitude of the wave. However, there is no single amplitude in electromagnetic radiation that requires the introduction of several new definitions with respect to the concept of intensity. Additionally, the surface the electromagnetic radiation flows through can be a plane or have a spherical geometry. This leads to a range of definitions with respect to the concept of "intensity."

Some of the concepts in the nomenclature of electromagnetic energy transport are defined as follows:

- Irradiance I : at a point of a surface, the radiant energy flux incident on an element of the surface divided by the area of that element (W/cm^2).
- Radiant energy W : energy emitted, transferred, or received as radiation (J).
- Radiant energy flux Φ : power emitted, transferred, or received as radiation (W).
- Radiant energy fluence rate Ψ : at a point in space the radiant energy flux incident on a small sphere divided by the cross-sectional area of that sphere (W/cm^2).
- Radiant energy density Ω : radiant energy in an element of volume divided by the volume of that element (J/cm^3).

These terms will be discussed in detail in the following sections.

2.2 Kirchhoff's Laws of Radiation

The radiation emitted from specific sources was empirically derived by John William Draper (1811–1882) and published by Gustav Kirchhoff (1824–1887) roughly in the year 1860. The findings are earmarked as Kirchhoff's laws of radiation, and are stated as follows:

- Kirchhoff's first law states that a hot, opaque, incandescent body, or thermal radiator emits radiation over a continuous range of wavelengths. The continuous range of wavelengths is called a "continuous spectrum." The brightness as a function of wavelength will only gradually change in this form of radiation.

- Kirchhoff's second law states that when the radiation emitted by the thermal source is viewed through an almost transparent cloud of cool gas, the continuous spectrum of the source is experienced as narrow, dark lines. Those dark lines in the transmitted spectrum are the result of discrete absorption at selected wavelengths due to the specific chemical configuration of the gas.
- The third law is related to the emission of radiation by a heated gas, which in turn only exhibits a discrete narrow line spectrum, of only a few spectral lines. These lines are specific and well defined for the state and molecular composition of the particular gas. If the chemical composition of the gas is same as used when deriving the second law the emission lines will occur at exactly the same wavelengths as the absorption lines in the cool gas.

A more specific description of the emission of electromagnetic radiation is given by the Planck law of radiation, which is discussed next.

2.2.1 Planck Function for BlackBody Radiation

Photons are emitted from any object that has energy, which can be converted into light energy. The emission is described by blackbody radiation and uses the premise of quantization of photon energy.

The energy (W) of a photon can be described by

$$W = h\nu \tag{2.14}$$

where h is Planck's constant, after the German physicist Max Planck (1858–1947), derived from the work by Wilhelm Wien and the Second Law of Thermodynamics. The Second Law of Thermodynamics states that the entropy of a system in equilibrium is always at its maximum value. Experimental observations showed that Wien's law was capable of describing the energy distribution of light for high frequencies, but broke down at low frequencies. After revising his perception of the Second Law of Thermodynamics as an absolute law of nature, and incorporating Ludwig Boltzmann's statistical explanation of the second law, he succeeded in deriving his formulation of the energy distribution of electromagnetic radiation in 1900, revealing the momentum of a photon, and the discreteness of the energy distribution of light (photon) energy. This phenomenon became more apparent when analyzing the energy distribution of black bodies, and the analogy with harmonic oscillators was proposed to explain the emission curve as a function of wavelength. The classical mechanics descriptions apparently failed at the atomic dimensions, and quantum mechanics was introduced.

The Planck function, derived by Max Planck in his theory of the blackbody, is defined as

$$B_\nu(T) = \frac{2h\nu^3}{c^2} \frac{1}{e^{(h\nu/KT)} - 1} \tag{2.15}$$

where K is Boltzmann's constant, c the speed of light, ν the frequency, and T the temperature in Kelvin. This equation describes the energy distribution of the different frequencies of light emitted by a blackbody at temperature T.

The direction of the oscillatory motion of the wave motion is what is called the polarization of the wave; to be more precise, for electromagnetic waves, it is the direction of the plane of oscillation of the electric field that determines the polarization. For a single wavefront with all electric fields oscillating in a single plane, we refer to this as linear polarization. In the case where two electric fields are at an angle with respect to each other, the combined effect can be threefold. Since both the electric and the magnetic field are vectors, vector addition of the respective fields is allowed. In the case where the electric fields of the two respective electromagnetic waves are perpendicular to each other, have the same amplitude, and are 90° out of phase with each other, the net result will be circular polarization. In this situation, both the net electric and the net magnetic field rotate around the direction of propagation in a harmonic fashion. Under similar conditions, with electromagnetic waves of different amplitude, the net result will be elliptical polarization. Finally, when there is no particular correlation between the phase and amplitudes of the two or more respective electric fields, the combined effect is random polarization.

The absolute index of refraction of a medium (n) is defined as the ratio of the speed of light in vacuum (c) divided by the speed of light in the medium (v):

$$n = \frac{c}{v} \tag{2.16}$$

When light travels from one medium (medium 1) into the next medium (medium 2) the speed of propagation of the electromagnetic radiation will most likely be different. As a result the wavefront will change its course and produce an angle with the normal (right angle locally to the interface of the two media) to the boundary(θ), which is different from the incident angle with the normal. This phenomenon was described by Willibrord Von Roijen Snell (1591–1626), who made his observations on the refraction of light beams at larger angles and proposed his empirical Law of Refraction in 1621 as

$$\text{Snells' law}: \quad \frac{\sin\theta_1}{\sin\theta_2} = \frac{n_2}{n_1} \tag{2.17}$$

2.3 Electromagnetic Wave Theory

In the case of electromagnetic radiation the waves propagate in all three dimensions with no particular preference, which warrants the wave equation (2.12) to be written in a Cartesian three-dimensional representation as

$$\frac{\partial^2 \Psi}{\partial x^2}+\frac{\partial^2 \Psi}{\partial y^2}+\frac{\partial^2 \Psi}{\partial z^2}=\frac{1}{c^2}\frac{\partial^2 \Psi}{\partial t^2} \tag{2.18}$$

which can be rewritten in vector format by using the vector **r** as the indicator of position in three-dimensional space as follows:

$$\frac{\delta^2 \Psi}{\delta \mathbf{r}^2}=\frac{1}{c^2}\frac{\delta^2 \Psi}{\delta t^2} \tag{2.19}$$

This can be proven with the help of Gauss's law, Faraday's law, and Maxwell equations for an electric field oscillating in the y-direction and a magnetic field perpendicular to the electric field oscillating in the z-direction. Both the electric and the magnetic field then propagate in the x-direction with $E(x,t) = E_0 \sin(\omega t-kx)$ and $B(x,t) = B_0 \sin(\omega t-kx)$, respectively. With E_x=0, E_z=0, E_y=E, and B_x=0, B_y=0, B_z=B, Figure 2.1 illustrates the electric field E in the x-direction perpendicular to the magnetic field B in the y-direction for an electromagnetic wave moving in the z-direction.

2.3.1 Gauss's Law

Based on the existence of an electric field several experimental and theoretical observations can be made. Consider that an electric field is in motion; an electric flux (Φ_E) can be defined as

$$\Phi_E = EA \tag{2.20}$$

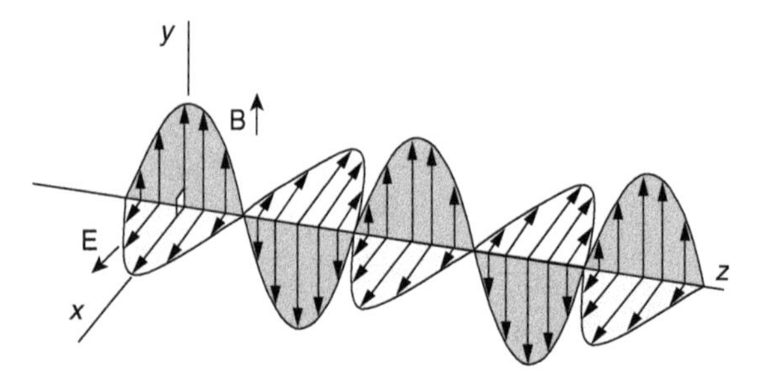

FIGURE 2.1
Electric and magnetic fields perpendicular to each other and the direction of propagation for an electromagnetic wave.

where the electric field E is passing through an area A. In the situation that the field makes an angle θ with the area of choice, only the field passing through the perpendicular projection of the area with the electric field migration contributes to the electric flux Φ_E. The equation becomes

$$\Phi_E = EA_\perp = EA\cos\theta \tag{2.21}$$

This can also be written in vector format as the vector dot product:

$$\Phi_E = \boldsymbol{E} \bullet \boldsymbol{A} \tag{2.22}$$

The only restriction is that the electric field E needs to be uniform over the entire area A. In case the electric field is not uniform, the area will need to be split into n number of sections where the electric field may be considered uniform, and subsequently all the electric fields E_i through each individual subarea ΔA_i will need to be added together to provide the total flux defined as

$$\Phi_E = \sum_{i=1}^{n} \boldsymbol{E}_i \bullet \Delta \boldsymbol{A}_i \tag{2.23}$$

When we let the individual areas approach a surface of 0, i.e., $\Delta A_i \to 0 = dA_i$, the summation can be replaced by an integral over the total area keeping the electric field distribution in consideration:

$$\Phi_E = \int \boldsymbol{E} \bullet \mathrm{d}\boldsymbol{A} \tag{2.24}$$

This was first postulated by the German mathematician and physicist Carl Friedrich Gauss (1777–1855) around 1821, and ties in with the Coulomb law.

Gauss's law considers the situation for a closed surface, such as a sphere, and the enclosed electric flux represents the charge Q from which this electric flux emanates. The electric field distribution can now be written as

$$\Phi_E = \oint \boldsymbol{E} \bullet \mathrm{d}\boldsymbol{A} = \frac{Q}{\varepsilon_0} \tag{2.25}$$

A moving charge provides a current and will generate a magnetic field as described by Ampere's law:

$$\mu_0 I = \oint \boldsymbol{B} \bullet \mathrm{d}\ell \tag{2.26}$$

Also consider this motion to be over a distance $d\ell$. Gauss now postulated the fact that magnetic field lines do not run between points, but are closed. This also means that there is nothing like a magnetic charge. This condition is formulated as

$$\oint B \bullet dA = 0 \tag{2.27}$$

Gauss's law in differential form gives the following two conditions for the electric and the magnetic field, respectively. When the electric field is propagating in a Cartesian coordinate system in the positive x-direction, with the field direction in the positive y-direction, and the magnetic field perpendicular to the electric field, also propagating in the positive x-direction, Gauss's postulates yield the following expressions for the electric and the magnetic field, respectively:

$$\frac{\partial E}{\partial y} = 0 \tag{2.28}$$

$$\frac{\partial B}{\partial z} = 0 \tag{2.29}$$

2.3.2 Faraday's Law

Similar to Gauss's derivation of the electric flux, Michael Faraday (1791 to 1867) proposed a definition of the magnetic flux around 1820/1821 as

$$\Phi_B = \int \mathbf{B} \bullet d\mathbf{A} \tag{2.30}$$

A magnetic flux changing over time was found to induce an electrical potential or "electromotive force" $\mathcal{E}$, which is equal to the work done on a positive test charge of 1 C magnitude, as expressed by

$$-\frac{d\Phi_B}{dt} = \varepsilon = \oint \mathbf{E} \bullet d\ell \tag{2.31}$$

Again, consider a Cartesian coordinate system with the electric field in the positive y-direction, and the magnetic field perpendicular to the electric field, in the z-direction, also propagating in the positive x-direction.

Faraday's law in differential form gives the following two conditions for the electric field and the relationship between the electric and the magnetic field, respectively:

$$\frac{\partial E}{\partial z} = 0 \tag{2.32}$$

and

$$\frac{\partial E}{\partial x} = -\frac{\partial B}{\partial t} \tag{2.33}$$

2.3.3 Maxwell Equations

From Equation (2.31), $\oint \mathbf{E}\cdot d\ell = -d\Phi_B / dt$; the following approximation can be made when solving this integral in infinitesimal steps:

$$y(x,t) = A \sin^2\left[\frac{3}{4}\pi(t+x)\right] \tag{2.34}$$

Assuming time to remain constant during this infinitesimal step, and using Faraday's law of induction, the left-hand side of Equation (2.31) can be written as

$$-\frac{d\Phi_B}{dt} = \varepsilon \tag{2.35}$$

Then Equation (2.33) can be rewritten as

$$\frac{d\Phi_B}{dt} = \ell\, dx \frac{dB}{dt}_{x=\text{constant}} = \ell\, dx \frac{dB}{dt} \tag{2.36}$$

and

$$\frac{d\Phi_E}{dt} = \ell\left(\frac{dB}{dt}\right) dx$$

by using the modified Ampere's law:

$$\oint \mathbf{B} \bullet d\ell = \mu_0 I = [B(x,t)]\ell - [B(x+dx,t)]\ell \approx -\ell\left(\frac{\partial B}{\partial x}\right) dx \tag{2.37}$$

The Maxwell extrapolation of Ampere's law, including the electric field in free space in an open circuit, is given by

$$\oint \mathbf{B} \bullet d\ell = \mu_0 I + \varepsilon_0 \mu_0 \left(\frac{\partial \Phi_E}{\partial t}\right) dx \tag{2.38}$$

After substitution of Equations (2.37) and (2.38) into Equation (2.31) this yields

$$-\frac{\partial B}{\partial x} = \varepsilon_0 \mu_0 \frac{\partial E}{\partial t} \tag{2.39}$$

where ε_0 is the electric permittivity and μ_0 the magnetic permeability.

The magnetic field follows Lenz's law with respect to the charged particle in oscillatory motion, representing an alternating current as expressed by

$$B = \frac{\mu_0 I}{2\pi \mathbf{r}} \tag{2.40}$$

When differentiating the Faraday equations with respect to the displacement x and the Maxwell equations with respect to the time t, the following conformity can be used:

$$\frac{\partial}{\partial x}\left(\frac{\partial E}{\partial x} = -\frac{\partial B}{\partial t}\right)$$

giving

$$\frac{\partial^2 E}{\partial x^2} = -\frac{\partial^2 B}{\partial t \partial x} \tag{2.41}$$

and

$$\frac{\partial}{\partial t}\left(-\frac{\partial B}{\partial x} = \varepsilon_0 \mu_0 \frac{\partial E}{\partial t}\right)$$

giving

$$-\frac{\partial^2 B}{\partial x \partial t} = \varepsilon_0 \mu_0 \frac{\partial^2 E}{\partial t^2} \tag{2.42}$$

EquatingEquations (2.41) and (2.42) gives

$$\frac{\partial^2 E}{\partial^2 x} = \varepsilon_0 \mu_0 \frac{\partial^2 E}{\partial t^2} \tag{2.43}$$

or

$$\frac{\partial^2 E}{\partial^2 x} = \frac{1}{V^2} \frac{\partial^2 E}{\partial t^2}$$

with $1/V^2 = \varepsilon_0\mu_0$, the reciprocal of the speed of light in vacuum squared, $c^2 = 1/\varepsilon_0\mu_0$ or

$$c = 1/\sqrt{\varepsilon_0\mu_0} \tag{2.44}$$

Equation (2.43) has as a solution $E = E_0 \sin(\omega t - kx)$.

Similarly, differentiating the Faraday equation with respect to time and the Maxwell equation with respect to direction of propagation x, gives a similar solution for the magnetic field:

$$\varepsilon_0\mu_0 \frac{\partial^2 B}{\partial^2 x} = \frac{\partial^2 B}{\partial t^2} \tag{2.45}$$

This also shows that $E = cB$.
Maxwell equations in differential form give the following two conditions for the magnetic field and the relation between the magnetic and the electric field, as illustrated in Equations (2.46) and (2.47) respectively:

$$\frac{\partial B}{\partial y} = 0 \tag{2.46}$$

Or in three-dimensional vector form $\nabla \cdot \mathbf{B} = 0$ in free space.

And vice versa the correlation between the space derivative for the magnetic field to the time derivative of the electric field is given by

$$-\frac{\partial B}{\partial x} = \varepsilon_0\mu_0 \frac{\partial E}{\partial t} \tag{2.47}$$

2.3.4 Energy and Momentum of Electromagnetic Waves

The energy density of the electric field of an electromagnetic wave is given by

$$w_E = \frac{1}{2}\varepsilon_0 E^2 \tag{2.48}$$

Analogously for the magnetic field the expression is given by

$$w_B = \frac{1}{2\mu_0}B^2 = \frac{1}{2\mu_0 c^2}E^2 = \frac{1}{2}\varepsilon_0 E^2 \tag{2.49}$$

The total energy density of the electric and magnetic fields combined is given by

$$w = w_E + w_B = \varepsilon_0 E^2 \tag{2.50}$$

The time average is obtained by

$$w_{\text{average}} = \varepsilon_0 \langle E^2 \rangle \tag{2.51}$$

The intensity can now be written as

$$I = cw_{\text{average}} = c\varepsilon_0 \langle E^2 \rangle \tag{2.52}$$

where

$$\langle E^2 \rangle = E_0^2 \sin^2(\omega t - kx) = \frac{1}{2} E_0^2 \tag{2.53}$$

which gives for the intensity

$$I = \frac{1}{2} c\varepsilon_0 E_0^2 \tag{2.54}$$

The cross product of the electric and magnetic fields, which are perpendicular to each other, is perpendicular to the wavefront and in the direction of propagation; thus,

$$|\mathbf{E} \times \mathbf{B}| = EB = \frac{1}{c} E^2 \tag{2.55}$$

The vector $c^2\varepsilon_0(\mathbf{E}\times\mathbf{B})$ thus has the magnitude of energy flow density, with the direction of the wave propagation. This vector expression is called the Poynting vector. The energy flux through a surface per unit time belonging to this Poynting vector is defined in

$$\int_S c^2 \varepsilon_0 (\mathbf{E} \times \mathbf{B}) \bullet \mathbf{e}_n \, dS = \frac{dW}{dt} \tag{2.56}$$

where $\mathbf{e}_n$ is the unit vector perpendicular to the surface.

2.3.5 Coherence of Electromagnetic Waves

Coherence of an electromagnetic wave is usually divided into two classifications: temporal and spatial.

2.3.5.1 Temporal Coherence

The average time interval during which the light wave oscillates in a predictable way is known as the coherence time Δt_c of the radiation. The longer the coherence time, the greater is the temporal coherence of the light. Since temporal coherence is given by the inverse of the frequency bandwidth of the source Δv, a large temporal coherence correlates to a small frequency bandwidth, and hence is a manifestation of spectral purity.

2.3.5.2 *Spatial Coherence*

Spatial coherence relates directly to a light wave's finite extent in space. A better way to think of spatial coherence is to find if the phase of the light wave at two laterally spaced points is correlated. If the phase at these two points is precisely correlated, then the radiation is spatially coherent. Spatial coherence is often measured using Young's experiment in which light is shown at two pinholes in an opaque screen and the resulting interference pattern observed. The more uncorrelated the light, the more closely the product of the two pinholes will average to zero.

2.3.6 Interference of Electromagnetic Waves

Owing to the fact that light is made up of electromagnetic waves another phenomenon of importance will need to be considered. When two waves with different phase but identical frequency meet, the wave patterns will add up. The addition will include the entire sine-wave function description, and not just the individual amplitudes of the respective waves.

In case wave 1 has the expression $A(t) = A_0\sin(\alpha t)$, and wave 2 satisfies $B(t) = B_0\sin(\beta t)$, and using the trigonometry fact that

$$\sin \alpha t + \sin \beta t = 2 \sin\left[\left(\frac{\alpha+\beta}{2}\right)t\right]\cos\left[\left(\frac{\alpha-\beta}{2}\right)t\right] \tag{2.57}$$

or in complex notation, $e^{it} = \cos(t) + i\sin(t)$, $A(t) = \mathrm{Re}(A^* e^{i\omega t})$, A^*=complex, $A^* = Ae^{i\phi}$, and $e^{i(a+b)} = e^{ia}\, e^{ib}$, we obtain the following expressions:

$$e^{i(a+b)} = \cos(a+b) + i\sin(a+b) \tag{2.58}$$

and

$$\sin(z) = \frac{\exp(iz) - \exp(-iz)}{2i} \tag{2.59}$$

From this we can see that when waves 1 and 2 have the same phase $\alpha=\beta$ and the same amplitude A, then the combined effect will be a wave with phase α and amplitude $2A$, i.e., $A_{tot} = 2A\sin(\alpha)$. When waves 1 and 2 are 180° out of phase, $\alpha = -\beta$, then $\sin(\alpha-\alpha)/2 = \sin(0)$ will ensure total extinction of the combined effect.

Complicated wave structures will result anywhere in between these two extremes.

2.3.7 Phase Velocity

As mentioned previously, the phase of the electromagnetic wave $\phi(x,t) = (\omega t - kx)$ is contained in the exponential term $A = A_0 \exp[i(\omega t - kx)]$. If we look at the partial derivative of the phase with respect to the time t, holding the position x constant, we obtain the rate of change of phase with time, or $(\partial\phi/\partial t)_x = \omega$, which is simply the angular frequency of the wave. Likewise, the rate of change of phase with distance, holding the time t constant, is $(\partial\phi/\partial x)_t = k$, and we obtain the wave number of the medium. Now, for the phase velocity v_p, we are interested in the speed of propagation for the condition of constant phase, which can be rewritten as follows:

$$\left(\frac{\partial x}{\partial t}\right)_\phi = \left(\frac{(\partial\phi/\partial t)_x}{-(\partial\phi/\partial x)_t}\right) = \pm\frac{\omega}{k} = v_{\mathrm{p}} \tag{2.60}$$

This is the speed at which the profile moves and is commonly known as the phase velocity.

2.3.8 Group Velocity

The medium in which the electromagnetic wave propagates determines the relationship among ω, k, and v. If in vacuum, the only truly nondispersive environment, the relationship is as in Equation (2.60), i.e., $v = \omega/k$. In a dispersive medium, every wave propagates at a speed that depends on its frequency. When a number of different frequency harmonic waves superimpose to form a composite wave, the resulting wave envelope will travel at a speed different from that of its constituent waves. This introduces the concept of the group velocity of the envelope v_g and its relationship with the phase velocity. The group velocity v_g is the rate at which the resultant wave envelope propagates, and $v_g = v = \omega/k$ applies specifically to a nondispersive medium in which the phase velocity is independent of wavelength so that the waves could have the same speed. In a dispersive medium, ω is dependent on λ, or equivalently on k. The dispersion relation for ω is given by $\omega = \omega(k)$, and if the frequency range is small and centered about the carrier ω_0, then the group velocity is approximately equal to the derivative of the dispersion relation evaluated at ω_0.

$$v_{\mathrm{g}} = \left(\frac{\Delta\omega}{\Delta k}\right)_{\omega_0} = \left(\frac{\partial\omega}{\Delta k}\right)_{\omega_0} \tag{2.61}$$

The resultant wave envelope propagates at a speed v_g that may be greater than equal to, or less than v, the phase velocity of the carrier.

2.3.9 Cauchy Theorem

Complex analysis is the study of functions of complex variables. Complex analysis is a widely used and powerful tool in optics, especially with regard to the index of refraction:

$$n = \sqrt{\varepsilon_r} \tag{2.62}$$

where Equation (2.62) represents the definition of the complex index of refraction as the square root of the relative dielectric permittivity ε_r, or in general, the square root of the real part of the relative dielectric permittivity for the real index of refraction. The full expression for the complex index of refraction is

$$n = n_r - in_i \tag{2.63}$$

The real and the imaginary parts of the index of refraction are defined as the index of refraction n_r and the index of extinction (i.e., absorption) n_i, respectively. The complex part of the index of refraction is considered to describe the attenuation of electromagnetic radiation.

In optics the French mathematician Augustin Louis Cauchy presented a mathematical treatment under the hypothesis that ether had the mechanical properties of an elastic medium. He formulated several theoretical analyses on wave propagation in liquids and elastic media.

The Cauchy integral theorem is primarily used in complex analysis and is an important statement on path integrals for holomorphic functions in the complex plane. The Cauchy theorem states that if two different paths connect the same two points, and a function is holomorphic in the mathematical space in between the two paths, then the two path integrals of the function will essentially be the same.

Technically, the Cauchy theorem defines the parameters under which it is permitted to integrate or differentiate a complex function.

Applying this criterion to the real and imaginary index of refraction, and substituting a suitable representation for the dielectric permittivity, the index of refraction (real part n_r) and the index of extinction (imaginary part n_i) can be expressed, respectively as a function of the real ε_{rr} and imaginary ε_{ri} parts of the relative permittivity using the following equations :

$$n_r = \sqrt{\frac{\sqrt{\varepsilon_{rr}^2 + \varepsilon_{ri}^2} + \varepsilon_{rr}}{2}} \tag{2.64}$$

and

$$n_i = \sqrt{\frac{\sqrt{\varepsilon_{rr}^2 + \varepsilon_{ri}^2} - \varepsilon_{rr}}{2}} \tag{2.65}$$

We define the relative permittivity as

$$\varepsilon_r = 1 + \chi_0 \tag{2.66}$$

where χ_0 signifies the dipole contribution of the medium, which becomes an integral part of the light–tissue interaction for anisotropic media. Here we have introduced the dimensionless variable

$$\chi_0 = \frac{\omega_p^2}{\omega_0'^2} \tag{2.67}$$

where

$$\omega_p^2 = \frac{Ne^2}{m_e \varepsilon_0} \tag{2.68}$$

is the natural frequency of the medium.

We further introduce ω'_0 as

$$\omega_0'^2 = \omega_0^2 - \frac{\omega_p^2}{3} \tag{2.69}$$

where the resonance frequency ω_0 of the medium is derived from the time-averaged mechanical energy E_m as follows:

$$\langle E_m \rangle = \frac{1}{2} m \omega_0^2 a_m^2 \tag{2.70}$$

where a_m is the dipole amplitude and m the dipole mass.

The final dimensionless variable

$$\Omega = \frac{\omega}{\omega_0'} = \frac{2\pi}{\omega_0'} \frac{c}{\lambda_0} \tag{2.71}$$

Combining all the definitions, the relationship between the relative dielectric permittivity and the introduced dimensionless variables yields the following expression for the real relative permittivity:

$$\begin{aligned} \varepsilon_{rr} &= 1 + \frac{c_0 \ (1-\Omega^2)}{(1-\Omega^2)^2 + \beta^2 \Omega^2} \approx 1 \\ &+ \chi_0 (1-\Omega^2)[1-(\beta^2-2)\Omega^2] \approx 1' \\ &+ \chi_0 [1-(\beta^2-1)\Omega^2] \end{aligned} \tag{2.72}$$

Equation (2.72) can be reduced to

$$\varepsilon_{rr} \approx (1+\chi_0)\left[1-\frac{(\beta^2-1)\chi_0\Omega^2}{(1+\chi_0)}\right] \tag{2.73}$$

with

$$\beta = \frac{1}{\omega_0'\tau} \tag{2.74}$$

where β is a dimensionless variable which uses the characteristic interaction time τ.

Thus, the index of refraction under the Cauchy theorem is expressed as

$$n \approx \sqrt{(1+\chi_0)}\left[1-\frac{(\beta^2-1)\chi_0\Omega^2}{2(1+\chi_0)}\right] = \sqrt{(1+\chi_0)} + \frac{(1-\beta^2)\chi_0\Omega^2}{2\sqrt{(1+\chi_0)}} \tag{2.75}$$

The obtained relation is called the Cauchy's relation $n=A+\frac{B}{\lambda_0^2}$ with

$$A = \sqrt{(1+\chi_0)}$$

and

$$B = \frac{(1-\beta^2)\chi_0}{2\sqrt{(1+\chi_0)}}\frac{4\pi c^2}{\omega_0'^2}$$

The introduction of the dimensionless variables in Equations (2.67), (2.71), and (2.74) allows the phase velocity and group velocity to be rewritten, respectively, as follows:

$$v_p = \frac{c}{\sqrt{1-\frac{\omega_p^2}{\omega^2}}} \tag{2.76}$$

and

$$v_g = c\sqrt{1-\frac{\omega_p^2}{\omega^2}} \tag{2.77}$$

The Cauchy theorem is used in Chapter 5 for the Taylor expansion of the light propagation. Additionally in Chapter 8 the Cauchy theorem is used in the description of index of refraction disparities resulting from tissue inhomogeneities and thermal effects applied to Schlieren imaging.

2.3.10 Electromagnetic Wave Spectrum

Moving electric charges produce magnetic fields, and possess an electric field of their own. Oscillatory motion of a charge is of special interest, since this produces a reproducible occurrence for an extended period of time. The oscillatory nature of the electric field as a function of both time and place provides an identical oscillatory generation of a magnetic field. This is known as an electromagnetic wave. The electromagnetic spectrum ranges across a wide spectrum, and is categorized by the source and the sensation. The convention is the following seven classifications: television and radio, microwaves, infrared, visible light, ultraviolet, x-rays, and gamma rays. These subcategories will be described in the following sections.

2.3.10.1 Television and Radio Waves

Television and radio waves operate in the long-, mid-, and short-wave range, with wavelengths in kilometers down to meters and frequencies ranging from tens of kilohertz to 10^6 kHz, these waves are not only generated in television and radio transmitters, but also occur naturally in the emission of stars.

2.3.10.2 Microwaves Waves

This part of the electromagnetic spectrum is used in radar and communication. In addition, it finds applications in investigating fundamental properties of atoms and molecules. Last but not the least, it is used to heat food. The wavelengths for microwaves range from approximately 1 m down to 1 mm and frequencies between 10^9 and 10^{12} Hz. The photon energy over this band runs from 10^{-5} to 10^{-3} eV.

2.3.10.3 Infrared Waves

Infrared radiation can be subdivided into far-, mid-, and near-infrared, with respective wavelengths ranging from 10^{-3} to 3×10^{-5} m, from 3×10^{-5} to 3×10^{-6} m, and for the near-infrared from 3×10^{-6} m to the visible 7.8×10^{-7} m, or 780 nm. Overall the infrared spectrum spans from 10^{-3} m to 7800 Å. The energy levels of infrared light range from 10^{-3} eV at the longest wavelength, up to 1.6 eV approaching the visible spectrum. The frequency range is from 3×10^{11} to 4×10^{14} Hz.

This is radiation recognized predominantly as emitted by black bodies or hot artifacts, such as stars or incandescent-light bulbs.

2.3.10.4 Visible Light Waves

This is electromagnetic radiation where the human eye is the most sensitive. The wavelength range spans roughly one octave from 780 nm down to 380 nm. This corresponds to a frequency range of 4×10^{14}–8×10^{14} Hz, respectively. The photon energy in the visible spectrum increases from

1.6 eV in the red to 3.2 eV in the blue–violet. Table 2.1 lists the color spectrum with respect to wavelength and frequency of the electromagnetic radiation.

Color vision is made possible by sensors in the eye called cones, while the eye also has black and white sensation by sensors referred to as rods. Both names are due to the cellular shapes of these respective sensors. The maximum spectral sensitivity of the eye is due to the rods and is at about 540 nm. All colors combined provide the perception of white, while the lack of colors or the lack of light in general is qualitatively represented as black.

This part of the electromagnetic spectrum is generally dealt with in the science of optics.

2.3.10.5 Ultraviolet Waves

This part of the electromagnetic spectrum runs from wavelengths of 380 nm down to 600 pm, and spans frequencies from 8×10^{14} to 5×10^{17} Hz. The photon energy of ultraviolet radiation is in between 3 and 2000 eV. The ultraviolet energy fits the nuclear energy of many atomic bonds and chemical reactions.

2.3.10.6 X-Ray Waves

X-rays, or Röntgen radiation, form a section of the electromagnetic spectrum ranging from approximately 10^{-9} to 6×10^{-12} m or 6 pm. The mechanism to produce x-rays was first discovered by Wilhelm C. Röntgen (1845–1923) in 1895. The X-ray frequencies are in between 3×10^{7} and 5×10^{19} Hz, with energies from 1.2×10^{3} to 2.4×10^{5} eV. X-rays are generally produced when fast-moving electrons hit a metal anode target and rapidly decelerate. Examples of metal targets are Cesium (Ce) and Tungsten (W).

2.3.10.7 Gamma Radiation Waves and Cosmic Rays

Gamma and cosmic rays originate from the charge motion in the nucleus of elements. As a rule there is no sharp demarcation between the high-energy

TABLE 2.1

Electromagnetic Spectrum in the Visible Light with Color Arranged with Respect to Wavelength and Frequency

Color	Wavelength λ (nm)	Frequency ν (´10^{14} Hz)
Violet	380–455	7.69–6.59
Blue	455–492	6.59–6.10
Green	492–577	6.10–5.20
Yellow	577–597	5.20–5.03
Orange	597–622	5.03–4.82
Red	622–780	4.82–3.84

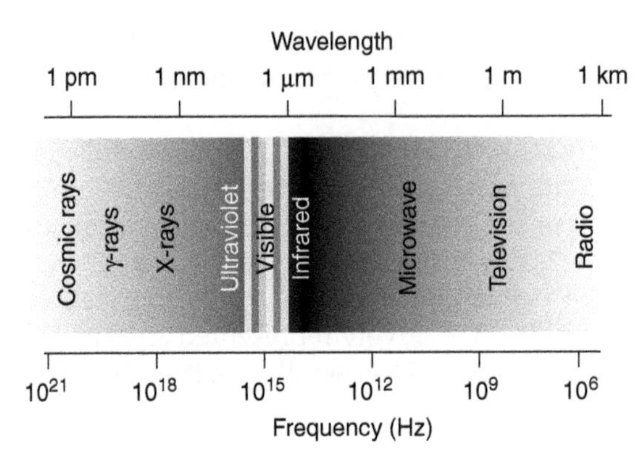

FIGURE 2.2
Electromagnetic spectrum.

x-ray and the beginning of the gamma-ray spectrum. Gamma rays have wavelength between 10^{-10} and 10^{-14} m, with the respective frequency spectrum of 3×10^{18} to 3×10^{22} Hz. The energy band runs from 10^4 to 10^7 eV. Cosmic radiation can be attributed to fast-moving elementary particles (mesons) from within the nucleus. Cosmic rays have even higher energy levels, and associated extremely short wavelengths. Cosmic rays cannot easily be stopped under ordinary conditions. Both gamma and cosmic rays are investigated in astronomy research.

Owing to the short wavelengths in the x-ray and gamma-ray groups the risk of ionization of organic materials is a serious concern. These are the wavelengths that people need to be shielded from to avoid cancerous exposure.

Figure 2.2 illustrates the range of the electromagnetic spectrum in both wavelength and frequency domain. The visible spectrum is colored with a few representative colors; however, the real visible spectrum is a continuously changing color array.

2.4 Light Sources

In general, light is emitted by an accelerating charge; the acceleration can be derived from many different sources.

When an atom or a molecule is in its ground state, all the energy levels of the particle are filled up from the lowest energy state, closest to the nucleus, upward to the energy level that can be filled with all the electrons present in the orbit around the nucleus in the Bohr atom model, after applying the Pauli exclusion principle, and the de Broglie orbital restrictions. The Pauli exclusion principle, devised by Wolfgang Pauli (1900 to 1958), states that at most two electrons can simultaneously occupy an orbit around a nucleus with the same energy; however, they will both have opposite spin. The de Broglie restriction will be discussed in Section 2.5 in greater detail, but summarizes to the following doctrine: All moving particles display a wave pattern, and for an electron in orbit, the wave pattern needs to form a closed wave to successfully occupy an orbit. This means that the wave pattern is a resonance, with energy correlating to the valence energy of that electron.

When subjecting the electrons to an external energy field, one or more electrons may be excited to higher energy states with respect to the nucleus within the electron orbit model.

Once the electrons are in a higher energy state, they will only remain in that state for a finite lifetime, before returning to their ground state. Once the electron returns to its ground state it is accelerated to facilitate the transition between the energy levels. The accelerated charge carries an electric field with it, and during acceleration, it produces a changing electric field in time and space. Based on the principles outlined in Ampere's law and the Maxwell interpretation of Ampere's finding, this changing electric field will induce a changing magnetic field, thus giving a combined changing electric and magnetic fields, electromagnetic radiation. The external energy field can thus be used to generate light. The energy delivered by the external field is in this way converted into photon energy when the excited electron returns to its ground state, and is lost from the atom or molecule.

When the external energy is provided repeatedly or continuously, there is a steady replenishment of the energy lost due to the emission of the photon energy. This principle provides the potential for light generation, and, under appropriate conditions, for the generation of laser light.

Temperature is linked to heat transport, and heat under classical mechanical conditions is directly proportional to the kinetic energy of an object such as an atom or a molecule in motion. The kinetic energy of elementary particles can be in translational, rotational, and vibrational motion. The rotational and vibrational motion both cause an oscillatory field to be generated when observed from the outside of the medium that holds these particles (far-field). This is where the blackbody radiation of Planck's function makes its introduction. Every object that is at a temperature above absolute zero (0 K) will emit light; thus, the higher the temperature, the more is the energy in the emitted light, until the visible spectrum, and even the ultraviolet, are reached. Electrical energy will raise the temperature as well when passing through a medium due to conversion into mechanical energy.

2.4.1 Broadband Light Sources

The initial light sources used in biomedical optics applications were incoherent, broadband sources such as sunlight, incandescent lamps, and noble gas arc lamps. Earlier to that only candlelight and oil lanterns were available for inspection of people suffering from medical afflictions. These light sources offered no real potential for diagnostic and treatment device development, because much of the broad spectrum was not useful. Once man learned to manage the sunlight and create high-output artificial light sources, the gateway for biomedical optics opened. Even still, it took a good part of two centuries to reach the point from the initial discovery of lenses to the development of medical apparatus based on light. All the light sources mentioned so far are called broadband sources. These sources emit light ranging across a wide band of the electromagnetic spectrum. They typically illuminate with the entire visible spectrum, including infrared. For the sun, the ultraviolet and shorter wavelengths are also integrated. These broadband sources have wavelengths that are not useful in visualization, and in the case of incandescent light or sunlight, there is a significant amount of infrared radiation, which simply adds heat energy. This heat energy was responsible for burning patients, and was one of the main reasons for the slow acceptance of endoscopes at the turn of the twentieth century. Later on, sources such as fluorescent light and light-emitting diodes (LEDs) were added to the spectrum of light-emitting devices, and in the invisible short-wavelength spectrum, the X-ray tube. The light emission bandwidth of these sources was narrower than what was available before then. Especially the LEDs offer a relatively narrow spectrum, however with only low available output power. In 1960, the LASER (light amplification by stimulated emission of radiation) was invented, offering high output and extremely narrow bandwidth, lending itself ideally for medical applications.

2.4.2 Laser Operation

There are three possible mechanisms for an atom to interact with light:

1. Spontaneous emission
2. Stimulated emission
3. Absorption

All three interactions are involved in the description of the production of the monochromatic light produced by the Laser.

With the advent of the Laser the selection of treatment or diagnostic wavelength became much more specific and selective.

Laser is produced by excitation of energy levels in the laser medium with external energy sources such as light or electrical current, where an electron is excited to a higher energy level when the appropriate external energy is delivered to an atom. The excited electron will return to its ground energy state after a finite time has elapsed, the duration depending on the stability of the excited state for that particular atom. When the excited electrons revert back to their original stable condition, they are accelerated. An accelerating electron emits an energy packet called a photon that can be explained by the theoretical works of Maxwell and Newton as described in Section 2.2. When one electron reverts back to its original energetic orbit, the photon that is released then stimulates the excited atoms in its path to revert back as well, emitting a photon that is identical in every aspect to the instigator of the process, and thus each photon has the same wavelength and phase.

Laser light is what is called a coherent light source of a single wavelength. Coherence is a relative concept. A single source, such as an oscillating single charge, will by definition produce a stable, single-wavelength electromagnetic wave with a perfect sinusoidal waveform with a single phase for an indefinite duration. In contrast, incoherence occurs when multiple oscillating charges produce electromagnetic waves that are not necessarily of the same frequency, and whose phase relations are completely random with respect to each other. Perfect coherence corresponds to an electromagnetic beam of waves of light each with perfectly identical frequencies and all waves traveling with indistinguishable phase of the sinusoidal expression of the wave.

The light emerging from the output mirror of a laser is monochromatic (single wavelength), coherent (in phase), and parallel.

For a laser light source, the photon energy emitted needs to be replenished on a continuous basis for the laser to continue to operate. For the laser to maintain stimulated emission there is a need for a continuous supply of excited electrons in the laser medium, ready to emit photons of equal frequency and phase. To establish the conditions for laser action to occur the following three conditions need to be satisfied:

1. There must be more atoms in an excited state than in the ground state. There must be more emission of energy, in the form of photons, than absorption of energy in the laser medium. This is called *population inversion*.
2. The excited state of the laser medium must be stable enough to maintain excitation before spontaneous emission can occur. This is what is referred to as a *metastable state*. Under these conditions the lifetime of the excited metastable state will exceed the typical excited state lifetime of 10^{-8} s. In this case stimulated emission will occur before spontaneous emission occurs.
3. The photons that are released under stimulated emission will need to remain confined to the excited medium long enough to provide stimulated emission of neighboring excited atoms.

The major components of a laser are an active medium where electrons can be excited, a resonant cavity where the excited energy is temporarily stored, and a pumping source to create the excited state of the laser medium.

We will begin by describing the concept of reaching an excited-state level population that will be capable of generating the laser conditions.

2.4.2.1 Einstein Population Inversion

The Einstein theory of population inversion describes the phenomenological relations concerning the interaction of an external electric or magnetic field with a group of atoms. This interaction is of interest in the generation of a situation where laser conditions can be accomplished. The primary condition to be satisfied to achieve laser action is population inversion. To establish this situation where there are a greater number of excited atoms in the laser medium than atoms in the ground state the following conditions need to be considered:

A minimum amount of energy will be required to raise the majority of atoms to the excited state if all energy injected into the laser medium when the pumping process of raising the electrons to the excited state is 100% effective. In that case only the energy lost due to radiation of light by stimulated emission will need to be replaced.

The excited state may not be a single energy band, but most likely it will be several bands that can be excited. Those multiple excitation bands can be further split into various modes. Each energy mode will have its own characteristic lifetime. When we call the energy modes within one single energy band N_m, and the average lifetime of these individual modes (τ_c), the following relation for the minimum energy required for pumping the laser medium to excited state can be derived.

Consider the idealized situation of a system with just two energy levels, the ground state and the excited state. Figure 2.3 illustrates the input of energy on a two-level system raising the ground-state condition to an excited state.

This system is subjected to an external electromagnetic field. The energy of a photon is expressed by the Planck relationship

$$W = h\nu = h\frac{c}{\lambda} \tag{2.78}$$

The photon energy W is the difference in energy between the excited energy level E_2, and the ground energy level E_1, and formulated as

$$E_2 - E_1 = h\nu_0 = h\frac{c}{\lambda_0} \tag{2.79}$$

Since this is a two-energy level system only one single wavelength λ_0 or frequency ν_0 will be generated in the transition.

This gives the conditions for the pumping power as expressed by

$$P = \frac{N_m hc}{\lambda_0 \tau_c} \tag{2.80}$$

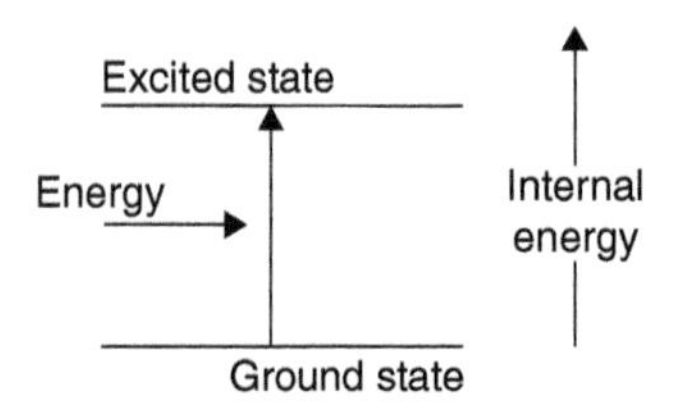

FIGURE 2.3
Energy added to a ground state raises the ground state to an excited state.

where $P = N_m/\tau_c$ with N_m being the necessary number of excited-state atoms.

The rate of stimulated emission now is directly proportional to the absolute difference between the number of atoms in excited state N_2 and the number of atoms in the ground state N_1; i.e., $N_2 - N_1$.

This difference between the excited and the ground states is directly proportional to the ratio of the average lifetime of the emission τ_e, to the average lifetime of the excited state as

$$\Delta N = N_2 - N_1 = \frac{N_m \tau_e}{\tau_c} \tag{2.81}$$

For spontaneous emission the rate of decay is directly proportional to the number of excited atoms as

$$\frac{dN_2}{dt} = -A_{21} N_2 \tag{2.82}$$

where A_{21} is the probability of spontaneous emission, or the Einstein coefficient A_{21}, and has the dimension of time^{-1}. Absorption of energy will take place according to the same principles as

$$\frac{dN_1}{dt} = -R_{12} N_1 = -B_{12} U(\nu_0) N_1 \tag{2.83}$$

where B_{12} is called the absorption coefficient, and has the dimension of time^{-1}. Similarly, in stimulated emission the equation can be rewritten as

$$\frac{dN_2}{dt} = -R_{21} N_2 = -B_{21} U(\nu_0) N_2 \tag{2.84}$$

where B_{12} is called the stimulation probability, and again has the dimension of time^{-1}.

The coefficient B_{12} depends only on the energy-level transition, and is a function of the atom in question. The decay of the population N_1 and N_2 also depends on the magnitude of the flux of the incident electromagnetic wave, the stimulus photons, and additionally the cross-sectional area of the stimulation target.

The stimulated emission can be formulated as

$$R_{21} = \sigma_{21} \Phi \tag{2.85}$$

where R_{21} is the stimulated emission rate, Φ the photon flux, and $\sigma_{21}$the stimulated emission cross-section.

The coefficients A_{12}, B_{21}, and B_{12} are called the Einstein coefficients.

2.4.2.2 Achieving a Metastable State

A laser medium needs to be constructed on the basis of the knowledge of the interaction of various elements in the presence of other elements, and how this atomic matrix will affect the energy structure of the neighboring atoms.

The decay of levels N_1 and N_2 as a function of time can be described by the following rate equations:

$$\frac{dN_2}{dt} = -A_{21}N_2 + B_{12}U(v_0)N_1 - B_{21}U(v_0)N_2 \tag{2.86}$$

$$\frac{dN_1}{dt} = -\frac{dN_2}{dt} \tag{2.87}$$

This can be done purely on the basis of the assumption that this system only has two energy levels.

Utilizing Planck's blackbody radiation equation for a blackbody at temperature T, emitting electromagnetic radiation with a spectral content of $U(v)$ is given by

$$U(v_0)dv = \frac{v^2}{c^3} hv \frac{1}{\exp\left[\frac{hv}{kT}\right] - 1} dv = e'hv\langle n\rangle dv \tag{2.88}$$

where e′ is the eigenvalue for the resonance energy density per photon, and ⟨n⟩ the average number of photons in each resonance energy density. The Planck's law will be described in detail in Chapter 14, section 14.12, under Thermographic imaging.

The occupation of the levels N_1 and N_2 in thermal equilibrium satisfies the expression

$$\frac{N_2}{N_1} = \frac{g_1}{g_2} \exp\left[\frac{-hv}{kT}\right] \tag{2.89}$$

where g_1 and g_2 are the radiation weights of the respective energy levels.

Under thermal equilibrium the decay rate from level 1 equals the decay rate from level 2, and they both will not experience decay. This condition is formulated as

$$\frac{dN_1}{dt} = \frac{dN_2}{dt} = 0 \tag{2.90}$$

This boundary condition results in the expression

$$-A_{21}N_2 + B_{12}U(v_0)N_1 = B_{21}U(v_0)N_2 \tag{2.91}$$

Now,

$$[A_{21} + B_{21}U(\nu_0)]N_2 = B_{12}U(\nu_0)N_1 \tag{2.92}$$

from which the conditions for the electromagnetic field can be solved as

$$U(\nu_0) = \frac{A_{21}}{B_{12}\left(\frac{N_1}{N_2}\right) - B_{21}} \tag{2.93}$$

Using the conditions for the ratio of occupation of the two levels in thermal equilibrium, in combination with the blackbody radiation, the Einstein coefficients can be correlated to each other as follows:

$$B_{12}\frac{g_1}{g_2} = B_{21} \tag{2.94}$$

and

$$\frac{\nu^2}{c^3}B_{21} = -A_{21} = U_e(\nu_0)B_{21} \tag{2.95}$$

where $U_e(\nu_0) = n^3/c^3$ represents the spectral energy density of one photon per eigenvalue oscillation when the average number of photons in each eigenvalue oscillation is one single photon.

This indicates that the stimulated emission in an external field equals the spontaneous emission, under conditions of a single photon in each energy configuration. This condition is satisfied due to the Pauli exclusion principle and the de Broglie orbital restrictions.

The fact that the energy levels are split up due to orbital restrictions in the Bohr atom model, the emission will not consist of a single narrow wavelength, but rather a spectrum with a linewidth. This linewidth will be dictated by the elements used to generate the laser medium.

When the average number of photons in each resonance energy density $\langle n \rangle < 1$ spontaneous emission will dominate, and when $\langle n \rangle > 1$, there is predominantly stimulated emission.

2.4.2.3 Harnessing the Photons in the Laser Medium

A laser is a cavity, with spherical or plane mirrors on opposite sides, which is filled with a laser medium. The medium is any material (glass, gas, crystal, dye, or semiconductors) that has atoms, which can be excited to a semistable energy state with greater energy content than their ground state. The mirrors or opposite sides of the laser medium reflect the emitted photons back and forth, creating an oscillation, and in some cases a resonance. While the flux of photons travels between these two mirrors with only a small portion allowed to escape through a single mirror of reflectivity <1 (the output mirror), the number of photons inside the cavity continually increases, if the gain is larger than the losses of the cavity, stimulating additional excited atoms to release their energy as identical photons.

2.4.2.4 Amplification

As a laser beam passes through an active medium, the rate of change of photon flux Φ is given by

$$\frac{d\Phi}{dz} = \frac{d\Phi}{c\,dt} = \frac{dN_p}{dt} = R_{21}N_2 - R_{12}N_1 = \sigma_{21}\Phi N_2 - \sigma_{21}\Phi N_2 = \sigma\Phi\,\Delta N \quad (2.96)$$

where

$$\Delta N = N_2 - N_1 \quad (2.97)$$

is the population inversion.

In Equations (2.96) and (2.97) spontaneous emission is neglected and for nondegenerate laser levels the cross-sections are equal, i.e., $\sigma_{12} = \sigma_{21}$. The solution is

$$\frac{d\Phi}{\Phi} = \sigma\,\Delta N\,dz \quad (2.98)$$

yielding

$$\Phi = \Phi(0)e^{\sigma \Delta N l} \quad (2.99)$$

where l is the length of the laser medium.

In thermal equilibrium, ΔN is negative. The negative population change leads to the following expression:

$$\left.\frac{N_2}{N_1}\right|_{\text{equilibrium}} = \exp-\left[\frac{E_2 - E_1}{kT}\right] \quad (2.100)$$

In the laser oscillator the pumping process makes the population inversion positive and the amplification of the photon flux can be determined on the basis of the threshold condition expressed by

$$R_1R_2\exp[2\sigma\,\Delta N l] = 1 \quad (2.101)$$

where R_1 and R_2 are the respective reflectance of the mirrors on both sides of the cavity when internal losses are neglected.

The critical population inversion is now given by

$$\Delta N_c = -\frac{\ln(R_1 R_2)}{2\sigma l} \tag{2.102}$$

2.4.3 Laser Light Sources

Laser light–tissue interaction and distribution of light energy in tissue depend on both laser characteristics and tissue properties. Lasers are available in the ultraviolet, visible, and near- and medium-infrared regime of the electromagnetic spectrum. Light is scattered and absorbed in the tissue. When light is absorbed, it is converted into another form of energy, mostly into heat, and to a smaller extent into light of another wavelength. Since conversion of light into heat occurs within tissue areas that the photons have managed to reach, heat generation by laser follows a pattern of the light distribution. In many cases, the incoherent nature and broad spectrum were not necessarily a disadvantage, since either merely the photochemical or photothermal effects were desired, or the wide bandwidth was taken in stride, while utilizing the beneficial spectrum. The fact that lasers are coherent light sources is still not necessarily used to its full potential, since the user is often generally interested in the photothermal capacity of the irradiation. In certain diagnostic applications only, the coherence properties of laser are capitalized on to deliver the unique performance of the instrumentation.

As far as the availability of laser sources is concerned there are generally five distinct types of lasers, in alphabetical order: chemical lasers, diode lasers, dye lasers, gas lasers, and solid-state lasers. A class by itself is the free-electron laser (FEL).

2.4.3.1 Chemical Laser

Chemical lasers are lasers that generate laser light from a chemical reaction. They typically operate on molecular transitions, although there is one that operates on an atomic transition, namely atomic iodine. Common chemical lasers are hydrogen fluoride (HF) and deuterium fluoride (DF). These lasers operate at wavelengths in the mid-infrared of 2.6–3.3 μm and 3.5–4.2 μm, respectively. Other chemical lasers include the hydrogen bromide (HBr) laser at 4.0–4.7 μm, the carbon monoxide (CO) laser at 4.9–5.8 μm, and the carbon dioxide (CO_2) laser at 10.0–11.0 μm. Atomic iodine produces laser irradiation at 1.3 μm, and is obtained via the generation of excited molecular oxygen by reacting molecular chlorine with hydrogen peroxide. The excited molecular oxygen transfers its energy to produce an excited state of atomic iodine. Table 2.2 lists several chemical lasers and the infrared emission wavelength.

TABLE 2.2
Various Chemical Lasers and the Emission Wavelength

Chemical Compound	Wavelength (μm)
HF	2.6–3.3
DF	3.5–4.2
HBr	4.0–4.7
CO and CO_2	4.9–5.8 and 10.0–11.0
Iodine (CF_3I)	1.315

2.4.3.2 Diode Laser or Semiconductor Laser

The class of semiconductor lasers is also a fairly new generation of light sources. A semiconductor is a material that has a resistance that falls between the resistance of the class identified historically as conductor, and that of the group of insulators. The semiconductor or diode lasers are made of semiconductor materials that emit light at the pn junction when a current is forced through this junction. The pn junction is the result of a contact between a semiconductor material doped with atoms carrying an excess of valence electrons (n-type semiconductors, due to the inherent surplus negative charge) and a semiconductor material with atoms lacking valence electrons (p-type semiconductors, due to the inherent deficient negative charge, thus carrying a positive charge). The charge depletion results in a charge migration, creating an electric field. An external current will force the recombination of charges, resulting in the emission of energy in the form of light as photons. Some examples are the gallium-arsenide (GaAs) laser emitting at 807–840 nm bands, aluminum-gallium-arsenide (AlGaAs) laser emitting at 760–790 nm, gallium-indium-arsenide-phosphorus (GaInAsP) laser emitting at 1300 nm, and the gallium-indium-arsenide-phosphorus (GaInAsP) laser emitting at 1550 nm. These diode lasers are primarily running on standard net voltage, and do not require elaborate cooling equipment. These specifications make them very desirable tools; in addition, they are at least an order of magnitude less expensive than the other categories of lasers, and have a significantly longer lifetime than most. One of the drawbacks of diode lasers, however, is the fact that the output power is often less than a comparable laser from any of the other classes. Other disadvantages are that they cannot be coupled into small fiberoptic core diameters, nor can the coupling to the fiber be accomplished at small numerical apertures. Also, the diode laser wavelengths are spotty on the spectral scale, often requiring sacrifices in operation modality compared with lasers in the other categories. The spectrum of diode lasers is expanding daily, with emissions gradually moving from the mid- to the near-infrared into the visible red and green, and recently into the visible blue. The new wavelengths become accessible when new combinations of semiconductor materials are combined to produce efficient laser conditions. Another emerging of age of diode lasers is the introduction of tunable diode lasers using chromium (Cr^{2+}) dopes for operation in the 2–3 μm waveband (CrZnSe: 2.1–2.8 μm), or thulium dopes (Tm^{2+}). Iron-doped (Fe^{2+}) diode lasers provide operation in the 3–4 μm wavelength region. A representative illustration of a diode laser is shown in Figure 2.4.

Lasers can be found in continuous-wave mode operation, or in pulsed laser delivery, or switchable between modes. Pulsed laser light delivery can be achieved by chopping, mode-locking, or Q-switching, depending on the desired pulse duration, pulse power, and repetition rate. Table 2.3 lists various diode laser systems with their respective wavelength.

2.4.3.3 Dye Laser

Dye lasers use mostly organic dyes as the active medium. The dye flows through the laser cavity to provide continuous regeneration of the active dyes. The dye is excited by light sources such as other lasers and high-power flash lamps.

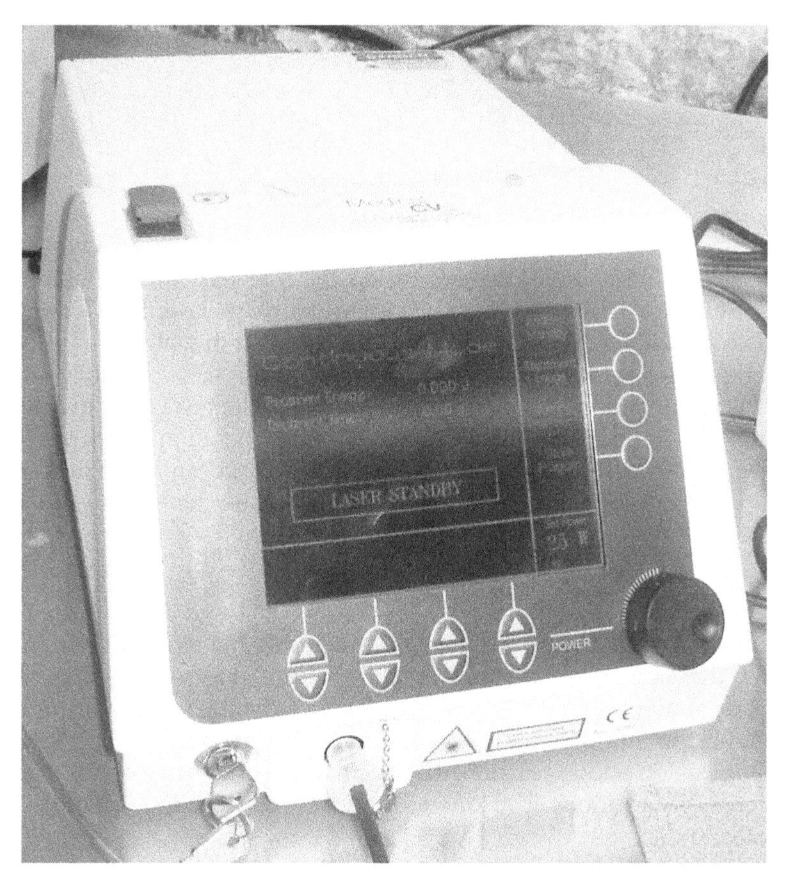

FIGURE 2.4
Photograph of a 25 W continuous wave air-cooled medical diode laser operating at 815 nm.

TABLE 2.3

Examples of Diode and Semiconductor Laser Media with the Specific Emission Wavelength of Each Medium

Laser Medium	Wavelength (nm)
Gallium-Nitride (GaN)	~403
Galium-arsenide (GaAs)	807–840
Aluminum-gallium-arsenide (AlGaAs)	760–790
Gallium-indium-arsenide-phosphorus (GaInAsP)	~1300
Gallium-indium-arsenide-phosphorus (GaInAsP)	~1550
Thulium doped (Tm^{2+})	3–4 μm
Iron doped (Fe^{2+})	3–4 μm
Chromium-zinc-selenide (Cr:ZnSe)	2100–2800

TABLE 2.4

Dye Lasers with Their Wavelength Range

Dye	Wavelength (nm)
Coumarin 1	440–475
Coumarin 30	450–500
Fluorescein	540–650
Rhodamine 6G	450–680
Stilbene	410–470
Tetracene	560–675

Dye lasers have been produced with virtually any liquid, producing visible- or near-infrared light, depending on the selected chromophores. One very popular dye laser is the rhodamine chromophore laser Rh6G at 600 nm. The main disadvantage of dye lasers is the fact that they need to be excited by another laser source. The main advantage, however, is that they are tunable. The combination of the two laser sources presents a rather awkwardly large instrument, not known for ease of use and operation. However, nitrogen-laser-pumped dye lasers are relatively small and can cover most of the visible spectrum from 360 to 720 nm. Table 2.4 shows the various dye lasers available and the tunable wavelength range.

2.4.3.4 Gas Laser

Gas lasers have a noble gas as the active medium. The most often used excitation to the higher energy states of the medium takes place by injection of energy from an electrical high-voltage discharge. Other forms of pumping the medium to a higher energy state are the use of chemical reactions and gas compression and expansion. The excited states mostly consist of vibrational energy transitions. The kinetic energy of the vibration can be released as photon energy.

Some of the gas lasers currently in use are helium-neon (HeNe) operating in the visible red at 632.8 nm; the multiwavelength argon ion laser in the blue–green spectrum, with laser lines at 351, 488, and 514.5 nm, and last but not the least the CO and CO_2 lasers. One other special category with the gas lasers is the excimer laser. Excimer is a contraction of the two words excited and dimer. The dimer refers to the fluoride, xenon chloride, or other halogen/inert gas combination molecules in the excited state. The dimer does not exist in nature in the unexcited state. The excited dimer consists of atoms of argon and halogens technologically bound together in a highly excited, temporary state to form a diatomic rare-gas halide. The decay of these unstable molecules (argon fluoride) to a stable state results in the emission of a highly energetic photon of ultraviolet light. The emissions wavelength of the argon fluoride excited dimer is 193 nm; xenon chloride (excimer) laser in the ultraviolet 308 nm or 375 nm. Another class of gas lasers is the metal-vapor laser, for instance copper-vapor laser with lines at 510 and 578 nm; krypton ion laser which emits laser radiation at 568 and 647 nm; gold-vapor laser at 628 nm; laser operating in the mid- and far-infrared at approximately 5 and 10.6 mm, respectively. The nitrogen laser emits in the ultraviolet spectrum at 337.1 nm. Table 2.5 has various gas lasers listed and the typical emission wavelengths.

2.4.3.5 Solid-State Lasers

Solid-state lasers are lasers made from a solid transparent medium (crystalline or glass) as the active medium. The transparent medium is doped with a specific element to provide the energy states that can be reached by

TABLE 2.5

Examples of Gas Laser Media with the Specific Emission Wavelength of Each Medium

Laser Medium	Wavelength (nm)
Excimer	
F_2	157
ArF	193
KrCl	222
KrF	248
XeCl	308
XeF	351
Gold vapor	312.4; 627.8
Helium-cadmium (HeCd)	325; 442
Nitrogen (N_2)	337.1
Argon ion	457.9; 476.5; 488[a]; 496.5; 501.7; 514.5[a]; 528
Copper vapor	510.6; 578.2
Helium-neon: HeNe (green, yellow, orange, red, infrared)	543.5, 594, 612, 632.8[a], 1152.3, 1523, 3391.2
Lead vapor	722.9
Carbon monoxide (CO)	5000–6000
Carbon dioxide (CO_2)	9,600; 10,600[a]

[a] Most frequently used.

excitation. The introduced impurity in the glass rod will determine the laser activity and thus the emission wavelength. In general this type of laser has the energy for excitation delivered by means of powerful light sources such as arc lamps or other lasers.

Examples of the solid-state lasers are ruby (ruby-glass) laser with sapphire as the host lattice and chromium as the active excitable ion. This laser emits at 694.3 nm in the visible red (this was the first working laser); the tunable titanium sapphire (680–1090 nm) and Alexandrite (700–900 nm) lasers; neodymium-vanadate (Nd:vanadate, Nd:YVO_4) laser, operating at 1342 nm, and frequency-doubled emitting at 671 nm; neodymium:yttrium-aluminum-garnet (Nd:YAG) laser, operating at the 1064 nm or 1334 nm wavelength, also often frequency-doubled to deliver light of 532 nm in pulsed fashion. An illustration of a 100 W medical Nd:YAG laser is shown in Figure 2.5. Other solid-state lasers are holmium:yttrium-aluminum-garnet (Ho:YAG) laser, operating at the 2060 nm wavelength; and erbium:yttrium-aluminum-garnet (Er:YAG) laser, operating in the mid-infrared at the 2900 nm wavelength. The majority of solid-state laser media are summarized in Table 2.6.

2.4.3.6 Free-Electron Laser (FEL)

An entirely different sixth class of laser is the free-electron laser that is broadly tunable, covering much of the electromagnetic spectrum from approximately 200 nm to 1 mm. The FEL is a unique laser source in that its electrons are not bound to a host medium, such as the gas or crystal or dye of the previous laser systems. The electrons are "free", hence FEL, and are generated, accelerated, and made to interact in an alternating magnetic array to produce light. This accelerator-driven technology is a pulsed system and, therefore, the FEL is a pulsed laser.

The journey starts at the source for the electrons, a cathode, typically a crystal of lanthanum hexaboride (LaB_6). The electrons are emitted from the cathode by heating (thermionic emission) and by placing the cathode in a strong, accelerating electromagnetic field (field emission). These free electrons are then accelerated to near relativistic energies in vacuum and made to interact with a periodic magnetic field, where their interaction generates light that depends on the energy of the electrons, and the periodicity and strength of the magnetic field. Figure 2.6 shows a typical layout of this interaction.

The high energy, free electrons are injected in the form of a beam in vacuum into this magnetic array called a wiggler, as it "wiggles" the electrons from their straight path. These highly energetic electrons are traveling at near the velocity of light $v = \beta c$, where c is the speed of light. The electrons execute N_{w} wiggles during the time t and are in the wiggler of length L_{w}. The time t is given by

$$t = \frac{L_{\mathrm{w}}}{v} \tag{2.103}$$

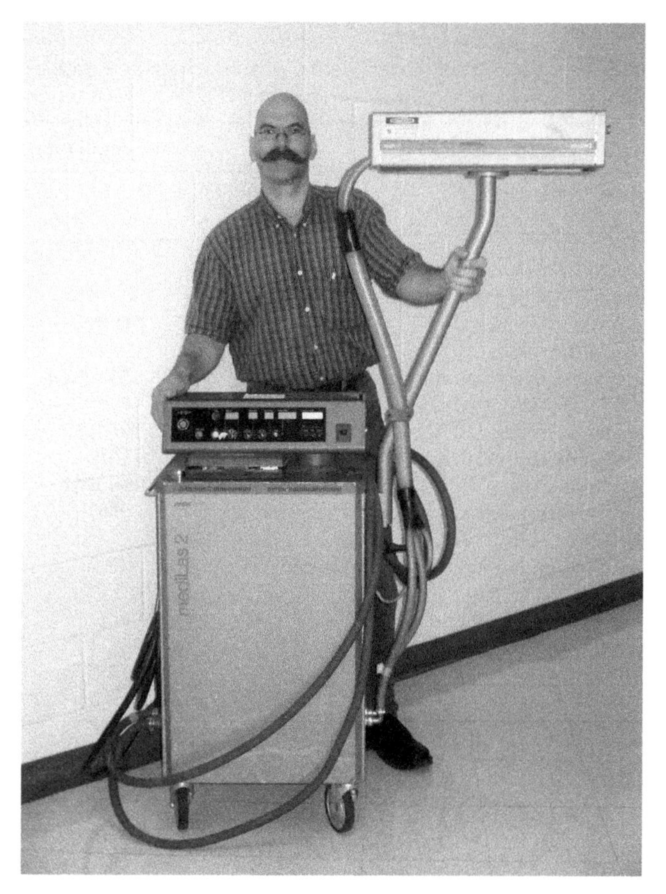

FIGURE 2.5
Photograph of a 100 W continuous wave water-cooled medical Neodymium:YAG laser operating at 1064 nm.

where $L_w = N_w\lambda_w$ and λ_w is the length of one period of the alternating wiggler magnetic field. At the end of the wiggler, a wavepacket of light emitted by the electrons at the beginning of the wiggler has moved a distance ct, while the back of the electron bunch is at vt. The total length of the wavepacket is then $(c-v)t$, and it contains N_w oscillations just as the wiggler does. The laser wavelength is then given by

$$\lambda_L = \frac{(c-v)t}{N_w} = \lambda_w \frac{1-\beta}{\beta} \tag{2.104}$$

Since the electrons are relativistic and $\beta \approx 1$ we can write:

$$\frac{1-\beta}{\beta} \approx \frac{(1-\beta)(1+\beta)}{2} = \frac{1-\beta^2}{2} \tag{2.105}$$

TABLE 2.6

Examples of Solid-State Laser Media with the Specific Emission Wavelength of Each Medium

Laser Medium	Wavelength (nm)
Krypton	337.5–799.3; 647.1–676.4[a]; 752.5[a]; 799.3[a]
Ruby (Cr^{3+}:Al_2O_3) (sapphire/chromium)	692.8; 694.3
Titanium-sapphire (Ti^{3+}:Al_2O_3)	680–1090
Alexandrite (Cr^{3+}:$BeAl_2O_4$)	700–900
Cr^{3+}:$LiCaAlF_6$	720–840
Neodymium:yttrium-aluminum-garnet (Nd:YAG) (Nd^{3+}:$Y_2Al_5O_{12}$)	353; 530; 1064
Cr-forsterite (Cr^{4+}:Mg_2SiO_4)	1150–1350
Neodymium:vandate (Nd:YVO)	1342
Neodymium-glass	1054–1062
Holmium-yttrium-aluminum-garnet (Ho:YAG)	2060
Erbium:yttrium-strontium-gallium-Garnet (Er:YSGG)	2797
Erbium:gallium-gadolinium-garnet (Er:GGG)	2821
Erbium:yttrium-aluminum-garnet (Er:YAG)	2937

[a] Most frequently used.

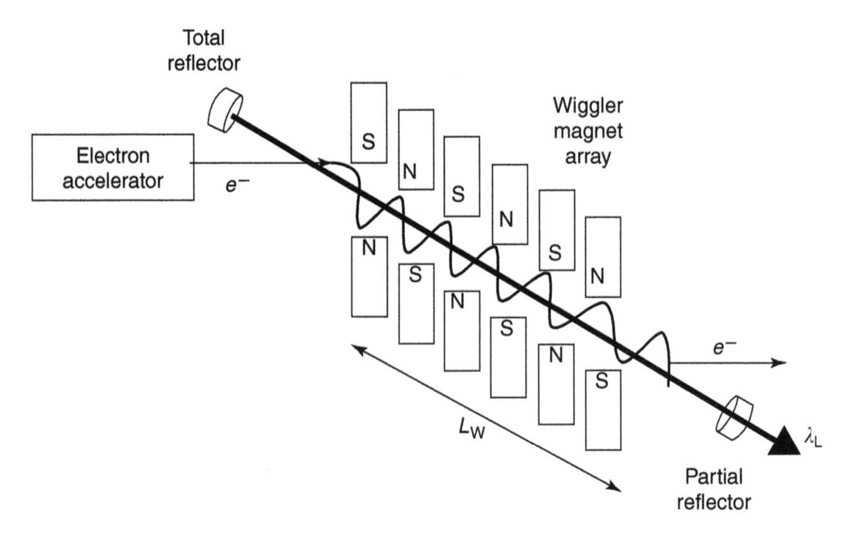

FIGURE 2.6
Diagram of mode of operation of free-electron laser.

We may describe this using the Einstein formula for the energy of a relativistic electron as

$$\gamma = \frac{1}{\sqrt{(1-\beta^2)}} \tag{2.106}$$

where γ is expressed in units of rest energy; $mc^2 / e = 0.511$ MeV with m being the electron rest mass, and e the electron charge. The laser wavelength is then given by

$$\lambda_L = \frac{\lambda_W}{2\gamma^2 \left[1 + \left(\frac{eB_W \lambda_W}{2\pi mc^2}\right)\right]} \tag{2.107}$$

where we have included the effect of strong wiggler magnetic fields B_W on the velocity of the electron. Strong magnetic fields can significantly increase the path length of the electrons in the wiggler and this increases the wavelength of the laser light. This formula is known as the resonance condition for an FEL. Typical values for the wiggler wavelength are in centimeters, electron beam energy values in the MeV to GeV range, and magnetic induction values in the kG range.

2.5 Applications of Various Lasers

The following section is a brief introduction to the various usages of the different laser media and wavelengths (see Table 2.7). More detailed descriptions will follow in subsequent chapters.

- Alexandrite lasers due to their tunable nature are ideal for removal of tattoos, pigmented lesions, and hair.
- Argon lasers with their main spectrum in the blue–green have a high absorption in blood, and are used exclusively for cutting in organs that are highly vascularized, in addition to tissue welding functions.
- The CO_2 laser operating at the 10.6 μm wavelength is used for skin resurfacing, transmyocardial revascularization (TMR) (drilling channels in the ventricular wall of the heart to redirect blood flow to sections of the heart muscle deprived of oxygen), and for general surgery cutting procedures.
- Diode lasers are used in general and microsurgery, dermatology, hair removal, ophthalmology, dentistry, urology, photodynamic therapy (PDT), for tissue welding, and for various diagnostic applications, with Optical Coherence Tomography (OCT) as the main field of interest. They have also found certain niche applications in photocoagulation procedures.

TABLE 2.7
Selected LASER applicatiions

Laser Medium	Wavelength (nm)	Biomedical Optics Application
Alexandrite	700–900	Dermatology
diode	400–2800	Dentistry
The Carbon dioxide (CO_2)	10600	TransMyocardial Revascularization (TMR) General surgery cutting procedures
Dye	440–680	Photo Dynamic Therapy (PDT) Treatment of vascular lesions
Er\ Ho\ Nd: YAG	1064, 1340, 2060, 2737	Ophthalmology
Excimer	157, 193, 222, 248, 308, 351	Transmyocardial Revascularization. Transdermal drug delivery
FEL	used in an amplifier configuration, generating short-pulse X-rays	Protein holography

- Dye lasers are also used in PDT, and additionally in the treatment of vascular lesions and the removal of varicose veins.
- Erbium:YAG and Er:YSGG lasers are used for skin resurfacing, in dentistry, in ophthalmology, and to allow needle-free blood-drawing and transdermal drug delivery.
- Excimer lasers have found applications in laser angioplasty, to remove plaque in the vasculature; they have been used for pacemaker lead removal and TMR. In addition, they are used in the reshaping of the cornea (LASIK) and in the treatment of glaucoma.
- Holmium:YAG is used in ophthalmology for various treatments; it is also used in laparoscopic surgery in orthopedics, as another alternative for TMR. Last but not least the Ho:YAG is used in cosmetic surgery for skin "rejuvenation".
- The neodymium:YAG laser in the near-infrared and the frequency-doubled version using a KTP crystal operating in the green are used in photocoagulation, and cutting applications respectively depending on the pertinent wavelength bands.
- The Ruby laser has been around the longest of all the lasers and is still finding its work cut out for it in certain aesthetic treatments and in ophthalmology, where it was first used.

The majority of optical principles developed over the centuries can also be applied in biomedical optics. The earlier idea of light as rays was the foundation for the theoretical formulation. The movement of photon rays can be illustrated by drawing the path of a photon on paper, which introduces basic geometric rules for the behavior of the ray. Some of the basic principles will be described in Chapter 3. We will also examine the various optical principles that apply to the interaction of light with tissue to gain a better understanding of how light can be used in diagnosis and therapy in surgery and medicine.

2.6 Summary

The basic electromagnetic laws have been postulated and the electromagnetic spectrum is described. The laser operation has been discussed in detail, and an introduction to the use of lasers in medical applications is given.

Further Reading

Brillouin, L., *Wave Propagation and Group Velocity*, Academic Press, Inc., New York, 1960.

Cauchy, A.L., Cours d'analyse de l'École Royale Polytechnique, 1ère partie: Analyse algébrique, Imprime Royal, Paris, 1821. Reprinted in OEuvres complètes, 2e série, Vol. 3, Gauthier-Villars, Paris, 1882.

Einstein, A., Zur Elektrodynamik der bewegter Körper, *Ann. Phys.*, 17, 891, 1905.

Fermann, M.E., Galvanauskas, A., and Sucha, G., *Ultrafast Lasers: Technology and Applications*, CRC Press, Boca Raton, FL, 2002.

Heaven, M.C., Chemical dynamics in chemical laser media, in *Chemical Dynamics in Extreme Environments*, Dressler R.A., Ed., Advanced Series in Physical Chemistry, World Scientific, Singapore, 2001, p. 138.

Rømer, O., Démonstration touchant le mouvement de la lumière, *Journal des Sçavans*, 7, 223–236, 1676.

Silfvast, W.T., *Laser Fundamentals*, Cambridge University Press, Cambridge, 1996.

Sir Isaac Newton, *Opticks*, Courier Dover Publications, New York, 1979.

Svelto, O., *Principles of Lasers*, 4th ed., Plenum Press, New York, 1998.

Townes, C.H., Production of coherent radiation by atoms and molecules (Nobel Lecture, December 11, 1964), in *Nobel Lectures, Physics 1963–1970*, Elsevier, Amsterdam, 1972. (http://nobelprize.org/physics/laureates/1964/townes-lecture.pdf).

Webb, C.E., and Jones, J.D.C, Eds., *Handbook of Laser Technology and Applications*, Institute of Physics Publishing Company, London, 2004.

Weber, M.J., *Handbook of Lasers Wavelengths*, CRC Press, Boca Raton, FL, 1998.

Problems

1. The description of a hypothetical wave propagating in the x-direction can be given as A= $A_0 \sin 2\pi(3x-10^{-6}t)$, where A_0=10 nm.
   a. What is the amplitude of this wave?
   b. What is the wavelength?

   c. What is the frequency?
   d. What is the propagation speed?
   e. Make a graphical representation of this wave, and illustrate the amplitude and the wavelength.
2. Describe the difference between a continuous spectrum and a line spectrum.
3. Prove that the magnetic field satisfies the wave equation.
4. Prove that $E = cB$.
5. What is the line spectrum of a Nd:YAG laser?
6. Prove that an electromagnetic wave is a transverse wave.
7. Place the infrared, radio, ultraviolet, visible, and x-ray regions of the electromagnetic spectrum in order of increasing photon energy (decreasing wavelength).
8. Interference can be seen predominantly with the use of coherent light sources. What are the two common ways by which coherent light sources can be obtained from one source?
9. A plane-wave and a spherical-wave interfere. Draw the pattern of the interference you would expect to see.
10. Incandescent light is produced when electric current flows through a metal wire.
    a. What happens in an incandescent-light bulb to produce light?
    b. Fluorescent lights are different from every other form of lighting that predates them. How are they different?
11. Can a water wave and a sound wave interfere? Give reasons.
12. At what wavelength does the Planck blackbody spectrum peak for a body temperature of 37°C?
13. Polarization by absorption is how polaroid film polarizes a light field. Polaroid film is made up of long-chain molecules with its associated electrons. Is the light that is absorbed along the direction or perpendicular to these long chains?
14. Describe the difference between spontaneous and stimulated emission.
15. After examining a copy of Kirchoff's laws, list other things in everyday life that might
    a. Give a continuum spectrum
    b. Provide line spectra
    c. Mention the characteristics of a continuous spectrum
16. The electrical field of a beam of electromagnetic radiation falling on a skin surface can be written as

$$E_x = \frac{E_0}{\sqrt{2}}\cos(kz - \omega t)e_x \text{ and } E_y = \frac{E_0}{\sqrt{2}}\cos(kz - \omega t)e_y \tag{2.108}$$

    where E_0 is electric field strength, k is the wave number, ω the angular frequency of the incident light, e_x and e_y the unit direction vector in the x and y direction, respectively. The beam of light hits the surface at a characteristic angle of $\theta = 33{,}750°$, assuming an index of refraction of the dermis of $n = 1.35$.
    a. Calculate the radiative flux density of the beam just on the inside of the skin.
    b. What can you conclude from this information with regard to the fact whether the light is polarized when entering the body at this angle?
17. Determine whether the following expressions are correct or incorrect:
    a. The shape of the waveform remains preserved when a wave propagates in a single medium.

b. The energy density of a wave is a function of the amplitude of the wave.
c. The energy density of a wave is a function of the phase of that wave.
d. The amplitude of a wave remains constant for a propagating wave in a single medium.
e. The phase of a wave remains constant for a propagating wave in a medium of constant index of refraction.

18. Provide the relationship between the following parameters (additional constants may be needed):
a. The wavelength, frequency, and wave-propagation velocity.
b. The amplitude of the electric field and the magnetic field of an electromagnetic wave.
c. The electric permitivity (ε_0), the magnetic permeability (μ_0), and the speed of light.
d. The energy density and the amplitude of an electromagnetic wave.
e. The irradiance (in W/m^2) and the amplitude of an electromagnetic wave.

19. Prove whether the following expressions are correct or incorrect:
a. When two electromagnetic waves interfere the result will be the vector sum of the two waves.
b. The energy density of a standing wave is zero.
c. A standing wave cannot be used to transport energy.
d. The total radiance of two independent light sources is always the sum of the individual values of the two sources.
e. A standing wave can be composed of two separate propagating electromagnetic waves, with one moving from left to right and one moving in the opposite direction from right to left.
f. The speed of light equals the phase velocity.
g. The speed of light equals the group velocity.
h. Information can be transported with:
   i. The speed of light
   ii. The phase velocity
   iii. The group velocity

20. A harmonic oscillation is represented by the following equation: $y=10 \sin(628.3x-6283t)$, with x and y in centimeters and t in seconds. Determine:
a. The angular frequency
b. The period
c. The velocity
d. The wavelength
e. The frequency
f. The wave number
g. The amplitude

21. The energy flux from the sun to the earth is approximately 1.4 kW/m^2 on the earth's surface.
a. Determine the maximum values of E and B for a harmonic plain electromagnetic wave of the same value energy flux.
b. Consider an electromagnetic wave as a stream of photons. Calculate how many photons hit the earth's surface per second per square meter, when the earth's diameter is in meters and the average wavelength of the photons is 550 nm).

22. Consider an electromagnetic wave propagating in the z-direction. The $\mathbf{E}$ vector of this wave can be written as follows: $\mathbf{E} = E_x e_x + E_y e_y$, with $E_x = \sin(kz + \omega t)$

and E_y=cos $(kz + \omega t)$, and e_x and e_y the unit vectors in the x and y direction, respectively.

a. For $t = 0$ create a three-dimensional illustration in the Cartesian coordinate system (x,y,z) of the **E**-field along the z-axis for the distance z=0,.......,λ.(Tip: first calculate the values at the locations: z=0, $\frac{1}{8}\lambda$, $\frac{1}{4}\lambda$, $\frac{1}{2}\lambda$, and λ.)

b. For z=0 create a three-dimensional illustration in the Cartesian coordinate system (x, y, z) of the **E**-field along the z-axis for the time interval t=0,...,T. (Tip: calculate the values at: $t = 0, \frac{1}{8}T, \frac{1}{4}T, \frac{1}{2}T$, and T.)

c. Create a graph for z=0 in the x–y plane (two-dimensional only), the position of the electric field vector at t=0, $\frac{1}{8}T, \frac{1}{4}T, \frac{1}{2}T$, and T, and also provide the values that the electric field can assume in this two-dimensional plane.

23. Consider the following three mathematical equations as function of time and place:

$$y(x,t) = A\sin^2\left[\frac{3}{4}\pi(t+x)\right] \tag{2.109}$$

$$y(x,t) = A(t-x)^2 \tag{2.110}$$

$$y(x,t) = \frac{A}{x^2\frac{3}{4}\pi - t} \tag{2.111}$$

a. Which equation(s) qualify as traveling waves?
b. Prove your conclusion.
c. Provide for the equations that qualify the magnitude and direction of the wave velocity.

24. Show that y = ln(a+x−vt) is a solution to the wave equation.

25. Two plane electromagnetic waves with the electric field oscillating in the z-direction can be represented by the following expressions:

$$E_x(x,t) = 4\sin\left(\frac{\pi}{3}x + 20t + \pi\right) \tag{2.112}$$

$$E_y(y,t) = 2\sin\left(\frac{\pi}{4}y + 20t + \pi\right) \tag{2.113}$$

a. Give the expression for the interfering wavefront at $x = 5$ and $y = 2$.

26 The dispersion curve of glass can be approximated by Cauchy's equation: n = A+(B/λ^2). .

a. Explain the difference between group velocity and phase velocity.
b. Determine the phase velocity and group velocity of light with wavelength = 500 nm for glass with $A = 1.40$ and $B = 2.5 \times 10^6$ Å^2?
c. Show that the sum of the phase velocity and the group velocity inside the glass medium is constant.

27. Two wavefronts with parallel electric fields are presented as expressed by

$$E_1 = 3 \sin\left(\mathbf{k}_1 \bullet \mathbf{r} - \omega t + \frac{\pi}{5}\right) \tag{2.114}$$

$$E_2 = 4 \sin\left(\mathbf{k}_2 \bullet \mathbf{r} - \omega t + \frac{\pi}{6}\right) \tag{2.115}$$

Consider the amplitude to be in kV/m. The wavefronts interact at a point where the phase difference due to the path difference between the two waves is $\pi/3$, and E_1 traversed a longer path length. Calculate the following parameters at the point were wavefront 1 is superpositioned on wavefront 2:

a. The radiance of the individual wavefront
b. The combined radiance as a result of interference
c. The average radiance
d. The fringe resolution

3

Review of Optical Principles: Classical Optics

In this chapter some of the basic definitions used in optics are presented, followed by an in-depth theoretical description of the electromagnetic theory; how electromagnetic waves are generated, what laws they obey, and the different classifications of electromagnetic waves from television waves to gamma radiation. The general area of optics in the visible light spectrum is covered, and the sources of electromagnetic radiation are discussed. In addition, the rules that were developed over the centuries regarding the modeling of light and image formation are presented and explained. The quantum theory of light is outlined, followed by several theories regarding absorption and how to model light scattering. At the end of this chapter follows a compendium of the common terms and definitions in the arena of optics principles.

3.1 Geometrical Optics

Geometrical optics describes the part of optics that involves image formation and related phenomena. The principles use geometry to track the path of the electromagnetic waves. These tracks can also be seen as rays of particles, consistent with the wave-particle duality of light. In Chapter 2 the wave characteristics of electromagnetic radiation were described. Generally, wave propagation in three dimensions is quite a complicated process, especially when there are multiple waves involved. In addition, one cannot always identify the point source of each single electromagnetic wave to apply the Maxwell equations to find the final result tied into the source. The description of the superposition of multiple waves traveling as a wavefront was formulated by Christiaan Huygens (1629–1695) in 1678, based on his observations on mechanical wave propagation, and was extrapolated in the early 1900s to fit the new description of electromagnetic waves. Christiaan Huygens was a contemporary of Sir Isaac Newton (1642–1727).

3.1.1 Huygens' Principle

Huygens' principle is used to predict the future position of a wavefront when an earlier position is known, and can be stated as follows: every point on a wavefront can be considered as a source of tiny wavelets directed in the same direction as the entire wave. The new wavefront is the envelope of all the wavelets. Since every oscillating charge is a point source, and multiple oscillating charges form a multipoint source, the overall light generation can present a rather complicated situation.

Huygens' theory of wave propagation used rudimentary mathematical concepts to track the path of a wave, or what is often referred to as the ray of energy of the wave. With this in mind the introduction of simple geometrical concepts was no quantum step. In electromagnetic wave theory every point in space turns into a secondary source, which is excited by the original source. As a result each point in space would be generating a spherical wave propagating in all the three dimensions. In Huygens' principle, only the components of the wavefront found by superposition of the entire collection of secondary sources form a tangent to the one and only Huygens' wavelet. The wavelets moving in the opposite direction of the original wavefront are neglected in the Huygens' postulate, without any justification. It later turns out that this is justifiable, since electromagnetic waves do not require a medium to travel in, and are entirely self-contained.

In the situation of an electromagnetic wave traveling through a boundary area between two media, the secondary source postulate forms a useful mechanism to explain some of the phenomena that will be described in the following sections.

3.1.2 Laws of Geometrical Optics

There are five laws that describe geometrical optics, and that help establish the concept of geometrical optics as a method of describing image formation. The five laws are as follows:

1. Law of rectilinear propagation
2. Law of reflection
3. Law of refraction
4. Law of diffraction
5. Law of conservation of energy

The first law describes the general concept of wave propagation in a homogeneous medium. The last four describe the interaction at the interface of two media. When light enters a tissue from another medium, it first interacts with the interface of the medium and tissue. Reflected light bounces back into the first medium (e.g., air), and refracted light enters the second medium (e.g., tissue). There are two types of reflection. Specular reflection occurs when the refractive indices of the medium and the tissue are different. In this case, light bounces back into the first medium from the interface. The other type of reflection, diffuse reflection, occurs when light entering the second medium scatters around, and finally scatters back into the first medium. Refraction depends on the angle of incidenct light and the refractive indices of the first and second medium. It will be shown that Snell's law defines the direction of refracted light.

The principles of the five foundations of geometrical optics will be described in detail next.

3.1.2.1 Law of Rectilinear Propagation

Light rays in a homogeneous medium travel in straight lines. A single light ray has no specific meaning in the context of Huygen's principle, since the wavefront of the Huygen's theoretical approach, however small, will always have a finite surface area. A finite area also entails that an infinite number of normals can be constructed to this surface. Figure 3.1 illustrates the Huygen's wavefront of every point on the path of a wave acting as a new source, reconstructing the wide beam wavefront.

3.1.2.2 Law of Reflection

In the case of reflection, the wavefront will not necessarily have the normal to the surface of interaction. The surface, namely, will be an interface between two uniquely characterized media. The "disturbance" causing the wavefront will approach the interface between the two media at an angle with the normal to the surface, and as such when a portion of the wavefront reaches the interface, the opposite end of the wavefront, by definition, may

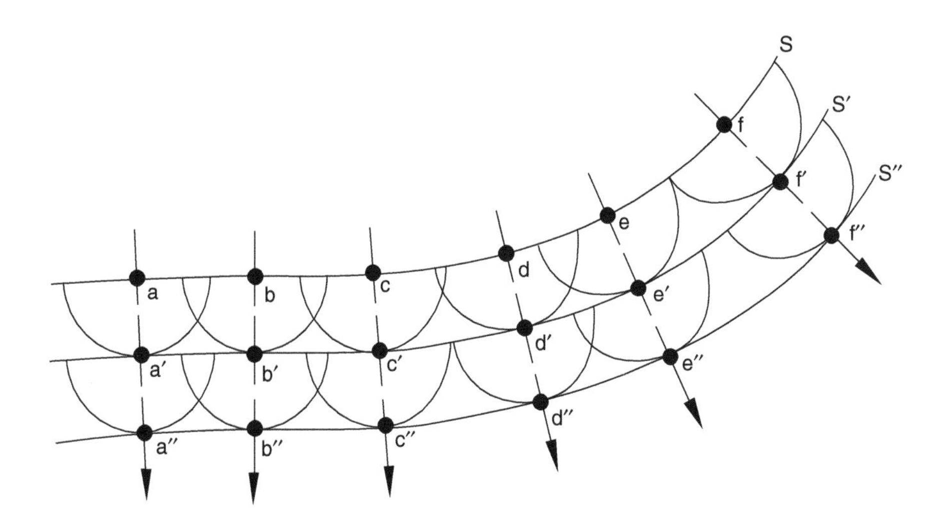

FIGURE 3.1
Illustration of the Huygen's principle

not have reached that interface yet. In this situation there will be a time difference between the generation of the disturbances recreating the wavefront from the opposite sides of the width of the wavefront. As a result the recombination of the wavefront will be re-emitted from the secondary sources on the interface in Huygens' model, resulting in a wavefront traveling in an opposite direction to the incident wavefront, but with an angle to the normal of the interface that has a different direction from the incident angle. This becomes clear in Figure 3.2.

When the interface between the two media is a plane surface, then it can be shown that the angle of incidence with the normal to the surface is the same as the angle of reflection, only in the opposite-side quadrant on the same side of the interface. As a rule the Huygens' diagram will be chosen so that all rays connecting the wavefronts are in a single plane.

3.1.2.3 Law of Refraction (Snell's Law)

Similarly in the case of a wavefront passing through an interface between two media, one side of the wavefront will pass through the surface separating the two media before the opposite boundary of the wavefront. In addition to the time difference between the crossover of the disparate extremes, the speed of propagation will be a different value depending on the characteristics of the second medium with respect to the first.

As a result of the combination of the time difference and the speed of propagation of the wavefront at either side of the wavefront, the distance traveled by the wavefront originating at the boundary will also be dissimilar using $x=vt$, as illustrated in Figure 3.2.

This phenomenon causes the wavefront to kink at the border separating the media. The rules that are obeyed at this intersection were derived by

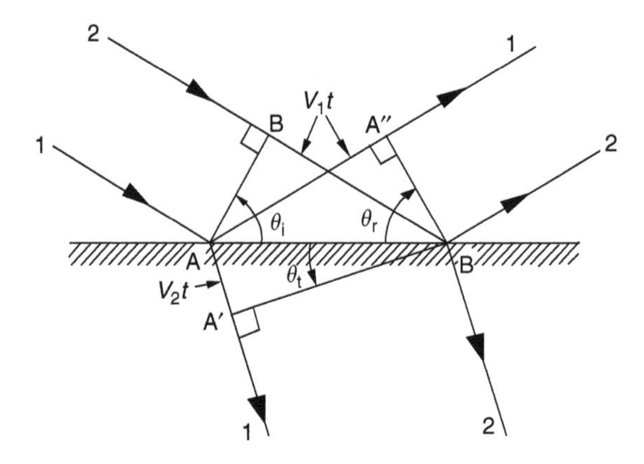

FIGURE 3.2
Refraction at surface with different indices of refraction on either side of the interface.

Willibrord Snell. Snell's law uses the nondimensional indicator for speed of propagation relative to the speed of propagation in vacuum:

$$n = \frac{c}{v} \tag{3.1}$$

where n is the relative index of refraction.

Snell's law of diffraction relating the incident and the refracted angle with the normal to the surface is now defined as

$$n_1 \sin\theta_1 = n_2 \sin\theta_2 \tag{3.2}$$

or

$$\frac{n_1}{n_2} = \frac{\sin\theta_2}{\sin\theta_1} \tag{3.3}$$

A list of the index of refraction of various media is presented in Table 3.1.

3.1.2.4 Law of Diffraction

In the mid-seventeenth century, a Jesuit priest, Fransesco Grimaldi (1618–1663), described his observations of sunlight passing through a finite diameter hole, producing a spot of illumination on the opposite wall that was larger than the size of the hole. He also observed that the edge of the spot was not sharply demarcated. Grimaldi attributed this fuzzy demarcation line to the same principles as observed by Huygens for water waves, that is, to diffraction based on the wave properties of light. In most cases,

TABLE 3.1

Index of Refraction of Various Media

Substance	State	Temperature (°C) at a Pressure of 100 kPa	Index of Refraction
Air	Gas	0	1.000293
Carbon dioxide	Gas	0	1.00045
Ice	Solid	0	1.309
Water	Liquid	20	1.333
Ethyl alcohol	Liquid	0	1.361
Fluorite (CaF2)	Solid	20	1.434
Fused quartz (SiO_2)	Solid	20	1.458
Benzene	Liquid	20	1.501
Glass, crown	Solid	20	1.52
Salt (NaCl)	Solid	10	1.544
Glass, flint	Solid	20	1.6–1.9
Diamond	Solid	20	2.419

however, the physical dimensions of an object are many orders of magnitude larger than the wavelength, thus minimizing the edge effects. This is why the ray description of light in geometrical optics is so useful in most of the cases. The smaller an aperture or, for that matter a lens, the more pronounced are the effects of the edges on the demarcation of the image formed by the limited size of the optical instrument. The aperture will produce a diffraction pattern, resulting from interference of waves generated at the various locations inside the span of the aperture (still using the Huygen's principle).

The distance traveled from one secondary source inside the aperture to the surface where the image is formed will have a finite difference with the wave emanating from the point adjacent to this point by the time both rays reach the surface where the image is formed: the projection screen. Since all the waves generated inside the aperture have the same phase, the waves will interfere with each other at the projection screen. When the path difference between two waves meeting at one place on the screen has a wavelength difference of a whole number between them, the phase difference will be zero, and there will be positive interference. In the other extreme, when the path lengths differ by an odd number times half a wavelength, the two waves will be out of phase by 180° or π, and there will be total extinction. In this situation the secondary sources also will generate wavefronts propagating as a spherical wave in 4π free space, thus providing path difference beyond the physical boundaries of the aperture. The single-slit configuration is illustrated in Figure 3.3. These conditions will produce a pattern of maximal and minimal interference at the location of the screen. This interference pattern is also known as the diffraction pattern of the aperture. The interference pattern is illustrated in Figure 3.4.

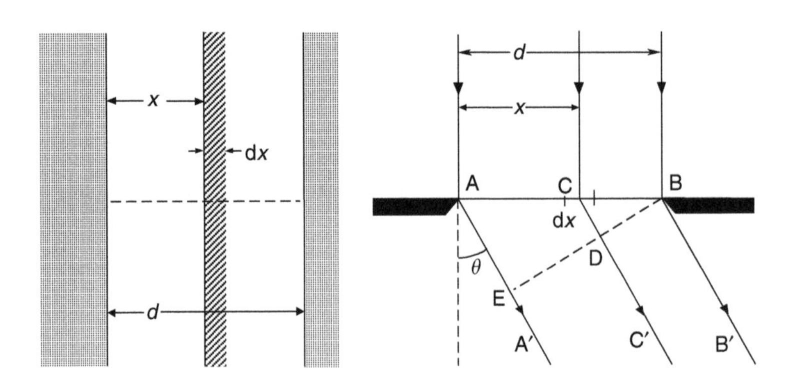

FIGURE 3.3
Rays of light passing through a slit where every point between the edges as well as the edges becomes a source of light. This pattern illustrates the diffraction from an aperture with width d. The figure on the left illustrates the view of the incident light, while the illustration the right shows the geometric outline for the ray pattern using section dx, providing an arbitrary secondary source at a distance x from one edge

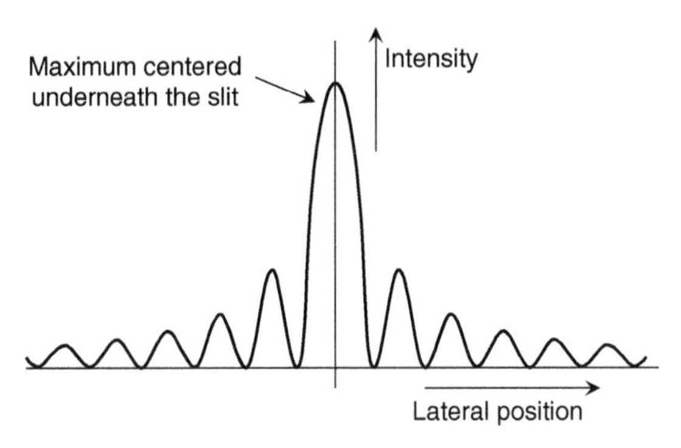

FIGURE 3.4
Diffraction through a slit with width d.

3.1.2.4.1 *Fresnel Diffraction*

As mentioned earlier, electromagnetic waves are just waves, and waves do not have finite dimensions in contrast to the particle theory for light. Waves will interfere with themselves and with objects in the path of the wave. The size of the obstruction and the wavelength of the interacting light will determine the total effect. Fresnel diffraction uses the fact that every location in space forms its own new point source, emitting spherical waves. This wave principle is similar to the ripples formed on the water surface when a stone is thrown into the water. This principle was introduced by the French mathematician and physicist Augustin Jean Fresnel (1788–1827). In geometrical optics, the Fresnel diffraction can describe the interaction of a light source with a narrow aperture, or several narrow apertures, with great ease and convenience. The aperture in general is divided into several Fresnel zones. For instance, when a beam of light hits a circular aperture, the aperture is divided into concentric circles with a subsequent spacing providing a path length s, from each edge of the concentric circle to the imaging screen giving a difference equivalent to half the wavelength of the incident light as

$$s_2 - s_1 = \frac{\lambda}{2} \tag{3.4}$$

The diffraction pattern generated by a single slit follows the rule that when the path length difference between two sources equals a whole number of wavelengths, the waves will line up and the amplitude will sum up to give the maximum amplitude of the two waves combined. In contrast, if the path length equals an odd number of half wavelengths, the amplitudes of the individual waves will cancel each other out to give minimum interference. The graphical representation of light passing through a slit is shown in Figure 3.4.

The equation for minimal interference follows from the condition for a single slit:

$$D \sin\theta = m\lambda \tag{3.5}$$

For a slit with width D and θ the angle with the normal to the plane of the slit. This is due to the fact that any point in between the edges of the slit also acts as a secondary source, and when the slit is split into half, there will be a counterpart for each source at a distance of $D/2$ starting from one edge, moving towards the center. An outline of this situation is presented in Figure 3.3.

The mathematical approach to solving the wave propagation using the Fresnel diffraction principle is not generally used. The integrals describing Fresnel diffraction in most cases are too difficult to even be considered for general use. Fresnel diffraction patterns have few analytical solutions, even in one dimension. Fraunhofer diffraction is a lot more forgiving, and is used more often in mathematical analyses.

3.1.2.4.2 *Fraunhofer Diffraction*

The description of light with simple geometrical objects was made by Joseph von Fraunhofer (1787–1826) as a field of interest to him resulting from his work as a spectacle maker (optician), and later on as part of the work in the fabrication of telescopes.

Boundary conditions can be limited to cases where the approaching light is parallel to the object in its path, and the light is monochromatic in nature. An additional restriction is the fact that the distance to the imaging plane needs to be much larger than the obstruction causing the light to diffract. This is known as imaging in the far-field, whereas Fresnel diffraction is in the near-field. Fraunhofer diffraction describes the interaction of coherent plane waves incident upon an obstruction. This principle resembles the interaction of landmasses and piers with the waves of an ocean.

3.1.2.4.3 *Fraunhofer Diffraction Equation*

Diffraction through a panel with two small slits will give an interference pattern on a screen at a distance behind the slits that obeys the following mathematical description:

$$I = 2I_0\left[\frac{\sin kb\theta/2}{kb\theta/2}\right]^2\left[1+\cos(k\,\Delta+kd\theta)\right] \tag{3.6}$$

where I_0 is the peak intensity for one slit, λ the wavelength of the incident light, $k = 2\pi/\lambda$ the wavenumber, d the distance between the slits, b the width of the slits, θ the angle of observation with the central axis to the screen, and the Δ difference of the optical lengths of interfering rays (in this case, for example, when the incident wave is not perpendicular to the screen or one slit is covered by glass).

An outline of the two-slit diffraction is shown in Figure 3.5.

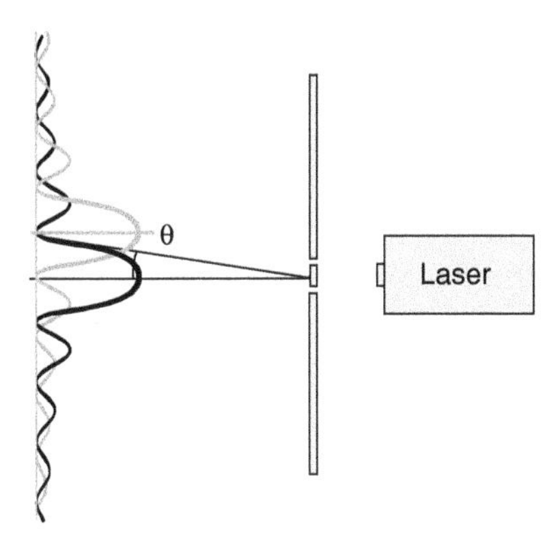

FIGURE 3.5
Diffraction pattern on a screen behind two slits.

Diffraction through a panel with N number of small slits will give an interference pattern on a screen at a distance behind the slits that obeys the following mathematical description:

$$I = I_0 \frac{\left[\sin^2(\beta)\sin^2(N\alpha)\right]}{\beta^2 \sin^2(\alpha)} \tag{3.7}$$

where

$$\alpha = \frac{\pi d \sin\theta}{\lambda} \tag{3.8}$$

and

$$\beta = \frac{\pi b \sin\theta}{\lambda} \tag{3.9}$$

3.1.2.4.4 Rayleigh Criterion

When two objects are projecting their image through the same aperture, both objects will generate a diffraction pattern that will overlap each other. Under the condition that the maximum of image 1 is adjacent to the maximum of image 2, and not overlapping, the two images can be separated or resolved. This condition is described by the Rayleigh criterion.

Using Equation (3.5) and the fact that at small angles $\sin(\theta) = \tan(\theta) = \theta$, Equation (3.5) gives the location where the maximum of the second aperture needs to be with respect to the maximum at 0° of the first aperture: $D\theta = m\lambda$, with $m = 1$, giving the expression

$$D\theta = \lambda \tag{3.10}$$

or

$$\theta = \frac{\lambda}{D} \tag{3.11}$$

The Rayleigh criterion for a circular aperture (e.g., lens) will require corrections for the nonuniform separation between the edges, and results in the angle that can be seen between two objects with respect to the aperture that needs to satisfy

$$\theta = \frac{1.22\lambda}{D} \tag{3.12}$$

where D is the opening of the aperture or diameter of the lens.

3.1.2.5 Law of Conservation of Energy

At the interface between two media the energy needs to be conserved. For the energy to be conserved the sum of the amplitudes of the incident and the reflected fields need to equal the transmitted amplitudes.

In addition, the phase of all the waves needs to be identical at the position of interface of the media. Only under these conditions will energy be conserved, since the energy of the electromagnetic radiation is the square of the electric and magnetic field waves combined. Also see the section on Poynting vector in Section 2.3.4 in Chapter 2.

3.1.3 Fermat's Principle

The Fermat principle is closely related to the action principle in classical mechanics. Analogous to other physics principles, it states all the pertinent physics laws pertaining to geometrical optics. Supposing there are two points in space, P_1 and P_2, and one wants to find the path of the ray of light that passes through both points. Fermat's principle states that any deviation of the path connecting both points by a true ray is of the first order in small quantities and will produce a deviation in optical path length that is at least of second order in small quantities. This was derived by the French mathematician Pierre de Fermat (1601–1665). Formulated differently, the time it takes for a light ray to travel from point P_1 to P_2 in space has an upper limit. In other words: *The actual path between two points taken by a beam of light is the one which is traversed in the least time*; the optical path length is either minimal or maximal, depending on the circumstances. The maximal path length will be achieved under gravitational lensing; for the conditions pertaining to the material in this book the propagation will strive to satisfy a minimal path length.

3.1.4 Ray Optics

Ray optics is the branch of geometrical optics that describes light as rays. The rays emanate from a source and are perpendicular to the wavefronts described in Section 2.2 in Chapter 2. The rays can illustrate the Laws of Rectilinear propagation, reflection, and refraction with great accuracy and ease. Examples of geometrical optics are presented in image formation with lenses and mirrors in the subsequent chapters. By definition, the light ray at a given point in space is along the gradient of the optical path function. The optical path function $L(\mathbf{r})$ is defined as the optical path length from a conveniently chosen reference wave surface Σ_0 to an arbitrary point P. The point P is identified by the vector $\mathbf{r}$, and this point can be reached by a ray emanating from the wave surface Σ0, and a point P_0 designates a location on the surface Σ0, with the vector $\mathbf{r}_0$ as position vector for P_0. The optical path function ΔL, traveling from point A to point B in three-dimensional space can be defined in a medium with speed of light v and index of refraction n as

$$\Delta L(\mathbf{r}) = \int_{A(\text{ray})}^{B} \frac{c}{v} \mathrm{d}s = \int_{A(\text{ray})}^{B} c\,\mathrm{d}t = \int_{A(\text{ray})}^{B} n(x,y,z)\,\mathrm{d}s \tag{3.13}$$

or in vector notation

$$L(\mathbf{r}) = \int_{\text{ray}}^{r} n\,\mathrm{d}s \tag{3.14}$$

When there is a disturbance in optical space occurring at time t_0, the wavefront Σ_0 is affected. At time t, when $t > t_0$, the wavefront will be at the two-dimensional surface Σ, consisting of all points that reach from Σ_0 after the time interval $(t - t_0)$, where the distance to the surface Σ can be defined as

$$L(\mathbf{r}) = c(t - t_0) \tag{3.15}$$

To obtain the path function $L(\mathbf{r})$, when $\mathbf{r}$ moves a distance $\mathrm{d}\mathbf{r}$ toward the location, P_1 the ray $P_0'P'P_1'$ will construct the new wave surface Σ_1 originating from Σ_0 at position P_1. The change in the optical path function $\mathrm{d}L$ is a direct result of the combined true path from P to P_1 and the index of refraction over that distance:

$$\mathrm{d}L = n \left| P'P_1 \right| = n(\mathrm{d}\mathbf{r} \cdot \Gamma) \tag{3.16}$$

where Γ is a unit vector at position P, which is tangential to the incoming ray $P_0 P$.

Since the length $\mathrm{d}L$ is in three-dimensional space, it can be written as

$$\mathrm{d}L = \frac{\partial L}{\partial x}\mathrm{d}x + \frac{\partial L}{\partial y}\mathrm{d}y + \frac{\partial L}{\partial z}\mathrm{d}z = \nabla L \cdot \mathrm{d}\mathbf{r} = n\Gamma \cdot \mathrm{d}\mathbf{r} \tag{3.17}$$

where the gradient in the optical path function can be represented by

$$\nabla L = n\Gamma \tag{3.18}$$

The absolute value of the gradient of the optical path function is known as the *eikonal equation*, and gives the expression

$$|\nabla L| = n \quad \text{and} \quad \left(\frac{\partial L}{\partial x}\right)^2 + \left(\frac{\partial L}{\partial y}\right)^2 + \left(\frac{\partial L}{\partial z}\right)^2 = n^2 \tag{3.19}$$

Equation (3.19) verifies that the unit vector Γ is along the gradient of the optical path function.

3.1.4.1 Critical Angle

Based on the Snell's equations (2.17) and (3.2) for diffraction at the interface between two media with different index of refraction, certain boundary conditions can be derived.

When light moves from medium 1, with an index of refraction greater than the index of refraction of the medium at the other side of the interface, i.e., medium 2, at a certain angle of incidence, the numerator will need to exceed the value 1, which is the maximum a sine or a cosine function can attain. At an angle of incidence where the incident light results in a value of the numerator of 1 based on the respective indices of refraction of the two media, the refracted light will pass along the surface of the interface between the two media. At any other angle of incidence greater than the aforementioned angle of incidence, the ray will be reflected and not refracted.

The angle θ_1, which satisfies the condition that the numerator $\sin(\theta_2)$ needs to be unity, is called the critical angle. At this angle the propagating wave has minimal penetration, and in addition, the amplitude of the wave degenerates rapidly. The refracted ray is confined to the surface layer of the second medium. At angles greater than the critical angle there will be total reflection only.

The polarization of the incident ray will need to be considered in greater detail, and Snell's law for horizontal and vertical polarized light is slightly different. The result is that unpolarized light may have a dominant component of the electric field parallel to the interface between the two media after interaction, and as such becomes horizontally polarized after reflection.

3.1.4.2 Brewster Angle of Polarization

The angle of incidence resulting in purely horizontally polarized reflected light is called the Brewster angle.

Consider a ray of light passing from air into glass. This ray can be considered to have an electric field parallel to the air–glass interface and one perpendicular to it. Both components have their respective coinciding magnetic fields perpendicular to the electric fields, to produce a vector addition that will explain elliptically polarized light. A special case of elliptically polarized light is circularly polarized light, which is universally the most frequently, naturally occurring polarization state. Here, the field vectors are of equal length, and hence form a circle. In unpolarized light, these two components will have infinitely more neighboring directions of the electric field vector, which again can be deconvolved into the vector sum of two perpendicular field vectors. However, if the angle of incidence with the air–glass interface is at a specific angle, known as the polarization angle θ_p, then the reflection coefficient (the amount of light reflected) of the parallel component of the electric field is zero. From experimental observation it is known that the polarization angle θ_p and the refraction angle θ_r, are at right angles to each other.

$$\theta_p + \theta_r = \frac{\pi}{2} \tag{3.20}$$

Substituting Equation (3.20) into Snell's law reveals

$$n_1 \sin\theta_p = n_2 \sin\theta_r = n_2 \sin\left(\frac{\pi}{2} - \theta_p\right) = n_2 \cos\theta_p \tag{3.21}$$

or

$$\theta_p = \tan^{-1}\left(\frac{n_2}{n_1}\right) \tag{3.22}$$

This is known as *Brewster's law*, and θ_p is the Brewster angle of polarization.

This phenomenon of polarization can be illustrated by solving the Maxwell equations for the electric field split up into a component parallel to the plane of incidence of the electric field propagation rays (sub-π), with the two adjacent media, and a component of the electric field perpendicular to the plane of incidence (sub-σ). In general, the latter would mean that the electric field is parallel to the interface between the media. The plane of incidence is the plane that contains the normal to the interface and the propagation vector of the incident light. The equations for reflection (R) and transmission (T) of the respective incident (sub-i), refracted (sub-f), and reflected (sub-r) components perpendicular to each other yield Equations (3.23)–(3.26). The reflection for the parallel field is given as

$$R_\pi = \frac{E_{r,\pi}}{E_{i,\pi}} = \frac{n_1 \cos\theta_f - n_2 \cos\theta_i}{n_1 \cos\theta_f + n_2 \cos\theta_i} \tag{3.23}$$

while the reflection with respect to the perpendicular field is given by

$$R_\sigma = \frac{E_{r,\sigma}}{E_{i,\sigma}} = \frac{n_1 \cos\theta_i - n_2 \cos\theta_f}{n_1 \cos\theta_i + n_2 \cos\theta_f} \tag{3.24}$$

The transmission for the parallel field is given as

$$T_\pi = \frac{E_{f,\pi}}{E_{i,\pi}} = \frac{2n_1 \cos\theta_i}{n_1 \cos\theta_f + n_2 \cos\theta_i} \tag{3.25}$$

while the transmission for the perpendicular field is

$$T_\sigma = \frac{E_{f,\sigma}}{E_{i,\sigma}} = \frac{2n_1 \cos\theta_i}{n_1 \cos\theta_i + n_2 \cos\theta_f} \tag{3.26}$$

The overall reflection coefficient average over all angles can be written as

$$R = \left(\frac{n_2 - n_1}{n_2 + n_1}\right)^2 \tag{3.27}$$

and the overall transmission coefficient is given by

$$T = \frac{4n_1 n_2}{(n_1 + n_2)^2} \tag{3.28}$$

3.1.4.3 Lens Maker's Equation

The lens maker's equation is the foundation for image formation via refraction. It not only applies to lenses but can also be used to derive the optic ray pattern inside a tissue.

The formation of an image by refraction from a curved surface with radius of curvature *R*, and the center of curvature indicated by C, separating two media with indices of refraction n_1 and n_2, can be derived using the ray propagation description derived from the Huygens' principles in geometrical optics. The equation describing the image formation relating the image distance to the object distance, the radius of curvature, and the indices of refraction will be derived by applying Snell's law to the incident and emerging rays from the curved surface and using the small-angle approximation. The small-angle approximation states that $\tan(\theta) = \sin(\theta) = \theta$ for small angles expressed in radians. The other tool is the goniometric expressions, which state that tan (θ) is the opposing edge of the triangle divided by the adjacent side of the triangle.

The angles of the ray with the normal to the surface θ_1 and θ_2 are related by Snell's law (Equation [2.17]), which for small angles can be written as $n_1\theta_1 = n_2\theta_2$(Figure 3.6). From the triangle ACP_2 it can be shown that

$$n_1\theta_1 = n_2\beta - n_2\gamma \tag{3.29}$$

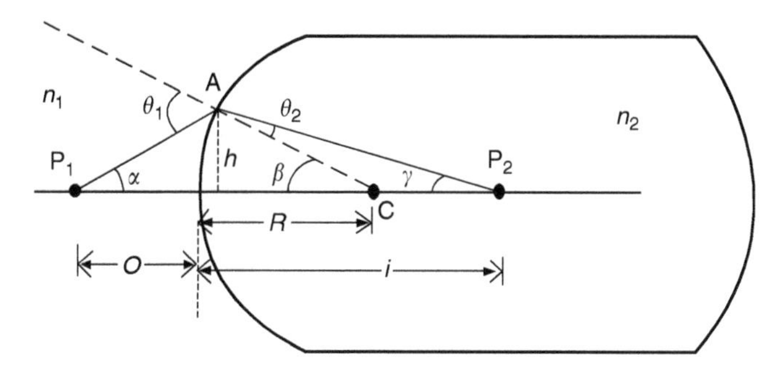

FIGURE 3.6
Geometry for determination of the power of a curved medium.

From the triangle P_1AC it can be shown that $\theta_1 = \alpha + \beta$. Eliminating θ_1 gives

$$n_1\alpha + n_2\gamma = (n_2 - n_1)\beta \tag{3.30}$$

When these angles are expressed in radians and the angles are small, they are related to the image distance, object distance, and radius of curvature by using the definition of radians as the arc length of a slice of a circle with radius 1; it follows that the respective angles α, β, and γ, when assuming that the incoming and exiting rays hit the curved surface at height h from the principal axis can be approximated as follows:
$\alpha \approx h/\mathrm{o}$, $\beta \approx h/\mathrm{R}$, and $\gamma \approx h/i$. Substituting these approximations for the angles in Equation (3.30) and eliminating h gives

$$\frac{n_1}{o} + \frac{n_2}{i} = \frac{n_2 - n_1}{R} \tag{3.31}$$

Note that for refraction of an object to the left of the surface, real images are formed to the right of the surface and virtual images to the left. Thus i_1 and R are taken to be positive if the image and center of curvature lie on the right of the surface.

In the process of image formation by a solid lens in a medium with a different index of refraction, commonly air with $n = 1$, this process will need to be repeated at the opposite face of the lens. Consider a lens of index of refraction n_2 embedded in a medium with index of refraction n_1 . The lens has a curved surface with radius R_1 in the direction of the incoming ray, and the other curved surface with radius R_2 on the exit side where the rays will emerge from. When an object is placed at a distance o from the first surface, application of Equation (3.29) gives the distance of the image due to refraction at the first surface as

$$\frac{n_1}{o} + \frac{n_2}{i_1} = \frac{n_2 - n_1}{R_1} \tag{3.32}$$

As a rule, a thin lens will not form an image inside the lens itself since the light is again refracted at the second surface. If the thickness of the lens is t, the distance from the image point i_1 to the second surface is $t-i_1$. The final image position is the result of the primary and secondary refraction at the front and back surface, respectively. By using the image distance from the front surface as the object distance for the second surface, it turns out that, for all possible values of the primary image distance i_1, the image formed by refraction at the second surface is at a distance i_2 from the second surface, which is the effective image distance i for the image formation by the total lens, where i again follows from the same goniometric relationships as were derived for the front surface :

$$\frac{n_2}{t-i_1}+\frac{n_1}{i}=\frac{n_1-n_2}{r_2} \tag{3.33}$$

For a *thin lens* the thickness t can be neglected in Equation (3.33) and i_1 can be eliminated by solving for n_2/i_1 as

$$\frac{1}{i}+\frac{1}{o}=\frac{n_2-n_1}{n_1}\left(\frac{1}{R_1}-\frac{1}{R_2}\right) \tag{3.34}$$

This gives the image distance i in terms of the object distance o, the incident medium index of refraction n_1 and the properties of the thin lens R_1, R_2, and n_2, which depends on the material the lens is constructed from. The radius of the aft surface of the lens will be negative for a convex surface as in Figure 3.6. Using Equation (3.35) for image formation we obtain

$$\frac{1}{f}=\frac{1}{i}+\frac{1}{o} \tag{3.35}$$

And the fact that the focal length of a lens is the location of the image from an object placed at infinite distance to the lens, Equation (3.34) combined with Equation (3.35) can be rewritten as

$$\frac{1}{f}=\frac{n_2-n_1}{n_1}\left(\frac{1}{R_1}-\frac{1}{R_2}\right)=P \tag{3.36}$$

Equation (3.36) is known as the *lens maker's equation*, where P is the power in diopter of the lens, and Equation (3.35) is the thin lens equation.

3.1.4.4 Optical Instruments

Geometrical optics covers a small but important aspect of the biomedical optics field. Several instruments have been developed over the ages whose operation can straightforwardly be described with the use of simple geometry and some knowledge of material properties, without the need to know about the theory of electromagnetic waves and quantum theory. Examples of some of the optical instruments that operate on the previously described principles that will be discussed in greater detail in Chapter 5 are the following: microscope, endoscope and other fiberoptic imaging devices, and ophthalmoscope.

3.2 Other Optical Principles

An additional topic in the theory of the interaction of electromagnetic radiation with turbid media is related to an accompanying aspect of the influence of speed of light of each particular media with respect to the wavelength of light. This dependence is known as dispersion and is described mathematically, as in Equation (2.61), by

$$v(\lambda) = \frac{c}{n(\lambda)}\lambda \tag{3.37}$$

where v is the wavelength-dependent speed of light in the medium, c the speed of light in vacuum, and n the wavelength-dependent refractive index of the medium.

3.2.1 Dispersion

The fact that the index of refraction has a correlation with the frequency/wavelength of the electromagnetic radiation has consequences regarding the propagation of different wavelengths through a single medium. In general, when white light, a combination of the entire visible spectrum in a single stream, enters a medium that has distinctly different indices of refraction for wavelengths ranging from blue through red, the colors will refract at different angles when encountering an interface at an angle different than zero with the normal. This phenomenon explains the creation of the rainbow when sunlight enters and exits a droplet of water.

3.3 Quantum Physics

The fact that electromagnetic radiation has the energy as if it were photon packages that are being transmitted, carrying energy, $W = hv$, has many implications.

An atom or a molecule absorbs electromagnetic radiation, especially when the frequency of the incident radiation matches the lines in the emission spectrum of the atom/molecule. This implies that the energy quanta that can be absorbed need to fit energy-level transitions inside the structure of the atom or molecule. These energy-level transitions will coincide with resonance spectra of vibrational or rotational motion, in addition to orbital energy transitions of charged particles in the configuration of the atomic structures, so-called excitation levels.

3.3.1 Schrödinger Equation

In the early 1900s experimental evidence was mounting to indicate that particles also exhibit wavelike characteristics. The initial link was the de Broglie condition for electron orbit. de Broglie stated that an electron in orbit around a nucleus had to satisfy a wave pattern that would form a closed wave to be able to establish an orbit. This means that the wave would repeat itself in orbit each time with no discontinuities.

In 1924 the Austrian physicist Erwin Schrödinger (1887–1961) formulated the wave behavior in an energy well as described next.

For a particle with mass m moving in one dimension (x-direction), with kinetic energy E and potential energy V, the Schrödinger wave equation reads as

$$\frac{\partial^2\psi}{\partial^2 x}+\frac{8\pi^2 m}{h^2}(E-V)\psi=0 \tag{3.38}$$

where h is the Planck's constant, and Ψ the wave function of the particle.

The Schrödinger equation gives the kinetic energy plus the potential (a sum also known as the Hamiltonian operator H) of the wave function Ψ, which contains all the dynamical information about a system. The wave function Ψ is a scalar function with complex values such as

$$(E-V)\psi=H\psi=-\frac{ih}{2\pi}\frac{\partial\psi}{\partial t}=\frac{-h^2}{8m}\nabla^2\,\psi+V(0,x)\psi \tag{3.39}$$

The Laplace operator is defined, in Cartesian or spherical coordinates, as

$$\nabla^2=\frac{\partial^2}{\partial x^2}+\frac{\partial^2}{\partial y^2}+\frac{\partial^2}{\partial z^2}=\frac{\partial^2}{\partial \theta^2}+\frac{\partial^2}{\partial r^2}+\frac{\partial^2}{\partial \varphi^2} \tag{3.40}$$

For the time-independent case, energy is written as the operator $-ih\partial/2\pi\partial t$, and kinetic energy as the square of the momentum operator $ih\nabla^2/8m$, given the potential at $t=0$, $V(0,x)$, and suitable boundary conditions. Solving this differential equation generates a wave function ψ, which contains all the properties of the system.

The Schrödinger equation can also be expressed in terms of the momentum of the particle as

$$P\psi(r,t) = -\frac{ih}{2\pi}\nabla^2 \psi(r,t) \quad (3.41)$$

Using the definition for energy

$$E = \frac{P^2}{2m} \quad (3.42)$$

where P equals the momentum operator $P = mv$, the particle mass multiplied by the velocity of the particle. The Schrödinger equation is used in Chapter 8 to calculate the wavefunction. It is also invaluable in matrix wave optics described next.

3.4 Gaussian Optics

Gaussian optics is a specific branch of geometrical optics. It was named after the same nineteenth-century mathematician Karl Friedrich Gauss we encountered earlier in the electric field theory. In Gaussian optics, the use of the Maxwell equations combined with simple geometry resulted in a set of vector equations that can very accurately describe the image formation by various optical media/instruments. They are the first-order approximation to the exact theory outlined in Sections 2.1 and 2.2 in Chapter 2. The only limitation is that the paths of the waves, the rays of light, and the normals to the interfaces between optical media make only small angles with the axis of symmetry, or the principal axis or optical axis. A ray is a straight line connecting points on the successive wavefronts in the direction of propagation, normal to the wavefront. The size, location, and orientation of an object are defined by light rays leaving from the object, identified as light ray vectors instead of the place and size of the object with respect to the optical instrument. The vector defines the coordinates of location z, and $\alpha = dz/dx$, an angle within a selected frame of reference in the phase space, in most cases the principal axis. The vector description of the object will appear like

$$\mathbf{x} = \begin{pmatrix} z \\ \alpha \end{pmatrix} \quad (3.43)$$

or in vector format as

$$\mathbf{x} = (z, \alpha) \quad (3.44)$$

Depending on the choice of the frame of reference, there will be a transformation involved to another frame of reference. In general, the image formation can be described in a Cartesian coordinate system with three dimensions x, y, and z.

3.4.1 Matrix Methods in Gaussian Optics

In Gaussian optics the light rays are imaged, rather than the object. In this geometry a light ray undergoes a transformation when passing through an image-forming structure while moving from point x_1 to point x_2 .This transformation can be described with the Matrix equation (3.44) shown as:

$$\mathbf{x}(x_2) = \mathbf{M}\mathbf{x}(x_1) \tag{3.45}$$

And if the image formation takes place through the interaction between various optical instruments, the entire system can be substituted by one single transformation matrix, which describes the cumulative effect of all the individual optical elements as shown in the following equation:

$$\mathbf{M} = \Pi_i \mathbf{M}_i = \begin{pmatrix} a & b \\ c & d \end{pmatrix} \tag{3.46}$$

For instance, when the optical element is a thin lens, with negligible thickness and focal length f, the matrix for the ray transformation becomes

$$\mathbf{M}_{\text{lens}} = \begin{pmatrix} 1 & 0 \\ -1/f & 1 \end{pmatrix} \tag{3.47}$$

On transposition over a distance d, the matrix becomes

$$\mathbf{M}_{\text{distance}} = \begin{pmatrix} 1 & d \\ 0 & 1 \end{pmatrix} \tag{3.48}$$

Refraction at a plane with the relative refractive index can be expressed in matrix format as

$$\mathbf{x}' = \mathbf{M}\begin{pmatrix} z_1 \\ \alpha_1 \end{pmatrix} = \begin{pmatrix} z_2 \\ n\alpha_2 \end{pmatrix} \tag{3.49}$$

with the matrix for the medium given by

$$\mathbf{M}_{\text{medium}} = \begin{pmatrix} 1 & 0 \\ 0 & 1/n \end{pmatrix} \tag{3.50}$$

A light ray passing through a medium with refractive index n and thickness d can be written in matrix format as

$$\mathbf{M} = \mathbf{M}_{\text{front surface refraction}} \times \mathbf{M}_{\text{thickness of medium}} \times \mathbf{M}_{\text{rear surface refraction}} \tag{3.51}$$

where the input and output planes lie just outside.
Or substituting the matrix elements, Equation (3.50) can be rewritten as

$$\begin{aligned} \mathbf{M} = \mathbf{M}_{\text{front}} \mathbf{M}_{\text{distance}} \mathbf{M}_{\text{rear}} &= \begin{pmatrix} 1 & 0 \\ 0 & 1/n \end{pmatrix} \begin{pmatrix} 1 & d \\ 0 & 1 \end{pmatrix} \begin{pmatrix} 1 & 0 \\ 0 & 1/n \end{pmatrix} \\ &= \begin{pmatrix} 1 & 0 \\ 0 & n \end{pmatrix} \begin{pmatrix} 1 & d/n \\ 0 & 1/n \end{pmatrix} = \begin{pmatrix} 1 & d/n \\ 0 & 1 \end{pmatrix} \end{aligned} \tag{3.52}$$

A light ray, by definition, is always perpendicular to the wavefront propagating from the electromagnetic wave source. The radius of curvature of the wavefront R at small distance to the principal axis ($z \ll R$) can be defined as $R = z / \alpha$. For an optical system with transformation matrix **M**, it can be shown that the transformation in Equation (3.53) is in effect for the radius of curvature of the wavefront:

$$R(x_2) = \frac{z(x_2)}{\alpha(x_2)} = \frac{az(x_1) + b\alpha(x_1)}{cz(x_1) + d\alpha(x_1)} = \frac{aR(x_1) + b}{cR(x_1) + d} \tag{3.53}$$

When the angles involved in the ray tracing of the optical paths are small, the sine of the angle becomes equal to the angle itself.

When a ray encounters a curved surface of medium 2 with index of refraction n_2 and radius of curvature R_1, originating in medium 1 with index of refraction n_1, the angle of refraction can be described as defined by the following equation (note these are the angles of the ray with the horizontal, they are not with respect to the normal to the surface of the medium):

$$\alpha_2 = \frac{n_1 \alpha_1}{n_2} + \left[\frac{(n_1 - n_2)}{R_1} \right] x_1 \tag{3.54}$$

Incorporating the angle between the normal to the surface and the horizontal, or a line parallel to the principal axis ϕ (which is the same for both sides of the interface), the angle of the ray with the normal θ becomes

$$\theta_1 = \phi + \alpha_1 \tag{3.55}$$

for medium 1, and

$$\theta_2 = \phi + \alpha_2 \tag{3.56}$$

for medium 2, while the location with respect to the principal axis does not change, i.e., $x_1 = x_2$.

The transformation matrix $\mathbf{S}_1$ at the refracting surface is given by

$$\begin{pmatrix} x_2 \\ \alpha_2 \end{pmatrix} = \mathbf{S}_1 \begin{pmatrix} x_1 \\ \alpha_1 \end{pmatrix} \tag{3.57}$$

The first curved surface is expressed in matrix form $\mathbf{S}_1$ as

$$\mathbf{S}_1 = \begin{pmatrix} 1 & 0 \\ -P_1/n_2 & n_1/n_2 \end{pmatrix} \tag{3.58}$$

where P_1 is referred to as the power of the surface, defined in the following equation :

$$P_1 \equiv \frac{n_2 - n_1}{R_1} \tag{3.59}$$

Thus, for the image formation of a thick lens with thickness d_{12}, the transformation matrix becomes

$$\begin{pmatrix} x_2 \\ \alpha_2 \end{pmatrix} = \mathbf{M}_1 \begin{pmatrix} x_1 \\ \alpha_1 \end{pmatrix} = \mathbf{S}_2 \mathbf{T}_{12} \mathbf{S}_1 \begin{pmatrix} x_1 \\ \alpha_1 \end{pmatrix} \tag{3.60}$$

where $\mathbf{S}_2$ is the second curved surface of the lens and $\mathbf{T}_{12}$ the translation matrix over the thickness d_{12}.

The order of matrix multiplication is important and cannot be reversed. However, matrix multiplication is associative, which means that the order in which multiplications are performed can be switched,

$$\mathbf{A}(\mathbf{BC}) = (\mathbf{AB})\mathbf{C} = \mathbf{ABC} \tag{3.61}$$

The impact of the Matrix method becomes evident when using software packages such as MathCad to model the light propagation solution. Additional refinements can be introduced by applying Fourier concepts in mathematical modeling as described next.

3.4.2 Fourier Optics

Fourier optics is a branch of physical optics, which provides a description of the propagation of light through an optical system using harmonic analysis and linear system theory. Fourier analysis offers the opportunity to represent an arbitrary two-dimensional wave equation as the sum of several harmonic functions of different spatial frequency and complex amplitude. The function $f(x,y)$, for example, can be written as a superposition of harmonic functions $\exp[-i2\pi(\nu_x x + \nu_y y)]$, where ν_x and ν_y respectively are the spatial frequency components in the x and the y directions of the harmonic function (cycles per unit length), and the complex amplitude of each function is determined from the Fourier transform of $f(x,y)$.

The plane wave is the simplest form of a three-dimensional wave. It can be used to analyze a traveling wave of arbitrary complexity. The optical disturbance of a plane wave, at a given instant in time, can be described by a sequence of equally spaced plane surfaces of constant phase called wavefronts. The wavefronts propagate through space in a direction parallel to the wavefront normal given by its wave vector $\mathbf{k} = (k_x, k_y, k_z)$.

An arbitrary function $f(x,y)$ can therefore be analyzed as a superposition of plane waves as described by the following equation:

$$f(x,y) = (2\pi)^{-2} \iint F(k_x, k_y) \exp[-i(k_x x + k_y y)] dk_x \, dk_y \tag{3.62}$$

where the complex amplitudes of the plane waves are determined by:

$$F(k_x, k_y) = \iint f(x,y) \exp[-i(k_x x + k_y y)] dx \, dy \tag{3.63}$$

The pair of Equations (3.62) and (3.63) should be recognized as the spatial two-dimensional Fourier transform pair. The plane wave is the spatial building block used to synthesize two-dimensional wavefronts of arbitrary complexity in the same way that the sinusoid is used to form one-dimensional temporal signals from single-frequency components.

In Chapter 5 we will examine the various interactions of light with tissue to gain a better understanding of how light can be used in diagnosis and therapy in surgery and medicine.

3.5 Summary

The elementary classical optics theories have been introduced and the relevance to biomedical optics have been outlined.

The concept of Fourier optics has particular value in Fraunhofer diffraction theory and in Huygen's-Fresnel spatial frequency domain calculations and will be outlined in subsquent chapters.

Further Reading

Born, M., and Wolf, E., *Principles of Optics: Electromagnetic Theory of Propagation, Interference and Diffraction of Light*, Cambridge University Press, Cambridge, 2003.

Gerrard, A, *Introduction to Matrix Methods in Optics*, Dover Publications, Mineola, NY, 1975.
Jackson, J.D., *Classical Electrodynamics*, 2nd ed., Wiley, New York, 1975; 3rd ed., 1998.
Mandel, L., and Wolf, E., *Optical Coherence and Quantum Optics*, Cambridge University Press, Cambridge, 1995.
Perlick, V., *Ray Optics, Fermat's Principle, and Applications to General Relativity*, Springer Verlag, Berlin, Heidelberg, 2000.

Problems

1. What is 'Gaussian optics'?
2. In terms of the interacting particle size D and the incident wavelength λ, what is the relationship between these for geometrical optics, and for wave optics?
3. Derive the equation for the power of a thick lens?
4. Describe the difference between diffuse and specular reflection?
5. What are the parameters that are unchanged when light propagates in a medium: wavelength, frequency, or speed?
6. The boundary conditions for the incident, reflected, and transmitted electric fields at an interface lead to relationships for the phase and amplitude of these fields.
   a. The phase relationship leads to which two laws?
   b. The amplitude relationship leads to which set of equations?
7. As the distance is increased from a diffracting aperture what does the pattern look like? State the three distinct diffraction patterns that can be distinguished.
8. In very dim light (moonlight at night), what is your resolution at 100 m? Assume that your pupil opens to 7 mm and the rods are most sensitive at 550 nm, while color vision with the cones is not possible under these conditions.
9. A 'hard' reflection from an interface is one where the light experiences a phase change. What one criterion must be met for the reflection at this interface to be 'hard'?
10. What happens to the angles of incidence, reflection, and refraction at the critical angle?
    a. Is there a critical angle for the interface between water and air, going from water to air?
    b. Is there a critical angle for the interface between air and water, going from air to water?
11. What are the characteristics of the focal length, angle of view, and image size for a thin lens?
12. Most of the polarized light around us is produced by which process?
13. Light can also be polarized by reflection from a surface. Which law accounts for this?
14. There are two types of diffraction, Fresnel and Fraunhofer:
    a. What is the difference between Fresnel and Fraunhofer diffraction?
    b. What are the similarities between the two types of diffraction?
15. Consider an optical system with two identical converging lenses. Both lenses are considered to be thin lenses, and each has a focal point f. The lenses are at a distance of L apart in air.
    a. Give the matrix expression for the ray transformation describing the passage from the entry plane V_1 to the exit plane V_2, as shown in Figure 3.7.

b. Describe the matrix with the use of the variables f and L.
c. An equivalent system describing the paraxial ray transformation between V_1 and V_2 can also be described with a single lens with focal length f_{eq}, and space on either side of the equivalent lens of l_1 and l_2, respectively. Express f_{eq}, l_1, and l_2 in terms of f and L.
d. For which value of L will the principal planes of the two lenses coincide, and where will this location be? Elaborate and express your answer in terms of L and f.
e. Assume now that $f = L$. Draw the location of the principal planes of the optical system using Figure 3.7, and show the path of a ray that has the entry and exit path running parallel to each other under these conditions.

16. Describe four principles that can transform unpolarized light into polarized light?
17. Describe the Huygens–Fresnel principle?
18. Give the Cauchy dispersion relationship for normal dispersion of an optical material, and explain how this is achieved?
    a. White light emerging from a point source is filtered by a filter with center wavelength 600 nm and line width of 0.1 nm. Determine the coherence length of the transmitted light.
    b. What are the conditions for a lens to be able to display two adjoining points as individually recognizable images?
19. A fisherman is standing on the side of the river and sees a trout that is at a distance of 2 m from the water's edge. Does the fisherman observe the trout as being closer or farther from the water's edge than the real distance?
20. Does an air bubble under water act as a converging or as a diverging lens?
21. Consider a microscope coverglass as shown in Figure 3.8 with index of refraction $n = 1.5$ and thickness $d = 5$ mm:

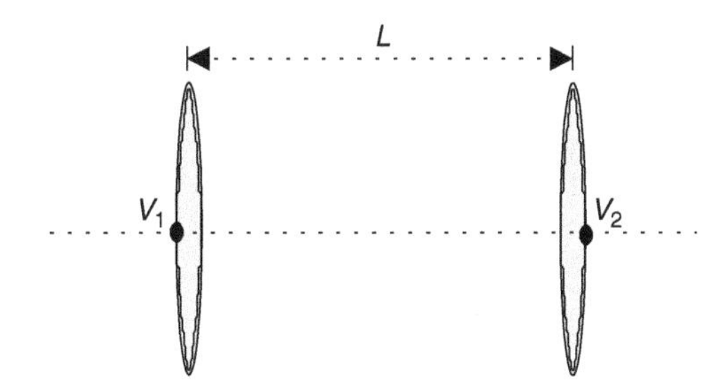

FIGURE 3.7
Set of two lenses.

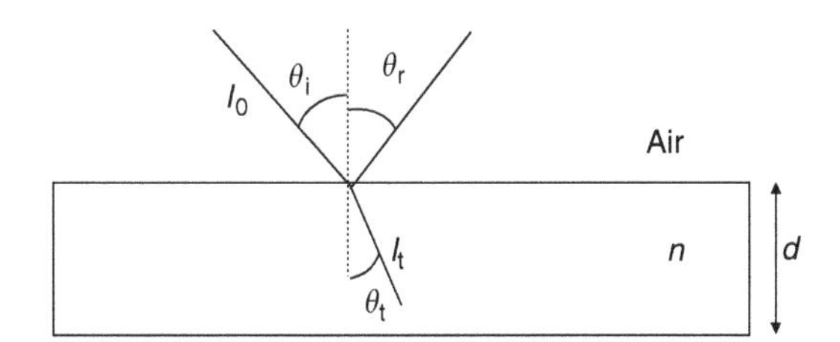

FIGURE 3.8
Diagram of rays of light interacting with a microscope cover glass.

a. Briefly describe the concepts of critical angle and Brewster angle.
b. Determine the critical angle and Brewster angle for the microscope coverglass in air. Repeat this calculation for the glass–air interface for internal reflection.

22. In a room illuminated by fluorescent lights no interference pattern can be observed that are generated by reflections from the upper and lower glass–air interfaces of the glass plate covering the light box. However, when placing two microscope covers on top of each other, an interference pattern will be observed. When placing the glass on a white sheet of paper, no interference pattern will be observed. Explain the conditions in each situation leading to the respective observations:
    a. The cover glass is irradiated by circular polarized laser light (λ = 500 nm), radiance I_0, hitting the glass surface at an angle of θ_i = 56.31° with the normal. Calculate the fraction of transmitted light under these conditions.
    b. The coverglass has wet sheet of paper placed on top of it (see illustration). As a result of the multiple internal reflections within the wet paper, the laser target area now acts as a point source. This results in a light ring on the paper. Explain this phenomenon, and derive the radius R of the lighted ring.
23. How does the electric field vector fluctuate in linearly polarized light?
24. How does the electric field vary in circularly polarized light?
25. What is the difference between diffraction and interference?
26. Which fundamental law of physics underwrites the requirement that $T + R = 1$?
27. What are the fundamental differences between the critical angle, Brewster angle, and the boundary angle?
28. Derive the lens makers' equation for a thick lens?

4

Review of Optical Interaction Properties

In this chapter some of the elementary definitions used in light–tissue optics are presented. In addition, the rules that were developed over the centuries regarding absorption and how to model light scattering are described. The conditions requiring the use of either Mie or Rayleigh scattering are also postulated. Some derived optical interaction processes are described, highlighting Raman scattering and the Doppler effect.

4.1 Absorption and Scattering

Absorption is the annihilation of photonic energy while interacting with electrons, atoms, or molecules and the conversion into heat or into photons with a much lower frequency, e.g., fluorescence or phosphorescence.

Scattering is the interaction between electromagnetic radiation and small particles, such as electrons, atoms, or molecules; however, no energy is lost in this case, and a more intricate mechanism redirects the incident wavefront or photon into a different direction, or more precisely, a probability of redirection into a solid angle. In general, a photon is first absorbed but immediately reemitted by the atoms or molecules, without loss of energy. The scattered photon can have any direction in 4π solid angle, but depending on the wavelength and the size of the interaction there are certain rules, which have been established in the distribution of the cone of preferential directions that can be expected.

There are three notable kinds of scattering that are considered in the context of this book: Rayleigh, Mie, and Raman scattering.

4.1.1 Light and Atom Interaction Overview

A summary of various light and atom interaction possibilities is illustrated in Figure 4.1. It is important to realize that several of these effects may occur simultaneously when a sample material is illuminated by a light source. The extent of each of the effects may also be lower compared to the total transmission of the light that passes through the sample material without experiencing any of the effects. Consequently, a brief description of these interactions is supplied so that the Raman scattering effect may be better understood in later sections.

The light resonance absorption effect shown in Figure 4.1a occurs when a photon, which is the particle characterization of the light wave, collides with an electron within an atom or molecule. The electron is excited to one of the available discrete energy levels by absorbing all of the photon energy. This occurs only if the photon energy precisely matches the change in electron energy at the new level. If it does not, the absorption does not take place and another type of photon-to-electron interaction may occur. Once the electron reaches the new energy level, it will remain at that higher level for a discrete amount of time before spontaneously falling back to the original energy level. During this transition, the total energy of the process must be maintained to satisfy the law of conservation of energy. Consequently, a new photon is emitted from the atom or molecule with the same frequency (i.e., energy) as the original photon.

The Rayleigh and Raman scattering effects shown in Figure 4.1b and Figure 4.1c, respectively, are similar in that the atoms in a molecule absorb a photon possessing the correct energy. Like resonance absorption, this causes the transition of the molecule to a higher energy level, which takes the form of vibrational and rotational energy changes of the molecule as well as possible electron excitation. However, for these effects the electron does not stay at the new energy level for long and quickly falls to a lower level. In the case of Rayleigh scattering the molecule returns to the original energy level and a photon is produced with the same frequency as the original one just like in the case of resonance absorption. The difference now is that there is no time delay

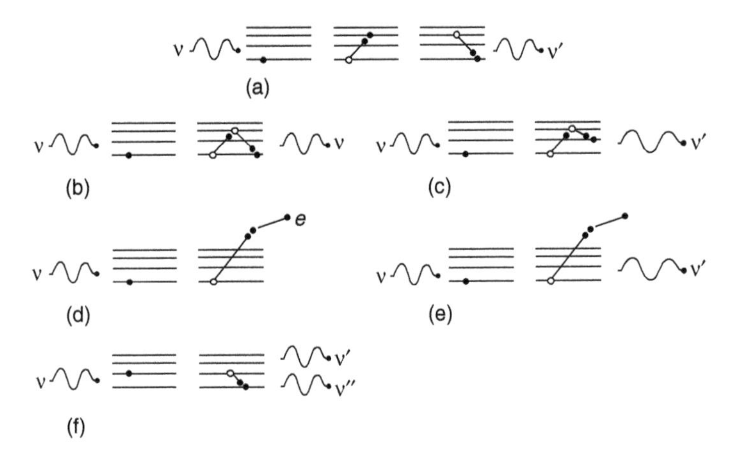

FIGURE 4.1
Summary of various light and atom interaction possibilities: (a) resonance absorption, (b) Rayleigh scattering, (c) Raman scattering, (d) photoelectric effect, (e) Compton scattering, and (f) stimulated emission.

between the energy transitions and the resulting emission of the photon. Since the emitted photon has the same frequency (i.e., energy) as the original photon, the Rayleigh scattering effect is also referred to as elastic scattering since no net energy is transferred to or from the molecule. In contrast, the Raman scattering effect is referred to as inelastic scattering since there is a net energy transfer. As a result, the molecule may provide a net gain of energy to the photon, resulting in an emitted frequency higher than the incident photon (not shown). If the incident photon supplies energy to the molecule that is not returned, then the atom will have a net gain in energy and the emitted photon will have a lower frequency as shown in Figure 4.1c.

The photoelectric effect and Compton scattering interactions shown in Figure 4.1d and Figure 4.1e, respectively, also begin with the absorption of a photon, but now the energy is sufficient to satisfy the work function, resulting in the removal of the electron from the atom. This ionizes the atom and results in a current flow in the presence of an external electric field. Visible and ultraviolet light sources provide the photons for the photoelectric effect while x-rays provide the photons for the Compton scattering effect. For the photoelectric effect, any energy that remains after the electron is freed is conserved through the transfer of kinetic energy to the electron. Energy is also conserved during the Compton scattering effect, but now a photon is released in addition to the transfer of kinetic energy to the electron. The emitted photon has a lower frequency than the original photon.

The stimulated emission effect shown in Figure 4.1f begins with an electron that has previously been excited to an energy level above its ground state, where the ground state is the resting energy level of the electron within the atom at a given temperature. The electron normally remains at the excited energy level for a period of time before spontaneously decaying to the ground state in the same way as the resonance absorption effect, which has been described earlier. When a photon with the correct energy interacts with the atom prior to the spontaneous emission event, it will not be absorbed but will stimulate the transition of the electron from the higher energy level to the ground state. This results in the emission of a photon with the same frequency and phase as the stimulating photon. Consequently, the incident photon is transmitted through the sample, causing the emission of another photon and resulting in light amplification.

4.1.2 Absorption

In general, absorption quantifies the conversion of light energy into other forms of energy, which turns out to be most particularly momentum or more precisely the impulse gained by atomic or molecular structures.

4.1.2.1 Classical Theory of Absorption

In the classical theory of absorption the assumption is made that the photon energy is converted into kinetic energy. This is in sharp contrast with relativistic absorption theory that needs to be taken into consideration only when the photon energy ($h\nu$) is of the same order as the rest energy (mc^2) of the particle it interacts with. This will be of interest only in the gamma ray region.

The kinetic energy stored in the atomic or molecular structure, however, may be released again as a new photon, with lower energy. In addition, the kinetic energy conversion may accelerate an electron to a higher energy band (excitation), which will eventually revert back to its ground state and accelerate in the process. As described earlier, an accelerated charge produces both an electric and a magnetic field fluctuation, which is a photon (electromagnetic wave). The absorption process is illustrated in Figure 4.2.

4.1.2.2 Lorentz Model

In terms of classical physics, Newton's second law describes the motion of a charged particle in an electromagnetic field as subjected to the Lorentz force $\mathbf{F}$:

$$\mathbf{F} = q(\mathbf{E} + \mathbf{v} \times \mathbf{B}) \tag{4.1}$$

where q and $\mathbf{v}$, respectively, are the charge and the velocity of the particle, and $\mathbf{E}$ and $\mathbf{B}$ are the electric and magnetic field vectors, respectively.

In the interaction of light with a medium, the interaction of the electromagnetic field with the nucleus is mostly neglected on the basis of the mass and size of the nucleus, and the motion imparted on this nucleus by the external fields. Only the individual proton and electron motion due to the Lorentz force is of interest.

In the dipole approximation ($x \ll \lambda$), the corresponding equation is given by

$$m\frac{\mathrm{d}^2 x}{\mathrm{d}t^2} = qE(\mathbf{x}, t) + F_\mathrm{b}(\mathbf{x}) \tag{4.2}$$

In the Lorentzian electron oscillator model, an electron in an atom responds to light as if it were bound to its atom by a simple spring just as a classical particle: $\mathbf{r} = \mathbf{r}_0 + \mathbf{x}$ in a stable circular orbit perturbed by an external force, as given by

$$F_\mathrm{b}(x) = k\mathbf{x} = -m\omega_0^2\mathbf{x} \tag{4.3}$$

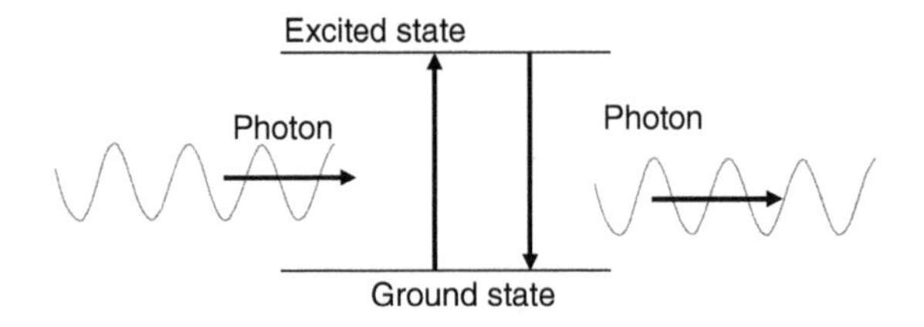

FIGURE 4.2
The light resonance absorption effect.

Since there are other external forces acting on the electron in orbit, neighboring charges, etc., an additional loss term $-b\mathbf{v}$ needs to be included, as shown in the following equation:

$$m\frac{\mathrm{d}^2x}{\mathrm{d}t^2} = qE(\mathbf{x},t) - m\omega_0^2\mathbf{x} - b\mathbf{v} \tag{4.4}$$

4.1.2.3 Beer–Lambert–Bouguer Law of Absorption

Absorption or the decline of irradiance of the amount $\mathrm{d}I$ of electromagnetic radiation, as with any other stream of energy or particles is directly proportional to the incident quantity I and the distance over which the absorption takes place $\mathrm{d}x$, with the proportionality factor μ, as expressed by

$$\mathrm{d}I = -\mu I\,\mathrm{d}x \tag{4.5}$$

The proportionality factor μ equals the absorption coefficient.

Solving Equation (4.5) gives Beer's law, or the Beer–Lambert law, or the Beer–Lambert–Bouguer law of absorption, as expressed in the following equation:

$$I(x) = I_0 \exp[-\mu x] \tag{4.6}$$

This is an empirical equation derived independently (in various forms) by Pierre Bouguer in 1729, Johann Heinrich Lambert in 1760, and August Beer in 1852.

4.1.3 Scattering

In scattering three modes of interaction can be distinguished. The distinction is made predominantly on the basis of the ratio of the wavelength with the scattering item. The three respective situations are Rayleigh, Mie, and Raman scattering.

4.1.3.1 Rayleigh Scattering

In case electromagnetic waves interact with particles that are smaller than the wavelength, the light is not necessarily absorbed; it will most likely be scattered. The elastic scattering of electromagnetic radiation on particles smaller than the wavelength is described by the work of the English mathematician John William Strutt Lord Rayleigh (1842–1919). Lord Rayleigh made an effort to explain the appearance of the blue sky and red sunset in his work published in 1871. His approximations were based on the assumption that electromagnetic waves interacted with items smaller than the wavelength, the atomic dimensions of the gases in the upper atmosphere.

These conditions can be expressed as

$$r \ll \lambda \tag{4.7}$$

and

$$r \ll \frac{\lambda}{n} \tag{4.8}$$

with r being the radius of the atomic dimension, λ the wavelength, and n the index of refraction.

The average scattering cross-section σ_R per particle as a function of wavelength is given by

$$\sigma_R(\lambda) = \frac{128\pi^5\alpha^2}{3\lambda^4}\frac{6+3\delta}{6-7\delta} \tag{4.9}$$

where, for n near unity,

$$\alpha = \frac{(n-1)}{2\pi N_0} \approx \frac{(n^2-1)}{4\pi N_0} \tag{4.10}$$

where n is the index of refraction, N_0 the density of molecules per unit volume at standard pressure and temperature, and δ the depolarization factor.

For n near unity, the Rayleigh scattering cross-section can be written as

$$\sigma_R = \frac{8\pi^3}{3}\frac{(n^2-1)^2}{\lambda^4 N^2} \tag{4.11}$$

This shows the $1/\lambda^4$ dependency of the scattering cross-section, which can be recognized in many tissue spectra.

Under these assumptions Rayleigh scattering can describe the interaction of ultraviolet and visible light with the molecules of air in the atmosphere, e.g., O_2, N_2, He, and CO_2. Also it can explain the interaction of infrared radiation with small aerosols, and microwave radiation with water droplets in clouds and rain. Some of the practical applications of Rayleigh theory are in weather radar and lidar for the remote detection of pollution in the air.

In matrix format, the Rayleigh scattering operator takes the following form:

$$P(\theta) = \begin{bmatrix} \frac{1}{2}(1+\cos^2\theta) & -\frac{1}{2}(1-\cos^2\theta) & 0 & 0 \\ -\frac{1}{2}(1-\cos^2\theta) & \frac{1}{2}(1+\cos^2\theta) & 0 & 0 \\ 0 & 0 & \cos\theta & 0 \\ 0 & 0 & 0 & \cos\theta \end{bmatrix} \tag{4.12}$$

4.1.4 Mie Scattering

When the dimension of the particle the electromagnetic radiation interacts with is of the order of magnitude of the wavelength of the radiation, the interaction closely approximates geometrical optics. This is the foundation of Mie scattering theory. A more precise definition involves the solution of the Maxwell equations applied to the interface with an isotropic, homogeneous, dielectric sphere. The Maxwell equations solved in spherical coordinates yields the scattering cross-section σ_M, and a scattering anisotropy factor g.

4.1.4.1 *Scattering Phase Function*

The scattering phase function is defined as the average scattering angle distribution weighted over the probability of each scattering angle, as expressed by

$$g = \int P(\theta)\cos\theta \, \mathrm{d}\omega \tag{4.13}$$

where $P(\theta)$ equals the probability of a photon being scattered in the direction θ.

In vector format, a photon coming in from direction $\mathbf{s}$ and redirected into direction $\mathbf{s}'$ is represented in the scattering phase function as

$$g = \int_{4\pi} P(\mathbf{s},\mathbf{s}')\cdot(\mathbf{s},\mathbf{s}')\mathrm{d}\omega \tag{4.14}$$

where the integration takes place over the full 4π steradian spherical geometry.

4.1.5 Raman Scattering

In 1921 Professor Chandrasekhara V. Raman and his students discovered that an incident light beam transmitted through a liquid resulted in a spectrum that included not only the original light but also secondary radiation at shifted wavelengths. The original experimental setup passed sunlight through an optical filter to produce a polarized incident light beam, which was then transmitted through the sample material. The emerging scattered light beam was then passed through a second optical filter and the resulting spectrum was observed visually. From these initial experiments it was found that the intensity of the resulting secondary radiation was dependent on both the frequency and the polarization of the light beam incident on the sample. The optical filters were used to control the polarization of the light beam. Later experiments replaced sunlight with a mercury arc lamp and a spectrograph was used to record the spectrums of various materials.

Professor Raman discovered that the secondary radiation effect also occurred when liquid samples were replaced with gases, solid crystals, and glasses. The experimental results revealed that each different material exhibited unique secondary radiation frequency shifts, and it was quickly realized that this was a new phenomenon that could be used to better understand the structure of matter.

Professor Raman announced his findings on March 16, 1928 at a meeting of the South Indian Science Association at Bangalore. Those attending the meeting realized the significance of the new phenomenon, and collaborative efforts continued to further define its nature. Professor Raman and K. S. Krishnan published a series of papers that reported the spectrums of various types of liquids, gases, and solids. From these works they concluded that the observed secondary radiation frequency shifts were the same as the oscillation frequencies of the chemical bonds of each atom within a molecule. This realization resulted in a new field within the study of molecular structures and provided an easy and revolutionary way of recording the vibrational and rotational spectra of chemical compounds.

4.1.5.1 Simple Atom and Molecule Models

The following section describes simple models that are often used to visualize the complex vibrations within atoms and molecules. While the applied forces within these models are mechanical, the actual forces are related to the interaction among charged particles, photons, and thermal energy.

The basics of a vibrating system are often illustrated using the simple mass and spring model shown in Figure 4.3a. One side of the spring is attached to a perfectly rigid structure and the other to an object with a constant mass. The idealized spring is assumed to have no mass so that energy losses due to gravity and momentum contributions from the spring may be ignored. To simplify the model, frictional energy losses due to the motion of the mass are also assumed to be zero. The spring follows Hooke's law of motion as described by Equation (4.15) with the necessary condition that the elastic range of the spring is never exceeded (i.e., the spring is not deformed by pulling it too far). When the mass is moved by an external force from the spring's equilibrium position ($x = 0$), a restoring force is exerted by the spring on the mass as shown in Figure 4.3b. This force is equal to the force constant of the spring (k), which is assumed to be constant over the range of motion of the mass, times the displacement distance:

$$f_{\mathrm{r}} = -kx \tag{4.15}$$

where x is the displacement distance from equilibrium and k is the spring constant. The negative sign given in Equation (4.15) simply indicates that the restoring force is in the opposite direction of the external force that causes the motion,

For a given displacement distance ($\pm x$), the mass and spring system will possess a discrete level of stored potential energy. Potential energy is proportional to the magnitude of the applied force and the resulting total distance the mass is moved. If the total distance is divided into incremental

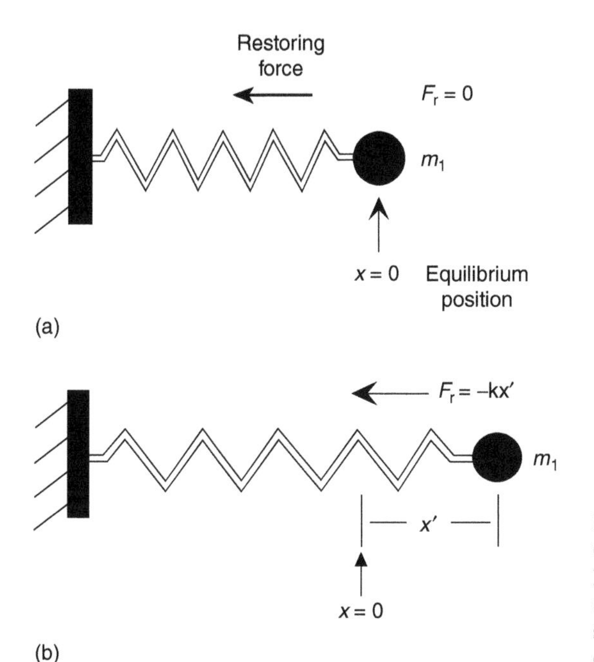

FIGURE 4.3
One-mass and spring system. (a) Equilibrium position of mass and (b) mass under influence of spring after being displaced by external force.

steps ($\mathrm{d}x$), then the corresponding incremental potential energy ($\mathrm{d}U$) is described as follows:

$$\mathrm{d}U = f_{\text{applied}}\,\mathrm{d}x = -f_r\,\mathrm{d}x \tag{4.16}$$

Substituting for the restoring force given by Equation (4.16) and performing the necessary integration, as shown in Equation (4.17), yields Equation (4.18), which describes the potential energy function for the mass and spring system:

$$\int \mathrm{d}U = \int -(-kx)\,\mathrm{d}x \tag{4.17}$$

$$U = \frac{1}{2}kx^2 \tag{4.18}$$

Modeling chemical bonds by approximating their potential using this energy function reduces an extremely complex system to a simple harmonic oscillator. While this yields good results, a better understanding of the bond interaction between two atoms within a molecule may be gained by the two mass and spring system shown in Figure 4.4a. The original mass and spring system is simply modified by removing the spring from the fixed wall and attaching a second mass. Hooke's law describing the restoring forces and all the assumptions for the original system still apply. For simplicity, it is also assumed that the masses only move along the line of the spring. This accounts for the stretching vibration of a molecular bond while ignoring the

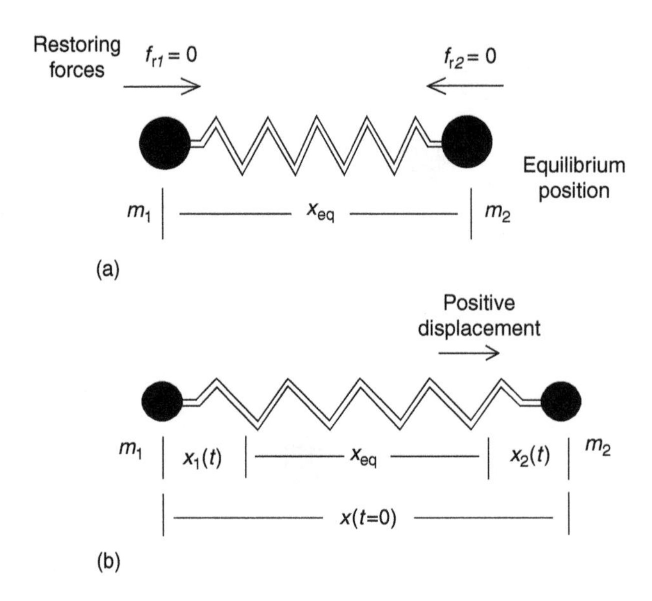

FIGURE 4.4
Two-mass and spring system: (a) equilibrium position and (b) one mass has been displaced by an external force.

bending vibration that is typically much smaller. To derive a general solution, the second mass will be assumed to be different than the first. However, the second mass may be equal to the first in the case of diatomic molecules such as O_2 or more complex molecules with similar symmetry.

From basic physics the kinetic energy of a particle is equal to one half of the mass time the velocity squared, as expressed by

$$\text{KE} = \frac{1}{2}mv^2 \tag{4.19}$$

For the two-mass system shown in Figure 4.4b the total kinetic energy is the sum of the kinetic energy of each mass as shown in the following equation:

$$\text{KE}_t = \frac{1}{2}m_1v_1^2 + \frac{1}{2}m_2v_2^2 \tag{4.20}$$

Knowing that velocity is simply the time rate of change in the distance traveled, the derivative with respect to time of the displacement functions $x_1(t)$ and $x_2(t)$ corresponding to masses m_1 and m_2, respectively, may be substituted for the velocity terms:

$$K_t = \frac{1}{2}\left[m_1\left(\frac{\mathrm{d}}{\mathrm{d}t}x_1(t)\right)^2 + m_2\left(\frac{\mathrm{d}}{\mathrm{d}t}x_2(t)\right)^2\right] \tag{4.21}$$

The total potential energy may be derived from Equation (4.22), where positive displacement is assumed to be toward the right in Figure 4.4b and negative displacement to the left. This convention accounts for the minus sign shown in the following equation:

$$U_t = \frac{1}{2}k[x_2(t) - x_1(t)]^2 \tag{4.22}$$

Lagrange's equation given by Equation (4.23) may be used to describe the motion of each particle, where $x_1(t)$ is the displacement function shown in Figure 4.4b for particles $i = 1$ and 2,

$$\frac{\mathrm{d}}{\mathrm{d}t}\left[\frac{\partial K_t}{\partial x_i(t)}\right] + \frac{\partial U_t}{\partial x_i(t)} = 0 \tag{4.23}$$

Substituting Equations (4.21) and (4.22) into Equation (4.23) and performing the partial derivatives for both $x_1(t)$ and $x_2(t)$ yields the second-order differential equations (4.24) and (4.26). That is, substitution using $x_1(t)$ yields

$$\begin{aligned}&\frac{\mathrm{d}}{\mathrm{d}t}\left\{\frac{\partial}{\partial x_1(t)}\left[\frac{1}{2}m_1\left(\frac{\mathrm{d}}{\mathrm{d}t}x_1(t)\right)^2 + \frac{1}{2}m_2\left(\frac{\mathrm{d}}{\mathrm{d}t}x_2(t)\right)^2\right]\right\}\\ &+\frac{\partial}{\partial x_1(t)}\left\{\frac{1}{2}k[x_2(t) - x_1(t)]^2\right\} = 0\end{aligned} \tag{4.24}$$

Equation (4.24) reduces to

$$m_1\frac{\mathrm{d}^2}{\mathrm{d}t^2}x_1(t) - k[x_2(t) - x_1(t)] = 0 \tag{4.25}$$

Substitution using $x_2(t)$ yields

$$\begin{aligned}&\frac{\mathrm{d}}{\mathrm{d}t}\left\{\frac{\partial}{\partial x_2(t)}\left[\frac{1}{2}m_1\left(\frac{\mathrm{d}}{\mathrm{d}t}x_1(t)\right)^2 + \frac{1}{2}m_2\left(\frac{\mathrm{d}}{\mathrm{d}t}x_2(t)\right)^2\right]\right\}\\ &-\frac{\partial}{\partial x_2(t)}\left\{\frac{1}{2}k[x_2(t) - x_1(t)]^2\right\} = 0\end{aligned} \tag{4.26}$$

Equation (4.26) reduces to

$$m_2\frac{\mathrm{d}^2}{\mathrm{d}t^2}x_2(t) + k[x_2(t) - x_1(t)] = 0 \tag{4.27}$$

Equations (4.25) and (4.27) are static equations stating that the two-mass and spring system does not move itself as a result of the applied forces. That is, the first term in each equation is the force due to the acceleration of each particle mass that follows Newton's second law of motion (F = ma). The second term in each equation is the restoring force following Hooke's law, which has been discussed earlier. The resulting parabolic potential energy plot about the spring's equilibrium point is shown in Figure 4.5. The negative sign in Equation (4.26) simply accounts for the positive and negative displacement convention shown in Figure 4.5.

Differential equations of the form given in Equations (4.25) and (4.27) may be solved by assuming periodic motion of the form given in Equations (4.28) and (4.29), where the vibration of each mass occurs at the same frequency and phase shift. The maximum oscillation amplitudes (A_1 and A_2) are different when there is a net vibration present. Substituting the equations for periodic motion into the differential equations results in two characteristic descriptions: Equations (4.25) and (4.27). Equations (4.25) and (4.27) may be solved simultaneously for the vibrational frequency using the following equations:

$$x_1(t) = A_1\cos(2\pi \nu t + \phi) \tag{4.28}$$

and

$$x_2(t) = A_2\cos(2\pi \nu t + \phi) \tag{4.29}$$

where ν is the frequency, t the time, and ϕ is defined as the phase shift.

Additionally, substituting Equations (4.28) and (4.29) into Equation (4.26) yields

$$\begin{aligned} m_1 \frac{d^2}{dt^2}[A_1 \cos(2\pi \nu t + \phi)] - k[A_2 \cos(2\pi \nu t + \phi) - A_1 \cos(2\pi \nu t + \phi)] = 0 \\ -4\pi^2\nu^2 m_1 A_1 \cos(2\pi \nu t + \phi) - kA_2 \cos(2\pi \nu t + \phi) + kA_1 \cos(2\pi \nu t + \phi) = 0 \\ \cos(2\pi \nu t + \phi)[(-4\pi^2\nu^2 m_1 + k)A_1 - kA_2] = 0 \end{aligned} \tag{4.30}$$

$$(-4\pi^2\nu^2 m_1 + k)A_1 - kA_2 = 0 \tag{4.31}$$

$$\frac{A_1}{A_2} = \frac{k}{(-4\pi^2\nu^2 m_1 + k)} \tag{4.32}$$

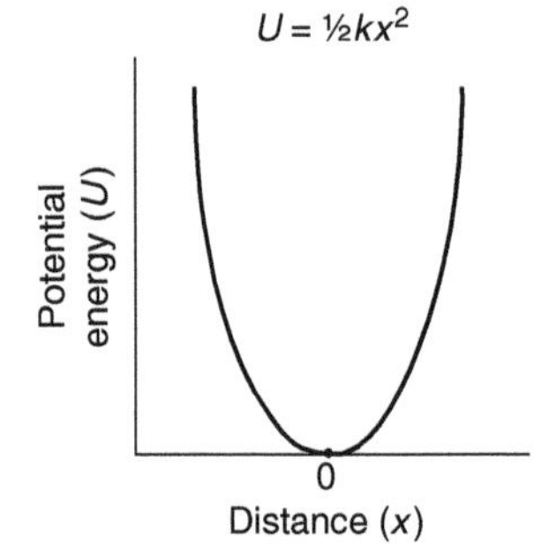

FIGURE 4.5
Displacement–energy well, parabolic potential energy plot of the mass for the position with respect to the spring's equilibrium point.

Substituting Equations (4.28) and (4.29) into Equation (4.27) yields

$$\begin{aligned} m_2 \frac{d^2}{dt^2}[A_2\cos(2\pi vt+\phi)]+k[A_2\cos(2\pi vt+\phi)-A_1\cos(2\pi vt+\phi)]=0 \\ -4\pi^2v^2m_2A_2\cos(2\pi vt+\phi)+kA_2\cos(2\pi vt+\phi)-kA_1\cos(2\pi vt+\phi)=0 \\ \cos(2\pi vt+\phi)[(-4\pi^2v^2m_2+k)A_2-kA_1]=0 \end{aligned} \quad (4.33)$$

The expression in Equation (4.33) is valid when the following condition is met:

$$-kA_1+(-4\pi^2v^2m_2+k)A_2=0 \quad (4.34)$$

or

$$\frac{A_1}{A_2}=\frac{(-4\pi^2v^2m_2+k)}{k} \quad (4.35)$$

Since both Equations (4.32) and (4.35) are equal to A_1/A_2, the right-hand side of each of these equations may be set equal to each other, yielding $k/(-4\pi^2\nu^2m_1+k) = (-4\pi^2\nu^2m_2+k)/k$.

Simplifying this expression yields

$$\begin{aligned} (4\pi^2v^2)^2m_1m_2-4\pi^2v^2k(m_1+m_2)+(k^2-k^2)=0 \\ 4^2\pi^4v^4m_1m_2-4\pi^2v^2k(m_1+m_2)=0 \\ v^2[4^2\pi^4v^2m_1m_2-4\pi^2k(m_1+m_2)]=0 \end{aligned} \quad (4.36)$$

Two solutions for the possible vibrational frequencies of the two-mass and spring model may be derived from Equation (4.36). The first solution, given by Equation (4.37), corresponds to no net vibration. That is, substituting $\nu = 0$ into either Equation (4.33) or Equation (4.35) yields the result. This means that the motion of each mass given by Equations (4.25) and (4.27) is the same, and that both particles always move the same amount and in the same direction. Consequently, there is no net vibration with respect to each mass but energy is present that results in the movement of the entire two-mass and spring system:

$$\nu^2=0 \quad (4.37)$$

$$\nu=\pm\sqrt{0}=0 \quad (4.38)$$

The second solution derived from Equation (4.36) is given by Equation (4.39). This equation corresponds to a net vibration between the two masses caused by differences in their movement over time. This is derived in the following equation:

$$
\begin{aligned}
4^2\pi^4 v^2 m_1 m_2 - 4\pi^2 k(m_1+m_2) = 0 \\
v^2 &= \frac{4\pi^2 k(m_1+m_2)}{4^2\pi^4 m_1 m_2} \\
v &= \pm\sqrt{\frac{k(m_1+m_2)}{4\pi^2 m_1 m_2}} \\
v &= \pm\frac{1}{2\pi}\sqrt{k\frac{(m_1+m_2)}{m_1 m_2}} \\
v &= \pm\frac{1}{2\pi}\sqrt{\frac{k}{\mu}}
\end{aligned} \tag{4.39}
$$

where μ represents the reduced mass of a two-mass system:

$$
\mu = \frac{m_1 m_2}{(m_1+m_2)} \tag{4.40}
$$

A relationship between the amplitude of the displacement functions is given by Equations (4.28) and (4.29). The masses of the particles may be found by substituting Equation (4.39) into either Equation (4.32) or Equation (4.35). The simplified result is shown in the following equation:

$$
\begin{aligned}
\frac{A_1}{A_2} &= \frac{-4\pi^2 m_2\left[\frac{1}{2\pi}\sqrt{\frac{k(m_1+m_2)}{m_1 m_2}}\right]^2 + k}{k} \\
\frac{A_1}{A_2} &= \frac{-4\pi^2 m_2\left[\frac{k(m_1+m_2)}{4\pi^2 m_1 m_2}\right] + k}{\mathrm{k}} \\
\frac{A_1}{A_2} &= \frac{-\frac{k(m_1+m_2)}{m_1} + k}{k} \\
\frac{A_1}{A_2} &= -\frac{(m_1+m_2)}{m_1} + 1 \\
\frac{A_1}{A_2} &= \frac{-m_1 - m_2 + m_1}{m_1} \\
\frac{A_1}{A_2} &= -\frac{m_2}{m_1}
\end{aligned} \tag{4.41}
$$

The ratio of the displacement functions also results in the same relationship between the particle masses as shown in the following equations:

$$\frac{x_1(t)}{x_2(t)} = \frac{A_1\cos(2\pi\nu t+\phi)}{A_2\cos(2\pi\nu t+\phi)} = \frac{A_1}{A_2} \tag{4.42}$$

$$\frac{x_1(t)}{x_2(t)} = -\frac{m_2}{m_1} \tag{4.43}$$

4.1.5.2 *Material Polarization*

A more precise way of describing the Raman scattering effect is to consider the wave nature of light and how it interacts with matter rather than treating light as a photon colliding with an atom. Only the electrical component of the electromagnetic wave interacts with molecules in a material to produce Raman scattering. Consequently, the effect the material has on the properties of the electric field (**E**) and conversely the effect the electric field has on the material as the light wave passes through the sample are important characteristics. Typically, the electric field may be assumed to be constant across a given molecule or chain since its size is much smaller than the wavelength of the incident light wave. This constant external electric field exerts a uniform force ($F = e^- \times \mathbf{E}$) on each electron ($1e^- = 1.6021 \times 10^{-19}$ C) within a molecule. This results in the displacement of the electrons from their average resting positions about the nuclei or simply the distortion of the original electron cloud shape. While all the electrons within a molecule are influenced to some degree, the electrons that are most affected typically reside in the outermost valence band. This movement of the negatively charged electrons relative to the positively charged nuclei results in an induced dipole moment ($\boldsymbol{\pi}$) within the molecule given by

$$\boldsymbol{\pi} = \alpha\mathbf{E}_{\text{loc}} = \varepsilon_0\alpha_{\text{el}}\mathbf{E}_{\text{loc}} \tag{4.44}$$

where α_{e1} is the electric polarizability of the molecule, $\mathbf{E}_{\text{loc}}$the electric field local to the molecule, and ε_0 the permittivity of free space.

The total electric polarizability (α_{e}) may be broken down into three distinct components as shown in Equation (4.45). The first component is referred to as the electronic polarizability (α_{e1}) and it is associated with the electron displacement or electron cloud distortion illustrated in Figure 4.6a. The ionic polarizability (α_{ion}) is the second component and it is related to the effect the electric field has on nonpolar molecules (e.g., ionic bonded CO_2) as shown in Figure 4.6b. In the absence of an external electric field, a

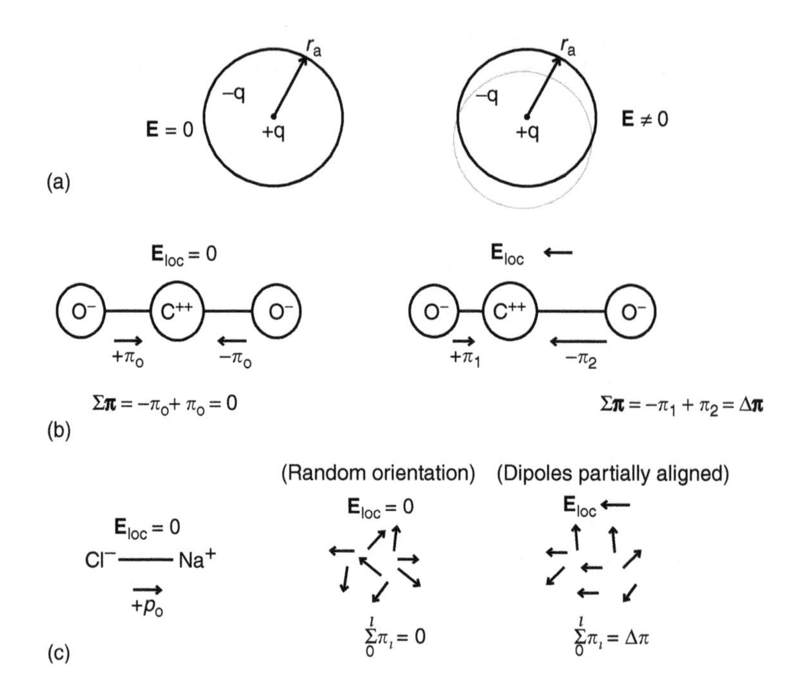

FIGURE 4.6
Component effects of electric polarizability: (a) electronic polarizability, (b) ionic polarizability, and (c) reorientation polarizability.

nonpolar molecule will not possess a net dipole moment due to the resting position of the ions within the molecule. That is, the effective positive and negative charge centers are at the same location within the molecule. However, when a sufficiently strong external field is applied the ionized atoms may move from their resting position within the molecular structure, resulting in a shift in the effective positive and negative charge centers. This shift results in a separation distance between the charge centers, which produces a net change in dipole moment per nonpolar molecule within the medium.

The third component of the total electric polarizability of a material is the reorientation polarizability (α_{reor}). This term deals with the tendency of polar molecules (e.g., NaCl) to align in the presence of an external electric field as shown in Figure 4.6c.

A polar molecule possesses a dipole moment even in the absence of an external field. However, a group of identical polar molecules will exhibit a zero net dipole moment when taken as a whole because of the random dipole orientation of each molecule caused by random thermal motion. When an external field is applied to the group of molecules, they will align with the field resulting in a net dipole moment. Each of the polarizability components is a function of the frequency of the incident light beam and the composition of the sample material. The total electron polarizability is given by

$$\alpha_{\mathrm{e}} = \alpha_{\mathrm{el}} + \alpha_{\mathrm{ion}} + \alpha_{\mathrm{reor}} \tag{4.45}$$

where α_{e1} is the electronic polarizability, α_{ion} the ionic polarizability, and α_{reor} the reorientation polarizability.

The local electric field vector ($\mathbf{E}_{\mathrm{loc}}$) shown in Equation (4.44) is the vector sum of the external electric field applied by the incident light wave and the fields created by the dipole moments from neighboring molecules in the medium. If the medium is viewed as a collection of N identical molecules with the same dipole moment, then the polarization (**P**) of the material, which is the net electric dipole moment per unit volume, may be described by

$$\mathbf{P} = \sum_{1}^{N} \boldsymbol{\pi} = N\boldsymbol{\pi} = N\varepsilon_0 \alpha_{\mathrm{e}} \mathbf{E}_{\mathrm{loc}} \tag{4.46}$$

$$\mathbf{P} = \varepsilon_0 \chi_{\mathrm{e}} \mathbf{E}_{\mathrm{loc}} \tag{4.47}$$

where $\chi_{\mathrm{e}} = N\alpha_{\mathrm{e}}$ is the electric susceptibility of the medium.

A molecule's total induced dipole moment is dependent not only on the direction of $\mathbf{E}_{\mathrm{loc}}$ but also on the direction in which it is applied to the molecule. That is, since molecules do not have a uniform surface their electric polarizability will be different depending on the side and angle at which the electric field is applied. Therefore, the induced dipole moment may be broken down into its vector components as shown in Equation (4.48). For simplicity, the Cartesian coordinate system is used:

$$\begin{aligned}
\pi_x &= \alpha_{xx}\varepsilon_0 E_x + \alpha_{xy}\varepsilon_0 E_y + \alpha_{xz}\varepsilon_0 E_z \\
\pi_y &= \alpha_{yx}\varepsilon_0 E_x + \alpha_{yy}\varepsilon_0 E_y + \alpha_{yz}\varepsilon_0 E_z \\
\pi_z &= \alpha_{zx}\varepsilon_0 E_x + \alpha_{zy}\varepsilon_0 E_y + \alpha_{zz}\varepsilon_0 E_z
\end{aligned} \tag{4.48}$$

This may be rewritten as

$$\begin{bmatrix} \pi_x \\ \pi_y \\ \pi_z \end{bmatrix} = \varepsilon_0 \begin{bmatrix} \alpha_{xx} & \alpha_{xy} & \alpha_{xz} \\ \alpha_{yx} & \alpha_{yy} & \alpha_{yz} \\ \alpha_{zx} & \alpha_{zy} & \alpha_{zz} \end{bmatrix} \begin{bmatrix} E_x \\ E_y \\ E_z \end{bmatrix} \tag{4.49}$$

The 3×3 matrix is referred to as the electric polarizability (α_{e}) tensor. Each of the components (α_{ij}) represents the polarizability that contributes to the induced dipole moment components in the $i = x$, y, and z directions produced by the electric field components in the $j = x$, y, and z directions.

If it is assumed that **E** is linearly polarized in the x-direction (i.e., $\theta = 0^0$) and propagates in the z-direction, then the vector components of the applied electric field will be described by the following conditions:

$$E_x = E_{max}\cos(2\pi\nu_0 t) \quad E_y = 0 \quad E_z = 0 \tag{4.50}$$

Under this applied electric field, Equation (4.48) simplifies to

$$\begin{aligned} \pi_x &= \alpha_{xx}\varepsilon_0 E_{max}\cos 2\pi\nu_0 t \\ \pi_y &= \alpha_{yx}\varepsilon_0 E_{max}\cos 2\pi\nu_0 t \\ \pi_z &= \alpha_{zx}\varepsilon_0 E_{max}\cos 2\pi\nu_0 t \end{aligned} \tag{4.51}$$

For simplicity, only the induced dipole moment in the x-direction (π_x) will be used during the remaining derivation. However, the induced dipole moments in the y and z directions may be investigated using the identical approach. Returning to the two-mass and spring system described earlier, assume that the electric polarizability of the molecule is linearly dependent on the difference between the mass displacements about their equilibrium positions; that is, there is linear relationship between the relative vibration between the two particles given by

$$x(t) = x_2(t) - x_1(t) \tag{4.52}$$

where displacement functions $x_1(t)$ and $x_2(t)$ are described by Equations (4.28) and (4.29), respectively. Under this assumption, the electric polarizability component α_{xx} may be described as

$$\alpha_{xx}(t) = \alpha_{xx}^{equil} + \left(\frac{d}{dx}\alpha_{xx}\right)x(t) \tag{4.53}$$

where the term α_{xx}^{equil} is the nonvibrating molecular polarizability and the term d α_{xx}/dx the polarizability change with respect to net displacement distance between the particles. Equation (4.53) can be simplified to

$$\alpha_{xx}(t) = \alpha_{xx}^{equil} + \left(\frac{d}{dx}\alpha_{xx}\right)A\cos 2\pi\nu t \tag{4.54}$$

where

$$A = A_2 - A_1 \tag{4.55}$$

Substituting Equation (4.51) into the π_x-induced polarizability component from Equation (4.51) yields

$$\pi_x = \left[\alpha_{xx}^{\text{equil}} + \left(\frac{\mathrm{d}}{\mathrm{d}x}\alpha_{xx}\right) A\cos 2\pi v t\right](\varepsilon_0 E_{\max}\cos 2\pi v_0 t) \tag{4.56}$$

$$\begin{aligned}\pi_x &= \varepsilon_0 E_{\max}\alpha_{xx}^{\text{equil}}\cos 2\pi v_0 t \\ &\quad + A\varepsilon_0 E_{\max}\left(\frac{\mathrm{d}}{\mathrm{d}x}\alpha_{xx}\right)\cos 2\pi v t\cos 2\pi v_0 t\end{aligned} \tag{4.57}$$

where v is the vibrational frequency of the particles and v_0 the frequency of the electric field from the incident light wave. Applying the trigonometry identity $\cos(A)\cos(B) = \frac{1}{2}[\cos(A+B) + \cos(A-B)]$, with $A = 2\pi v_0 t$ and $B = 2\pi v_0 t$ in the second term of the above equation yields

$$\begin{aligned}\pi_x &= \varepsilon_0 E_{\max}\alpha_{xx}^{\text{equil}}\cos 2\pi v_0 t + \frac{1}{2}A\varepsilon_0 E_{\max}\left(\frac{\mathrm{d}}{\mathrm{d}x}\alpha_{xx}\right)\cos 2\pi(v_0 + v)t \\ &\quad + \frac{1}{2}A\varepsilon_0 E_{\max}\left(\frac{\mathrm{d}}{\mathrm{d}x}\alpha_{xx}\right)\cos 2\pi(v_0 - v)t\end{aligned} \tag{4.58}$$

Using the same derivation methods, the induced dipole moments for the y and z directions are

$$\begin{aligned}\pi_y &= \varepsilon_0 E_{\max}\alpha_{yx}^{\text{equil}}\cos 2\pi v_0 t + \frac{1}{2}A\varepsilon_0 E_{\max}\left(\frac{\mathrm{d}}{\mathrm{d}x}\alpha_{yx}\right)\cos 2\pi(v_0 + v)t \\ &\quad + \frac{1}{2}A\varepsilon_0 E_{\max}\left(\frac{\mathrm{d}}{\mathrm{d}x}\alpha_{yx}\right)\cos 2\pi(v_0 - v)t\end{aligned} \tag{4.59}$$

and

$$\begin{aligned}\pi_z &= \varepsilon_0 E_{\max}\alpha_{zx}^{\text{equil}}\cos 2\pi v_0 t + \frac{1}{2}A\varepsilon_0 E_{\max}\left(\frac{\mathrm{d}}{\mathrm{d}x}\alpha_{zx}\right)\cos 2\pi(v_0 + v)t \\ &\quad + \frac{1}{2}A\varepsilon_0 E_{\max}\left(\frac{\mathrm{d}}{\mathrm{d}x}\alpha_{zx}\right)\cos 2\pi(v_0 - v)t\end{aligned} \tag{4.60}$$

The significance of Equations (4.57)–(4.60) is that they describe how the linearly polarized incident light wave interacts with a vibrating diatomic molecule.

The oscillation of each dipole moment component (π_x, π_y, and π_z) may be broken down into three discrete parts corresponding to each of the terms on the right-hand side of each equation, where each oscillates over time at unique frequency. From electromagnetic theory, it is known that an oscillating dipole radiates energy in the form of scattered radiation with a frequency equal to the oscillation frequency. Consequently, each of the three discrete dipole components emits scattered light at three unique frequencies. The first term on the right-hand side of the equations corresponds to the Rayleigh scattering effect, where the emitted light has the same frequency but different direction than the incident light wave. The second term in each equation corresponds to the anti-Stokes line in the Raman spectrum. The frequency of the emitted light is greater than that of the incident light due to the addition of the vibrational frequency ($\nu_0+\nu$). The third term in each equation corresponds to the Stokes line in the Raman spectrum where the frequency of the emitted light is less than that of the incident light due to the subtraction of the vibrational frequency ($\nu_0-\nu$).

4.1.5.3 Biochemistry Applications

This section provides a brief description of some applications of Raman scattering spectroscopy in the field of biochemistry. The information is provided for reference purposes and little attempt is made to explain the complex biochemistry concepts.

There is great interest in non-native states of proteins since they provide insight into the stability, function, and folding of proteins. Non-native states range from ideal random coils at one extreme to molten globules at the other. Detailed characterization of these states has proven to be extremely difficult due to their heterogeneity. Raman spectroscopy has emerged as a valuable new tool to probe chiral elements of the peptide backbone in aqueous solution that has led to valuable new insights into the complexity of molten globule states. One application is the study of hen egg-white lysozyme and bovine ribonuclease A in unfolded denatured states.

Many anticancer drugs react at the DNA level of cancerous tissue. The transfer of drug effects is related to the size and shape of each constituent molecule, which depends on the electron distribution of the component atoms in each molecule. The bond polarizability between partner molecules within the complex drug structure also impacts its effectiveness. Surface-enhanced Raman scattering spectroscopy, which is not discussed within this report, has been used to investigate the nature of the interactions in supramolecular complexes of enzymes, DNA, and antitumor drugs.

Shifted-excitation Raman difference spectroscopy (SERDS) has been used to observe the biochemical processes within living systems. SERDS appears to use the same experimental setup as the basic Raman technique (Figure 4.7); only the laser source is tunable. Two spectra are obtained for a given sample by shifting the second excitation laser frequency by 10 cm^{-1} relative to the first. The resulting spectra are then subtracted to yield the

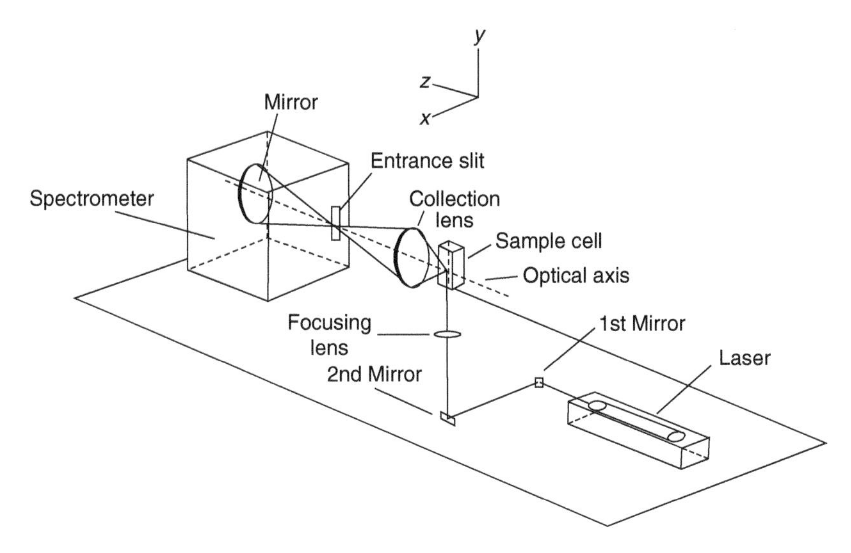

FIGURE 4.7
Configuration of basic Raman spectroscopy setup.

SERDS spectrum. One example is the investigation and comparison of the photosynthetic reaction centers between green bacterium (*Chloroflexus aurantiacus*) and purple bacterium (*Rhodobacter sphaeroides*), where the green bacterium is an evolutionary ancestor of the purple bacteria. The reaction centers of each bacterium were isolated and suspended in a buffer solution before obtaining the spectra. It was found that certain characteristics of the spectra from the two species are nearly identical. These similarities between two very different bacterial species have led to the suggestion that the common nuclear and electronic dynamics are characteristic of all photosynthetic reaction centers.

4.2 Doppler Effect

When light interacts with moving objects or when the light source itself moves there will be a frequency shift. This frequency shift has been described in various textbooks for both sound and light. The underlying principle to the frequency shift with regard to a moving source or observer is the fact that the perceived wavelength will be caught at progressively earlier (respectively later) phase intervals when source and observer are moving towards (respectively away from) each other. This phenomenon was first described by the Austrian mathematician Christian Doppler (1803–1853) in 1842. We leave the derivation of the expression to the basic physics textbooks and will only show the final statement for a frequency shift resulting from both moving source and observer. The following equation gives the perceived frequency by the observer in relative motion to the source:

$$f' = \left(\frac{1 \pm \frac{u_o}{c'}}{1 \pm \frac{u_s}{c'}} \right) f = f + \Delta f \tag{4.61}$$

where f' is the Doppler shifted frequency, f the laser or Super Luminescent Diode (SLD) frequency of light, u_0 observer velocity or the speed of motion of the Piezoelectric Transducer (PZT) in the detector arm, u_s source velocity or in our case the speed of the modulated mirror in the reference arm of the interferometer, and c′ the speed of light in the medium. The signs on top, for the numerator $\pm$ and for the denominator $\mp$, are for the situation where the source is moving towards the detector and the signs on the bottom mean that the source is moving away from the observer. Since both the mirror and the PZT in the detection arm are modulated, they will both reverse directions simultaneously within one period.

In case a scattering or reflecting object is moving at an angle with respect to the incident ray only, the motion in the direction of the connecting line between the primary source and the reflector is relevant. The object (e.g., blood cell) that is capable of scattering the light back in the direction it came from moves at presumably constant velocity, or at least with an average velocity u_s, and in case of a moving mirror the motion is predetermined. The motion of the object is at an angle θ with the normal to the surface of the light source.

The device used to measure the red blood cell flow velocity as a function of place in the blood vessel is called a laser Doppler flow anemometer. The laser targeting allows for specific location isolation to successively measure the entire velocity profile across the diameter of the vessel. For the larger blood vessels the pulsatile character of the beating heart will be visible and the measurements will need to be trigged to obtain flow characteristics of the exactly same segment of the flow pattern of the heart stroke.

Over a given time frame the object moves away from the detector; thus, each part of the wave reflected at any successive time interval will be released farther from the detector. The compound effect is that the wave is detected in a stretched form, or with a longer wavelength. The reflective object has become a secondary source that will be the new source in the description of the frequency shift. The frequency of the light collected by a still-standing detector can now be described by

$$f' = f \frac{1}{\left[1 + (u_s \cos\theta / c')\right]} \tag{4.62}$$

Equation (4.62) can be developed in a series to yield the following expression:

$$\frac{f'}{f} = \frac{1}{[1+(u_s \cos\theta/c')]} = 1 - \frac{u_s \cos\theta}{c'} + \left(\frac{u_s \cos\theta}{c'}\right)^2 + \cdots \tag{4.63}$$

where it can be shown that the second- and higher-order terms can be discarded when the speed of moving reflector is much less than the speed of light. As a result the descriptions for either a moving source or a moving detector have the same expression as given by

$$\frac{f'}{f} = 1 - \frac{u_s \cos\theta}{c'} \tag{4.64}$$

The frequency of emitted light by the particle/mirror f is perceived by the detector as f'. The reflected signal in reference to the original source frequency f for the situation where the red blood cell or mirror is moving away from the detector is now given by rewriting Equation (4.62) to yield the following equation:

$$f' = f\left(1 - \frac{u_s \cos\theta}{c'}\right) \tag{4.65}$$

This allows us to rewrite Equation (4.65) in a way that shows the relative Doppler shift for the reflected frequency as given by

$$\frac{f' - f}{f} = \frac{u_s \cos\theta}{c'} = \frac{\Delta f}{f} \tag{4.66}$$

When light is reflected by a moving mirror placed at an angle or by a particle with irregular topography and the collected light is detected at an angle φ, with respect to the incoming ray, the reflected or, respectively, scattered light will be Doppler shifted by an amount Δf given by

$$\Delta f = \frac{u_s f}{c'}(\cos\theta + \cos\varphi) \tag{4.67}$$

When collecting scattered light from a sample in either optical coherence tomography or Doppler flow measurements all these conditions need to be taken into account for a proper theoretical description.

In this chapter we have described the defining principles of optics, from the classical geometric optics to quantum optics. We have discussed electromagnetic wave theory and the different light sources that are used in the field of biomedical optics. Finally, absorption, scattering, and polarization were introduced.

In Chapter 5 we will examine the various interactions of light with tissue to gain a better understanding of how light can be used in diagnosis and therapy in surgery and medicine.

4.3 Summary

This chapter defined the basic optical parameters that apply to optics and to biomedical optics in particular. In addition, several theoretical phenomenological descriptions have been presented, such as the scattering process and the Doppler effect.

Further Reading

Bemporad, A., and Puiseux, P., VII. 2. Réfraction et Extinction, in *Encyclopédie des Sciences Mathématiques Pures et Appliquées*, Gauthier-Villars, Paris, 1913, pp. 14–67.

Doppler, C., Über das farbige licht der Doppelsterne und eininger anderer Gestirne des Himmels, *Abh. K. Böhm. Ges. Wiss., Prague, Series V*, 2, 465–482, 1842.

Michelson, A., Influence of motion of the medium on the velocity of light, *Am. J. Sci.*, 3, 31, 185, 1886.

Waynant, R.W., *Lasers in Medicine*, CRC Press, Boca Raton, FL, 2001.

Weber, M.J., *Handbook of Lasers*, CRC Press, Boca Raton, FL, 2000.

Problems

1. The photoelectric effect could not be described by classical physics and ushered in the era of modern physics. Describe the experiment and how it led to quantization of the light field.
2. A series of linear polarizers are set up with polarized light I_0 at the input. The polarizers can be used to attenuate a linearly polarized light beam.
   a. What angle should be chosen for the first polarizer if you want an optical density OD 1.0 filter, an attenuation of $10^{1.0} = 10$?
   b. What angle should be chosen for the first polarizer if you want an OD 2.0 filter, an attenuation of 100?
   c. What angle should be chosen for the first polarizer if you want an OD 4.0 filter, an attenuation of 10,000?
3. The photoelectric effect is measured on an unknown material. A pulse of light lasting 1 μs is shown on the unknown material and a current of 10 mA is measured.
   a. What is the total charge generated by this pulse of light?
   b. The photon energy necessary for the current to be nonzero is 2.2 eV. What cutoff frequency does this correspond to?

4. Consider some of the people involved in the development of the theories of scattering.
   a. Name three pioneers in scattering theory.
   b. What were the contributions of each of these scientists?
5. How many photons are required for the energy to be equivalent to that of the rest energy of an electron?
6. Is it a prerequisite that linearly polarized light should be coherent?
7. What is the difference between diffraction and interference?

5

Light–Tissue Interaction Variables

This chapter describes the theoretical and practical variables of both the light source and the turbid medium involved in the irradiation process.

The interaction of light with tissue is the combined effect of the properties of the light source and the characteristics of the tissue. In general the light source is monochromatic, limiting the variables involved in the light–tissue interaction process. A monochromatic light source can be produced by laser as discussed in Section 2.3 in Chapter 2. The variables affecting the light–tissue interaction are described first for the source and subsequently for the target.

5.1 Laser Variables

Several key factors that influence the light–tissue interaction are briefly described below, and the correlation with the actual light distribution inside the tissue due to the influence of these parameters is subsequently explained in detail in the following sections.

5.1.1 Laser Power

The total photon energy per second sent out by the source described by the Poynting vector, as discussed in Section 2.3.4 in Chapter 2, times the area of the beam cross-section is the emitted power of the source.

5.1.2 Light Delivery Protocol

The light emitted from the laser cavity can be used as a bare beam or it can be coupled into a fiberoptic cable for transportation over greater distances. The bare beam can also be redirected to a point not in the direct line of sight of the output mirror of the laser with the use of mirrors, either in free space or within an enclosed tube system: the articulating arm.

Frequently the bare beam is coupled into a light guide, which may carry the light over extended distance to the point of delivery without altering the properties of the light, except for transmission losses. Virtually all fiberoptic light guides have a circular geometry, specialty fibers exempt. While irradiating through a fiber optic, the impact zone is circular when directed perpendicular to the tissue surface. However, when the circular profile is hitting the target at an angle, the profile will become oval, with a nonuniform power density as a function of position on the target, especially because the region closest to the output aperture will have spread the least compared to the more distal positions. This is the result of either the waist of the bare beam or the numerical aperture of the fiberoptic light guide.

A nominal wave package (photon) has minimal spread in the Heisenberg uncertainty principle expressed in

$$\delta x \delta p_x \geq \frac{\hbar}{2} \tag{5.1}$$

where h is Planck's constant, $\hbar$ Planck's constant divided by 2π, x the position, and p_x the momentum, making δ_x and δ_{px} the uncertainties in position and momentum, respectively.

By using the following substitution: $p_x = k_x h/2\pi$, Equation (5.1) can be rewritten as

$$\delta x \delta k_x \geq \frac{1}{2} \tag{5.2}$$

where k represents the wave number, i.e., $\mathbf{k} = \omega/c = 2\pi/\lambda$, with c the speed of light and λ the wavelength.

This criterion reformats the solution to the wave equation (Equation [3.37] in Chapter 3) with an electric field in all three dimensions as shown in

$$E(x,y,z,t) = E_0 \exp[i(-k_x x - k_y y - k_z z)] \exp[i(-\omega t)] \tag{5.3}$$

In three dimensions the position vector $\mathbf{s}$ is used to indicate the location of the electric field in space as $\mathbf{s} = (x, y, z)$, and the wavevector $\mathbf{k}$ in a dispersive medium is defined as $\mathbf{k} = (k_x, k_y, k_z) = (2\pi/\lambda_1, 2\pi/\lambda_2, 2\pi/\lambda_3)$. This means that the speed of propagation in the three Cartesian directions is not necessarily the same for all directions. This presents the definition of a dispersive medium. In a dispersive medium the primary wave front can be deformed due to inhomogeneities.

Using the vector description, Equation (5.3) can also be written as

$$E(\mathbf{r},t) = E_0 e^{i\mathbf{k} \bullet \mathbf{s}} e^{-i\omega t} \tag{5.4}$$

where E_0 represents the magnitude of the electric field.

Now assume that this wave travels in the positive z-direction, with the electric field perpendicular to the direction of propagation. This means by definition that the polarization of the wave is perpendicular to the wave vector $k_z = \omega/c$.

The electric field still satisfies the wave equation (Equation [3.38]) or

$$\nabla^2 E + k^2 E = 0 \tag{5.5}$$

Equation (5.5) can be shown to have a solution in the form of

$$E(x,y,z,t) = E_0 \exp[i(-k_z z)]\varphi(x,y,z)\exp[i(-\omega t)] \tag{5.6}$$

To satisfy the requirement of a collimated beam, this needs to be a solution that does not change dramatically over distance z. This implies for $\varphi(x, y, z) = \varphi(\mathbf{s})$ that over a distance $\Delta z = \lambda/2\pi$, the following condition holds true: $\varphi \ll 1$, or

$$\left|\frac{\partial \varphi}{\partial z}\right| \ll |k\varphi| \tag{5.7}$$

Developing Equation (5.7) in a Taylor series gives

$$\left|\frac{\partial^2 \varphi}{\partial z^2}\right| \ll \left|k\frac{\partial \varphi}{\partial z}\right| \tag{5.8}$$

Applying the condition expressed in Equation (5.8) to Equation (5.6) after substitution in Equation (5.5) gives

$$\frac{\partial^2 \varphi}{\partial x^2} + \frac{\partial^2 \varphi}{\partial y^2} = -\frac{2k}{i}\frac{\partial \varphi}{\partial z} \tag{5.9}$$

Out of the various possible solutions to Equation (5.9) the still-standing wave-package is the solution that satisfies the requirements for minimal divergence. This solution will be chosen to be at $z = 0$. Using Equation (5.2) we obtain the following as a possible solution:

$$\varphi(x,y,0,t) \propto \exp\left\{-\frac{1}{2}\left[\left(\frac{x}{x_0}\right)^2 + \left(\frac{y}{y_0}\right)^2\right]\right\}\exp(-i\omega t) \tag{5.10}$$

While looking in the direction $y = 0$ we find that the dispersion of δ_x with respect to the average position $\langle x \rangle = 0$ can be written as

$$\delta x = \sqrt{\langle x^2 \rangle} = \frac{\sqrt{\int x^2 |\varphi|^2 \, dx dy}}{\sqrt{\int |\varphi|^2 \, dx dy}} = \sqrt{\frac{1}{2}} x_0 \tag{5.11}$$

Similarly the x-component of the wave vector: k_x satisfies (using Equation (5.2))

$$\delta k_x = \sqrt{\langle k_x^2 \rangle} = \sqrt{\frac{1}{2}} \frac{1}{x_0} \tag{5.12}$$

Combining Equations (5.12) and (5.11) provides the conditions for a minimum wave packet since

$$\delta x \delta k_x = \frac{1}{2} \tag{5.13}$$

Assume the x and y components to be equal in size: cylindrically polarized ($x_0 = y_0$), then the equations can be converted into cylindrical coordinate system using

$$r^2 = x^2 + y^2 \tag{5.14}$$

Now the distance to the axis of propagation where the amplitude of φ has dropped to e^{-1} can be defined as the waist of the beam w_0, meaning the narrowest part of the beam at position $z = 0$. This can be written in energy terms as

$$|\varphi|^2 \propto |E|^2 \propto e^{-2} \tag{5.15}$$

To find the length over which the waist can be considered to be minimal the following definition needs to be introduced:

$$\alpha = \frac{x(z)}{z}; \quad \alpha = \frac{dx}{dz} \text{ or } \alpha_{z=0} = \frac{dx}{dz}_{z=0} = \alpha_0 \tag{5.16}$$

From here the waist-length L can be defined as

$$L = w_0 / \alpha_0 \tag{5.17}$$

with

$$w_0 = \sqrt{\left(\frac{L\lambda}{\pi}\right)} = \sqrt{\left(\frac{2L}{k}\right)} \tag{5.18}$$

and

$$\alpha_0 = \sqrt{\left(\frac{\lambda}{L\pi}\right)} = \sqrt{\left(\frac{2}{kL}\right)} \tag{5.19}$$

The waist-length of the bare beam can now be written as

$$L = \frac{kw_0^2}{2} \tag{5.20}$$

All these factors combined show the limitations of the initially assumed infinitely parallel beam. Every laser beam will diverge; however, the amount of divergence will depend on the geometry of the laser cavity. The waist will be the smallest at the output mirror, and diverges from there. Focusing can produce a smaller waist size; in return the waist-length will also decrease as expressed by Equation (5.20).

5.1.3 Power Density Profile of the Beam

A normal radial beam profile of laser light exiting from a laser will have a greater energy in the center than towards the edge of the beam; the square of the electric field will drop with the square of the distance from the optical axis. This intensity profile is called a Gaussian profile. The Gaussian beam profile is illustrated in Figure 5.1. On the other hand, the energy profile may be uniform across the entire cross-section of the beam; this profile is often called a "top-hat" profile.

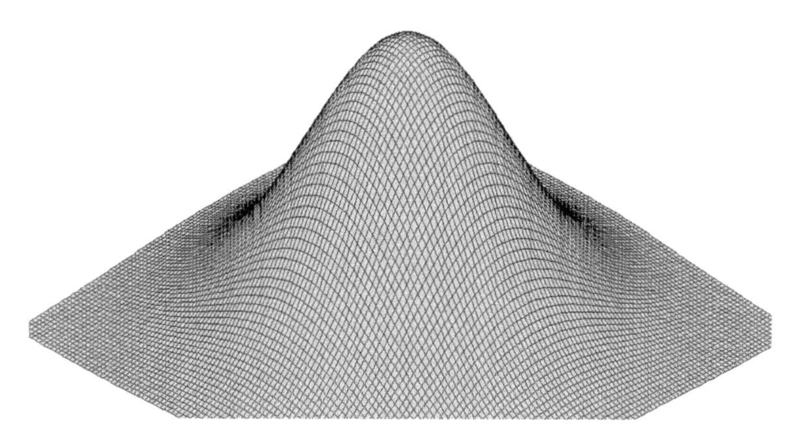

FIGURE 5.1
Illustration of the Gaussian beam profile.

5.1.4 Gaussian Beam Profile

The output profile of most laser cavities will be Gaussian shaped. The nature of the Gaussian profile gives a probability of 0.683 of being within one standard deviation (in case of the laser beam, the waist of the beam is w_0) of the mean. The Gaussian distribution is a continuous function that approximates the exact binomial distribution of events. The binomial distribution function specifies the number of times (ξ) an event occurs in n independent trials where "p" is the probability of the event occurring in a single trial. For the number of photons released by a laser the number of events can easily be taken to be infinity.

The binomial distribution reads as follows:

$$f_b(\xi) = \frac{n!p^{\xi}(1-p)^{n-\xi}}{\xi!(n-\xi)!} \tag{5.21}$$

With the mean of the number of events given by

$$\langle \xi \rangle = np \tag{5.22}$$

The standard deviation of the number of events is defined by

$$\sqrt{np(1-p)} \tag{5.23}$$

It is an exact probability distribution for any number of discrete trials. If n is very large, it may be treated as a continuous function.

The power density of a laser beam after passing through a fiber optic will depend on the category and quality of the fiber and the fiber preparation.

The power density is usually defined as the luminous irradiance I. The irradiance is the rate of flow of electromagnetic radiation or radiant power, in W, incident on a surface. The irradiance is hence expressed in W/m^2 or in $J/s\ m^2$.

The default Gaussian distribution of a power profile is bell shaped, with the maximum power density at the center of the spot, and dropping off towards the periphery of the beam. For the circular output the irradiance as function of radius drops with the square of the radius and is called a Gaussian profile, given by

$$I(\mathbf{r},\mathbf{s}) = I_0 e^{-2r^2/w_0^2} e^{\mathbf{k} \bullet \mathbf{s}} \tag{5.24}$$

where w_0 is the waist of the beam, r the radius from the center of the beam, and I_0 the irradiance on the center of the beam. It is clear that the irradiance profile is symmetric around the principal axis.

For the rectangular profile of selected lasers, both axes describing the rectangular configuration have a Gaussian distribution,

$$I(x,y,\mathbf{s}) = I_{x_0}\left(\frac{\sqrt{2}x}{w_0}\right) I_{y_0}\left(\frac{\sqrt{2}y}{w_0}\right) e^{-(x^2+y^2)/w_0^2} \mathrm{e}^{\mathbf{k}\bullet\mathbf{s}} \tag{5.25}$$

Irradiation profiles can range from the Gaussian distribution to the so-called "top-hat" profile. The top-hat profile will be described next.

5.1.5 Top-Hat Beam Profile

The top-hat beam profile has a constant irradiance I_0 over the transverse plane (x, y) for $r \leq a$ compared with Equation (5.24) and no irradiance outside its boundary at $r > a$.

One of the most popular models for describing top-hat (or flat-topped) profiles is the so-called super-Gaussian (SG) function given by superimposed complex Gaussian functions:

$$\mathrm{SG}(r) = \sum_{\lambda=1}^{N} A_{\gamma} \exp(-|r|^{\gamma}/B_{\gamma}^{2}) \tag{5.26}$$

where r is the radial coordinate and γ the SG power. The constant B_γ is a dimensional and provisional coefficient, and A_r a normalization coefficient. Usually, γ is an even integer and Equation (5.26) reduces to the Gaussian case when $\gamma = 2$, and A is a characteristic length of the phenomenon in question.

The SG function consists of a sum of superimposed complex Gaussian functions. The sum of all Gaussian functions approaches a rectangular distribution. The SG function has the advantage of being expressed explicitly. However, the function often turns out to be unfriendly in analytical terms, as the study of propagation features of SG beams has to be handled by a numerical method. One recent model to overcome some of these problems was F. Gori's flattened Gaussian (FG) profiles. His sequentially infinite set of profiles is given by

$$\mathrm{FG}(r) = \sum_{l=0}^{N} \frac{r^{21}}{l!} \exp(-r^2/B_l^{\,2}) \tag{5.27}$$

where the flatness of the FG profile is related to the integer $N = 0, 1, 2, \ldots$, and Equation (5.27) reduces to the Gaussian case when $N = 0$. Again the constant B_l represents a dimensional constant.

A functional form for flat-topped beam profiles takes advantage of the root cause of flatness of SG and FG beams at the center of their profiles; that is, their first, second, and higher derivatives with respect to r are identical to

0 at the point $r = 0$. This can be accomplished by introducing a function F (r) that represents the profile of an axially symmetric flat-topped profile at a certain transverse plane:

$$F(r) = \frac{\{1-[1-\exp(-r^2/w^2)]^M\}}{M} \tag{5.28}$$

where $M = 1, 2, 3, \ldots$, and $F(r)$ reduces to the circular Gaussian function exp $(-r^2/w^2)$ when $M = 1$. The flat-topped beams are obtained when $M > 1$. The only difference between these profiles is the steepness of the edges and this is a very small difference. Equation (5.28) can also be used to describe beams with spike-shaped and triangular profiles.

Finally, one of the simplest expressions for flat-topped beam profiles is the super-Lorentzian (SL) profile,

$$\text{SL}(r) = \frac{1}{1+\left|r/w\right|^M} \tag{5.29}$$

where $M = 2$ for a Lorentzian distribution and a flat-topped profile is rapidly approached with an M of 50. However, a numerical technique is required for studying beam propagation of beams with SL profiles.

Next we will discuss what is involved in defining the beam spot size.

5.1.6 Irradiation Spot Size

The spot size of a bare beam will be discussed first. The spot size can have various geometric configurations depending on the construction of the laser cavity. The majority of lasers will have a circular output profile; however, in particular the excimer lasers often have a rectangular-shaped output profile. The light delivery from a fiber optic presents a more diverse approach. The light guide relies on Snell's law for internal reflection to provide a means for transporting the light by increasing the critical angle explained in Chapter 2, and remaining confined within the material designed to guide the light, until Snell's law is no longer valid. The point where Snell's law seizes to confine the light rays by reflection is the distal open end of the fiber optic. Two situations can be discriminated when delivering light through a fiber optic:

(1) Near-field irradiation: in this case the face of the fiber is toughing the biological medium, or close to toughing. The reflections from the wall of the fiber are totally scrambled in this situation, and the Gaussian beam profile of the bare beam coupled into the fiber optic is lost. The output power profile is constant across the entire face of the fiber, which is often referred to as a top-hat profile.

(2) Far-field irradiation: at a distance of several fiber diameters away the radiance drops off towards the edge of the spot, creating a Gaussian profile again or a Lorentzian profile.

5.1.7 Power Density and Energy Density of a Light Source

Combining the power of the source with the spot size provides the output power distributed over an area. The power per unit area is referred to as the power density. The area in question can be the cross-sectional area of the bare beam, for instance impacting a target, or it can be the cross-sectional area of a fiberoptic light guide. Depending on the particular application, the light will exit from the fiber optic and travel through free space for a certain distance before encountering the target biological medium. The amount of light power coupled into the medium under irradiation will be a function of the laser power output. At a constant power density, the radiant fluence rate in the tissue at the core of the spot size will increase with spot size, as shown in Figure 5.2 At constant laser power, the power density at the irradiation site will decrease with the equivalent of the square of the radius of the laser spot with increasing spot size. At any power density, the depth of coagulation will increase as a function of duration of exposure. Lesion maximum width and lesion depth are a direct function of the temperature generated inside the tissue and thus directly correlated to the amount of heat deposited

FIGURE 5.2
Spot size as a function of distance to a fiberoptic source. Light emitted from the fiberoptic distal end diverges out. The spot size increases with distance from the fiber tip; as also, the beam profile changes from "top-hat" to Gaussian within a few hundred micrometers, depending on the fiber diameter.

in the tissue. The amount of heat deposited in each volume of the medium is the local fluence rate times the local absorption of light. The power and energy density will determine the rate and the range of thermal damage.

5.1.8 Local Beam Angle of Incidence with the Tissue

The maximum energy penetration into the tissue occurs at a normal incidence. When laser irradiation reaches the tissue surface under a local angle, a progressively higher percentage of photons will be reflected off the surface at an increasing angle of oblique incidence. The total reflectance in percentage as a function of the angle of incidence with the normal to the tissue for a smooth surface and for a rough surface is depicted in Figure 5.3. The reflectance can be calculated from Fresnel's Arrhenius equations described in Chapter 3, section 3.1.4.2:

$$R = \left(\frac{n_1 \cos\theta_1 - n_2 \cos\theta_2}{n_1 \cos\theta_1 + n_2 \cos\theta_2} \right)^2 \tag{5.30}$$

In combination with Snell's law: $\sin\theta_1 / \sin\theta_2 = n_2 / n_1$, where n is the index of refraction of the medium, θ the angle the beam makes with the normal to the surface, and 1 and 2 the first and the second medium, respectively, in the order the light passed through them.

In the next section the difference between collimated irradiation and diffuse illumination will be described.

5.1.9 Collimated or Diffuse Irradiation

The beam profiles from the previous sections are usually delivered to the tissue in a collimated geometry, where the light interacts with the tissue at normal incidence along the z-axis. Diffuse irradiance, in contrast, presents light at many angles to the tissue.

The radiance L(**r**,**s**) describes the spatial extent of the cone of light, where **r** denotes the position along the cone and **s** the unit vector in the direction of the cone. The two extremes describe completely diffuse light that is isotropic in a 4π solid angle at one extreme and that propagates straight along the z-axis as a collimated beam, as described in the previous sections. Figure 5.4 shows the difference between the collimated and diffuse irradiance incident at an air–tissue interface.

Collimated irradiation presents a greater challenge than diffuse irradiation in deriving an expression for the light distribution inside a turbid medium that is irradiated by a laser beam, as explained in Section 3.2 in Chapter 3.

As part of the description of light delivery the issue of the wavelength needs to be addressed as well. The following sections will discuss the wavelength issues involved with irradiation of biological samples.

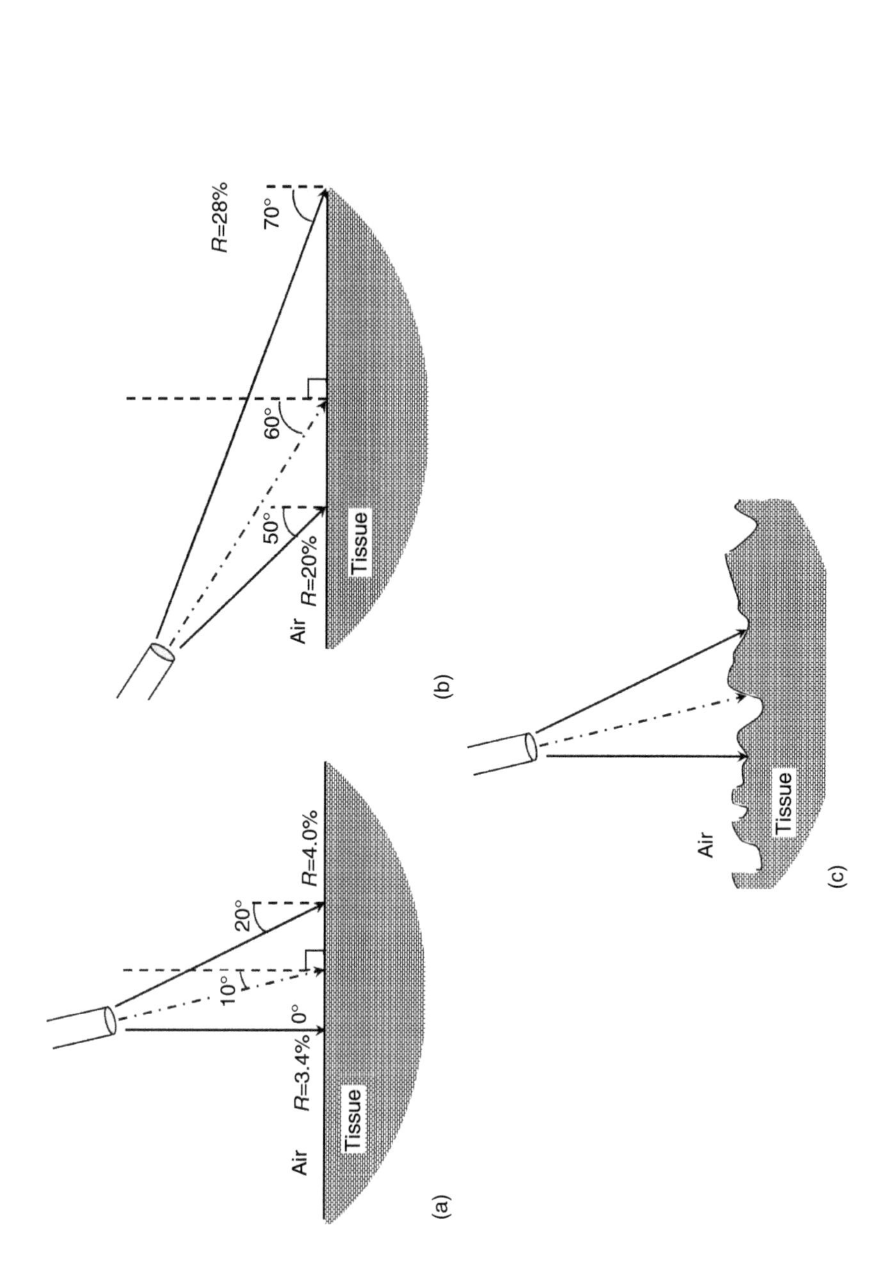

FIGURE 5.3
Reflection as a function of angle of incidence with a surface. Three different types of surface configurations are illustrated. The surface contour with respect to the light delivery protocol will largely depend on the spot size and target organ: (a) close to perpendicular incidence, (b) grazing angle of incidence, and (c) trabeculated surface with many angles of incidence simultaneously.

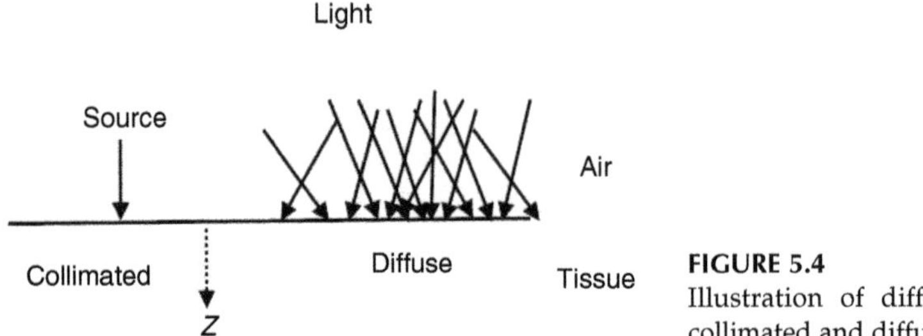

FIGURE 5.4
Illustration of difference between collimated and diffuse irradiation.

5.1.10 Light Wavelength

The fundamental interaction of electromagnetic radiation with matter depends in part on the ratio of the wavelength of the light and the dimensions of the "particles" involved in the interaction. These particles can be of atomic size up to red blood cell size (several micrometers) in the case of biological materials.

Other issues related to light delivery involve the duration or mode of delivery, whether it is pulsed or continuous and how frequent the delivery takes place. The implication of delivery frequency will be described briefly in the next section.

5.1.11 Repetition Rate

When a pulsed laser system is used, the laser–tissue interaction phenomena will also depend on the repetition rate, the pulse duration, and the pulse modulation, in addition to the dependence on the peak power and spot size. When the pulse repetition rate is high, the light can appear to be continuous, even though it is pulsed. The amount of time that the laser is on versus off, known as the duty cycle, is very important in photothermal interactions. The theory of pulsed laser systems will be discussed in detail in Chapter 8.

5.1.12 Pulse Length

For a pulsed laser system, the energy is typically delivered in short pulses. It is desirable to have a laser pulse duration that is shorter than the thermal relaxation time of the tissue or the time it takes for the tissue to diffuse the heat energy. This allows one to confine the laser energy to a small area near the irradiation site, thus confining the thermal energy.

5.1.13 Light Delivery Pulse Modulation

The pulse repetition rate and pulse width are factors directly related to the type of pulse modulation. A pulse can result from simply turning the laser on and off at regular intervals, resulting in pulse durations of milliseconds to seconds with repetition rates often smaller than 20 Hz. Faster repetition rates and shorter pulse widths can be achieved by interrupting the delivery of the laser beam by mechanically blocking the laser beam at regular intervals. High repetition delivery schemes can be realized by the mode-locking and Q-switching techniques described in Chapter 2, Section 2.4.3.2 and Chapter 8.

5.2 Tissue Variables

The interaction of light with biological media for treatment or diagnostic purposes depends on several factors, with the optical properties of the tissue as the primary influence. The optical properties of tissue are a series of parameters that characterize light propagation in tissue. The optical properties of importance are typically the absorption and scattering coefficients, the scattering anisotropy factor, and the refractive index.

5.2.1 Optical Properties of the Tissue

Once light enters tissue, photons may be annihilated, redirected, or reduced in energy. This translates in absorption or scattering events. The principles of absorption and scattering are described in detail in Chapter 2. The ability of tissue to absorb photons is represented by the parameter μ_a with units per length (per mm or per cm), which is defined as the probability per unit path length that a photon will encounter an absorption event. When an absorption event occurs, the incident photons terminate and their energy turns into other forms of energy, such as heat. In those events where the energy is absorbed by fluorescent dye, part of it turns into photons emitted at a longer wavelength (lower energy).

In a scattering event, the photons do not lose their energy but change the direction of propagation. The scattering coefficient μ_s, also with units per length (per mm or per cm), represents the probability per unit path length that a photon will encounter a scattering event. Rayleigh scattering, in general as shown in Chapter 4, can describe scattering for particle size smaller than the wavelength of the incident photon flux. For particle size equal to or greater than the wavelength the scattering events can be analyzed by Mie theory, also described in Chapter 4. When considering the cell anatomy involved in light–tissue interaction many different-size particles are implicated.

Color and composition of the tissue will determine the optical and thermal properties relevant for laser–tissue interaction. Optical properties are different for distinctive tissues, and these differences are not always distinguishable by visual examination. The optical parameters are in fact a function of the delivery protocol of the light source, tissue composition, and temperature and water content of the tissue. Different tissues and tissue constituents will interact with light according to their specific optical parameters, and require individual attention in the classification of their optical parameters. Photocoagulation, dehydration, and carbonization resulting from laser irradiation will furthermore change the optical properties of the tissue.

When light strikes a particle with an index of refraction n that has a different value than the surrounding environment, the light is refracted. The angle at which the light is refracted can theoretically be explained either by Rayleigh or by Mie scattering; however, the basis is the fact that the scattering angle is a function of the size and shape of the particle as well as the wavelength of the incident light, in addition to the angle at which the incident light reaches the particle interface surface. This introduces a third parameter to describe the fate of a photon when interacting with a turbid medium such as biological tissue. This parameter indicates the anisotropy of the medium: the mean cosine of scattering angle or scattering anisotropy factor (g).

5.2.1.1 Tissue Index of Refraction

The index of refraction for tissue is usually assumed to be that of water, since tissue generally constitutes minimally 70% water. For fat tissue (adipose tissue) the story is slightly different, since there is significantly less water in adipose tissue. Both for water and tissue the index of refraction depends on the wavelength of the incident light. Studies by Bausch and Lomb indicate that a heuristic relationship between the water content and the refractive index n can be derived. The Bausch and Lomb report suggests the following correction factor at the helium-neon laser wavelength of 632.8 nm:

$$n = 1.53 - 0.2W \tag{5.31}$$

where W represents the water content of the tissue in grams of water per total tissue weight. Further corrections are needed to account for the wavelength dependency of the index of refraction.

Additionally temperature adjustments are also needed to find n as the refractive index of the tissue with respect to vacuum, using the dependence on temperature, under T absolute temperature (in K), with λ as the wavelength of light and ρ mass density. Next introduce the reference temperature $T^* = 273.15$ K and the reference density $\rho^* = 1000$ kg/m^3 all in reference to the center wavelength $\lambda^* = 0.589$ μm.

By introducing the following dimensionless variables: Temperature $\overline{T} = T/T^*$, density $\overline{\rho} = \rho/\rho^*$, and wavelength $\overline{\lambda} = \lambda/\lambda^*$, the heuristic expression for the refractive index n as a function of wavelength can be represented by the following empirically derived equation:

$$\begin{aligned} \frac{n^2-1}{n^2+2}\frac{1}{\overline{\rho}} &= a_0 + a_1\overline{\rho} + a_2\overline{T} + a_3\overline{\lambda}^2\overline{T} \\ &\quad + \frac{a_4}{\overline{\lambda}^2} + \frac{a_5}{\overline{\lambda}^2 - \overline{\lambda}_{\mathrm{UV}}^2} + \frac{a_6}{\overline{\lambda}^2 - \overline{\lambda}_{\mathrm{IR}}^2} + a_7\overline{r}^2 \end{aligned} \tag{5.32}$$

Using the following coefficients: $a_0 = 0.244257733$, $a_1 = 9.74634476 \times 10^{-3}$, $a_2 = 3.73234996 \times 10^{-3}$, $a_3 = 2.68678472 \times 10^{-4}$, $a_4 = 1.58920570 \times 10^{-3}$, $a_5 = 2.45934259 \times 10^{-3}$, $a_6 = 0.900704920$, $a_7 = -1.66626219 \times 10^{-2}$, we obtain $\lambda_{UV} = 0.2292020$ and $\lambda_{IR} = 5.432937$.

These temperature dependent optical characteristics are used for diagnostic applications such as Schlieren imaging, as discussed in Chapter 8, section 8.4.

5.2.2 Surface Contour

Since most biological tissues have a rugged surface, accidental grazing angles of incidence will affect the light distribution inside the medium as a result of gradual loss across the surface due to increased reflection at larger angles of incidence, following Snells' law of reflection. In addition the interface between the two media will be subject to Snell's law of refraction, which will determine the direction of the rays of light at the boundary of the various tissues in an organ, each with its own index of refraction. The initial encounter with a different index of refraction will be when entering the biological medium itself, and subsequently the inhomogeneities within the medium. A view of an optical fiber is exemplified in Figure 5.5a, showing a special configuration with feedback in Figure 5.5b.

5.2.3 Tissue Temperature

In most cases the laser is used to heat the tissue to therapeutic temperature levels, for example hyperthermia, coagulation, or evaporation. Temperature rise is accomplished as a result of the conversion of light energy into heat due to a certain amount of absorption. Since phase transitions take place at specific and unique temperature levels (coagulation at approximately 60°C, and water evaporates at 100°C), the baseline temperature is of critical importance to the laser parameters required to reach the target result. The most striking example of temperature effects on the optical properties of biological media is the denaturization of albumin. When the contents of an egg are heated above 60°C, in addition to the fact that the albumin turns from liquid into solid, a change in optical properties is expressed as a change in appearance: change from clear to white. Note that the temperature effect is not as well demarcated as the previous one, which seems to indicate that there exists a time–temperature effect that determines the ultimate coagulation outcome. This time–temperature effect will be discussed in detail in Chapter 7.

5.2.4 Thermodynamic Tissue Properties

Once the tissue is heated as a result of the absorption of light energy and conversion of this energy into vibrational, rotational, and translational energy, the tissue will experience a temperature rise locally. Tissue temperature

(a)

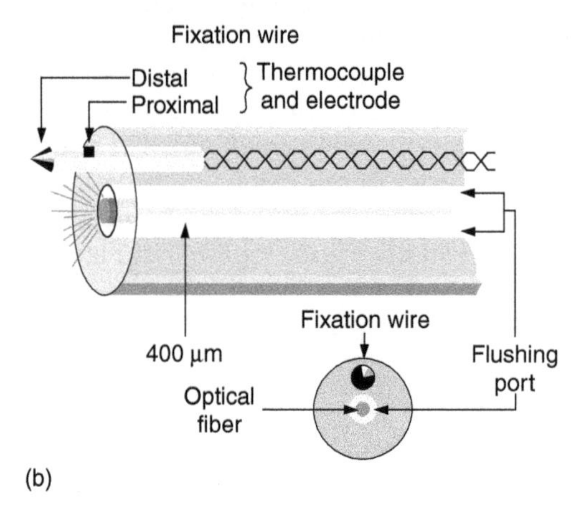

(b)

FIGURE 5.5
(a) Fiberoptic bare tip and (b) fiber optic inside a catheter with thermocouples for temperature feedback and electrodes for cell depolarization feedback information.

rise and the rate of heat transfer to surrounding tissues depend, among others, on the specific heat, thermal diffusivity, and thermal conductivity of the tissue constituents. These three parameters are the main thermodynamic properties in the principal equations describing heat generation and heat transfer in a medium: the First Law of Thermodynamics—net heat input equals change in internal energy, plus work; however, no work is generated in this particular case.

5.2.5 Tissue Blood Flow and Blood Content

The presence or absence of blood in the tissue is the primary discrimination factor between in vivo and in vitro observations. Blood itself can be distinguished into oxygenated and de-oxygenated form. The heme group in hemoglobin has its own characteristic absorption and scattering spectrum, while the presence of oxygen significantly alters this spectrum. A benefit

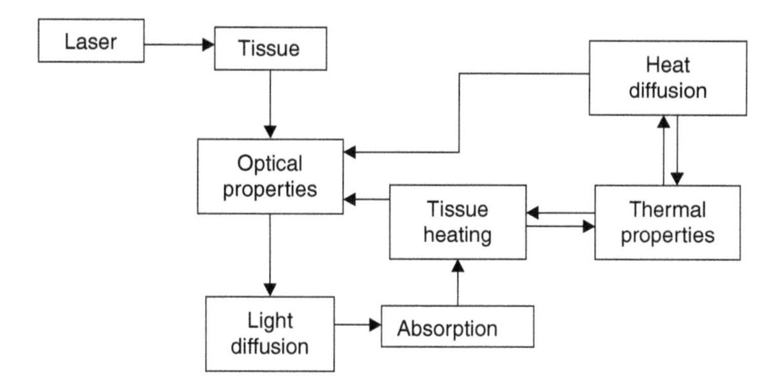

FIGURE 5.6
General impression of the interrelations of the optical properties and the thermodynamic parameters of a tissue under laser irradiation.

resulting from this characteristic spectrum of, respectively, oxy- and deoxyhemoglobin is the ability to measure the oxygen saturation of blood noninvasively by optical means. The optical blood-gas analyses will be discussed in detail in Chapter 16.

Tissue blood content influences the optical properties of the tissue, as well as the thermodynamic properties. Tissue blood flow will serve as a heat sink resulting in tissue cooling, depending on the perfusion rate and blood temperature.

A general impression of the inter-relations of the optical properties and the thermodynamic parameters of a tissue under laser irradiation is illustrated in Figure 5.6.

5.3 Light Transportation Theory

The field of radiative transfer is aimed at the theoretical modeling of the transport of particles through a dispersive medium. Radiative transfer theory is also called transport theory. The foundation for radiative transfer theory comes from two disciplines in the field of physics: astrophysics and neutron physics. Astrophysics deals with the light propagation of distant galaxies through galactic dust and planetary debris. Neutron physics, in contrast, is the study of the propagation of elementary particles and has the added element of a continuous energy spectrum, whereas light has a discrete spectrum.

In general the foundation of transport theory is the assumption of migrating particles that do not interact with each other, and conservation of energy applies.

In the case of radiative transfer in biomedical optics, the particles are photons, which are probability electromagnetic wave packets, interacting with atomic structures and with optical geometries.

In order for the transport theory to be applicable, the photon's localization in space needs to be small with respect to the dimensions of the medium it interacts with, and the spread in momentum needs to be small as well. Under these conditions the photons will not interact with each other. Even when starting out with coherent light from a laser source, this coherence is quickly lost after several scattering events have taken place. Since the photon is small, numerous scattering events will occur during the lifetime of a photon in biological media, and the chance for interference effects is virtually negligible.

In the theoretical description of biomedical optics, the main concern is the energy transfer within the biological media, and the final distribution of the deposited energy in the tissue resulting from absorption.

Radiative transport theory provides the equation of radiative transfer, which is in many aspects similar to the Boltzmann equation used in kinetic gas theory, only for photons.

The equation of radiative transfer can also be used to describe the interaction of light with inorganic media such as the irradiance and radiance of photographic emulsions, paint layers, or dye solutions in liquid or solid form, respectively, for the description of the light penetration from an external source, or the radiation by the scattered light, or even from fluorescence. The equation of radiative transfer is also used to describe the light scattering in seawater for oceanographic applications, the light scattering in planetary atmospheres in astronomy, and the neutron migration in reactor physics.

First, several definitions in light transport theory need to be established.

5.3.1 The Time-Dependent Angular and Spatial Photon Energy Rate Distribution

The time-dependent angular and spatial photon energy rate distribution is described by the light power incident on a cross-sectional area flowing within a solid angle, and is designated by the parameter radiance: $L(\mathbf{r}, \mathbf{s}, t)$ (W/sr cm^2). In this notation vector $\mathbf{r}$ denotes the position of the location where the radiance is quantified and $\mathbf{s}$ the direction vector for the photon migration, all as a function of time t.

The radiance follows from the Poynting vector : $\mathbf{S} : \mathbf{S} = (\mathbf{E} \times \mathbf{B})/\mu$ (Section 2.3.4 in Chapter 2). The radiance is defined as the time average of the length of the Poynting vector, averaged over one or more cycle-lengths,

$$L(\mathbf{r},\mathbf{s},t) = \left\langle \frac{\mathbf{E} \times \mathbf{B}}{\mu} \right\rangle \tag{5.33}$$

where μ is the dielectric permeability, relative permeability times the permeability of free space: $4\pi \times 10^{-7}$ H/m.

Since there are no external forces acting the energy in the system is conserved; a general balance equation (such as those used to model chemical reactions) can be used to help develop an illumination model. For example, the general energy balance follows the following rules:
Accumulation within a system = flow-in through system boundaries − flow-out through system boundaries + generation within system − consumption within system.

The energy balance results in a description of the dissemination of the light, which is expressed as

$$\frac{1}{c}\frac{\partial L(\mathbf{r},\mathbf{s},t)}{\partial t} = -\mathbf{s}\bullet\nabla L(\mathbf{r},\mathbf{s},t) - [\mu_a(\mathbf{r}) + \mu_s(\mathbf{r})]L(\mathbf{r},\mathbf{s},t) + \mu_s(\mathbf{r})\int_{4\pi} P(\mathbf{s},\mathbf{s}')L(\mathbf{r},\mathbf{s}',t)\mathrm{d}\omega' + S(\mathbf{r},\mathbf{s}',t) \tag{5.34}$$

The terms μ_a, μ_s, and $P(\mathbf{s}.\acute{\mathbf{s}})$ are the three basic microscopic optical properties that characterize the light propagation in tissue. One of the derived optical parameters is the total attenuation coefficient μ_t, which is equal to $\mu_a + \mu_s$ and represents the probability per unit path length that a photon will encounter either an absorption or a scattering event. When the anisotropy coefficient g is not known, the reduced scattering coefficient μ'_s is useful to describe the scattering event. The μ'_s is defined as $(1-g)\mu_s$. Similarly, the reduced total attenuation coefficient μ'_t equals $\mu'_s + \mu_a$.

Equation (5.34) can be seen as the composite effect of the following components of radiation transfer within a small but finite volume V of turbid medium. The first term in this equation is the change in radiance per unit volume:

$$\frac{1}{c}\frac{\partial L(\mathbf{r},\mathbf{s},t)}{\partial t} \tag{5.35}$$

The next term, the first term after the equal sign, signifies the radiance lost through the boundaries of the volume:

$$-\mathbf{s}\bullet\nabla L(\mathbf{r},\mathbf{s},t) \tag{5.36}$$

The following term indicates the loss due to absorption and scattering into a different direction:

$$-[\mu_a(\mathbf{r}) + \mu_s(\mathbf{r})]L(\mathbf{r},\mathbf{s},t) \tag{5.37}$$

The subsequent term identifies the recovery of radiance into the original direction as a result of scattering from direction $\acute{\mathbf{s}}$ into direction s:

$$+\mu_s(\mathbf{r})\int_{4\pi} P(\mathbf{s},\mathbf{s}')L(\mathbf{r},\mathbf{s}',t)\mathrm{d}\omega' \tag{5.38}$$

The dot product of $\mathbf{s}'$ and $\mathbf{s}$: $(\mathbf{s}\cdot\mathbf{s}') = \omega$ yields the solid angle of scattering, which reduces to $(\mathbf{s}\cdot\mathbf{s}') = \theta$ in two dimensions. The term $d\omega'$ defines the differential solid angle about the direction $\mathbf{s}'$.

The final term in Equation (5.34) is a source term, the laser source either within the medium or incident on the surface of the medium, or even fluorescence.

In Chapter 7, section 7.2.2.7 and Chapter 14, section 14.13 the specific conditions for time-dependent light–tissue interaction will be discussed in detail. For now the steady-state light–tissue interaction will provide the basis of the theoretical description of the photon migration through turbid media.

5.3.2 Steady-State Angular and Spatial Photon Energy Rate Distribution

The time-independent equation of radiative transfer will be used from here on to derive the principal optical process of light–tissue interaction.

The stationary radiance equation applies to continuous irradiation only; however, pulsed irradiation for long pulse lengths also satisfies the conditions for steady-state approximation:

$$\begin{aligned}\mathbf{s}\bullet\nabla L(\mathbf{r},\mathbf{s}) &= -[\mu_a(\mathbf{r})+\mu_s(\mathbf{r})]L(\mathbf{r},\mathbf{s})\\ &\quad +\mu_s(\mathbf{r})\int_{4\pi} P(\mathbf{s},\mathbf{s}')L(\mathbf{r},\mathbf{s}')d\varpi'+S(\mathbf{r},\mathbf{s}')\end{aligned} \tag{5.39}$$

5.3.3 Boundary Conditions

The boundary conditions to the equation of radiative transfer will need to be defined at each interface between two media within the tissue, which have significantly different optical properties, and at the interface between the tissue and the medium the light source is located in.

5.3.4 Radiation Definitions for Turbid Media

In the light distribution theory the following definitions have been adopted: power of delivered light, radiance, fluence rate, power of light inside turbid medium, radiative flux, and radiation pressure. A derived parameter associated with the absorption and scattering properties of a tissue under irradiation at a specific wavelength is the penetration depth. The radiation terms as mentioned will be defined in the following paragraphs.

5.3.4.1 Photon Power Related to Radiance

The radiance $L(\mathbf{r}, \mathbf{s})$ can be measured directly with a standard small-aperture photodetector and is calculated from the measured power $P(\mathbf{r})$ as

$$P(\mathbf{r}) = \iint_{\text{detector surface}} L(\mathbf{r},\mathbf{s})\mathbf{n}\bullet\mathbf{s}\,\partial\Omega\,\partial A = \iint_{\text{detector surface}} L(\mathbf{r},\mathbf{s})\cos\theta\,\partial\Omega\,\partial A \tag{5.40}$$

5.3.4.2 Radiance Incident on a Volume (Fluence Rate)

The radiance $L(\mathbf{r}, \mathbf{s})$ integrated over 4π solid angle is called the fluence rate $\Psi(\mathbf{r})$. The fluence rate signifies the power incident on a surface of a volume of medium with finite but small dimensions at the location $\mathbf{r}$. The unit for fluence rate is W/cm^2. Fluence rate is expressed as

$$\Psi(\mathbf{r}) = \int_{4\pi} L(\mathbf{r}, \mathbf{s}) \mathrm{d}\varpi \tag{5.41}$$

This term is commonly used to describe the light distribution inside a turbid medium.

5.3.4.3 Photon Power Density in Medium

The fluence rate $\Psi(\mathbf{r})$ multiplied by the local absorption coefficient provides the power density $P(\mathbf{r})$ in the medium in W/cm^3, and thus the absorbed energy per unit time at position $\mathbf{r}$:

$$P(\mathbf{r}) = \mu_a \Psi(\mathbf{r}, \mathbf{s}) = \mu_a \int_{4\pi} L(\mathbf{r}, \mathbf{s}) \mathrm{d}\varpi' \tag{5.42}$$

5.3.4.4 Photon Radiative Flux

The radiance at any given point per unit cross-sectional area is defined as the radiative flux $F(\mathbf{r})$:

$$F(\mathbf{r}) = \int_{4\pi} \mathbf{s} \bullet L(\mathbf{r}, \mathbf{s}) \mathrm{d}\varpi' \tag{5.43}$$

The radiative flux will be shown to play an important role in several approximations to the equation of radiative transfer under specific boundary conditions.

5.3.4.5 Photon Radiation Pressure

The radiation pressure P_r exerted by the photon flux is taken with respect to the normal $\mathbf{n}$ to the surface, as is customary in pressure, and is defined as

$$P_r(\mathbf{r}) = \frac{1}{c} \int_{4\pi} (\mathbf{n} \bullet \mathbf{s})^2 L(\mathbf{r}, \mathbf{s}) \mathrm{d}\varpi' \tag{5.44}$$

This parameter is not frequently used in biomedical optics theory, but is supplied to provide a more complete overview of all conditions that apply in optical radiation therapy and diagnostics.

5.3.4.6 *Penetration Depth*

The penetration depth of light in biological tissue is determined by the distance over which the diffuse energy fluence rate drops to e^{-1} of its initial value and is given by

$$\delta = \frac{1}{\mu_{eff}} \tag{5.45}$$

where $\mu_{eff}{}^2 = 3\mu_a[\mu_a + (1-g)\mu_s]$ is the square of the effective attenuation coefficient for diffuse light (see Chapter 6, section 6.1.2).

5.4 Light Propagation under Dominant Absorption

The most ideal and simple situation to solve the light distribution inside a turbid medium is the situation where absorption is much greater than scattering. Under these conditions the light propagation simply reduces to the Beer–Lambert law. The light propagation can be described by a one-dimensional attenuation proportional to the distance, with respect to absorption only,

$$\psi(z) = \psi_0 e^{-\mu_a z} \tag{5.46}$$

Combining the Gaussian beam profile of Section 5.1.4 with the Beer–Lambert law for attenuation based on predominant absorption, the two-dimensional attenuation is described by

$$\psi(r,z) = \psi_0 e^{-\mu_a z} e^{-2r^2/w_0{}^2} \tag{5.47}$$

Generally the light distribution can be described by the so-called equation of radiative literature, especially for the case where the absorption is smaller than the scattering. Otherwise the Beer–Lambert law is more convenient when absorption is more than an order of magnitude greater than scattering.

Most of light propagation theory is a development of methods that are compromises between physical accuracy and mathematical feasibility.

To quantify the spatial localization of optical radiance, the most likely distribution of photon trajectories needs to be calculated.

5.4.1 Description of Angular Distribution of Scatter

The cumulative effect of all scattering events throughout the entire volume of the particle results in a scattering profile. The scattering profile is a representation of all possible angles of scattering and the likelihood of each scattering direction. This profile is unique for every particle, since there will be subtle differences in shape, size, and orientation of each scattering core with respect to the incoming light. The scattering profile is called the phase function. The name of phase function has a historical background that will not be elaborated on here, but it has no correlation with the phase of the electromagnetic wave, a more appropriate name would be scattering function.

The phase function $P(\mathbf{s},\mathbf{s}')$ (per sr) represents the probability that a photon incident from the $\mathbf{s}$ direction will leave in a unit of solid angle in the $\mathbf{s}'$ direction. Generally the phase function is not the same for each particle that the photon encounters. It depends on the size and shape of the particles. For simplicity, an average phase function is used in most models of light propagation.

Usually the average phase function also assumes that the probability of scattering from one direction to another is only a function of the angle θ between two directions. The angle θ ranges from 0° to 180°, with $\theta = 0°$ and 180° corresponding to forward and backward directions, respectively. Therefore, $P(\mathbf{s},\mathbf{s}') = P(\theta)$. Notice $\cos(\theta)$ has a one-to-one relationship with the angle θ when θ is in the range [0°, 180°]. So, the average phase function $P(\mathbf{s},\mathbf{s}')$ can also be expressed as a function of $\cos(\theta)$, i.e., $p[\cos(\theta)]$, which additionally implies that each interaction takes place in a plane.

The phase function is a probability distribution; consequently, normalization condition requires that the integral of the phase function over all angles equals unity:

$$\int_{4\pi} P(\mathbf{s},\mathbf{s}')\mathrm{d}\omega = 1 \tag{5.48}$$

Equation (5.48) implies a lossless scatterer.

From Equation (5.48) it is obvious that the isotropic phase function (first-order) $P(\mathbf{s},\mathbf{s}')$ can be expressed as

$$P(\mathbf{s},\mathbf{s}') = \frac{1}{4\pi} \tag{5.49}$$

This entails that the probability for scatter in every 4π solid angle has equal probability, and this is referred to as isotropic scattering.

When introducing potentially different probabilities for scattering over the various quadrants of space, a new definition will be required. Since the phase function considers the average scattering angles, no discontinuities can be expected in the phase function. The next logical step is to introduce the mean scattering angle. This parameter can implicate anisotropy, but still assumes a smooth distribution of angles, however indicating a preferential net direction:

$$\int_{4\pi} P(\mathbf{s},\mathbf{s}')(\mathbf{s},\mathbf{s}')\mathrm{d}\omega = g \tag{5.50}$$

where g is the average cosine of the scattering angle.

When trying to capture the scattering process in a mathematical description for radiative transfer analysis, the anisotropic behavior in terms of the scattering angle relative to the direction of the incident photon is best characterized by the mathematical expression of the Henyey–Greenstein phase function in three-dimensional form as

$$g_{HG}(\cos\theta) = \frac{1}{4\pi}\left[\beta + (1-\beta)\frac{1-g^2}{(1+g^2-2g\cos\theta)^{3/2}}\right] \tag{5.51}$$

The two-dimensional form is given as

$$g_{HG}(\cos\theta) = \frac{1}{2\pi}\left[\beta + (1-\beta)\frac{1-g^2}{(1+g^2-2g\cos\theta)}\right] \tag{5.52}$$

where θ is the scattering angle and β an indicator for the fraction of isotropic scattering associated with the anisotropic component. Equation (5.49) again is normalized. In actuality this equation reduces to the formal Henyey–Greenstein phase function when β=0.

Slightly forward scattering can be represented by the first two moments of the phase function, expressed in the mean cosine of the scattering angle $g = \langle\cos(\theta)\rangle$:

$$p(\theta) = \text{Const}[1+3g\cos(\theta)] \text{ or } p(\cos\theta) = \text{Konst}[1+3g\cos(\theta)] \tag{5.53}$$

Equation (5.53) can be normalized as

$$p(\cos\theta) = \frac{1}{4\pi}[1+3g\cos(\theta)] \tag{5.54}$$

A value of g = 0 represents the condition associated with isotropic scattering, while a value of g near 1 corresponds to predominantly forward-directed scattering.

Another common proposition for the scattering phase function is the delta-Eddington approximation, $p_{\delta\text{-E}}(\cos\theta)$,

$$p_{\delta-\text{E}}(\cos\theta) = \frac{1}{4\pi}\{2f\delta(1-\cos\theta) + (1-f)[1+3g\cos(\theta)]\} \tag{5.55}$$

where f represents the fraction of light scattered in a forward peak; when f approximates 1, the phase function becomes a true delta function, and when f goes to 0, it approximates the first-order phase function of Equation (5.54).

The μ_a, μ_s, and $\langle\cos\theta\rangle$ terms are the three basic microscopic optical properties that characterize the light propagation in tissue. Several other microscopic parameters derived from these root parameters are also commonly seen in the literature. The first derived optical parameter is total attenuation coefficient μ_t. This equals $\mu_a + \mu_s$ and represents the probability that a photon will encounter either an absorption or a scattering event per unit path length. Since this coefficient represents the probability of encountering either an absorption or a scattering event, the reciprocal of this is the path length of a photon before it encounters an event; thus, the mean free optical path length D can be introduced as

$$D = \frac{1}{\mu_a + (1-g)\mu_s} \tag{5.56}$$

When the anisotropy coefficient is not directly known, the reduced scattering coefficient μ_s' is useful to describe the scattering event. The reduced scattering coefficient is defined as $\mu_s' = (1-g)\mu_s$ and signifies the forward-only-scattered portion. Similarly, the total reduced attenuation coefficient μ_t' equals $\mu_s' + \mu_a$.

The microscopic optical properties describe single absorption or scattering events. They can be measured directly from samples thin enough that multiple scattering events are negligible. This requires that the sample thickness is less than the mean free path of a photon (i.e., $1/\mu_t'$). The microscopic properties can be measured by indirect methods, in which the macroscopic parameters such as diffuse reflectance, diffuse transmittance, and collimated transmittance are measured. Then by applying light propagation models, the microscopic optical properties are derived from the macroscopic parameters.

Additional methods have been developed that offer the potential for in vivo determination of optical properties as a result of a specific interpretation of the light transport equation approximations under certain boundary conditions. These experimental methods will be discussed in Section 7.2 in Chapter 7.

Many recent advances regarding light propagation in tissue are based upon transport theory. It deals directly with the transport of energy through a medium containing particles and was initiated by Schuster in 1905 in his study of radiation in foggy atmospheres. Schuster used the Boltzmann equation as a basis for his derivations. Two other popular theories are the diffusion theory and Kubelka–Munk theory; both are approximations of the Maxwell equations and the equation of radiative transfer.

Generally it can be seen from Equation (5.34) that the radiance $L(\mathbf{r},\mathbf{s})$ [W/m^2 sr] of light at position $\mathbf{r}$ traveling in a direction of the unit vector $\mathbf{s}'$ is decreased by absorption and scattering and increased by the light scattered from all the other directions into the direction of $\mathbf{s}$. The exact solution to the equation of radiative transfer, however, is usually impossible to handle analytically for the majority of boundary conditions. Therefore, approximations need to be made. One of the methods used to solve the equation of radiative transfer described in this chapter is based on polynomial expansion of the unknown radiance. Because analytical solutions for the transport equation are only available for cases with specific boundary conditions, it is usually solved numerically. Moreover, several approximate theories are available.

In Chapter 6 several approximation techniques for solving the equation of radiative transfer are presented with respect to specific boundary conditions.

5.5 Summary

Both the laser and the tissue variables affecting the light–tissue interaction process were described and correlated to the biological medium interaction. The equation of radiative transfer was introduced, as derived from astrophysics. Additionally the vocabulary of light delivery was defined.

5.6 Nomenclature

Absorption — Loss or attenuation of light energy due to material properties. In general the change in energy due to absorption (dW) is directly proportional to the incident energy (W_0) and to the distance over which the amount of absorption is considered to take place (dx), with the proportionality factor μ_a; i.e., $dW = \mu_a W_0 dx$. Solving for the amount of absorption as a function of the path length μ traveled gives the following solution: $W(x) = W_0 e^{-\mu_a x}$.

a_0 — Albedo: ratio of scattering over scattering plus absorption, $\mu_s/(\mu_a + \mu_s)$(–).

a_1 — Single scattering albedo: ratio of scattering over scattering plus absorption limited to a single scattering event (constricted by choice of boundary conditions) $\mu_s/(\mu_a + \mu_s)$ (–).

Attenuation — Amount of light energy lost when propagating over a distance, often expressed in dB (decibels) per kilometer for fiberoptic attenuation.

Bandwidth — Range of frequencies, or wavelength, over which a light source operates or a light guide propagates without loss.

Chromatic dispersion — Separation of a broadband source into its various wavelength components when traveling through a medium.

Diffuse reflection	Reflection from a surface with random normal incidences (coarse surface) (–).
g	Scattering anisotropy factor: mean cosine of photon scattering angle relative to the direction of the incident photons (–).
Frequency	Number of oscillations per second (Hz).
Index of refraction	Ratio of the velocity of electromagnetic radiation in vacuum to the velocity of electromagnetic radiation in a refractive medium at a given wavelength n.
I	Intensity: amplitude squared.
I	Irradiance: Flow of electric magnetic radiation through a unit area (W/m^2) (also space irradiance). At a point on a surface, the radiant energy flux incident on a surface element divided by the area of that element.
Flux density	Time-averaged Poynting vector (W/m^2).
Light	Electromagnetic radiation that can be perceived by the human eye.
L	Radiance: at a point of a surface and in a given direction, the radiant energy fluence rate of an element of the surface divided by the area of the orthogonal projection of this element on the plane perpendicular to the given direction (W/cm^2 sr).
Mean free optical path	Over this distance D, in one direction the radiance falls to e^{-1} of its value: $D = 1/\mu_a + (1-g)\mu_a$ (m).
Numerical aperture	NA: Sine of the maximum acceptance half angle of an optical device (e.g., fiberoptic lens) θ_{max} times the index of refraction of the device. The larger the numerical aperture, the greater is the amount of light that can be accepted into the device $(-)$.
Reflectance	Reflectance: ratio of incident flux on surface of a sample relative to the reflected flux from the surface. Value range from 0 to 1 $(-)$.
R	Refletion: ratio of light redirected in the opposite direction of entry with respect to the incident-light, obeys snell's law of reflection $(-)$.
R	Radius: fixed radius (mm).
r	Radius (mm).
Scattering phase function $P(\mathbf{s},\mathbf{s}')$	Probability of light being scattered from direction $\mathbf{s}'$ into direction $\mathbf{s}$, which is normalized so that the chance of a photon being scattered over 4π steradian solid angle equals 1: $\int_{4\pi} P(\mathbf{s},\mathbf{s}')d\omega = 1$.

Specular reflection	Reflection from a flat surface or finite flat areas of surface (–).
t	Time (s).
Transmission	The percentage of light or energy passing through a medium, with respect to the incident amount (–).
Transmittance	The fraction of light or energy passing through a medium, with respect to the incident amount (–).
W	Radiant energy: energy emitted, transferred, or received as radiation (J).
w_0	Laser waist.
Wavelength	Length of a wave traveling a full 2π phase angle (m).
Wave number (k)	Frequency of the wave divided by the velocity of propagation (per m).
z	Depth (m).
δ	Penetration depth: distance over which perfectly diffuse light falls to e^{-1} of its initial value, defined as the reciprocal of the effective attenuation coefficient (mm).
μ_a	Absorption coefficient: the distance over which the radiant energy flux falls to e^{-1} of the initial value in the path of the incident beam, due to conversion of specific photon energy into other forms of energy [per mm].
μ_{eff}	Effective attenuation coefficient $\mu_{eff} = \sqrt{3\mu_a[\mu_a + (1-g)\mu_s]}$ (per mm).
μ_s	Scattering coefficient: the distance over which the radiant energy flux falls to e^{-1} of the initial value in the path of the incident beam, due to diversion of photons of light diffracting particles (per mm).
ρ	Density (kg/m^3).
Φ	Radiant energy flux: power emitted, transferred, or received as radiation (W).
ϕ	Polar angle (rad).
Ψ	Radiant energy fluence rate: at a point in space, the radiant energy flux incident on a small sphere divided by the cross-sectional area of that sphere (W/cm^2) (also radiant flux density).
Ω	Radiant energy density: radiant energy in an element of volume divided by that element (J/cm^3).

Additional Reading

AAPM Report No. 57, *Recommended Nomenclature for Physical Quantities in Medical Applications of Light*, American Institute of Physics, College Park, 1996, http://www.aapm.org/pubs/reports/rpt_57.pdf.

Abramowitz, M., and Stegun, I.A., *Handbook of Mathematical Functions*, Dover Publications, New York, 1972.

Bausch and Lomb report S-5195, 0675, Refractive index and percent dissolved solids scale, Analytical Systems Division.

Born, M., and Wolf, E., *Principles of Optics*, Pergamon Press, London, 1959.

Case, K.M., and Zweifel, P.F., *Linear Transport Theory*, Addison-Wesley Publishing Co., Reading, 1967.

Chance, B., Optical method, *Annu. Rev. Biophys. Biophys. Chem.*, 20, 1–28, 1991.

Chandrasekhar, S., *Radiative Transfer*, Oxford University Press, Oxford, 1950.

Frisch, U., *Wave Propagation in Random Media, Probabilistic Methods in Applied Mathematics*, Bharuda-Reid, A.T., Ed., Academic Press, New York, 1968, pp. 75–198.

Gori, F., Flattened Gaussian beams, *Opt. Commun.*, 107, 335–341, 1994.

Hansen, J.E., and Travis, L.D., Light scattering in planetary atmospheres, *Space Sci. Rev.*, 16, 525–610, 1974.

Henyey, J.C. and Greenstein, J.L., Diffuse radiation in the galaxy, *Astrophys. J.*, 93, 70–83, 1941.

Hovenier, J.W., Lumme, K., Mishchenko, M.I., Voshchinnikov, N.V., Mackowski, D.W., and Rahola, J., Computations of scattering matrices of four types of non-spherical particles using diverse methods, *J. Quant. Spectrosc. Radiat. Transf.*, 55(6), 695–705, 1996.

Ishimaru, A., *Wave Propagation and Scattering in Random Media*, Vols. 1 and 2, Academic Press, Burlington, 1978.

Jacques, S.L., Origins of tissue optical properties in the UVA, Visible, and NIR regions, *OSA TOPS on Advances in Optical Imaging and Photon Migration*, Vol. 2, Alfano, R.R., and Fujimoto, J.G., Eds., Optical Society of America, Washington, 1996.

Klier, K., Absorption and scattering in plane parallel turbid media, *J. Opt. Soc. Am.*, 62, 7, 882–885, 1972.

Li, Y., Propagation and focusing of Gaussian beams generated by Gaussian mirror resonators, *J. Opt. Soc. Am. A*, 19, 9, 1832–1843, 2002.

Lynch, D.K., and Livingston, W., *Color and Light in Nature*, Cambridge University Press, Cambridge, 2001.

Mie, G., Beiträge zur Optik trüber Medien,speziell kolloidaler Metallösungen, *Ann. Phys.*, 4, 25, 377–392, 1908.

Pope, R.M., and Fry, E.S., Absorption spectrum (380–700 nm) of pure water. II. Integrating cavity measurements, *Appl. Optics*, 36(33), 8710–8723, 1997.

Purcell, E.M., and Pennypacker, C.R., Scattering and absorption of light by non-spherical dielectric grains, *Astrophys. J.*, 186, 705–714, 1973.

Rutily, B., and Bergeat, J., The solution of the Schwarzschild-Milne integral equation in an homogeneous isotropically scattering plane-parallel medium, *J. Quantitative Spectroscopy Radiat. Transfer*, 51, 823–847, 1994.

Schiebener, P., Straub, J., Levelt Sengers, J.M.H., and Gallagher, J.S., Refractive Index of Water and Steam as Function of Wavelength, Temperature and Density, *J. Phys. Chem. Ref. Data*, 19, 3, 677–717, 1990.

Schwarzschild, K., Über Diffusion and Absorption in der Sonnenatmosphäre, *Sitz. Ber. Preuss. Akad. Wiss. Berlin*, 1183–1196, 1914.

Schwarzschild, K., Über das Gravitationsfeld eines Massenpunktes nach der Einsteinischen Theorie, *Sitz. Ber. Preuss. Akad. Wiss. Berlin*, 189–196, 1916.

Van der Hulst, H.C., *Multiple Light Scattering*, Vols. 1 and 2, Academic Press, Burlington, 1980.

Problems

1. An Ho:YAG laser beam at a wavelength of 1.9 μm irradiating the plaque in a blood vessel, the absorption coefficient is estimated to be 25/cm and scattering is negligible. The beam has a Gaussian profile. Find the radial profile of light intensity and the rate of heat generation at the tissue surface, $z = 0$, and its axial profile along the center axis of the beam, $r = 0$. Graph $L(r, 0)$ for r ranging from $-2w$ to $2w$ and $L(0,z)$ for $z = 0$ to $z = 5 \times \mu_{eff}$. For simplicity, assume $L_0 = 1$ W/cm^2.

2. Describe the decay of ballistic light after emerging from a turbid tissue sample with a thickness of 30 mean free optical path lengths. If the scattering coefficient of the tissue is 170/cm, calculate the corresponding thickness in centimeters.

3. Explain the similarities and differences among radiance, irradiance, (radiant energy) fluence rate, absorption, and scattering.

4. A laser beam is incident from air on the surface of a tissue (i.e., water) ($n = 1.33$); calculate the transmission angle for a beam incident on the tissue at 45°. Calculate the relative loss due to reflection at this angle.

5 Liver tissue is irradiated by the pulsed Ho:YAG laser. Absorption dominates over scattering for this combination of wavelength and tissue. Assume the absorption coefficient at the Ho:YAG laser wavelength is 30/cm and the thermal diffusivity of the tissue is the same as that of water. If the threshold temperature of ablation is 150°C, determine the threshold fluence (J/cm^2) for ablation.

6. In the visible region, tissue scattering dominates absorption in an experimental setting using an HeNe laser operating at 632.8 nm, with an absorption coefficient of 0.05/cm, a scattering coefficient of 143/cm, and a tissue thickness of 5 cm. To detect on average a single ballistic photon transmitted through the tissue per second, what is the required energy in Joules for the incident light? Compare this energy of the incident light with the solar radiance incident on the earth's surface over the entire hemisphere?

7. A fiberoptic probe is used to perform a diagnostic and therapeutic procedure on a cancerous lesion in the esophagus. The probe first determines the boundary of the suspected cancerous lesion by measuring the reflectance of the light at the interface, where it is known that the refractive index of the cancer tissue is 1.2. The normal esophageal tissue refractive index is 1.4. The fiber is polished at an angle of 30° (angle of incidence). Helium-neon (HeNe) laser light is propagated down a fiberoptic probe. The power density at the fiber tip is 1 W/m² at the interface with the tissue. The optical properties of the cancerous tissue at this wavelength are given as μ_a = 3/cm and μ_a = 150/cm. The anisotropy factor g is 0.9. Assume a semi-infinite geometry. Calculate the reflectance for the two tissues, normal and cancerous, at the interface between the sapphire fiber optic (n = 1.7) and the tissue.

8. You are asked to model the following endoscopic laser therapy in lung tissue with refractive index n_2 = 1.7. The radiant energy fluence rate required to excite a particular photosensitizer in a tumor in the lung is 0.5 W/cm². The laser light is diffusely transmitted out of the tip of a fiber optic (n_1 = 1.3) in a semi-infinite geometry. The tumor is located 1 mm from the fiber tip. The optical properties of the lung at this wavelength are given as μ_a = 3/cm and μ_s' = 3/cm.

   a. Determine r_{21}. Assume the laser light radiance $L(\mathbf{r},\mathbf{s})$ to be given by diffusion theory and a step function for Fresnel reflection.
   b. Determine the radiant energy fluence rate at the tumor.
   c. Will the photodynamic therapy be successful?

9. If the radiant energy fluence rate Ψ and the net flux **F** have the same units (W/m²), how are they different?

10. What does the reduced scattering coefficient μ_s' describe?

11. A pulsed laser has 50 mJ of energy in a 1 μs pulse. The laser operates at a repetition rate of 10 Hz. The laser is focused on an area of 0.001 cm².

    a. What is the time-averaged power?
    b. What is the power in the pulse?
    c. What is the irradiance of the laser beam?
    d. What is the fluence?

12. A multi-fiberoptic probe is used to perform a diagnostic and therapeutic procedure on a cancerous lesion in the brain. The first fiber in the probe determines the boundary of the suspected cancerous lesion by measuring the reflectance of the light at the optic–tissue interface, where it is known that the refractive index of the cancer tissue is 1.1. The normal brain tissue refractive index is 1.45. The fiber is polished at

an angle of 30° (angle of incidence). Assume that the light is s-polarized within the fiber.

a. Calculate the reflectance R for the two tissues, normal and cancerous, at the interface between the sapphire fiber optic ($n = 1.7$) and the tissue.
b. Calculate the transmittance T of the laser light into the cancerous tissue.

13. You are asked to model an endoscopic laser therapy. The radiant energy fluence rate required to excite a particular photosensitizer in the cancerous lesion is 0.75 W/cm^2. The laser light is diffusely transmitted out of the tip of a second sapphire fiber in a semi-infinite geometry. The tumor is located 1 mm from the fiber tip. The optical properties of the cancer tissue at this wavelength are given as $\mu_a = 3/\text{cm}$ and $\mu_s' = 15/\text{cm}$.

    a. Determine r_{21}? Assume the laser light radiance $L(\mathbf{r},\mathbf{s})$ to be given by diffusion theory and a step function for Fresnel reflection.
    b. Determine the radiant energy fluence rate at the tumor in units of W/cm^2?
    c. Will the photodynamic therapy be successful?

14. You are asked to model endoscopic laser therapy. The radiant energy fluence rate required to excite a particular photosensitizer in the cancerous lesion is 0.75 W/cm^2. The laser light is diffusely transmitted out of the tip of a second sapphire fiber in a semi-infinite geometry. The tumor is located 1 mm from the fiber tip. The optical properties of the cancer tissue at this wavelength are given as $\mu_a = 3/\text{cm}$ and $\mu_s = 15/\text{cm}$. You do not want to ablate the tissue for fear of spreading the cancer. So, the ablation threshold of this tissue must not be reached. Instead, you want to do selective photodynamic therapy (PDT), as such you want the photochemical interaction to take place *before* the photothermal interaction. Use the normalized fluence rate Ψ at $z = 1$ mm as calculated in problem 13. For an exposure time of 1 s, will the final tissue temperature be above 100°C? Consider the temperature rise to be solely due to conduction.

15. The Duke University Mark III Infrared Free-Electron Laser (FEL) generates an energy per pulse of 20 mJ in a "macropulse" duration of 2 µs (full width half maximum, FWHM) at 10 Hz. Within this macropulse is a "comb" of 1 ps pulses FWHM at a repetition rate of 2.857 GHz. What is the

    a. Time-averaged power P_{ave}?
    b. Average power in the macropulse P_m?
    c. Peak power in the micropulse P_μ?
    d. Micropulse energy E_μ?
    e. Peak irradiance L_μ for a diffraction limited Gaussian beam at 3 µm?
    f. Number of photons in a micropulse N_μ?

16. When monochromatic light passes from air into another material, which of the following does not change, regardless of the material: its speed, frequency, or wavelength?

17. Plot the reflectance R_p (p-parallel to the plane of incidence) and the transmittance T_p versus angle (0°–90°) at the interface between two media of refractive index $n_1 = 2.53$ and $n_2 = 1.32$.

    a. What is the value of the critical angle?
    b. What is the value of the Brewster's angle?

18. In the limit of particle size much less than the wavelength of the incident light, scattering is generally referred to as Rayleigh scattering. The intensity of the scattered light in Rayleigh scattering is proportional to d^6/λ^4, where diameter is d and λ is the wavelength.

    a. What happens to the scattered light signal if the particle size is reduced by half?
    b. What does this say about the wavelength of the incident light if we wish to measure even smaller particles?

19. Plot on a computer the specular reflectances R_s (s-polarized signifies normal to the plane of incidence) and R_p (p-polarized is parallel to the plane of incidence) versus incident angle θ_i for an optic–tissue interface for two cases:

    a. The tissue does have absorption ($k_2 \neq 0$).
    b. The tissue does not have absorption ($k_2 = 0$).

20. The Henyey–Greenstein scattering phase function has been shown to be a good approximation of the true phase function of biological tissue and is given by $P_{HG} = (1/4\pi)[(1-g^2)/(1-2g\mu+g^2)^{3/2}]$, where $\mu = \mathbf{s}\cdot\mathbf{s}' = \cos(\theta)$. Plot P_{HG} on a polar diagram for g = 0, 0.1, 0.3, 0.5, 0.7, and 0.9. Plot on a computer and normalize each plot's amplitude to 1.

21. The reflection factor r_{21} is for diffuse light in medium 2 reflecting off medium 1 and is given by $F_{n+}(\mathbf{r})/F_{n-}(\mathbf{r})$. Determine the boundary condition for r_{21} for a refractive index mismatched boundary. Assume isotropic radiance $L(\mathbf{r},\mathbf{s}) = L_0(\mathbf{r},\mathbf{s})$, which is permitted far from boundaries and sources. Additionally assume that the Fresnel reflection is given by step functions $R = 0$ for $\theta < \theta_{critical}$ and $R = 1$ for $\theta \geq \theta_{critical}$. Express the answer in terms of the refractive index ratio $n = n_2/n_1$.

6

Light–Tissue Interaction Theory

This chapter describes the available fundamental theoretical resources of light–tissue interaction on the basis of optical parameters of both the source and the tissue.

6.1 Approximations of the Equation of Radiative Transport

For a scattering-dominated medium, the integrodifferential equation of transport theory can be simplified to the differential equation of diffusion theory. However, diffusion theory cannot provide accurate estimation of the radiance of light near the boundary of the tissue.

Another method of solving the equation of radiative transfer uses a numerical approach. In this situation a computer simulation recursively calculates numerical values for potential photon paths to reach a solution that fits the equation. This method is referred to as the Monte Carlo technique.

6.1.1 Spherical-Harmonics Substitution Method to Solve the Equation of Radiative Transport

In the mathematical approach for solving the rather complex equation of radiative transfer, different series are introduced as a technique to reach a solution that is analogous to the exact solution. The radiance has an angular dependence that can be developed in a truncated series of polynomials. In most cases the function that is part of the differential equation can be developed in a series of power series solutions, Legendre polynomials, Bessel functions, or Laguerre polynomials.

The spherical-harmonics development as a solution method for the equation of radiative transfer is based on a Taylor expansion.

6.1.1.1 Taylor Expansion of Radiance

Using Cauchy's formula and the Taylor theorem any function can be developed in a series of orthogonal polynomials:

$$f(x)=\sum_{k=0}^{\infty} A_k x^k \tag{6.1}$$

The Taylor expansion can be developed in as many terms as desirable or as found useful, or as determined accurate. The angular distribution of the radiance L(**r**, **s**) is subsequently multiplied by a series of orthogonal polynomials $Y_n^m(s)$ for every m and n of the series:

$$\begin{aligned} L(\mathbf{r},\mathbf{s}) &= \sum_{n=0}^{1}\sum_{m=-n}^{n} Y_n^m(\mathbf{s})L_n^m(\mathbf{r}) = L_0^0(\mathbf{r})Y_0^0(\mathbf{s})+L_1^{-1}(\mathbf{r})Y_1^{-1}(\mathbf{s}) \\ &+L_1^0(\mathbf{r})Y_1^0(\mathbf{s})+L_1^1(\mathbf{r})Y_1^1(\mathbf{s})+L_2^0(\mathbf{r})Y_2^0(\mathbf{s})+\cdots \end{aligned} \tag{6.2}$$

The angular distribution as developed in a series of spherical-harmonic functions can be expressed as

$$Y_n^m(s)=Y_n^m(\theta,\varphi)=(-1)^m\sqrt{\frac{(2n+1)(n-m)!}{4\pi(n+m)!}}\mathrm{e}^{im\varphi}P_n^m(\cos\theta) \tag{6.3}$$

where $m=-n,\ldots, n$, and thus n is always smaller than m.

In Equation (6.3) P_n^m is an associated Legendre polynomial, which is defined as follows:

$$P_n^m(x)=\frac{(1-x^2)^{m/2}}{2^n n!}\frac{\mathrm{d}^{m+n}}{\mathrm{d}x^{m+n}}(x^2-1)^n \tag{6.4}$$

where $x = \cos(\theta)$.

Under cylindrical symmetry all terms with $m \neq 0$ are zero, and can be eliminated from the series, yielding

$$P_n(x)=\frac{1}{2^n n!}\frac{\mathrm{d}^n}{\mathrm{d}x^n}(x^2-1)^n \tag{6.5}$$

The first four Legendre polynomials are given as

$$P_0[\cos(\theta)]=1 \tag{6.6}$$

$$P_1[\cos(\theta)] = \cos(\theta) \tag{6.7}$$

$$P_2 = \frac{1}{2}\{3[\cos(\theta)]^2 - 1\} \tag{6.8}$$

$$P_3 = \frac{1}{2}\{5[\cos(\theta)]^3 - \cos(\theta)\} \tag{6.9}$$

The first sets of spherical-harmonic functions are as follows:

For $m = 0$,

$$Y_0^0(\theta, \varphi) = \sqrt{\frac{1}{4\pi}} \tag{6.10}$$

For $m = 1$,

$$Y_1^0(\theta, \varphi) = \sqrt{\frac{3}{4\pi}} \cos(\theta) \tag{6.11}$$

$$Y_1^1(\theta, \varphi) = -\sqrt{\frac{3}{8\pi}} e^{i\varphi} \sin(\theta) \tag{6.12}$$

For $m = 2$, and so on.

$$Y_2^0(\theta, \varphi) = -\sqrt{\frac{5}{16\pi}} [3\cos^2(\theta) - 1] \tag{6.13}$$

$$Y_2^1(\theta, \varphi) = -\sqrt{\frac{15}{8\pi}} e^{i\varphi} \sin(\theta)\cos(\varphi) \tag{6.14}$$

$$Y_2^2(\theta, \varphi) = -\sqrt{\frac{15}{32\pi}} e^{i2\varphi} \sin^2(\theta) \tag{6.15}$$

The phase function P (**s, s′**) is developed in the same way.

The compiled expression is subsequently integrated over solid space, 4π, to derive an exact solution. Generally this process is very elaborate and detailed. In general, the series can be truncated to three or four terms to derive an accurate solution.

The lowest order approximation assumes a uniform distribution of the scattering angles: the P_0 approximation. This case describes the scattering probability to be uniform in all directions, which is the highly unlikely case of isotropic scattering. The second-order approximation assumes a uniform base distribution with as forward directed second term. This approximation will work in a wide variety of applications, and is called the diffusion approximation.

6.1.2 Diffusion Approximation of the Equation of Radiative Transfer

Another approximation that can be derived from the polynomial expansion is the first-order diffusion approximation, which truncates the expansion after the first term. The diffusion theory is a more pragmatic approach and has a more empirical foundation. It is based on the general transport theory, which can be used to describe many types of transport in physical systems, e.g., mass, heat, electromagnetic energy, or sound. It is essentially the approximation where no particular direction of transport is preferred for an infinitesimal element of the property being transported. In the case of photon scattering, this means that every time a photon changes its direction, due to scattering, it may take off in a new direction with a uniform probability. It also means that if a detector were to be placed in a point inside the tissue, it would detect an equal amount of photons regardless of which direction it is pointed in and where the light source is located. This is a simplification, which is not always true, but for many important purposes it is valid. The diffusion theory approximation gives an analytical solution to the light distribution.

The first-order approximation of the Taylor expansion of the equation of radiative transfer is called the P_1 approximation. This is also called the diffusion approximation. This approximation is valid only if

$$\frac{\mu_a}{\mu_s} \ll (1-g) \tag{6.16}$$

The cylindrical symmetric two-dimensional diffusion equation is now written as

$$D\nabla^2 \psi(r,\theta,z) - \mu_a \psi(r,\theta,z) = -S(r,\theta,z) \tag{6.17}$$

where the diffusion coefficient D is defined as the effective attenuation coefficient for diffuse light,

$$D = \mu_{eff} = \sqrt{3\mu_a[\mu_a + (1-g)\mu_s]} \tag{6.18}$$

The two-dimensional diffusion equation can be solved for a variety of source functions, $S(r, \theta, z)$, and various boundary conditions.

Consider a monochromatic bundle of light energy $I(\mathbf{r}, \mathbf{s})$ radiating into a solid angle $d\omega'$, incident on a volume V at position $\mathbf{r}$ in the medium with total attenuation cross-section ($\alpha + \sigma$). This is the combined scattering and absorption cross-section, and length d**s**, in the direction **s**.

The decrease in radiance dI(**r**, **s**) for the volume V is expressed as

$$\mathbf{s}\bullet\text{grad}[I(\mathbf{r},\mathbf{s})] = -(\mu_a + \mu_s)I(\mathbf{r},\mathbf{s}) \tag{6.19}$$

At the same time the radiance increases because light incident from direction **s**′ scatters into direction **s** and adds to the radiance. If we consider a wave incident on a scattering particle, the scattered wave is governed by the scattering amplitude f(**s**,**s**′).

The scattered radiance in the direction **s** from **s**′ is therefore

$$|f(\mathbf{s},\mathbf{s}')|^2\, I(\mathbf{r},\mathbf{s})\mathrm{d}\omega \tag{6.20}$$

Adding the incident flux from all directions **s** leads to

$$d\, I(\mathbf{r},\mathbf{s}')_{\text{gain}} = |f(\mathbf{s},\mathbf{s}')|^2\, I(\mathbf{r},\mathbf{s}')\mathrm{d}\,\omega \tag{6.21}$$

If θ is the angle between **s** and **s**′, the phase function P(**s**, **s**′) can be defined as

$$P(\mathbf{s},\mathbf{s}') = \frac{1}{\mu_a + \mu_s}|f(\mathbf{s},\mathbf{s}')|^2 \tag{6.22}$$

with

$$\frac{1}{4\pi}\int P(\mathbf{s},\mathbf{s}')\mathrm{d}\,\omega = \frac{\mu_s}{\mu_a + \mu_s} = a_1 \tag{6.23}$$

The variable a_1 is called the albedo of single scattering.

The phase function expresses the statistical distribution of scattering angles with respect to the incident ray of light.

A third contribution in radiance may come from a source within the medium (e.g., blackbody irradiation) ε(**r**, **s**). With this information the total change in irradiance I(**r**, **s**) can be written as

$$\begin{aligned}\mathbf{s}\bullet\text{grad}[I(\mathbf{r},\mathbf{s})] = &-(\mu_a + \mu_s)I(\mathbf{r},\mathbf{s}) \\ &+(\mu_a + \mu_s)\int I(\mathbf{r},\mathbf{s})P(\mathbf{s},\mathbf{s}')\mathrm{d}\omega' + \varepsilon(\mathbf{r},\mathbf{s})\end{aligned} \tag{6.24}$$

The space irradiance I(**r**, **s**) can be divided into a collimated part I_c(**r**, **s**) and a diffuse part I_d(**r**, **s**),

$$I(\mathbf{r},\mathbf{s}) = I_c(\mathbf{r},\mathbf{s}) + I_d(\mathbf{r},\mathbf{s}) \tag{6.25}$$

The collimated part satisfies the equation:

$$\mathbf{s}\bullet\text{grad}[I(\mathbf{r},\mathbf{s})] = -(\mu_a + \mu_s)I_c(\mathbf{r},\mathbf{s}) \tag{6.26}$$

while the diffuse part satisfies the equation:

$$\begin{aligned}\mathbf{s}\bullet\text{grad}[I_d(\mathbf{r},\mathbf{s})] = &-(\mu_a + \mu_s)I_d(\mathbf{r},\mathbf{s}) \\ &+(\mu_a + \mu_s)\int I_d(\mathbf{r},\mathbf{s})P(\mathbf{s},\mathbf{s}')\mathrm{d}\omega' \\ &+\varepsilon_{ri}(\mathbf{r},\mathbf{s}) + \varepsilon(\mathbf{r},\mathbf{s})\end{aligned} \tag{6.27}$$

where $\varepsilon_{ri}(\mathbf{r}, \mathbf{s})$ is the equivalent source function,

$$\varepsilon_{ri}(\mathbf{r},\mathbf{s}) = (\mu_a + \mu_s)\int P(\mathbf{s},\mathbf{s}')I_c(\mathbf{r},\mathbf{s})\mathrm{d}\omega' \tag{6.28}$$

In the diffusion approximation the following assumption is made: The diffuse radiance consists of a major isotropic part and a slightly anisotropic part to account for a net flux of light.

This assumption can describe the diffusion approximation by extending $I_d(\mathbf{r}, \mathbf{s})$ in a Taylor's expansion in terms of the power of $\mathbf{s}\cdot\mathbf{s}_f$.

The following useful solution can be derived by truncating the Taylor expansion after the second term. The limitation is that this solution satisfies only a selective group of boundary conditions:

$$I_d(\mathbf{r},\mathbf{s}) = U_d(\mathbf{r}) + cF_d(\mathbf{r})\bullet\mathbf{s} \tag{6.29}$$

where

$$U_d(\mathbf{r}) = \int_{4\pi} I_d(\mathbf{r},\mathbf{s})\mathrm{d}\omega' \tag{6.30}$$

represents the isotropic space irradiance and

$$F_d(\mathbf{r}) = \int I_d(\mathbf{r},\mathbf{s})\mathrm{d}\,\omega = F_d(\mathbf{r})\bullet\mathbf{s}_f \tag{6.31}$$

is the diffuse net flux.

Hence "diffusion approximation," describing an isotropic diffuse component with equal probability of scatter in all 4π angles of solid space, combined with a net flux. The diffusion approximation resembles Fick's law known in biochemistry.

Substitution of Equation (6.29) into Equation (6.31) will provide the constant

$$c = \frac{3}{4\pi} \tag{6.32}$$

Therefore, the diffuse radiance $I_d(\mathbf{r}, \mathbf{s})$ is given by

$$I_d(\mathbf{r},\mathbf{s}) = U_d(\mathbf{r}) + \frac{3}{4\pi} F_d(\mathbf{r}) \bullet \mathbf{s} \tag{6.33}$$

We now integrate Equation (6.27) over all 4π of solid angle and obtain

$$\text{div}\left[F_d(\mathbf{r})\right] = -\mu_a U_d(\mathbf{r}) + \mu_s U_c(\mathbf{r}) + E(\mathbf{r}) \tag{6.34}$$

where

$$U_c(\mathbf{r}) = \int_{4\pi} I_c(\mathbf{r},\mathbf{r}) \mathrm{d}\omega \tag{6.35}$$

and

$$E(\mathbf{r}) = \int_{4\pi} \varepsilon(\mathbf{r},\mathbf{s}) \mathrm{d}\omega \tag{6.36}$$

Next substitute Equation (6.33) into Equation (6.27) under the assumption that the phase function $P(\mathbf{s}, \mathbf{s}')$ is a function of the angle between $\mathbf{s}$ and $\mathbf{s}'$,

$$\mathbf{s} \bullet \mathbf{s}' = \cos\theta \tag{6.37}$$

and write

$$P(\mathbf{s},\mathbf{s}') \quad \text{as} \quad P(\upsilon,\upsilon') \tag{6.38}$$

using

$$\upsilon = \cos\theta \tag{6.39}$$

where θ is the angle with the positive z-axis.
Substitution leads to

$$\begin{aligned} \mathbf{s} \bullet \text{grad}[U_d(\mathbf{r},\mathbf{s})] + \frac{3}{4\pi}\text{grad}[F_d(\mathbf{r},\mathbf{s})] = {} & -(\mu_a + \mu_s) U_d(\mathbf{r}) - \frac{3}{4\pi}(\mu_a + \mu_s) F_d(\mathbf{r}) \cdot \mathbf{s} \\ & + U_d(\mathbf{r}) + \frac{3}{4\pi}(\mu_a + \mu_s) F_d(\mathbf{r}) \bullet \mathbf{s} a_1 \\ & + \varepsilon_{ri}(\mathbf{r}) + \varepsilon(\mathbf{r}) \end{aligned} \tag{6.40}$$

Introducing the albedo of multiple scattering events, a_0, in relation to the albedo of single scatter a_1:

$$a_1 = \int_{4\pi} P(\mathbf{s},\mathbf{s}') \mathbf{s} \bullet \mathbf{s}' \mathrm{d}\omega = a_0 g \tag{6.41}$$

where the scattering anisotropy factor g is defined as

$$g = \frac{\int_{4\pi} P(\mathbf{s},\mathbf{s}')\mathbf{s}\bullet\mathbf{s}'\mathrm{d}\omega}{\int_{4\pi} P(\mathbf{s},\mathbf{s}')\mathrm{d}\omega} \tag{6.42}$$

Equation (6.40) represents the averaged forward scattering $\mathbf{s}\bullet\mathbf{s}'>0$ minus the backward scattering $\mathbf{s}\bullet\mathbf{s}'<0$ of a single particle.

Now multiply Equation (6.40) by $\mathbf{s}$ and integrate over all 4 π of solid angle to obtain

$$\begin{aligned} \mathrm{grad}[U_\mathrm{d}(\mathbf{r},\mathbf{s})] = {} & \frac{3}{4\pi}(\mu_\mathrm{a}+\mu_\mathrm{s})(1-a_1)F_\mathrm{d}(\mathbf{r}) \\ & + \frac{3}{4\pi}\int_{4\pi} \varepsilon_{ri}(\mathbf{r})\mathrm{d}\,s + \frac{3}{4\pi}\int_{4\pi} \varepsilon(\mathbf{r})\mathrm{d}\,s \end{aligned} \tag{6.43}$$

The quantity $(\mu_\mathrm{a}+\mu_\mathrm{s})(1-a_1)$ is called the transport coefficient (often referred to as transport cross-section; however, it does not have the dimensions of cross-section):

$$\mu_\mathrm{tr} = (\mu_\mathrm{a}+\mu_\mathrm{s})(1-w_1) = \mu_\mathrm{s}(1-g)+\mu_\mathrm{a} \tag{6.44}$$

By eliminating $F_\mathrm{d}(\mathbf{r})$ from Equation (6.44) it reduces to the following:

$$\begin{aligned} \nabla^2 U_\mathrm{d}(\mathbf{r},\mathbf{s}) - \kappa_\mathrm{d}^2 U_\mathrm{d}(\mathbf{r}) = {} & -\mu_\mathrm{tr}U_\mathrm{c}(\mathbf{r}) - \frac{3}{4\pi}\mu_\mathrm{tr}E_\mathrm{d}(\mathbf{r}) \\ & + \frac{3}{4\pi}\Delta\bullet\int_{4\pi} \varepsilon_{ri}(\mathbf{r},\mathbf{s})\mathrm{d}\,\mu + \frac{3}{4\pi}\nabla\bullet\int_{4p} \varepsilon(\mathbf{r},\mathbf{s})\mathbf{s}\,\mathrm{d}\,\omega \end{aligned} \tag{6.45}$$

where

$$\kappa_\mathrm{d}^2 = 3\mu_\mathrm{a}\mu_\mathrm{tr} \tag{6.46}$$

If space has cylinderical symmetry and if only a plane parallel medium is being considered with the incident light coming from one side, the description can be rewritten as follows. Note that under these conditions all directions are related to the inward normal.

The plane parallel medium is described by x and y ordinates on the surface and by z perpendicular to the surface, $z=0$, at the upper air–medium interface, and the positive z-direction is the inward pointing normal to this surface.

The phase function P(υ,υ') is a function of the scattering angle θ only, and can be expanded in a series of Legendre functions to the order n,

$$P(\upsilon,\upsilon') = 4\pi \sum_{0}^{\infty} \frac{1}{2n+1} w_n P_n(\upsilon) P_{n'}(\upsilon') \tag{6.47}$$

Using the definition from Equation (6.39) the equation of radiative transfer, as expressed in Equation (6.27), is now written as

$$\begin{aligned} \mu \frac{\partial I_{\mathrm{d}}(z,\upsilon)}{\partial z} &= -(\mu_{\mathrm{a}} + \mu_{\mathrm{s}}) I_{\mathrm{d}}(z,\upsilon) \\ &\quad + (\mu_{\mathrm{a}} + \mu_{\mathrm{s}}) \int_{4\pi} I_{\mathrm{d}}(z,\upsilon) P(\upsilon,\upsilon') \mathrm{d}\mu + S(z,\upsilon) \end{aligned} \tag{6.48}$$

where $S(z, \upsilon)$ is the source function inside the medium and will be discarded from here on:

$$S(z,\upsilon) \equiv \varepsilon_{ri}(\mathbf{r},\mathbf{s}) + \varepsilon(\mathbf{r},\mathbf{s}) \tag{6.49}$$

The phase function is described by $P(\upsilon, \upsilon')$ from Equation (6.47),

$$\int_{-1}^{+1} P_n(\upsilon) P_{n'}(\upsilon') \mathrm{d}\upsilon = \frac{2\delta_{nn'}}{2n+1} \tag{6.50}$$

which can be rewritten as

$$\sum_{0}^{\infty} P_n(\upsilon) P_{n'}(\upsilon') \frac{2n+1}{2} = \delta(\upsilon - \upsilon') \tag{6.51}$$

This describes the delta-Dirac approximation for an infinitely short pulse in an infinite homogeneous medium, in which case the source can be written as $q(r', t') = \delta(r', t')$:

$$\int_{-1}^{+1} P_n(\upsilon,\upsilon') \mathrm{d}\upsilon = 2a_0 \tag{6.52}$$

Using the P_n approximation equation (6.47), and multiplying Equation (6.48) by $P_{n'}$, and subsequently integrating over 4 π solid angle leads to

$$\frac{\partial}{\partial z}\int_{-1}^{+1} \upsilon I_d(z,\upsilon)d\upsilon = (\mu_a+\mu_s)\left\{\int_{-1}^{+1} I_d(z,\upsilon)d\mu + \int_{-1}^{+1}\frac{1}{2}d\upsilon\int_{-1}^{+1}P(\upsilon,\upsilon')d\upsilon' + \int_{-1}^{+1} I_c(z,1)P(\upsilon,1)d\upsilon\right\}$$
$$= \alpha\int_{-1}^{+1} I_d(z,\upsilon)d\upsilon + \sigma I_c(z,1) \tag{6.53}$$

Developing Equation (6.53) in the Taylor series for $P_n(\upsilon, \upsilon')$ gives

$$\frac{1}{\mu_a+\mu_s}\frac{\partial}{\partial z}\int_{-1}^{+1}\upsilon^2 I_d(z,\upsilon)d\upsilon = -\int_{-1}^{+1}\upsilon I_d(z,\upsilon)d\upsilon + w_1\int_{-1}^{+1}\upsilon I_d(z,\upsilon)d\upsilon + w_1 I_c(z,1) \tag{6.54}$$

$$\frac{1}{\mu_a+\mu_s}\frac{\partial}{\partial z}\int_{-1}^{+1}\frac{1}{5}(3P_3+2P_1)I_d(z,\upsilon)d\upsilon = -\int_{-1}^{+1}P_2 I_d(z,\upsilon)d\upsilon + w_2\int_{-1}^{+1}P_2 I_d(z,\upsilon)d\upsilon + w_2 I_c(z,1) \tag{6.55}$$

and so on. In the diffusion approximation of the equation of radiative transfer, in general, only the first two orders are used, giving

$$\frac{\partial}{\partial z}\int_{-1}^{+1} I_d(z,\upsilon)d\upsilon = -\mu_a\int_{-1}^{+1} I_d(z,\upsilon)d\upsilon + I_c(z,1) \tag{6.56}$$

$$\frac{\partial}{\partial z}\int_{-1}^{+1}\upsilon^2 I_d(z,\upsilon)d\upsilon = -(\mu_a+\mu_s)(1-w_1)\int_{-1}^{+1}\upsilon I_d(z,\upsilon)d\upsilon + (\mu_a+\mu_s)w_1 I_c(z,1) \tag{6.57}$$

Using Equation (6.33) the concept of radiative flux can be introduced.

In the Kubelka–Munk approach the stream of photons is seen as a one-dimensional particle diffusion or flux, where the forward flux is defined as

$$F_{d+}(z) = 2\pi\int_{0}^{+1}\upsilon I_d(z,\upsilon)d\upsilon = 2\pi\left\{\frac{1}{2}U_d(z) + \frac{1}{3}F_d(z)\right\} \tag{6.58}$$

In the same configuration the backward flux is defined as

$$F_{d-}(z) = -2\pi \int_{-1}^{0} \upsilon I_d(z,\upsilon) d\upsilon = 2\pi \left\{ \frac{1}{2} U_d(z) - \frac{1}{3} F_d(z) \right\} \tag{6.59}$$

On substituting Equations (6.58) and (6.59) in Equations (6.46) and (6.48), respectively, they reduce to

$$\frac{d}{dz}\{F_{d+}(z) - F_{d-}(z)\} = -2\mu_a \{F_{d+}(z) + F_{d-}(z)\} + \sigma F_c(z) \tag{6.60}$$

and

$$\begin{aligned} \frac{d}{dz}\{F_{d+}(z) + F_{d-}(z)\} = -\frac{3}{2}(\mu_a + \mu_s)(1 - w_1)\{F_{d+}(z) \\ - F_{d-}(z)\} + \frac{3}{2}(\mu_a + \mu_s) w_1 F_c(z) \end{aligned} \tag{6.61}$$

Solving Equations (6.60) and (6.61) for F_c, F_{d+}, F_{d-} yields (see also Equation [6.45])

$$\frac{dF_c(z)}{dz} = -(\mu_a + \mu_s) F_c(z) \tag{6.62}$$

and

$$\begin{aligned} \frac{dF_{d+}(z)}{dz} = &-\left\{2\alpha + \frac{3}{4}(\mu_a + \mu_s)(1 - w_1) - \mu_a\right\} F_{d+}(z) \\ &+ \left\{\frac{3}{4}(\mu_a + \mu_s)(1 - w_1) - \mu_a\right\} F_{d-}(z) \\ &+ \left\{\frac{\mu_a}{2} + \frac{3}{4}(\mu_a + \mu_s) w_1\right\} F_c(z) \end{aligned} \tag{6.63}$$

and, respectively,

$$\begin{aligned} \frac{dF_{d-}(z)}{dz} = &-\left\{2\mu_a + \frac{3}{4}(\mu_a + \mu_s)(1 - w_1) - \mu_a\right\} F_{d-}(z) \\ &+ \left\{\frac{3}{4}(\mu_a + \mu_s)(1 - w_1) - \mu_s\right\} F_{d+}(z) \\ &+ \left\{\frac{\mu_a}{2} - \frac{3}{4}(\mu_a + \mu_s) w_1 F_c(z)\right. \end{aligned} \tag{6.64}$$

Which is in essence equivalent to a three-flux model. The "original" three-flux approach will be outlined in Section 6.1.3.2.

Note that the P_1 diffusion approximation is valid under the condition $\mu_a \ll \mu_s'$ only.

6.1.2.1 Isotropic Source Solution

The solution to the diffusion approximation for an isotropic source inside an infinite medium can be shown to satisfy a Green's function in the form

$$\psi(r) = \frac{\psi_0}{4\pi D} \frac{e^{-r/\partial}}{r} \tag{6.65}$$

where the fluence rate depends only on the radius to the source and the diffusion coefficient (Equation [6.18]).The diffusion coefficient is correlated to the effective penetration depth as

$$\partial = \frac{1}{\sqrt{\mu_a D}} \tag{6.66}$$

6.1.3 Discrete-Ordinate Method for Solving the Equation of Radiative Transport

The solid angle ω itself can also be developed in discrete directions. By selecting the discrete directions in entire quadrants of space, the equation of radiative transfer is split into fluxes of photons flowing in specific solid angles. This angular discretion gives a finite number of fluxes and is therefore called the N-flux method.

6.1.3.1 Two-Flux Theory

Based on Beer's law, or the Beer–Lambert law, or the Beer–Lambert–Bouguer law of collimated attenuation, Paul Kubelka and Franz Munk derived a similar phenomenological description of diffuse light passing through an absorbing and scattering medium. They introduced a scattering term that redirects a portion of the incoming net flux of light in the opposite direction, providing a net flux in the direction directly opposing the incident flux of photons.

If the medium is a one-dimensional slab without reflections on the boundaries and the incident radiance is diffuse light, the transport theory can also be simplified to the Kubelka–Munk theory. In this theory, the scattering and absorption coefficients of medium may be expressed directly by the measured diffuse reflectance and transmission. This made the Kubelka–Munk theory very popular in the design of experimental procedures to measure the optical properties of tissues.

A schematic diagram illustrating single and multiple scattering, absorption, and reflection from a photon beam entering an imaginary tissue is shown in Figure 6.1.

The Kubelka–Munk diffuse absorption and scattering coefficients are, respectively, defined as K and S. The incident flux is defined as i, and the redirected flux as j.

Consider diffuse light traveling from one interface to another parallel interface layer in the positive z-direction over a distance dz. The path length of the light will not be dz but $\mathrm{d}z/\cos(\theta)$, where θ is the angle with the normal to the surface boundary. For homogeneously diffuse illumination many rays of light will travel from one interface to the next. The average path of the light passing through the layer dz towards the unilluminated edge of the slab of turbid medium $\mathrm{d}\xi_i$ will be

$$\mathrm{d}\xi_i = \mathrm{d}z \int_0^{\pi/2} \frac{\partial i}{i\partial\theta} \frac{\mathrm{d}\theta}{\cos\theta} \equiv u\,\mathrm{d}z \tag{6.67}$$

where $\partial i/\partial\theta$ is the angular distribution of the radiance i.

Similarly, the backscattered light traveling toward the illuminated boundary has the following average path length $\mathrm{d}\xi_j$:

$$\mathrm{d}\xi_j = \mathrm{d}z \int_0^{\pi/2} \frac{\partial j}{j\partial\theta} \frac{\mathrm{d}\theta}{\cos\theta} \equiv v\,\mathrm{d}z \tag{6.68}$$

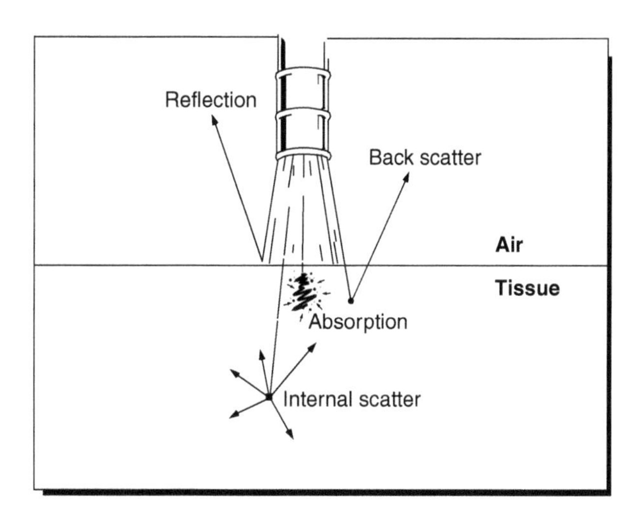

FIGURE 6.1
Schematic diagram illustrating single and multiple scattering, absorption, and reflection from a photon beam entering an imaginary tissue.

Assume that initially the infinitely thin layer dz is only illuminated by I, resulting in loss due to scattering "cross-sections" σ and absorption "cross-sections" α; the attenuation through layer dz can be expressed as

$$(\alpha+\sigma)i\,\mathrm{d}\xi_i=(\alpha+\sigma)ui\,\mathrm{d}z \tag{6.69}$$

where the absorbed portion is expressed as

$$\sigma i\,\mathrm{d}\xi_i=\sigma ui\,\mathrm{d}z \tag{6.70}$$

and the scattered portion traveling in the opposite direction is represented by

$$\alpha i\,\mathrm{d}\xi_i=\alpha ui\,\mathrm{d}z \tag{6.71}$$

In analogous form an expression can be derived for illumination from the backside by j only,

$$(\alpha+\sigma)j\,\mathrm{d}\xi_j=(\alpha+\sigma)vj\,\mathrm{d}z \tag{6.72}$$

where the backscattered portion is expressed as

$$\alpha j\,\mathrm{d}\xi_j=\alpha vj\,\mathrm{d}z \tag{6.73}$$

The true representation would have the incident and backscattered fluxes occur simultaneously, each contributing to the other, giving

$$-\mathrm{d}i=-(\alpha+\sigma)ui\,\mathrm{d}z+\alpha vj\,\mathrm{d}z \tag{6.74}$$

$$\mathrm{d}j=-(\alpha+\sigma)vj\,\mathrm{d}z+\alpha ui\,\mathrm{d}z \tag{6.75}$$

Substituting the Kubelka–Munk scattering and absorption cross-section by the Kubelka–Munk scattering coefficient $S=u\sigma$ and absorption coefficient $K=u\varepsilon$, respectively, it can be shown that $u=v=$constant, yielding for the diffuse forward flux:

$$-\mathrm{d}i=-(S+K)i\,\mathrm{d}z+Sj\,\mathrm{d}z \tag{6.76}$$

and for the diffuse backward flux the expression reads

$$\mathrm{d}j=-(S+K)j\,\mathrm{d}z+Si\,\mathrm{d}z \tag{6.77}$$

These equations signify the loss of the incident flux i, due to scattering and absorption while moving in the direction dz, and gain from the scattered "backscattered" flux j. Also the backscattered flux j is the result of scattered incident flux into the opposite direction, again with loss due to both scattering in the opposite (now forward) direction, and absorption of the backward flux. The definition of the various light flux propagations is illustrated in Figure 6.2.

Perfectly diffuse light will have a homogeneous distribution of radiance in all directions; thus, when hitting an arbitrary plane inside the medium the angular distribution can be represented as

$$\frac{\partial i}{\partial \theta} = i \sin 2\theta \tag{6.78}$$

Substituting Equation (6.78) back into Equation (6.67) gives for perfectly diffuse light the following condition:

$$u = \int_0^{\pi/2} \sin 2\theta \frac{\mathrm{d}\theta}{\cos\theta} = 2 \tag{6.79}$$

Introducing the substitution $a=(S+K)/S$, Equations (6.76) and (6.77) can be written as

$$-\frac{\mathrm{d}i}{S\mathrm{d}z} = -ai + j \tag{6.80}$$

and, respectively,

$$\frac{\mathrm{d}j}{S\mathrm{d}z} = -aj + i \tag{6.81}$$

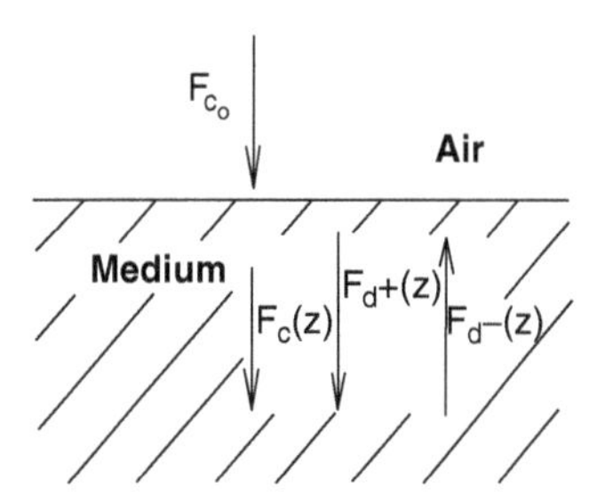

FIGURE 6.2
Illustration of the light flux model.

Next we introduce the reflection coefficient $r=j/i$. It can now be shown that the following holds true:

$$\frac{dr}{S\,dz}=r^2-ar+1 \tag{6.82}$$

This is integrated over a finite depth z' within the turbid medium, assuming index matching on the top surface $z=0$, providing the boundary condition $r_{z=0}=0$, and tissue with black background at $z=z'$, $r_{z=z'}=r_0$, yielding

$$z=\frac{1}{S\sqrt{(a^2-1)}}\operatorname{arctgh}\left(\frac{1-ar_0}{\sqrt{(a^2-1)}r_0}\right)=\frac{1}{bS}\operatorname{arctgh}\left(\frac{1-ar_0}{br_0}\right) \tag{6.83}$$

Using the following substitution $b=\sqrt{(a^2-1)}$, the expression for the ratio of the backward over the forward flux at any point inside the medium can be shown to satisfy

$$r_0=\frac{1}{a+b[\operatorname{ctgh}(bSz)]} \tag{6.84}$$

Thus the backward and forward fluxes, respectively, can be expressed as $j=r_0i_0$ and $j=j_0/r_0$:

$$j=j_0=r_0j_0=\frac{i_0}{a+b[\operatorname{ctgh}(bSz)]} \tag{6.85}$$

$$i=i_0=j_0/r_0=j_0\{a+b[\operatorname{ctgh}(bSz)]\} \tag{6.86}$$

The solutions to Equations (6.76) and (6.77) are, respectively,

$$\begin{aligned} i_0 &= C_1\{a[\sinh(bSz)]+b[\cosh(bSz)]\} \\ &= I\frac{a[\sinh(bSz)]+b[\cosh(bSz)]}{a[\sinh(bSZ)]+b[\cosh(bSZ)]} \end{aligned} \tag{6.87}$$

and

$$\begin{aligned} j_0 = C_2\sinh(bSz) &= I\frac{1}{a+b[\operatorname{ctgh}(bSZ)]}\frac{\sinh(bSz)}{\sinh(bSZ)} \\ &= I\frac{\sinh(bSZ)}{a[\sinh(bSZ)]+b[\cosh(bSZ)]}\frac{\sinh(bSz)}{\sinh(bSZ)} \\ &= I\frac{\sinh(bSz)}{a[\sinh(bSZ)]+b[\cosh(bSZ)]} \end{aligned} \tag{6.88}$$

where I is the radiance incident on the slab and Z the total thickness of the slab of medium.

From this the black backing reflection R_0 can be defined as

$$R_0 = \frac{1}{a + b\left[\text{ctgh}(bSZ)\right]} = \frac{\sinh(bSZ)}{a\left[\sinh(bSZ)\right] + b\left[\cosh(bSZ)\right]} \tag{6.89}$$

In the Kubelka–Munk theory the total transmitted and reflected amount of light is of primary interest, not so much the light distribution inside the medium. The transmittance is the ratio of the radiance leaving the backside of the medium I_g over the incident radiance I_0:

$$T = I_g / I_0 \tag{6.90}$$

The reflectance is the ratio of the backscattered radiance leaving the medium on the side of the incident beam J over the incident radiance I_0:

$$R = J / I_0 \tag{6.91}$$

A special case to be considered is the bulk tissue reflectance, where the tissue is assumed to be of infinite thickness (which usually is achieved at a thickness of > 40 optical free path lengths),

$$R_\infty = \lim_{r \to \infty} r = a - \sqrt{(a^2 - 1)} = a - b = 1 + \frac{s}{k} - \sqrt{\left(\frac{S^2}{K^2} + 2\frac{S}{K}\right)} \tag{6.92}$$

This is sometimes a useful relationship to determine the order of magnitude of light loss due to backscatter out of the tissue.

The Kubelka–Munk transmittance follows from Equations (6.85), (6.86), and (6.90):

$$T = \frac{b}{a\left[\sinh(bSZ)\right] + b\left[\cosh(bSZ)\right]} \tag{6.93}$$

Using conservation of energy, the reflectance, transmittance, and absorption inside the medium need to balance out to the delivered radiance, giving the following balance equation using Equations (6.89) and (6.93):

$$T^2 + b^2 = (a - R_0)^2 \tag{6.94}$$

Solving Equation (6.94) for a gives

$$a = \frac{1 + R_0^2 - T^2}{2R_0} \tag{6.95}$$

The general solutions to Equations (6.85) and (6.86) are, respectively,

$$i(z) = A\exp[-\alpha z] + A'\exp[+\alpha z] \tag{6.96}$$

$$j(z) = B\exp[-\alpha z] + B'\exp[+\alpha z] \tag{6.97}$$

$$\alpha^2 = K^2 + 2KS \tag{6.98}$$

These equations are however limited to the description of diffuse incident light, since collimated incidence will provide an additional boundary condition, requiring a three-flux model, described in the next section.

Equations (6.85) and (6.86) assume perfectly diffuse light and homogeneous scattering conditions. Homogeneous scattering is the case where the light has an equal probability to be scattered in the forward or backward direction, since these are the only two directions that are considered.

Further refinement of this theory includes a probability distribution of the scattering angle expressed as the likelihood that the photons are scattered preferentially within a solid angle cone pointing in one direction over any other direction:

$$-\mathrm{d}i = -1/2(S+K)ui\,\mathrm{d}z + 1/2\,Svj\,\mathrm{d}z \tag{6.99}$$

$$-\mathrm{d}i = -1/2(S+K)vi\,\mathrm{d}z + Suj\,\mathrm{d}z \tag{6.100}$$

with

$$u \equiv \int_0^{\pi/2} \frac{\partial i}{i\partial\theta}\frac{\partial\theta}{\cos\theta} \tag{6.101}$$

$$v \equiv \int_0^{\pi/2} \frac{\partial j}{j\partial\theta}\frac{\partial\theta}{\cos\theta} \tag{6.102}$$

The Kubelka–Munk absorption coefficient (K) and scattering coefficient (S) can be linked to the respective absorption and scattering cross-section and hence to the absorption and scattering parameters μ_a, μ_s, and g as follows:

$$\frac{\mu_a}{\mu_s} \approx \frac{3K}{8S} \tag{6.103}$$

and furthermore

$$\mu_s(1-g)=\frac{4}{3}S+\frac{1}{6}K \tag{6.104}$$

$$\mu_a=\frac{1}{2}K \tag{6.105}$$

In the next section the three-flux model for collimated irradiance is described.

6.1.3.2 Three-Flux Theory

A more useful and appropriate description involves the illumination of a turbid medium with a collimated beam of light i_c, which is subsequently scattered throughout the medium, rendering the collimated beam into a diffuse forward flux i_d, while concurrently providing a backward diffuse flux j_d, resulting from the same scattering phenomena.

The definition of the incident collimated flux i_c is as follows: $i_c(0)=I_0$, at the interface between air/vacuum and the turbid medium. The diffuse flux in the forward direction is defined as $i_d(0)=r_{id}j_d$, and the diffuse backward flux as $j_d(d)=r'_{id}i_d$, with r_{id} and r'_{id} the Fresnel reflection coefficients from the interface on the incident side of the medium and from the interface between the the first medium and the second medium, respectively,

$$\frac{di_c}{dz}=-(S+K)i_c \tag{6.106}$$

$$\frac{di_d}{dz}=-[\alpha K+\alpha(1-\beta)S]i_d+\alpha(1-\beta)Si_d+\beta Si_c \tag{6.107}$$

$$-\frac{dj_d}{dz}=-[\alpha K+\alpha(1-\beta)S]j_d+\alpha(1-\beta)Sj_d+(1-\beta)\,Si_c \tag{6.108}$$

The additional coefficients α and β are correction factors for, respectively, a stronger diffuse absorption with respect to collimated irradiation (α) and the fraction of the forward scattered light with respect to the total scatter envelope (β).

Substitution of the correction factors in the original definition for Kubelka–Munk absorption and scattering gives

$$K' = \alpha K \tag{6.109}$$

$$S' = \alpha(1-\beta)S \tag{6.110}$$

which can be filled in Equations (6.107) and (6.108) to give the following expressions for the collimated flux, diffuse forward flux and diffuse backward flux, respectively,

$$\frac{\mathrm{d}i_\mathrm{c}}{\mathrm{d}z} = -(S+K)i_\mathrm{c} \tag{6.111}$$

$$\frac{\mathrm{d}i_\mathrm{d}}{\mathrm{d}z} = -(K'+S')i_\mathrm{d} + S'i_\mathrm{d} + \beta S i_\mathrm{c} \tag{6.112}$$

$$-\frac{\mathrm{d}j_\mathrm{d}}{\mathrm{d}z} = -(K'+S')j_\mathrm{d} + S'j_\mathrm{d} + (1-\beta)S i_\mathrm{c} \tag{6.113}$$

Next the following substitutions are proposed:

$$a = \frac{S'+K'}{S'} \tag{6.114}$$

$$b = \sqrt{a^2-1} \tag{6.115}$$

This leads to the following solutions of Equations (6.111)–(6.113), respectively,

$$i_\mathrm{c}(z) = I_0 \exp\left[-(K+S)z\right] \tag{6.116}$$

$$i_\mathrm{d}(z) = c_1\exp(\beta S'z) + c_2\exp(-\beta S'z) \tag{6.117}$$

$$i_\mathrm{d}(z) = c_1(a+b)\exp(\beta S'z) + c_2(a-b)\exp(-\beta S'z) \tag{6.118}$$

Introducing the attenuation coefficient $q'=K+S$, the particular solution to Equations (6.112) and (6.113) can be shown to be

$$i_{\mathrm{d}}^{\mathrm{p}}(z)=d_1I_0\exp[-(K+S)z]=d_1I_0\mathrm{e}^{-q'z} \tag{6.119}$$

$$j_{\mathrm{d}}^{\mathrm{p}}(z)=d_2I_0\exp[-(K+S)z]=d_2I_0\mathrm{e}^{-q'z} \tag{6.120}$$

Substituting Equation (6.119) and (6.120) back into Equations (6.112) and (6.113), respectively, yields

$$\begin{aligned}-\frac{\mathrm{d}\,i_{\mathrm{d}}^{\mathrm{p}}}{\mathrm{d}z}=-q'd_1I_0\mathrm{e}^{-q'z}&=-(K'+S')d_1I_0\mathrm{e}^{-q'z}+S'd_2I_0\mathrm{e}^{-q'z}+\beta SI_0\mathrm{e}^{-q'z}\\&=\{-(K'+S')d_1+S'd_2+\beta S\}I_0\mathrm{e}^{-q'z}\end{aligned} \tag{6.121}$$

and

$$\begin{aligned}-\frac{\mathrm{d}\,j_{\mathrm{d}}^{\mathrm{p}}}{\mathrm{d}z}=q'd_2I_0\mathrm{e}^{-q'z}&=-(K'+S')d_2I_0\mathrm{e}^{-q'z}+S'd_1I_0\mathrm{e}^{-q'z}\\&\quad+(1-\beta)SI_0\mathrm{e}^{-q'z}\\&=\{-(K'+S')d_2\end{aligned} \tag{6.122}$$

Further refinement gives the following solutions:

$$i_{\mathrm{c}}(z)=I_0\mathrm{e}^{-q'z} \tag{6.123}$$

$$i_{\mathrm{d}}(z)=c_1\mathrm{e}^{bSz}+c_2\mathrm{e}^{-bSz}+d_1I_0\mathrm{e}^{-q'z} \tag{6.124}$$

$$i_{\mathrm{d}}(z)=c_1(a+b)\mathrm{e}^{bSz}+c_2(a-b)\mathrm{e}^{-bSz}+d_2I_0\mathrm{e}^{-q'z} \tag{6.125}$$

The coefficients c_1, c_2, d_1, and d_2 will need to be determined from the boundary conditions particular to the situation.

6.1.3.3 Special Situation for Infinite Slab

Considering the Kubelka–Munk approach for backscatter inside an infinitely thick slab under perfectly diffuse illumination presents some insight into the fluence rate directly below the boundary interface of the turbid medium with another medium that has the source.

If there is an index mismatch between the two media, there will be reflection on the basis of Fresnel's law of reflection (section 2.4.2.3). In contrast, the tissue itself will have a bulk reflection coefficient as defined by Equation (6.92). This condition will create a circumstance comparable with a Fabry–Perot interferometer, or a laser cavity, resulting in photon accumulation directly beneath the surface of the medium.

The diffuse light initially entering the tissue will penetrate into it, and a reverse flux will result as a consequence of the scattering. At the interface with the first medium, due to a mismatch in index of refraction, another forward flux will be generated, to be combined with the initial incident beam of diffuse light.

At any point directly below the interface surface, the total fluence will be the result of three fluxes, the incident diffuse beam I_{0d}, the backscattered flux J, and the forward scattered flux I:

$$L_{\text{total}}(z) = 2I_{0d} + 2[I + J] \tag{6.126}$$

However, since the reflection process repeats itself an infinite number of times the following situation arises:

$$I_{z=\delta}/I_0 = 1 + R_\infty + R_\infty r_{id} + R_\infty^2 r_{id} + R_\infty^2 r_{id}^2 + \cdots \tag{6.127}$$

This can be seen as the addition of two recursive series:

$$\begin{aligned} I_{z=\delta}/I_0 &= 1 + 2\left[\sum_{n=1}^{\infty} R_\infty^n r_{id}^n + \sum_{n=1}^{\infty} R_\infty^n r_{id}^{n-1}\right] \\ &= \frac{1}{1 - R_\infty r_{id}} + \frac{1}{r_{id}} \frac{R_\infty r_{id}}{1 - R_\infty r_{id}} = \frac{R_\infty + 1}{1 - R_\infty r_{id}} \end{aligned} \tag{6.128}$$

Equation (6.128) shows an increase in light energy fluence rate directly below the interface of the second medium. The location of this accumulation is in the order of one free optical path length depth.

There is one additional condition of the diffuse reflectance at the entry resulting from the difference in index of refraction of the two media as well. This initial reflectance will reduce the amount of light actually entering the tissue on the basis of Fresnel's law of reflection.

6.1.4 Light Distribution under Infinite Wide Beam Irradiation

Even for absorption smaller than scattering there are certain boundary conditions that will allow for a simple solution. Again the solution for light distribution inside the medium may be approximated by a simple exponential attenuation with depth, still under the assumption of absorption smaller than scattering. Namely, for a wide beam incident on a scattering and absorbing medium (beam diameter minimal 40 times the penetration depth δ) the on-axis light attenuation at depths greater than minimal three times the length of the penetration depth can be approximated according to

$$\psi(z) = \psi_0 e^{-\mu_{\text{eff}} z} \tag{6.129}$$

The validity of Equation (6.129) can be shown from the Kubelka–Munk approximation in Section 6.1.2.

6.1.5 Numerical Method to Solve the Equation of Radiative Transport

The problem of calculating the distribution of light in tissue can be approached in several ways. Two of the most common methods are diffusion theory and Monte Carlo simulation. In the Monte Carlo simulation each photon is considered as a game of chance. The propagation of the photon packet continues until it escapes the medium or its weight is below certain threshold. In the latter case, a technique called Russian roulette is normally used to avoid the violation of the energy conservation rule, which would be violated by simply terminating all the packets below the limit. During the simulation process, the fluence rate distribution in tissue, diffuse reflectance, and transmittance can all be recorded. After simulating sufficient photon packets, these recorded values approach their realistic values. Given enough simulation time, the Monte Carlo model is believed to be the best estimation of light distribution in tissue.

The Monte Carlo simulation technique will be discussed in detail in Chapter 7. Additionally, Chapter 7 will outline the available methodology to derive the optical characteristics of tissues.

6.2 Summary

This chapter presents several techniques that are currently available to solve the equation of radiative transfer under well-defined approximation conditions. The kernel of the irradiation conditions offering acceptable theoretical validation approximations was illustrated for several clinically and investigative relevant protocols.

Further Reading

Bellman, R.E., Kalaba, R., and Prestud, M.C., *Invariant Imbedding and Radiative Transfer in Slabs of Finite Thickness*, Am. Elsevier, New York, 1963.

Chandrasekhar, S., *Radiative Transfer*, Oxford University Press, Oxford, 1950.

Groenhuis, R.A.J., Ferwerda, H.A., and Ten Bosch, J.J., Scattering and absorption of turbid materials determined from reflection measurements, 1: Theory, *Appl. Optics* 22, 2456-2462, 1983.

Ishimaru, A., *Wave Propagation and Scattering in Random Media*, Vols. 1 and 2, Academic Press, Burlington, 1978.

Kottler, F., Turbid media with plane-parallel surfaces, *J. Opt. Soc. Am.*, 50, 5, 483–490, 1960.

Kubelka, P., New contribution to optics in intensely scattering media, Part I, *J. Opt. Soc. Am.*, 38, 448–457, 1948.

Kubelka, P., New contribution to optics in intensely scattering media, Part II, *J. Opt. Soc. Am.*, 44, 5, 330–335, 1954.

Rutily, B., and Bergeat, J., The solution of the transfer equation in a homogeneous isotropically scattering plane-parallel medium, *J. Quant. Spectrosc. Radiat. Transfer*, 52, 857–885, 1994.

Saidi, I.S., Jacques, S.L., and Tittel, F.K., Mie and Rayleigh modeling of visible light scattering in neonatal skin, *Appl. Optics*, 34, 7410–7418, 1995.

Star, W.M., Diffusion theory of light transport, in *Optical-Thermal Response of Laser-Irradiated Tissue*, Welch, A.J. and Van Gemert, M.J.C., Eds., Plenum Press, New York, 1995, pp. 131–206,

Van Gemert, M.J.C., Jacques, S.L., Sterenborg, H.J.C.M., and star, W.M., Skin optics, *IEEE Trans. Biomed. Eng.*, 36, 1146–1154, 1989.

Problems

1. Derive the accumulation of light energy fluence rate directly beneath the surface of a semi-infinitely large turbid medium under collimated irradiation? Note that $L_{\text{total}}(z) = F_c + 2(F_{d+} + F_{d-})$.
2. When measuring the concentration of an absorber in a scattering medium with known scattering coefficient, the following assumptions can be made. If the reduced scattering coefficient μ'_s is known and the radiance at a distance r from an isotropic source with radiance L_0 can be measured, an algebraic equation may be solved for the absorption coefficient on the basis of diffusion approximation given that the radius r is large enough for diffusion approximation to be valid. Show the approach to determine the absorption coefficient and the concentration of the absorber if the undiluted absorber (e.g., Indian Ink) has an absorption coefficient of 80 cm^{-1} at the wavelength of the light source used.
3. Using diffusion theory the light penetration in biological tissue can be approximated to the first order of accuracy. A 50-mW HeNe laser is launched into the tissue through a fiberoptic light guide with a diffusing tip at the end,

inserted into the tissue at 10 cm depth. Assume that the absorption coefficient is 0.1 cm^{-1}, the scattering coefficient of the tissue is 100 cm^{-1}, and the scattering anisotropy factor 0.9 at a wavelength of 632.8 nm. Calculate the light fluence rate 5 cm from the source.

4. Use diffusion theory to estimate solar luminescence light transmission in biological tissues. Assume that the average absorption coefficient over the entire solar spectrum hitting the skin to be 0.1 cm^{-1}, the average scattering coefficient 100 cm^{-1}, with a scattering anisotropy factor of 0.9. Calculate the light fluence rate at 3 mm depth using perpendicular incidence and solar luminance.
5. State the three criteria for deriving the diffusion approximation of light transport in tissue.
6. No matter how complicated the relations become for scattered light, the fluence rate distribution of collimated light with tissue depth is governed by which fundamental law that accounts for both absorption and scattering in the tissue?
7. A 1-mm fiberoptic probe is used to do a therapeutic procedure on a cancerous lesion in the brain. The radiant energy fluence rate required to excite a particular photosensitizer in the tumor is 0.5 W/cm^2. The laser light is directed out of the tip of a fiber optic ($n_1 = 1.7$) to the tumor, located 2 mm from the fiber tip, and is 1 mm in diameter. The optical properties of the normal brain tissue and tumor are given as $n_{norm} \approx 1.1$, $\mu_{a-norm} = 5cm^{-1}$, and $\mu'_{a-norm} = 10cm^{-1}$, and $n_{tumor} \approx 1.1$, $\mu_{a-tumor} = 25cm^{-1}$, and $\mu'_{a-tumor} = 10cm^{-1}$, respectively.
   a. Assuming negligible loss in the fiber, what does the incident irradiance need to be for activating the photosensitizer in the tumor?
   b. With this threshold incident irradiance, what is the expected temperature rise in both the intervening tissue and the tumor for a 3 s exposure? Consider the temperature rise to be solely due to conduction.
8. Which two balance (conservation) equations are used to derive the steady-state diffusion equation from the transport equation?
9. Write the general equation for the total radiant energy fluence rate in the tissue for a collimated irradiance E_0 incident at the interface between air and tissue.
10. How can the irradiance just below an air–tissue interface be larger than the incident irradiance? Assume a tissue that has scattering.

7

Numerical and Deterministic Methods in Light–Tissue Interaction Theory

This chapter describes the available fundamental theoretical resources of light–tissue interaction on the basis of the optical parameters of tissues.

7.1 Numerical Method to Solve the Equation of Radiative Transport

The problem of calculating the distribution of light in tissue can be approached in several ways. Two of the most common methods are diffusion theory and Monte Carlo simulation. The Monte Carlo method uses a numerical approach to calculate the light energy fluence rate in each individual point in the tissue and is exact, but tedious.

Monte Carlo modeling assumes that the photons in a medium are neutral particles, and their wave phenomenon can be neglected. But it is easy to be implemented and flexible enough to model complex tissue. In this model, each photon packet of incident light goes through a random walk in the medium. It will meet absorption and scattering events. The possibility of these events happening depends on the medium's optical properties. In absorption events, a part of the photon packet is absorbed. In scattering events, the photon packet changes its propagation direction as described by the phase function. The propagation of the photon packet continues until it escapes the medium or its weight is below certain threshold. In the latter case, a technique called Russian roulette is normally used to avoid the violation of the energy conservation rule, which would be violated by simply terminating all the packets below the limit. During the simulation process, the fluence rate distribution in tissue, diffuse reflectance, and transmittance can all be recorded. Given enough simulation time, Monte Carlo model is believed to be the best estimation of light distribution in tissue.

7.1.1 Monte Carlo Simulation

As a justification for solving the equation of radiative transfer, computer models have been developed. These models follow one photon at a time and calculate the probability of scattering or absorption of this photon, and if scattered find the direction, according to the optical parameters entered into the program. Where diffusion theory relies on solving a differential equation to calculate the photon density in the scattering medium, the Monte Carlo method explicitly simulates individual photons as they move through the medium. Scattering is a highly stochastic process, and thus a random number generator must be utilized to accomplish the simulation. A large number of photons are simulated in this manner, and eventually a distribution will form. The restrictions of the diffusion theory do not apply to the Monte Carlo method. This process is repeated for a certain predetermined number of photons. The end result is either a cylindrically symmetric two-dimensional or true three-dimensional graphical display of a concentration distribution of absorbed photons in a medium and the corresponding radiant energy fluence rate. Owing its name from the famous casino, it is a statistical method based on a sophisticated type of random photon walk, generally referred to as "random walk." It is possible to determine the probability distributions from which the random numbers are drawn, e.g., to favor a certain direction of scattering for the photons. The Monte Carlo simulation technique provides a numerical solution to the light distribution conundrum. The propagation pattern of photons inside the medium can be described by connecting coordinates with corresponding radiant energy fluence rate, thus forming isodose curves representing the light distribution in a two-dimensional plane. Since this process resembles a chance game, this simulation is called the Monte Carlo method.

7.1.1.1 Simulation Process

A photon packet, with an initial weight, is launched into the simulated tissue. In each interaction point, the scattering direction, the absorption, and the distance to the next interaction point are computed as stochastic variables on the basis of the optical properties of the tissue. The weight is then updated and the photon packet moved to the next interaction point. The physical parameters of the photon packet can be logged during this walk, and the behavior of a large number of photon packets results in a statistical estimation of the macroscopic property of interest.

Monte Carlo simulations have proved to be a very powerful way of modeling light propagation. The method does not suffer from the inaccuracies of the diffusion equation. It is also possible to simulate any geometry and still obtain valid results. However, the method has one major drawback that makes it less attractive for practical use: it is computationally very time-consuming. Since it is a statistical method, the more photon packets added, the lower is the variance and the better the result. Often many hours even on relatively fast computers are required to obtain good statistics.

Treating the boundary conditions in the case of Monte Carlo simulation is a simple task: just make use of Fresnel's and Snell's laws explicitly when the photon packet hits the boundary.

A fundamental problem when constructing the reconstruction algorithm is knowing where the photons have been on their way from the source to the detector. An absolute knowledge of this is of course impossible to obtain because of the stochastic nature of the propagation process, but what can be calculated is the distribution of photon paths. As a measure for identifying the distribution of the quantity of the photon paths the concept of photon hitting density has been introduced. The photon hitting density is a measure of the expected local time that photons spend at different positions when migrating from source to detector within a given time interval.

Consider a point source applied at time $t=0$ in $\mathbf{r}_1$, and the light is detected at time $t=\tau$ in $\mathbf{r}_2$. The probability that a photon has been at a point $\mathbf{r}$ somewhere between the source and the detector will depend on two factors: primarily, on the photon density ρ_p in $\mathbf{r}$ during a time frame from the time the first light reaches $\mathbf{r}$ to the time the last light has to leave $\mathbf{r}$ to have a chance to reach $\mathbf{r}_2$ at time $t=\tau$; and secondary, on the probability $\mathbf{P}$ that a photon in $\mathbf{r}$ at a time $t=\tau'$ within the time frame will reach $\mathbf{r}_2$ at time $t=\tau$. Another factor, the escape function E is also involved. The probability $\mathbf{P}$ and the escape function together determine the photon hitting density in $\mathbf{r}$ for the given source-detector configuration for time $t=\tau$ when integrated over the time window. The integration interval will depend on t, and thus the integration is a convolution. These are the basics of the approach when determining the photon hitting density utilizing the diffusion equation.

Monte Carlo simulation lends itself to a simple way of determining the photon hitting density. Since the simulation logs the position of the photon packets, the photon paths that traverse from the source to the detector in time $t=\tau$ directly build up the photon hitting density distribution.

The first stage of the Monte Carlo photon distribution will be to calculate the photon hitting density.

Let us first consider the simplest case: a single source in an infinite, homogeneous medium, Dirac-distributed in both space and time, given by

$$q_0(\mathbf{r},t) = \delta(\mathbf{r})\delta(t) \tag{7.1}$$

The diffusion equation is then easily solved by using Green's functions:

$$\rho_\mathrm{p}(\mathbf{r},t) = \upsilon(4\pi D)^{-3/2} t^{-3/2} \exp(-\mu_\mathrm{a}\upsilon t) \exp\left[-\frac{r^2}{4D\upsilon t}\right] \tag{7.2}$$

Equation (7.2) is the Green's function for free diffusion.

For other geometries, as long as they are not too complicated, the standard method is to introduce mirrored image sources, so that the boundary conditions are met and the calculation can be performed over infinity. The solution is then generalized accordingly to

$$\rho_{\mathrm{p}}(\mathbf{r},t)=\upsilon(4\pi D)^{-3/2}t^{-3/2}\exp(-\mu_{\mathrm{a}}\upsilon t)\sum_{k}\left\{\exp\left[-\frac{r_{\mathrm{pk}}^{2}}{4D\upsilon t}\right]-\exp\left[-\frac{r_{\mathrm{nk}}^{2}}{4D\upsilon t}\right]\right\} \quad (7.3)$$

where r_{pk} and r_{nk} represent the distances to the components of the kth-order image source dipole. Fortunately, in the cases usually at interest, the higher order terms in Equation (7.3) will never affect the result, so the sum can be truncated at relatively low values of k.

In an even more general form, the photon density may be written as

$$\rho_{\mathrm{p}}(\mathbf{r},t)=\int G(\mathbf{r},\mathbf{r}';t)\rho(\mathbf{r}',0)\mathrm{d}^{3}\mathbf{r}' \quad (7.4)$$

where $G(\mathbf{r},\mathbf{r}';t)$, the Green's function for the geometry, can be seen as an impulse response function. Note that the $\mathbf{r}'$ in this expression only means that the impulse source is allowed to be applied anywhere, not just in the origin as assumed previously in Equations (7.1) and (7.2). The photon hitting density can now be written as

$$\upsilon(\mathbf{r};\mathbf{r}_1,\mathbf{r}_2;t)=\frac{1}{G(\mathbf{r}_1,\mathbf{r}_2;t)}\int_{|\mathbf{r}-\mathbf{r}_1|/\upsilon}^{t-(|\mathbf{r}_2-\mathbf{r}|/\upsilon)} G(\mathbf{r}_1,\mathbf{r}_2;t')G(\mathbf{r},\mathbf{r}_2;t-t')\mathrm{d}t' \quad (7.5)$$

Equation (7.5) may need some explanation. The inverted Green's function outside the integral is merely a normalization factor. The integration limits are the same as were discussed earlier. The first Green's function in the integral represents the photon density in $\mathbf{r}$ due to the impulse source in $\mathbf{r}_1$. The second represents the probability that the photons in $\mathbf{r}$ will reach $\mathbf{r}_2$ given the specific time criteria, the escape function. That this probability is also given by a Green's function may not be obvious, but viewing the point $\mathbf{r}$ as the position of a new source and $\mathbf{r}_2$ as the position where the photon density from this source is to be calculated might be helpful. The difference is that in the first case (from $\mathbf{r}_1$ to $\mathbf{r}$) the source was an impulse in both space and time, and in the second case (from $\mathbf{r}$ to $\mathbf{r}_2$) the source is only an impulse in space. The spreading in time is taken care of by the convolution integral.

This is the definition of the photon hitting density. Here the first Green's function represents ρ_{p} in $\mathbf{r}$, and the second, the escape function, represents ρ_{p} in $\mathbf{r}_2$ due to a source in $\mathbf{r}$. One might argue that the escape function should instead be defined by the photon current density multiplied by a normal vector, $nJ(\mathbf{r}_2,t-t')$, when the detector is located outside the medium. This quantity would then be a measure of the photon flux over a small area $\mathrm{d}A$ on the boundary at $\mathbf{r}_2$, due to a source in $\mathbf{r}$. Nevertheless, using the photon density ρ_{p} as the escape function has the advantage that some degree of symmetry is evident in the problem, which makes calculation simpler.

After this theoretical survey, we conclude that the photon hitting density in which we are interested is proportional to

$$v(\mathbf{r};\mathbf{r}_1,\mathbf{r}_2;t) \propto \int_{\mathbf{r}_1}^{t-\mathbf{r}_2} G(\mathbf{r}_1,\mathbf{r}_2;t')E(\mathbf{r},\mathbf{r}_2;t-t')\mathrm{d}t' \tag{7.6}$$

where the escape function E is given by either ρ as in Equation (7.2) or J as in Fick's law, and t_1 and t_2 represent the time spans given in Equation (7.6)

Log the photon packets as they traverse through the medium, and if the packet hits the position that is appointed as the detection point, add the photon path to a result matrix. Eventually, when an adequate number of photons have been filed, the sum of photon paths forms a distribution that approximately equals the photon hitting density.

In practice, however, a number of difficulties become apparent, all of which have their origin in the computationally time-consuming nature of the Monte Carlo method. Since we are only interested in the photons that actually hit the detector, a lot of computation done by the computer is a waste. The fraction of photons that contribute depends on the optical properties of the simulated medium and the geometry, primarily the distance between the source and the detector, but even under favorable conditions this fraction is no more than perhaps 1:1,000 or 1:10,000. This means that the total number of photon paths that has to be calculated is typically several millions. As the distance to the detector increases, the number of detected photon packets decreases rapidly. At some distance it is no longer useful to perform calculations because the statistics become too poor for any reasonable computation times.

Closely connected to the problem with the distance to the detector are the optical properties of the medium. The scattering mean free path here acts as a sort of scaling factor. Increasing the mean free path by a factor of 10 (i.e., decreasing μ_S by the same amount) means that only a tenth of the calculations is needed for a photon packet to reach a certain point. This opens up a possibility of performing faster simulation, by varying the optical properties so that μ_S has a minimum value while μ_S' still has the same value. This can be accomplished by setting $g=0$.

Another problem is the finite nature of the method. The result has to be presented as some sort of discrete matrix or array. The pixel size must be a compromise between, on the one hand, small pixels, which provide high accuracy, and on the other large pixels, to obtain acceptable statistics within each pixel (by pixel, in this context, is meant the smallest discrete element, which not necessarily has to be of two dimensions).

For certain simple geometries, such as the homogeneous slab, the statistics can be significantly enhanced by taking advantage of symmetries. In the case of the slab, introducing cylindrical coordinates reduces the problem from three spatial dimensions to two when it comes to plotting the result. In the general case, when the distribution of photon paths is a function of x,

y, z, and t, determining the pixel size (i.e.Δx, Δy, Δz, and Δt) may be a delicate matter.

The way Monte Carlo simulation is normally implemented, the photon packets are not actually logged along the whole of the traversed path, but rather merely in the interaction points. This implies that there is not much point in having the (spatial) pixels much smaller than the mean free path length, because otherwise the distribution would be built from scattered points instead of connected paths, requiring yet more computation to get good statistics. Larger pixels thus smooth the curve. As far as accuracy is concerned, larger pixels are allowed as long as their size is small compared to the overall variation of the distribution. The problem might here arise near the source and detector, where the distribution may be expected to vary rather sharply. The same goes for the time steps. Since a complete spatial distribution is required for each time step, these steps must be large enough to give good statistics, i.e., a sufficient number of photons detected within the time interval. The upper limit of the step size is determined by the condition that it must be small compared to the overall variation in detected intensity.

The finite size of the detector also has to be considered. The simplest way of implementing detection is to assign detector status to a certain pixel, and thus make the detector size equal to the pixel size. It is tempting to make the detector larger to obtain a higher fraction of detected photons. In reality, the detector size would be determined by the diameter of an optical fiber or in the future perhaps the pixel size of an arrayed semiconductor detector, in any case not much larger than around 1 mm^2. Also, when comparing the results to those obtained by the diffusion approximation method, one must keep in mind that the method assumes an infinitely small detection point. Setting the detector size equal to the pixel size seems sensible for the following reason: the photon-hitting-density at the source and at the detector must be the same because, as per definition, every photon packet that has contributed to the photon-hitting-density distribution has been at the source and at the detector. Now, in the result matrix, the photon hitting density at the source will be represented by the number of photon interactions within the pixel that contains the source. To get the same value at the detector, the detector area should then be of the same size as the source pixel.

7.1.1.2 Light–Tissue Interaction Model

The tissue is generally modeled using a cylindrical coordinate system to represent the assumed cylindrical symmetry of the light distribution in a homogeneous medium.

The commonly used model applies the cylindrical symmetry and reduces the tissue to a plane in ($\mathbf{r}$, z). The rationale behind it is that the cylinder can be reproduced by rotating this plane along the z-axis. This holds good if the tissue is homogeneous, but if it is not, it fails. It provides no way of modeling nonuniform tissue.

7.1.1.3 Random Walk Model

The photon is assumed to move with its own spherical coordinate system. This coordinate system is used to represent the current direction of the photon.

At every stage of the random walk, a random walk length is generated. Using this length the new position of the photon is calculated. Mean cosine of scattering (g) is used to calculate the new direction of photon, which is a function of current direction.

7.1.1.3.1 Simulation Model

The simulation of light propagation in turbid media is a simulation of laser light and tissue interaction. So, the two most important objects, which we have already come across, are the tissue and the laser light. In fact, from the level of abstraction that we are concerned with here, these two are the main objects. As the objective of this simulation is to simulate the light distribution and photocoagulation in tissue as a result of laser light irradiation, we need to identify the various interactions that result in this phenomenon. The key issue here is to model how laser light and tissue interact. To do that, we need to have a good understanding of how light travels in a scattering and absorbing medium and what light consists of. There are various theories for this, but for our purpose the theory of random walk of photons is most suited to model light and light propagation.

Laser light is decomposed into photons, so the propagation of light is constructed from the random walk of these photons. In each step of their random walk, photons lose energy and scatter. The amount of energy loss and scattering depends on the local optical media, in this case a specific tissue. As a result of energy deposition the tissue gets heated and may change its optical properties; this is known as photocoagulation. One of the objectives of the simulation model is to account for the optical changes in the tissue as a result of photocoagulation. To achieve this, we need to model heat dissipation and photocoagulation.

The simulation model has the following objects:

- Tissue
- Laser beam
- Photon

The relevant interactions between them are as follows:

- Random walk
- Scattering of light
- Absorption of light
- Reflection of light
- Heat generation and dissipation

7.1.1.3.2 Tissue Model

This simulation project treats the tissue as an optical medium. Modeling light by the random walk of photons involves modeling absorption and scattering of light or photon. The amount of scattering and absorption of light is different in various media. We need to incorporate absorption and scattering through some parameters in the tissue model. Our aim is to model the tissue as real as possible. The real tissue has different optical properties in its various parts, due to various inhomogeneities in the tissue. Our model will be able to represent the tissue inhomogeneities. Further, this implies that there will be more than one medium for light to travel in. When light travels from one medium to another, there is reflection and refraction of light. To accommodate this phenomenon we need to include the index of refraction into our model. Absorption of light in tissue increases the temperature of the tissue. This increase in temperature can alter the tissue phase, resulting in coagulated tissue.

The objectives of the tissue model are as follows:

1. Model optical abstraction of real tissue
2. Model absorption and scattering of light
3. Specify different inhomogeneities of real tissue, such as blood vessels, fat, and muscle
4. Model tissue temperature and be able to adopt to the change in the tissue properties due to the changes in the temperature of the tissue.

7.1.1.3.3 Light Propagation Model

Random walk is one of the most critical processes to be modeled in simulation of light propagation for obtaining the light distribution in a turbid media. Light propagation in a medium can be thought of as the random walk of all the photons through the medium. There are many different activities happening to make this apparently simple random walk. Depending on the optical properties of the medium light is traveling through, photons lose energy (absorption of light), they get scattered into different directions (scattering of light), and if there is change in the medium, they get reflected or refracted. All these phenomena of light are associated with the random walk, although they are independent of each other. The random walk model represents the interaction between light and the medium light travels through; here: tissue.

The objectives of the light propagation model are as follows:

1. Model light propagation through the random walk of a photon
2. Model scattering of light
3. Model absorption of light
4. Model reflection and refraction of light.
5. Model thermodynamic effects of the interaction of light with biological media
6. Model heat generation and dissipation due to absorption of light
7. Model photocoagulation

7.1.1.3.4 The Actual Light Propagation Model

The process of simulating light propagation involves the following steps:

1. Calculation of walk path length or displacement in the location of photon
2. Calculation of the new direction of the photon in the tissue media

The random walk length L_{rnd} is calculated using the function

$$L_{\mathrm{rnd}} = -\ln \frac{(1-\mathrm{RND})}{\mu_t} \tag{7.7}$$

where RND is a random number such that $0<\mathrm{RND}<1$ and it is from a uniform distribution.

If the previous location of the photon is $\mathrm{Loc}_{\mathrm{prev}} = [x\ y\ z]$, the current location n is given by

$$\mathrm{Loc}_{\mathrm{curr}} = [x + L_{\mathrm{rnd}} y + L_{\mathrm{rnd}} z + L_{\mathrm{rnd}}] \tag{7.8}$$

The scattering model is in the Cartesian coordinate system. The Monte Carlo simulation calculations are based on a Henyey–Greenstein angular scattering distribution in spherical coordinates.

The scattering of photon in a spherical coordinate system is calculated using the following functions:

$$\varphi_{\mathrm{s}} = \arccos\left\{\frac{1}{2g}\left[1+g^2-\left(\frac{1-g^2}{1-g+2g\,\mathrm{RND}}\right)^2\right]\right\} \tag{7.9}$$

and

$$\psi_{\mathrm{s}} = 2\pi\,\mathrm{RND} \tag{7.10}$$

with g the scattering anisotropy factor.

These directions are the deflection from the previous direction of the photon. These are used to obtain the net direction by transforming the old direction vector into these new directions.

The photon moves with a coordinate system, which has the z-axis in the same direction as that of the photon. Figure 7.1 shows a photon with its local moving coordinate system. We keep track of this using a direction vector in the tissue coordinate system.

The following composition of three-dimensional rotation matrix gives the scattered direction of the photon:

$$D_{\mathrm{scattered}} = D_{\mathrm{old}} R_z(-\psi) R_y(\varphi_{\mathrm{s}}) R_z(\psi + \psi_{\mathrm{s}}) \tag{7.11}$$

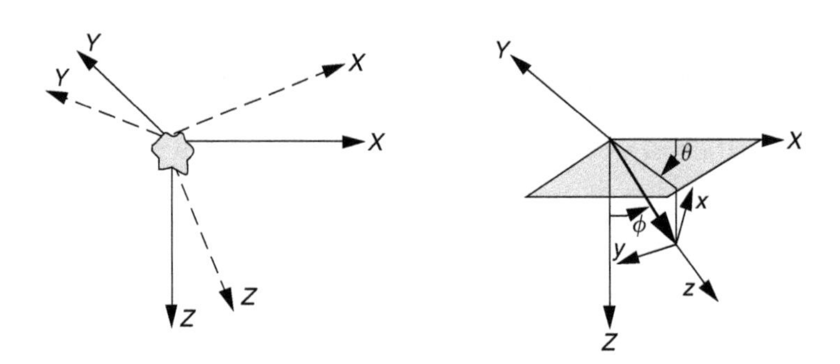

FIGURE 7.1
Photon with its local moving coordinate system.

Reflection is calculated using Fresnel's law of reflection when the photon moves from tissue to air or at the boundary when a photon moves from one tissue medium to another tissue medium.

7.1.1.4 Light Propagation Model Components

The light propagation model involves modeling the incident beam as well as the light–tissue interaction mechanism.

7.1.1.4.1 Infinitely Narrow Beam Excitation

Monte Carlo simulation begins with the simulation of an infinitely narrow photon beam normally incident on a multilayered tissue. The response of tissue is regarded as an impulse response. Later, this impulse response will be used to get the response of tissue to a finite-size photon beam by convolution.

Figure 7.2 shows the flowchart of Monte Carlo simulation. The simulation of a photon packet propagating in tissue includes the following steps: launch, interaction, absorption and scattering, boundary conditions, photon termination, and finally convolution of multiple photons. These six phases of the simulation will be discussed in the following sections.

7.1.1.4.2 Launching

A photon packet starts with unity weight w. Its location and direction are specified by the coordinates (x, y, z) and the directional cosines (μ_x, μ_y, μ_z), respectively. The directional cosines are defined as

$$\mu_x = \mathbf{r} \cdot \mathbf{x} \tag{7.12}$$

$$\mu_y = \mathbf{r} \cdot \mathbf{y} \tag{7.13}$$

$$\mu_z = \mathbf{r} \cdot \mathbf{z} \tag{7.14}$$

where **r** is the unit vector of the direction of the photon propagation and **x**, **y**, **z** are unit vectors along each axis. The initial location of the photon packet is set to (0,0,0) and its initial directional cosines are (0,0,1).

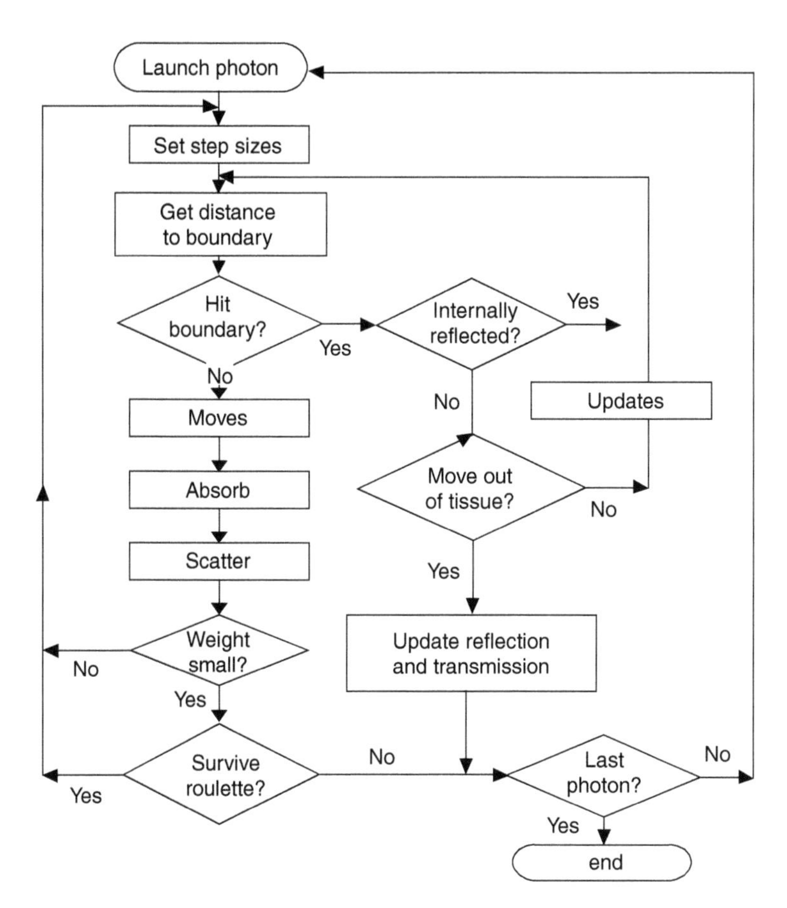

FIGURE 7.2
Flowchart of Monte Carlo simulation.

If mismatched boundary is present when the photon packet is launched, the effect of specular reflection will count in by reassigning weight w with $w–R_{sp}$. The specular reflectance at the tissue surface R_{sp} at normal incidence is equal to (see Chapter 2)

$$R_{sp} = \frac{(n_1 - n_2)^2}{(n_1 + n_2)^2} \tag{7.15}$$

where n_1 and n_2 are refraction indices of outside medium and tissue, respectively.

7.1.1.4.3 *Simulation Steps*

One step in tissue can be described as follows. The step size s of a photon packet can be regarded as the free path of the packet. At the end of each step, the packet will encounter absorption and scattering events. The value s is generated randomly according to the total attention coefficient μ_t of the tissue. As mentioned above, μ_t represents the probability per unit path length that a photon will encounter either an absorption or scattering event. On the basis of the definition of step size, we have

$$P(s \geq s' + \mathrm{d}s') = P(s \geq s', \text{no interactions in ds}') \tag{7.16}$$

As we know, μ_t is a constant and does not depend on the previous propagation history of the photon packet. So, $s \geq s'$ is independent of whether there are any interactions at location ds′:

$$P(s \geq s' + ds') = P(s \geq s')P(\text{no interactions in ds}') \tag{7.17}$$

We also have

$$P(\text{no interactions in ds}') = 1 - \mu_t \, \mathrm{d}s' \tag{7.18}$$

Substitution of Equation (7.18) into Equation (7.17) yields

$$P(s \geq s' + \mathrm{d}s') = P(s \geq s')(1 - \mu_t \, \mathrm{d}s') \tag{7.19}$$

Rearranging Equation (7.19) gives the following expression:

$$\frac{P(s \geq s' + \mathrm{d}s') - P(s \geq s')}{P(s \geq s')\mathrm{d}s'} = \frac{\mathrm{d}P(s \geq s')}{P(s \geq s')\mathrm{d}s'} = -\mu_t \tag{7.20}$$

Next we integrate both sides of Equation (7.20) from 0 to s_1 to obtain

$$P(s \geq s_1) = \exp(-\mu_t s_1) \tag{7.21}$$

$$P(s < s_1) = 1 - \exp(-\mu_t s_1) \tag{7.22}$$

$$P(s_1) = \mu_t \exp(-\mu_t s_1) \tag{7.23}$$

Based on its distribution, the random variable s_1 is generated from a uniformly distributed variable ξ :

$$s_1 = -\frac{\ln(\xi)}{\mu_t} \tag{7.24}$$

where $0 \leq \xi \leq 1$ and is generated by the computer simulation.

Once the step size s is determined, the program will check if the packet hit the boundary of layered tissue. If yes, the program will hit boundary routine, which will be introduced in the following discussion. If not, the photon packet will move to a new location $(x,y,z,)_{new}$:

$$x_{new} = x + \mu_x s \tag{7.25}$$

$$y_{\text{new}} = y + \mu_y s \tag{7.26}$$

$$z_{\text{new}} = z + \mu_z s \tag{7.27}$$

7.1.1.4.4 *Absorption*

At the new location, the packet first encounters an absorption event. A weight that is equal to $(\mu_a/\mu_t)w$ will be deducted from the packet weight w before interaction. This weight will also be stored in an array $A(r,z)$, which represents the total absorbed light at $A(r,z)$. A cylindrical coordinate system is used here because of the cylindrical symmetry properties of $A(r,z)$. This coordinate system shares the same origin and z-axis with the Cartesian $A(x,y,z)$ coordinate system.

After an absorption event,

$$w_{\text{new}} = w - \left(\frac{\mu_a}{\mu_t}\right) w \tag{7.28}$$

$$A(r,z)_{\text{new}} = A(r,z) + \left(\frac{\mu_a}{\mu_t}\right) w \tag{7.29}$$

7.1.1.4.5 *Scattering*

After coming to the new location and having experienced an absorption event, the likelyhood of encountering either a scattering or absorption event will be calculated. We next assume that the photon packet will be scattered. During this event, the packet will change its propagation direction without losing any weight. The new direction is described by a deflection angle θ ($0 \le \theta < \pi$) and an azimuthal angle ψ ($0 \le \psi < 2\pi$). The cosine of the deflection angle θ is decided by the phase function, while the azimuthal angle ψ is distributed uniformly from 0 to 2π. Based on their distributions, the random variables $\cos(\theta)$ and ψ are generated from a uniformly distributed variable ξ:

$$\cos\theta = \begin{cases} \dfrac{1}{2g}\left[1 + g^2 - \left(\dfrac{1-g^2}{1-g+2g\xi}\right)^2\right] & \text{if } g \neq 0 \\ 2\xi - 1 & \text{if } g = 0 \end{cases} \tag{7.30}$$

$$\psi = 2\pi\xi \tag{7.31}$$

Given the directional cosine of the current direction, the value of $\cos(\theta)$ and ψ, the new directional cosines will be

$$\mu_{x,\text{new}} = \frac{\sin\theta}{\sqrt{1-\mu_z^2}}(\mu_x \mu_z \cos\psi - \mu_y \sin\psi) + \mu_x \cos\theta \tag{7.32}$$

$$\mu_{y,\text{new}} = \frac{\sin\theta}{\sqrt{1-\mu_z^2}}(\mu_y\mu_z\cos\psi + \mu_x\sin\psi) + \mu_y\cos\theta \tag{7.33}$$

$$\mu_{z,\text{new}} = -\sin\theta\cos\psi\sqrt{1-\mu_z^2} + \mu_z\cos\theta \tag{7.34}$$

7.1.1.4.6 *Hitting Boundaries*

If the photon packet hits the boundary of current layer during one step, it will be either reflected back or transmitted outside the current layer. The probability of internal reflection depends on the angle of incidence, α_i, and is determined by the Fresnel reflection $R(\alpha_i)$:

$$R(\alpha_i) = \frac{1}{2}\left[\frac{\sin^2(\alpha_i - \alpha_t)}{\sin^2(\alpha_i + \alpha_t)} + \frac{\tan^2(\alpha_i - \alpha_t)}{\tan^2(\alpha_i + \alpha_t)}\right] \tag{7.35}$$

Note: The following angles are defined: the angle of incidence $\alpha_i = \cos^{-1}(|\mu_z|)$ and the angle of transmission $\alpha_t = \sin^{-1}(n_i\sin(\alpha_i)/n_t)$.

A uniformly distributed random variable $\xi \in [0,1]$ is then generated to determine whether the photon packet is internally reflected or transmitted. If $\xi \le R(\alpha_i)$, the packet is internally reflected, which means $\mu_{z,new} = -\mu_z$. If $\xi > R(\alpha_i)$, the packet is transmitted outside the current layer.

For the transmitting situation, if the packet is transmitted to a new layer of tissue, the program will continue propagation with the updated step size and direction. If the packet is transmitted to the outside of the tissue, the propagation of the packet will stop and its weight will be counted in diffuse reflectance or transmittance.

7.1.1.4.7 *End of a Photon Packet*

The photon packet may end its propagation by transmitting outside the tissue. If the packet's weight is very small and it is still propagating in the tissue, only little information can be yielded. But if the packet is stopped directly by assigning its weight to zero, the energy conservation rule is violated. A technique called Russian roulette is used here to void this violation. In this technique, the photon packet has a probability of $1/m$ to survive with weight mw. A uniformly distributed random variable $\xi \in [0,1]$ is generated. If $\xi \le 1/m$, the packet will survive with weight mw and continue its propagation. Otherwise, the packet's weight is set to zero and the packet is ended, i.e.,

$$w_{\text{new}} = \begin{cases} mw & \text{if } \xi \le 1/m \\ 0 & \text{if } \xi > 1/m \end{cases} \tag{7.36}$$

After a photon packet is terminated, a new packet will be launched unless the total simulation packets' number is met.

7.1.1.4.8 *Convolution to Obtain the Light Distribution for a Finite-Size Light Beam*

There are two ways to get the response of tissue to a finite-size photon beam. One is to launch photons with initial locations distributed spatially on the tissue–beam interaction plane. This way is simple but it needs large amount of simulation to reach an acceptable variance. However, the system with a finite-size photon beam incident normally on a multilayered tissue is linear and space-invariant. Hence, the responses of tissue to a finite-size photon beam can also be obtained by convolution, which is much more efficient than the first approach. The linearity means if the intensity of the incident photon beam is multiplied by a constant, the responses also need to be multiplied by the same constant. It also means if several photon beams are incident simultaneously on the tissue, the total response is the summation of the responses to each photon beam. The space-invariance means if the incident photon beam shifts in the surface of the tissue, the response will also shift correspondingly. An illustration of the light distribution simulation is shown in Figure 7.3.

If $G(x,y,z)$ is the impulse response of the multilayered tissue to an infinite narrow incident beam and the intensity profile of the finite size photon beam is $S(x,y)$, then the response of tissue to this beam, $C(x,y,z)$, is

$$C(x,y,z) = \int_{-\infty}^{\infty}\int_{-\infty}^{\infty} G(x-x',y-y',z)S(x',y')\mathrm{d}\,x'\mathrm{d}\,y' \tag{7.37}$$

The integration is solved numerically by the extended trapezoidal rule.

However, the location of the fluence rate recording grid, (r_0, z_0), is in a cylindrical coordinate system, while the location of the packet, (x_0, y_0, z_0), is in a Cartesian coordinate system. During initialization process, (x_0, y_0, z_0), is set to $(x_0, 0, z_0)$.

These routines can be implemented in various computer programming languages. The visualization of the simulated light propagation inside a two- or three-dimensional block of homogeneous or inhomogeneous fictitious tissue, however, is an entirely different topic (Figure 7.4).

7.2 Measurement of Optical Parameters

The measurement of optical properties can be achieved by either invasive or noninvasive techniques. The invasive techniques will provide greater detail, since the medium can be investigated for anatomical anomalies as well as tissue diversity. In addition, these techniques provide several more degrees of freedom of tissue interrogation compared with the noninvasive techniques.

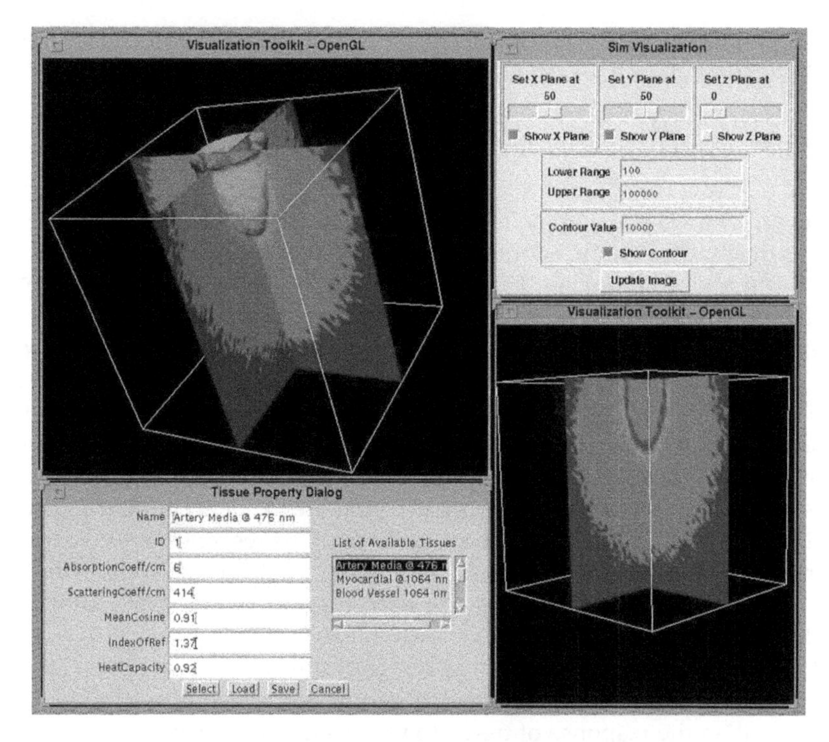

FIGURE 7.3
Graphic representation of light distribution generated by Monte Carlo simulation.

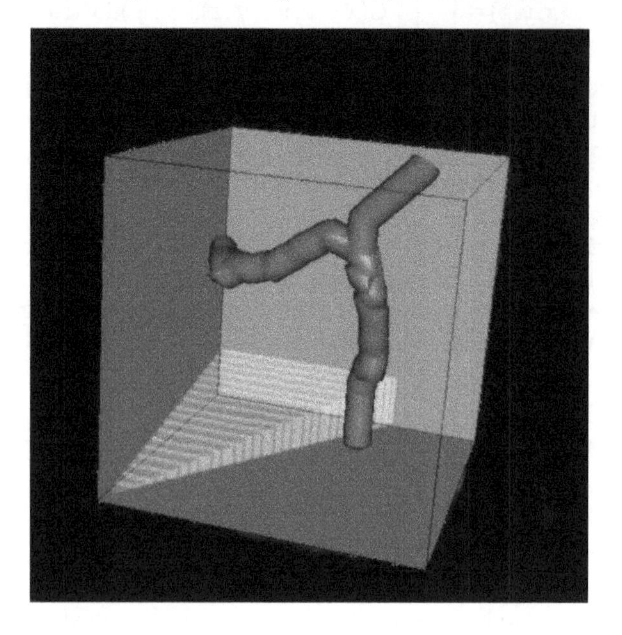

FIGURE 7.4
Computer visualization of an inhomogeneous three-dimensional tissue configuration, with muscle (bulk part), blood vessel, and fat.

Invasive techniques usually come in two forms: in one form the tissue is removed similar to a biopsy punch, and in the other fiber optics are inserted into the tissue. Fiber measurements can be relatively minimally invasive, but are nonetheless invasive.

Noninvasive determination of optical characteristics is often preferable; however, due to its nature it may not always be able to determine all individual optical parameters. In noninvasive measurements the medium will remain anatomically undisturbed, thus giving the opportunity to avoid tissue changes resulting from loss of blood circulation and degeneration.

The invasive techniques will be discussed first on the basis of both the documented accuracy and the historical background.

7.2.1 Invasive Techniques

Two different invasive techniques can be distinguished: excision of a slab of tissue and subsequently slicing this slab into thinner slabs of tissue approximating one optical free path length thickness. Since the optical free path length is not known initially, several preparations and measurements will be needed. The second method of invasive determination of optical properties is by inserting a probe into the tissue and measuring the light distribution inside the medium under illumination with a monochromatic source.

7.2.1.1 Integrating Sphere Measurements

Based on the diffusion approximation of the equation of radiative transfer adapted to a modified Kubelka–Munk (three-flux) theory, it is possible to compute diffusion optical properties from measurements of reflection and transmission.

The combination of diffusion theory and Kubelka–Munk theory is used in this method to derive the absorption coefficient μ_a, scattering coefficient μ_s, and scattering anisotropy factor g from measurements of diffuse reflectance, diffuse transmittance, and collimated transmittance through thin sliced tissues. The Kubelka–Munk approach thus provides three measurements to derive three optical parameters.

7.2.1.1.1 Double Integrating Sphere Methods

For measuring diffuse reflectance R_d, a thin slice of tissue is placed on a hole in a sphere that has been coated with a diffuse reflective coating appropriate for the wavelength range of the measurements. The sphere has two additional holes. One hole is to allow the light source to enter the sphere, which subsequently reflects from the diffuse inner wall of the sphere an infinite number of times, resulting in perfectly diffuse light falling on each surface section of the wall. With a section of the wall removed for the sample, this sample will receive the diffuse light needed for measuring the diffuse reflection from the sample, and the diffuse transmission through the sample. The measurement is performed by the placement of a light sensor in a third opening in the sphere. The third hole holds the light sensor. The surface areas occupied by these holes cannot be too large, since each hole will act as a loss for the light source used for probing. Additional precautions involve shielding both the sensor and the sample from light coming directly from the source hole, since this will not be diffuse light, in addition to the fact that the incoming collimated source will have a greater radiance than the diffuse illumination by the entire wall. If the sensor sees the source directly, this light will drown out any diffuse light from the sample. For convenience in the double integrating sphere configuration the sample is initially placed on top of a sphere for measuring diffuse reflection R_d. A schematic representation of this experimental setup is shown in Figure 7.5.

Three individual measurements need to be performed, which are the calibration of the sphere and the coating, by measuring the reflection from a standard white plate on the sample port—R_m; another calibration measurement involve the reflection from the sphere with an empty sample port to determine the loss by the missing reflective surface area—R_0; and finally a measurement with the sample on the port—R_s. R_d is calculated using the following formula:

$$R_d = \frac{R_s - R_0}{R_m(100/m) - R_0} \tag{7.38}$$

The factor $(100/m)$ in the formula is a calibration factor for the reflection coating of the integrating sphere and the standard plate. The value of m normally ranges from 95 to 99, depending on the coating material and the wavelength used to measure the diffuse reflection and diffuse transmission of the biological sample.

For diffuse transmittance measurement, i.e., T_d, two spheres are needed and two measurements need to be performed, which are the transmission through an empty sample port, collected by the empty sample port in the

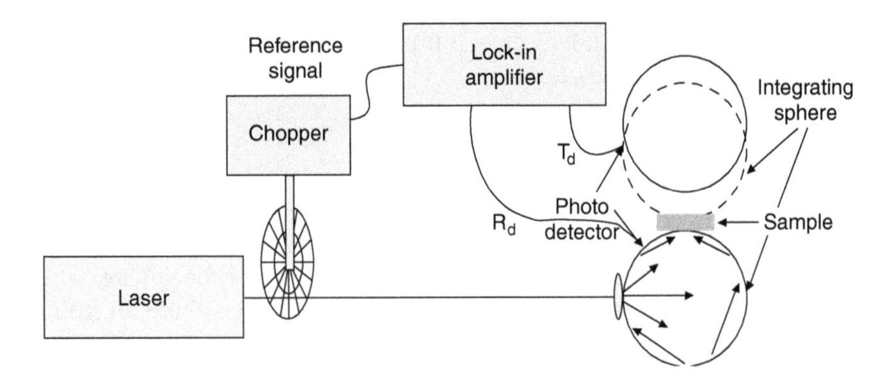

FIGURE 7.5
Double-integrating-sphere configuration to measure diffuse reflection and diffuse transmission of a biological sample.

second sphere, i.e., T_{100}, equivalent to 100% transmission; and next a measurement needs to be made with the sample on the port of the source sphere, with the collection sphere sample hole in contact with the opposite side of the sample. The collection sphere collects all the light remitted by the sample, and the second hole in this sphere also has a light sensor to collect the diffusely transmitted light through the sample. The diffuse transmission T_d can now be calculated as

$$T_d = \frac{T_s}{T_{100}} \tag{7.39}$$

In this formula it is assumed that $T_0=0$, which means no transmission when the hole is closed, and no additional light losses other than by the sample.

However when the sample is more than one optical free path length thick, light may escape through the edge of the sample, thus introducing an error in the measurements.

These measurements are in accordance with the Kubelka–Munk two-flux theory for a forward and a backward diffuse photon flux. The Kubelka–Munk scattering coefficient S and absorption coefficient K are, respectively, shown in the following equations:

$$S = \frac{1}{bd} \ln\left[\frac{1 - R_d(a-b)}{T_d}\right] \tag{7.40}$$

$$K = (a-1)S \tag{7.41}$$

with

$$a = \frac{1 + R_d^2 - T_d^2}{2R_d} \tag{7.42}$$

$$b = \sqrt{a^2 - 1} \tag{7.43}$$

The third measurement will require the measurement of collimated attenuation. Figure 7.6 shows the experimental setup for measuring collimated transmittance.

Several apertures are needed in the path of the laser beam, especially after leaving the sample, to ensure that only the collimated unscattered light is captured by the photo sensor. For convenience, an aperture of the same diameter as the aperture in front of the sensor can be placed to warrant a one-to-one conversion of the incident to the transmitted beam. Additional apertures, e.g. placed directly after the sample can be used to block the scattered components of the laser beam passing through the sample. The placement of the sensor/aperture combination will need to be sufficiently far away to ensure a

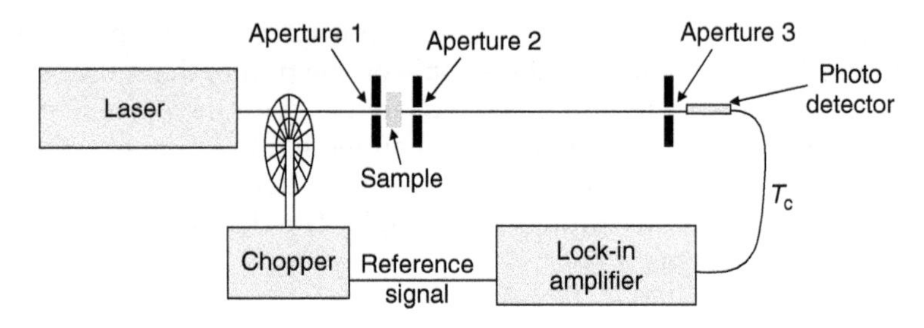

FIGURE 7.6
Illustration of the configuration for measuring collimated transmittance.

relatively small numerical aperture for collection, which again depends on the selection criteria used to size the initial apertures. For collimated transmittance T_c, two measurements need to be performed to calibrate the transmission. First the transmission through two glass slides, sandwiched together to eliminate the reflection loss from the air–glass interfaces, needs to be made: V_{ref}. The second light transmission measurement needs to be with the sample of thickness d between the glass windows: V_c. The collimated transmission is now calculated by

$$T_c = V_c - V_{ref} \tag{7.44}$$

where the collimated transmission satisfies the Beer–Lambert law,

$$T_c = I_0 \exp(-qd) \tag{7.45}$$

The total attenuation coefficient q is the combined loss due to absorption and scattering,

$$q = \mu_a + \mu_s = \mu_t \tag{7.46}$$

From Equation (7.45) total attenuation coefficient μ_t follows as

$$\mu_t = -\frac{\ln T_c}{d} = q \tag{7.47}$$

In this measurement, the laser beam is incident perpendicular to the surface of the sample after passing through an aperture.

The absorption coefficient μ_a (mm^{-1}), scattering coefficient μ_s (mm^{-1}), and the scattering anisotropy factor g follow from T_d, R_d, and T_c, according to

$$\mu_a = \frac{K}{2} \tag{7.48}$$

$$\mu_s = \mu_t - \mu_a \tag{7.49}$$

$$g = 1 - (\mu_a + 4S)/3\mu_s \tag{7.50}$$

This technique is attractive because simple analytic relations exist to determine uniquely the transport (μ_a, μ_s, g) parameters from macroscopic measurements of the reflection and transmission. Three measurements are required that can be rapidly obtained in one single experimental setup. Dehydration is therefore minimized. The disadvantages of the approach are, first, the empirical difficulty of separating the on-axis unscattered from scattered irradiance. Any problems with the primary beam divergence can be eliminated by beam expansion and collimation by lenses before entering the tissue.

In the second place, errors caused by index mismatching at the tissue boundaries have not been accounted for. This discrepancy is greatest at the air–glass slide interface; it is assumed that tissue and glass have similar refractive indices. The measurement of the value T_0 eliminates this error; a fourth reflection measurement using black backing on the sample is needed to implement a correction mathematically, which was suggested for diffuse incidence. However, this approach requires a known value for the total internal reflectance.

In the third place of items of importance is the fact that the thickness of the tissue must be sufficiently thin to enable low-noise detection of the heavily attenuated light.

7.2.1.2 Fiberoptic Probing to Determine the Optical Properties

With the use of an isotropic diffuse probe tip at the end of a fiberoptic light guide the actual energy fluence rate Ψ (r, θ, z) can be measured directly because the isotropic probe will capture all 4π steradian of light incident on a sphere (a small systematic error will be introduced by the surface area of the probe occupied by the fiber optic).

To derive the optical properties of the medium under illumination by a collimated beam, the probe will be inserted into the medium to measure the local energy fluence rate. The diffusion approximation will need to be rewritten in spherical coordinates for the theoretical derivation of the measured parameters. Again this can be shown to reduce to a two-flux arrangement with one-photon flux moving in the positive axial direction: $F_+(r)$, and one flux moving in the negative axial direction: $F_-(r)$. Both fluxes technically are a function of the radial distance from the center of the source. This theoretical description of the light energy fluence rate is represented by

$$\frac{\partial}{\partial r}\left\{r^2[F_+(r)-F_-(r)]\right\}=-\mu_a r^2[F_+(r)+F_-(r)] \tag{7.51}$$

$$\frac{\partial}{\partial r}[F_+(r)+F_-(r)]=-\frac{3}{2}\mu_{tr}[F_+(r)-F_-(r)] \tag{7.52}$$

There is no source term because there is no isotropic point source in this geometry.

For the positive and negative axial directions, the net fluxes are, respectively, defined as follows:

$$F_+(r) = 2\pi \int_0^{+1} L(r,\mu)\mu \, d\mu \tag{7.53}$$

$$F_-(r) = -2\pi \int_{-1}^{0} L(r,\mu)\mu \, d\mu \tag{7.54}$$

where $L(r, \mu)$ is the radiance and μ the cosine of the angle of the ray with the optical axis $\mu=\cos(\theta)$. The optical axis technically can be defined as the normal to the surface.

The solutions to Equations (7.51) and (7.52) can, respectively, be written as follows:

$$F_+(r) + F_-(r) = \frac{[c_1 \exp(-\mu_{\text{eff}} r) + c_2 \exp(\mu_{\text{eff}} r)]}{r} \tag{7.55}$$

$$F_+(r) - F_-(r) = \frac{[c_1(1-\mu_{\text{eff}} r)\exp(-\mu_{\text{eff}} r) + c_2(1-\mu_{\text{eff}} r)\exp(\mu_{\text{eff}} r)]}{\frac{3}{2} r^2 \mu_{\text{tr}}} \tag{7.56}$$

This will yield for the fluence rate

$$\psi(r) = 2[F_+(r) + F_-(r)] \tag{7.57}$$

and the radiance can be represented as

$$L(r,\mu) = \frac{\left\{F_+(r) + F_-(r) + \frac{3}{2}\mu[F_+(r) - F_-(r)]\right\}}{2\pi} \tag{7.58}$$

An additional concern is the difference in index of refraction between the isotropic beam and the turbid medium. Reflection at the interface will result in loss of light measurement, and hence introduces a systematic error. This error can be eliminated when the indices are known. Finally, the probe cannot have a significant absorption coefficient in the range of wavelengths of the source.

To find the optical parameters a curve fit will be required, and the only two parameters that can be resolved are the reduced scattering coefficient $[\mu_s'=(1-g)\mu_s]$ and the absorption coefficient.

7.2.2 Noninvasive Techniques

Noninvasive techniques allow for measurement of the tissue in its living, undisturbed state. These techniques could provide in situ real-time values for the absorption, scattering, and index of refraction properties. The main advantage would be the independence of mechanical, thermal, and preservation influences. Additionally, this type of measurement can be applied to diagnostic imaging to reveal pathological tissue conditions.

These techniques are more difficult, as they are constrained to outer surface measurements, through fiber endoscopes in body openings or through reasonably thin tissue sections. The depth to which the tissue can be interrogated is a function of the wavelength of the incident light.

The dependence of the optical properties on wavelength allows for deep light penetration in the near-infrared and shallow penetration in the ultraviolet and mid-infrared. In these latter two wavelength ranges the attenuation is largely due to absorption being much greater than scattering. When the optical properties are derived from external observation of the light distribution, it is considered noninvasive. There are at least four ways of measuring the optical characteristics noninvasively. The four experimental methods all rely on either backscatter or diffuse transmission. Additionally, collected induced fluorescence can be used in combination to provide more deep tissue details. The backscatter can be collected by camera or fiberoptic probe, and similarly for diffuse transmission. The number of data points used to collect the signal will determine the maximum degree of accuracy attainable for the number of optical parameters that can be derived. In most cases there is only a single degree of freedom: radius to the source, which means that only one optical parameter can be determined with accuracy. This single parameter turns out to be the diffuse transport attenuation μ_{tr}.

7.2.2.1 Specular Reflectance Measurement

One relatively simple optical measurement relies on the specular reflection of bulk tissue. The specular reflection is solely a function of the index of refraction of the tissue target as discussed in Section 3.1.4 in Chapter 3. A complication, however, is introduced by surface roughness.

7.2.2.2 Diffuse Backscatter Measurement by CCD Camera Imaging

A narrow beam (typically a laser beam) incident on turbid medium and subsequent collection of the backscattered light from the medium by Charged Coupled Device (CCD) camera will give the light distribution as a function of the radius to the incident light. An experimental representation is given in Figure 7.7. The specular reflection from the incident beam will need to be eliminated to avoid saturation of the detector. Depending on the sensitivity of the CCD array [8 bit, 12 bit, 16 bit], the accuracy of the backscattered radiance can be interpreted with increased perceptiveness. Curve fitting of the radial distribution of the backscattered radiance reveals an exponential

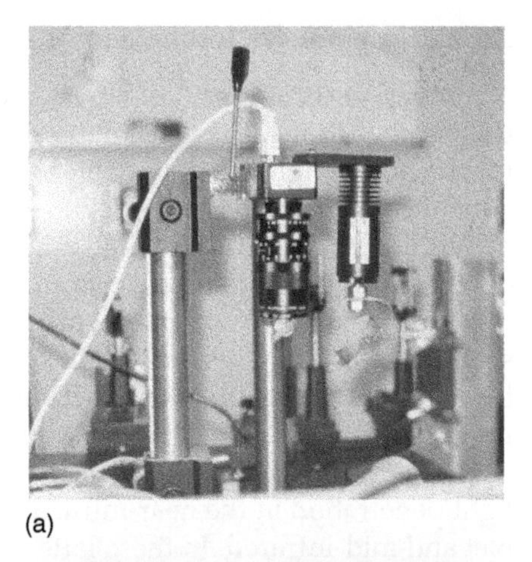

(a)

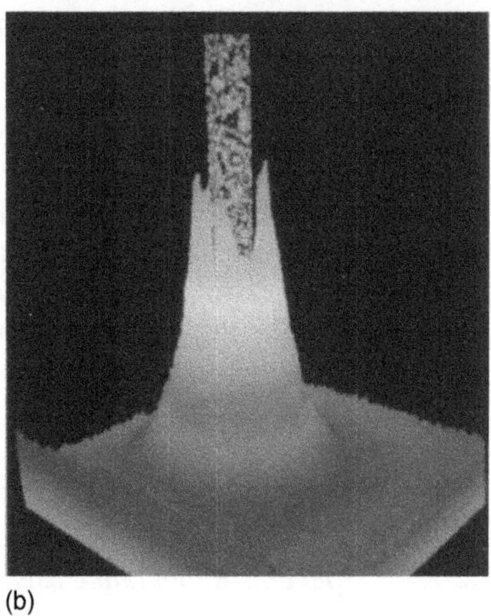

(b)

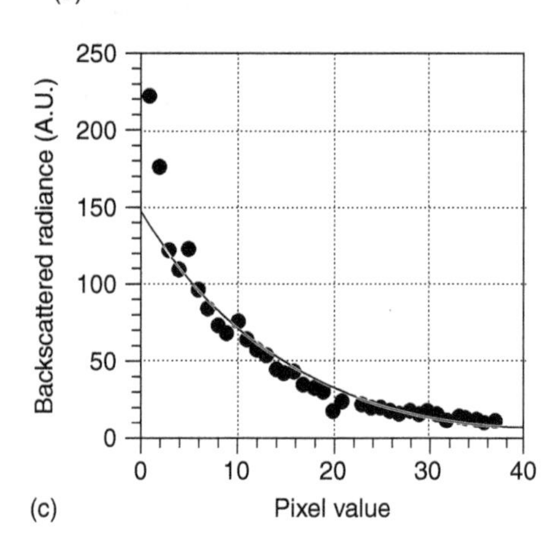

FIGURE 7.7
(a) Illustration of CCD measurement of optical properties; (b) representative image of backscattered light where both color and height are in indication of the collected fluence; and (c) normalized radial decay of backscattered radiance for coagulated myocardial muscle.

coefficient (μ_r). The backscatter radiance $R(r)$ as a function of radius r on the surface of the cylinder coordinates relative to the incident narrow beam. This can be shown to satisfy the following empirical relationship, primarily based on diffusion theory:

$$R(r) = \frac{a_0'}{4\pi}\left\{\frac{1}{\mu_t'}\left[\mu_{\text{eff}} + \frac{1}{c_1}\right]\frac{e^{-\mu_{\text{eff}}c_1}}{c_2^2} + \left[\mu_{\text{eff}} + \frac{1}{c_2}\right]\left(\frac{1}{\mu_t'} + \frac{4R_{\text{int}}}{\mu_t'}\right)\frac{e^{-\mu_{\text{eff}}c_{21}}}{c_2^2}\right\} \tag{7.59}$$

where $a_0' = [(1-g)\mu_s]/[\mu_a + (1-g)\mu_s]$ is the transport albedo; $\mu_t' = \mu_a + (1-g)\mu_s$ the reduced total attenuation coefficient; μ_s the effective attenuation coefficient; and c_1 and c_2 the relative locations with respect to an isotropic point source at approximately one optical mean free path length depth and a negative isotropic source at a strategic location above the surface, respectively, to satisfy all other boundary conditions. The locations c_1 and c_2 are defined as

$$c_1 = \sqrt{\left[\left(\frac{1}{\mu_t'}\right)^2 + r^2\right]} \tag{7.60}$$

and

$$c_2 = \sqrt{\left[\left(\frac{1}{\mu_t'} + \frac{4R_{\text{int}}}{3\mu_t'}\right)^2 + r^2\right]} \tag{7.61}$$

In a first-order approximation this can be characterized as

$$R(r) = C_1 \exp(-\mu_r r) \tag{7.62}$$

where C_1 is the maximum bit value of the camera, defined as 256 for an 8-bit camera. This exponential decay with radius can be proven to be the transport coefficient when the backscattered light is collected from a distance greater than three optical free path lengths away from the source. A combination of predetermined μ_a, μ_s, and g from in vitro measurements and in situ backscatter measurements by the CCD camera can be used for estimation of in situ optical properties. The optical effective attenuation coefficient, or transport coefficient μ_{eff} (in mm^{-1}) is defined as $\mu_{\text{eff}} = \sqrt{3\mu_a[\mu_a + (1-g)\mu_s]}$, thus giving $\mu_r = \mu_{\text{eff}}$.

Close to the source the radially measured reflectance is primarily the result of scatter, with low contributions from absorption ($D < 3\delta$, with D the mean free optical path length), thus providing mainly information on the reduced scattering coefficient, while at greater distances from the source the measurements are dominated by the effective attenuation coefficient. The assumption made is that the tissue is relatively homogeneous and sufficiently large to be considered as a bulk tissue with infinitive boundaries in terms of diffusion theory.

The absorption coefficient can now be expressed in terms of effective attenuation coefficient obtained from the proximal and distal radial decaying reflectance and the total diffuse reflectance. This way the reduced scattering coefficient and the absorption coefficient can be derived. In diffusion theory these two values are sufficient to provide a first-order approximation of the light distribution inside the tissue and perform dosimetry calculations, as well as rudimentary optical biopsy. Note that diffusion theory requires the reduced scattering coefficient to be greater than the absorption coefficient $\mu_s' \gg \mu_a$.

Additionally numerical solutions generated by Monte Carlo computer calculations of the radial light propagation, for various combinations of optical properties, can be used to find which sets of absorption, scattering, and g value will result in a theoretical decay coefficient (e.g., μ_C) of remitted light backward perpendicular to the surface of the medium, which is identical to the measured value. This may result in a finite number of combinations, which can be narrowed down based on prior invasive measurements of similar tissue. This, however, is not an exact method, but it will give a first-order approximation of the treatment protocol.

7.2.2.3 Diffuse Backscatter by Fiberoptic Area Measurement

A multiple fiber-based optical probe is used to measure optical properties rapidly over the wavelength range of interest. Figure 7.8 shows a schematic of the probe that is composed of five separate fibers. The contact surface of the probe is painted in black. The resulting boundary conditions allow the use of the diffusion theory for semi-infinite medium assuming air–tissue boundary. The fiber #1 is the source fiber through which the white spectrum is delivered to the tissue. The diffusely reflected photons are collected by fibers 2, 3, 4, and 5 at increasing distances from the source fiber, respectively. The data from the detector fibers are delivered to the spectrograph where the attenuation spectrum is plotted as a function of wavelength. After calibration of the data using phantoms with known optical properties, the optical properties can be deduced by fitting the data to diffusion theory for each wavelength, similar to the method described in the preceding section. This

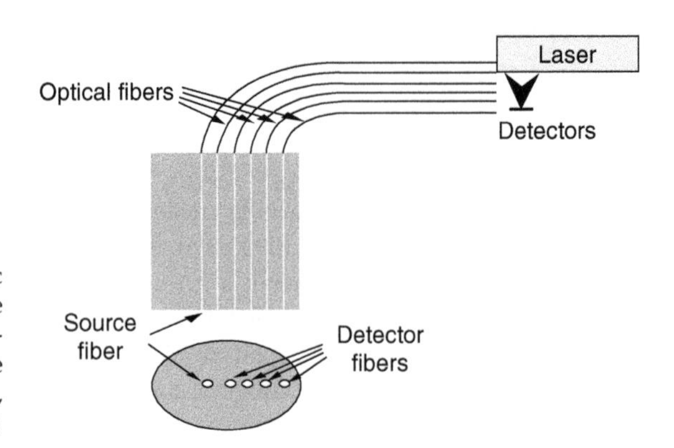

FIGURE 7.8
Probe design for fiberoptic reflectometry to measure tissue optical properties of tissue volume spanning the width of the fiber array. (After Bob-Min Kim, Yonsei University, Seoul, Korea.)

method provides the optical information relevant to the immediate surface area; however, no depth information can be obtained.

7.2.2.4 Diffuse Backscatter Measurement by Interferometric Probing

A method similar to optical coherence tomography can be used to obtain the light decay as a function of depth inside the tissue, as outlined in Figure 7.9. Optical coherence tomography will be discussed in detail in Chapter 14. The light decay curve directly underneath the delivery fiber is measured as a function of depth by adjusting the length of the reference arm of the interferometer. This light decay curve is directly dependent on the combinations of optical properties in the path of the beam.

The normalized radiant intensity as a function of depth measured with a single-mode fiber interferometer needs to be correlated to computer simulations of various combinations of optical properties to find the best curve fit, and hence the closest matching combination of optical properties. This is an elaborate method; usually the starting parameters are obtained from in vitro measurements.

7.2.2.5 Diffuse Forward Remittance Measurement

When a thin enough tissue section is available, a diffuse transmission measurement can be applied yielding the Beer–Lambert law of attenuation. This method can directly provide the effective attenuation coefficient from the input and output signal from the tissue combined with the sample thickness. Other applications of this principle are the use of cuvettes for the measurement of biological fluids and the initial invasive measurement on a slab of tissue described in Section 7.2.1.1.

7.2.2.6 Diffuse Forward Scatter Measurement by CCD Camera Imaging

A Mach–Zehnder interferometer can be used for relative refractive index measurement as illustrated in Figure 7.10. The changes in refractive index of

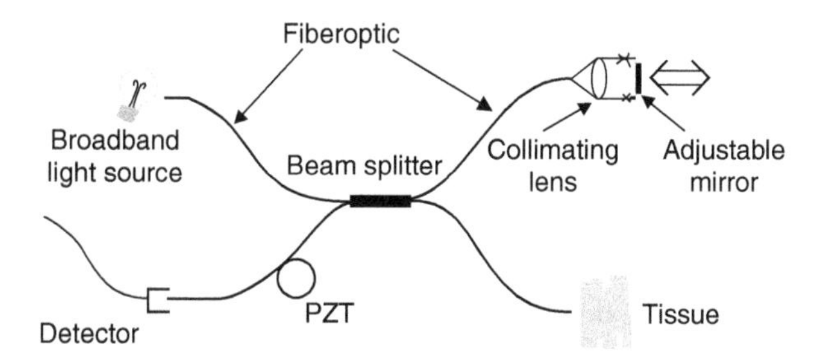

FIGURE 7.9
Schematic of fiberoptic Michelson interferometer used to measure the optical decay pattern as a function of depth resulting from the passage through tissues with characteristic sets of optical properties. Amplitude modulation is acchieved by wrapping the fiber optic around a piezoelectric transducer (PZT).

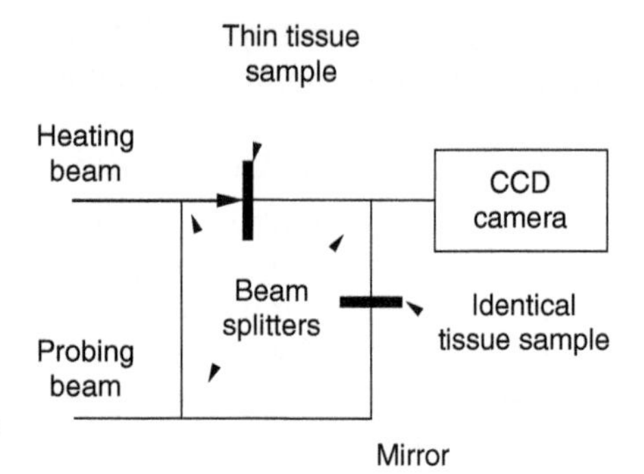

FIGURE 7.10
Mach–Zehnder interferometer for dynamic refractive index measurement.

the tissue by temperature rise will induce the phase changes that are directly correlated to the light velocity and refractive index changes. The interferometric fringe patterns can be quantified using a CCD camera so that two-dimensional phase distribution could be analyzed. The time delay between the heating beam and the probing beam can be adjusted using a delay generator to monitor the time-dependent phase changes. An identical tissue sample will be placed on the other arm of the interferometer so that the phase is matched before laser heating.

7.2.2.7 Time-Resolved Reflectance and Transmittance Measurement

Elaborating on the solution to a finite light source and the diffuse spatial light distribution, the time element can be used as an additional discriminator.

Using the time-dependent diffusion equation in spherical geometry with an isotropic point source, a solution can be found:

$$\frac{1}{c}\frac{\partial L(\mathbf{r},t)}{\partial t} - \frac{\nabla^2 L(\mathbf{r},t)}{3[\mu_a + (1-g)\mu_s]} + \mu_a L(\mathbf{r},t) = S(\mathbf{r},t) \tag{7.63}$$

where c is the speed of light in the medium.

Equation (7.63) has a solution in the form of a Green's function for an infinite medium as given by

$$L(\mathbf{r},t) = c\left[\frac{3[\mu_a + (1-g)\mu_s]}{4\pi ct}\right]^{3/2} \exp\left(-\frac{r^2 3[\mu_a + (1-g)\mu_s]}{4ct} - \mu_a ct\right) \tag{7.64}$$

using the assumption that photons incident on the outside of the medium are initially scattered at the depth given by

$$z = \left[\frac{1}{(1-g)\mu_s}\right] \tag{7.65}$$

The term $\mu_a+(1-g)\mu_s$ can be substituted by the definition of the transport coefficient:

$$\mu_{tr}=\mu_a+(1-g)\mu_s \tag{7.66}$$

Equation (7.64) can then be rewritten as

$$L(\mathbf{r},t)=c\left[\frac{3\mu_{tr}}{4\pi ct}\right]^{3/2}\exp\left(-\frac{r^2 3\mu_{tr}}{4ct}-\mu_a ct\right) \tag{7.67}$$

The source term in Equation (7.63) reduces in this case to a delta function at the origin:

$$S(\mathbf{r},t)=\delta(0,0) \tag{7.68}$$

A second assumption is that the boundary conditions force the solution (Equation [7.64]) to be zero at the interface of the medium to the outside as shown in the following equation:

$$L(\mathbf{r},t)=0 \quad \text{at} \quad z=0 \tag{7.69}$$

This condition can be met by introducing a negative source in the geometry at the mirror location of the first source.

Reformatting Equation (7.67) in cylindrical coordinates, as a substitute of spherical coordinates, including both sources (primary and negative mirrors) will give access to use this solution for external measurements of optical properties:

$$\begin{aligned}L(\mathbf{r},\mathbf{z},t)=c\left[\frac{3\mu_{tr}}{4\pi ct}\right]^{3/2}\exp(-\mu_a ct)\left\{\exp\left[\frac{3\mu_{tr}[(z-z_0)^2+r]}{4ct}\right]\right.\\ \left.-\exp\left[\frac{3\mu_{tr}[(z+z_0)^2+r]}{4ct}\right]\right\}\end{aligned} \tag{7.70}$$

Using Fick's law to find the total photon flux $J(\varsigma, 0, t)$ reaching the surface of the medium through backscatter gives

$$J(\mathbf{r},\mathbf{0},t)=-\frac{1}{3\mu_{tr}}\nabla L(\mathbf{r},\mathbf{z},t)_{z=0} \tag{7.71}$$

The backscattered light leaving the tissue surface can now be defined as

$$R(\mathbf{r},t)=|J(\mathbf{r},\mathbf{0},t)|=-\left(\frac{3\mu_{tr}}{4\pi c}\right)^{3/2} z_0 t^{-5/2}\exp(-\mu_{tr}ct)\exp\left(-\frac{(r^2+z_0^2)3\mu_{tr}}{4ct}\right) \tag{7.72}$$

When $r^2 \gg z_0^2$ the backscatter equation can be approximated for infinite time as

$$\frac{\mathrm{d}}{\mathrm{d}t}\ln[\mathrm{R}(\mathbf{r},t)] = -\frac{5}{2t} - \mu_\mathrm{a}c - \left(\frac{3\mu_\mathrm{tr}r^2}{4ct^2}\right) \tag{7.73}$$

and

$$\lim_{t\to\infty}\frac{\mathrm{d}}{\mathrm{d}t}\ln\left[R(\mathbf{r},t)\right] = -\mu_\mathrm{a}c \tag{7.74}$$

Equation (7.74) shows that the absorption coefficient will follow from the slope of the $\ln[R(\mathbf{r},t)]$ versus time curve. The reduced scattering coefficient follows from the same graph by solving the following equation:

$$(1-g)\mu_\mathrm{s} = \frac{1}{3r^2}(4\mu_\mathrm{a}c^2t_{\max}^2 + 10ct) - \mu_\mathrm{a} \tag{7.75}$$

using the fact that the logarithmic curve of the backscatter signal levels off after a finite time period: $t_{\max}$.

In a simplified approximation Equation (7.74) can be written as

$$R(\mathbf{r},t) = K_1 t^{-5/2}\exp(-\mu_\mathrm{a}ct/n)\exp(-r^2/r_0^2) \tag{7.76}$$

or in the time domain:

$$R(t) = K_2 t^{-3/2}\exp(-\mu_\mathrm{a}ct/n) \tag{7.77}$$

The time release of photons will have an initial peak, after which it decays according to $t^{-5/2}\exp(-\mu_\mathrm{a}ct/n)$ or $\exp(-\mu_\mathrm{a}ct/n)$.

For the time period after the peak, Equation (7.77) can be rewritten in μ_a, after differentiation, to yield

$$\mu_\mathrm{a} = -\frac{n}{c}\left[\frac{\mathrm{d}\ln(R(t))}{\mathrm{d}t} - \frac{3}{2t}\right] \tag{7.78}$$

Similarly,

$$\mu_\mathrm{a} = -\frac{n}{c}\left[\frac{\mathrm{d}\ln(R(\mathbf{r},t))}{\mathrm{d}t} - \frac{5}{2t}\right] \tag{7.79}$$

In contrast, the peak rate of photon escape will occur at the measured time $t_{\max}$, yielding the following expression for the reduced scattering coefficient:

$$\mu_\mathrm{s}' = \frac{1}{3\zeta^2}\left[\frac{4\mu_\mathrm{a}c^2t_{\max}}{n^2} + \frac{10ct_{\max}}{n}\right] \tag{7.80}$$

Combining all the information shows that the absorption coefficient and the reduced scattering coefficient can be derived from time-resolved measurement of backscatter sufficiently far away from the source to eliminate source contributions compared with the values obtained close to the incident light, under the condition that only diffuse light is measured.

A similar derivation can be applied for diffuse transmission measurements.

7.3 Temperature Effects of Light–Tissue Interaction

At each step of the random walk, the photon loses energy in the tissue. The amount of energy loss is based on the ratio of the absorption coefficient and the total attenuation coefficient. This energy is converted into heat using the bioheat equation.

The heat transfer Q at any given point in the tissue is directly proportional to the local temperature gradient $\Delta T/\text{length}=[(d/dx)+(d/dx)+(d/dx)]T=\nabla T$, as shown in the following equation:

$$Q=-k_{\mathrm{T}}\left(\frac{\mathrm{d}}{\mathrm{d}x}+\frac{\mathrm{d}}{\mathrm{d}x}+\frac{\mathrm{d}}{\mathrm{d}x}\right)T=-k_{\mathrm{T}}\,\nabla T \tag{7.81}$$

where T is the temperature in K and k_{T} the thermal conductivity, on average 0.5 W/m K for soft tissues.

The process of heat transfer is time-dependent and the following energy balance can be written:

$$\rho c_{\mathrm{v}}\frac{\partial T}{\partial t}=\nabla(k_{\mathrm{T}}\,\nabla T) \tag{7.82}$$

where ρ is the tissue density and for most tissues, it is equal to 1.04×10^{3} kg/m^3, and c_{v} is the heat capacity, approximately, on average, 3.7×10^{3} J/kg K.

Including a heat source q in Equation (7.82) resulting from absorption of light gives

$$\rho c_{\mathrm{v}}\frac{\partial T}{\partial t}=\nabla(k_{\mathrm{T}}\,\nabla T)+q \tag{7.83}$$

where q equals the rate of energy deposited in a volume of tissue cells, which is the light distribution (the local fluence rate) times the local absorption coefficient, i.e., $\mu_{\mathrm{a}}\psi(r,\theta,z)$.

Additionally, heat transfer resulting from blood perfusion usually needs to be included, since there are a significant number of blood vessels in all living tissues. Including blood perfusion in the energy balance for heat will need to account for the perfusion flow rate of the blood ω_b in m^3/kg s, the density of the blood ρ_b, the specific heat of blood c_b, and the temperature difference between the blood and the surrounding tissue $T-T_b$, to yield a heat loss expressed as

$$-\omega_b \rho_b c_b \rho (T - T_b) \tag{7.84}$$

7.3.1 The Bioheat Equation

Combining all the features described, the following function that describes the tissue temperature evolution as a function of time expression can be obtained:

$$\begin{aligned} \frac{\partial T}{\partial t} &= \frac{1}{\rho c_v}\left[q + k_T\left(\frac{\partial^2}{\partial x^2} + \frac{\partial^2}{\partial y^2} + \frac{\partial^2}{\partial z^2}\right)T - \omega_b \rho_b c_b \rho (T - T_b)\right] \\ &= \frac{1}{\rho c_v}\left[k_T \nabla^2 T - \omega_b \rho_b c_b \rho (T - T_b) + q\right] \end{aligned} \tag{7.85}$$

In the process of tissue heating other phenomena will take place, such as coagulation and evaporation, which are both heat sinks as well.

The process of coagulation encompasses a phase transition that has a latent heat h_{coag}, expressed in J/kg, and is tissue-specific.The heat loss due to coagulation, Q_{coag}, depends on the coagulated tissue volume V_t and the tissue density ρ_t. These on combination provide the coagulated tissue mass, which is expressed as

$$Q_{coag} = h_{coag} \rho_t V_t \tag{7.86}$$

The amount of heat loss due to evaporation can be described primarily by the rate of water loss.

The heat loss due to evaporation is now the rate of evaporation times the latent heat of evaporation; for water it is equal to $h_{ev,w} = 2.26 \times 10^6$ J/kg:

$$Q_W = V_W(\rho_{W,\infty} - \rho_{W,S}) h_{ev,W} \tag{7.87}$$

where V_w is the volume of evaporated water, $\rho_{w,s}$ the mass density of saturated water vapor at the surface of the vapor bubble, and $\rho_{w,\infty}$ the vapor mass density outside the tissue or cell.

One noteworthy fact is that the water vapor is usually created inside a cell membrane, which causes an increased pressure, ultimately changing the parameters in the equation.

Generally the metabolic heat generation can be neglected. Integration over a volume of tissue gives for the bioheat equation

$$V\rho c_{\mathrm{v}}\frac{\partial T}{\partial t}=\iint_{s}k_{\mathrm{T}}\,\nabla^{2}T\times\mathbf{n}\cdot\mathbf{ds}-V\omega_{\mathrm{b}}\rho_{\mathrm{b}}c_{\mathrm{b}}\rho(T-T_{\mathrm{b}})+qV-\dot{Q}_{\mathrm{W}}-\dot{Q}_{\mathrm{coag}} \quad (7.88)$$

where the volume of heated tissue has a surface area s, and $\mathbf{n}$ is the normal to the surface of the area, and $\dot{Q}_{\mathrm{w}}$ and $\dot{Q}_{\mathrm{coag}}$ are the rate of heat loss due to evaporation and coagulation, respectively.

7.3.2 Computer Modeling of Absorption, Heat Dissipation, and Photocoagulation

The bioheat equation can be solved numerically by integrating it over time to produce the temperature distribution over a period of laser irradiation.

For a volume element, voxel v_{ijk} in the tissue, the temperature calculation will be as described by the following equations for the three Cartesian directions:

$$\frac{\partial^{2}T}{\partial x^{2}}=\frac{(T_{i+1}-T_{i}+T_{i-1})}{(i+1)-i[i-(i-1)]} \quad (7.89)$$

$$\frac{\partial^{2}T}{\partial y^{2}}=\frac{(j_{i+1}-T_{j}+T_{j-1})}{(j+1)-j[j-(j-1)]} \quad (7.90)$$

$$\frac{\partial^{2}T}{\partial z^{2}}=\frac{(T_{k+1}-T_{k}+T_{k-1})}{(k+1)-k[k-(k-1)]} \quad (7.91)$$

Following the numerical process through will provide the change in temperature due to energy deposition by laser irradiation in tissue while accounting for all other loss processes.

7.3.3 Damage Integral

Photocoagulation or irreversible tissue damage can be calculated using the so-called damage integral. The damage integral calculates the loss of normal tissue to the denatured state as expressed in the following equation, which is a standard reaction equation:

$$\Omega(t)=\ln\left(\frac{[N_{\mathrm{total}}]}{[N_{\mathrm{total}}]-[N_{\mathrm{denatured}}]}\right)=A\int_{0}^{t}\exp\left(\frac{-E}{RT}\right)\mathrm{d}t \quad (7.92)$$

where R is the universal gas constant (=8.314 × 10^3 J/kmol K), A the rate constant (=3.1 × 10^{98} s^{-1}), E the activation energy for burn injury (=6.28 × 10^5 J/mol), T the temperature of the tissue obtained from the absorption of light, $[N_{total}]$ the total number of cells in the tissue volume being irradiated, and $[N_{denatured}]$ the number of denatured cells after laser irradiation.

7.3.4 The Damage State

The damage state D can be determined for exposure to laser irradiation integrated over various discrete time intervals. Equation (7.93) illustrates the damage calculation for time intervals of 0.1 s:

$$D = \Omega(0.1) = A\exp\left(\frac{-E}{RT}\right)(0.1) \tag{7.93}$$

The total amount of damage follows from repeating the summation successively over the total time of laser irradiation, taking into account the cumulative effect up to the stage before the next summation. The cumulative effect means that the damage integral needs to be adjusted for the temperature evolution at each discrete time frame.

The following situations can generally be distinguished:

- $0 \leq D \leq 0.53$: Tissue is in normal state, unharmed by the effects of heat.
- $0.53 < D < 1$: Tissue is coagulated.
- $D \geq 1$: Tissue will be carbonized.

The thermal damage process regarding therapeutic laser applications will be discussed in detail in Chapter 8.

7.4 Summary

As a follow-up on Chapter 6 the numerical method to solve the equation of radiative transfer was presented and the methodology available to find the optical characteristics that are needed to describe the light propagation inside turbid tissue media was outlined. Both invasive and noninvasive techniques for the determination of the optical parameters were discussed. Additionally, the technique for calculating the tissue damage resulting from light exposure was presented.

Further Reading

Carter, L.L., Horak, H.G., and Sandford, M.T. II, An adjoint Monte Carlo treatment of equations of radiative transfer for polarized light, *J. Comp. Phys.*, 26, 119–138, 1978.

Egan, W.G., and Hilgeman, T., Integrating sphere for measurements between 0.185 μm and 12 μm, *Appl. Optics*, 14, 5, 1137–1142, 1975.

Groenhuis, R.A.J., Ferwerda, H.A., and Ten Bosch, J.J., Scattering and absorption of turbid materials determined from reflection measurements, 2: Measuring method and calibration, *Appl. Optics*, 22, 2463–2467, 1983.

Miller, O.E., and Sant, A.J., Incomplete integrating sphere, *J. Opt. Soc. Am.*, 48, 11, 828–831, 1958.

Nichols, M.G., Hull, E.L., and Foster, T.H., Design and testing of a white-light, steady-state diffuse reflectance spectrometer for determination of optical properties of highly scattering systems, *Appl. Optics*, 36(1), 93–104, 1997.

Niemz, M.H., *Laser-Tissue interaction*, Springer, Heidelberg, 1996.

Patterson, M.S., Schwartz, E., and Wilson, B.C., Quantitative reflectance spectrophotometry for the noninvasive measurement of photosensitizer concentration in tissue during photodynamic therapy, *Proc. SPIE*, 1065, 115–122, 1989.

Welch, A.J., van Gemert, M.J.C., Star, W.M., and Wilson, B.C., *Definitions and Overview of Tissue Optics, Optical-Thermal Response of Laser-Irradiated Tissue*, Welch, A.J. and Van Gemert, M.J.C., Eds., Plenum Press, New York, 1995, pp. 15–46.

Zaidi, H., *The Monte Carlo Method: Theory and Computational Issues*, Institute of Physics Publishing Company, London, 2003.

Problems

1. A collimated laser beam with a radiance of 25 mW is sent through a scattering medium with negligible absorption at the wavelength used (e.g., Intralipid). The laser beam is sent through an aperture of 1 mm for consistency, while at 50 cm from the sample another 1 mm pinhole is placed in front of a power meter. The sample is 3.5 cm thick assuming no scattering takes place in the container wall; however, there is a glass–air interface both at entry and exit. The collimated transmission was measured through a small aperture to be 0.0025 mW. Calculate the scattering coefficient.
2. A bare neodimium:YAG laser operating at 1064 nm with a Gaussian beam is to be used for performing a portwine stain treatment, using the first-order approximation solution to the radiance as $L(r, z)=L_0\exp(-\mu_t z)\exp[-2(r/\omega_b)^2]$, where $\mu_t=\mu_a+\mu_s$. Under what conditions will this equation hold true? In case of portwine stain treatment there are several tissue layers involved as outlined in Figure 7.11. At the center of the beam ($r = 0$) give the expression of the energy fluence rate as a function of depth and graph L and the rate of heat generation Q in the tissue by the laser, $L(r = 0, z)$ and $Q(r = 0, z)$. Assume $L_0 = 100$. The absorption and scattering coefficients of each skin layer and blood vessel for this wavelength are given in Table 7.1.
3. A biological tissue is given with an absorption coefficient $\mu_a = 0.1\ \text{cm}^{-1}$, since the scattering coefficient and scattering anisotropy factor are not known, these two are combined as reduced scattering coefficient. Using the diffusion approximation, draw in a single graph the logarithm of relative radiance versus radius for four different values of reduced scattering coefficient μ_a', 10, 50, and 100 cm^{-1}, for r ranging from 0.1 to 2 cm.
   a. What conclusions can be drawn from the effects of scattering coefficient and scattering anisotropy factor on the light distribution profile and the effect of the reduced scattering coefficient on the light distribution?

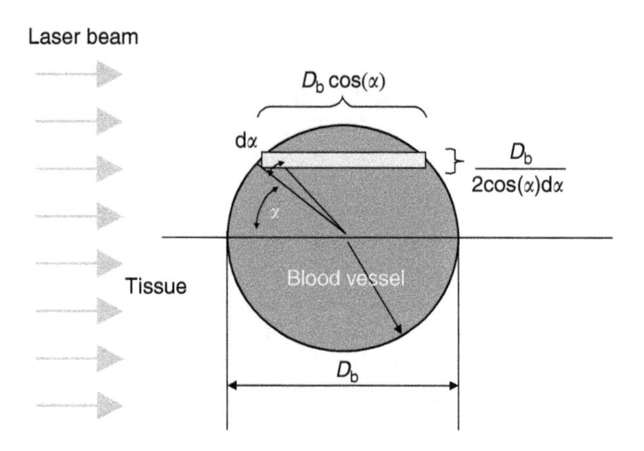

FIGURE 7.11
Illustration of blood vessel under skin layers representing the situation for portwine stain treatment.

TABLE 7.1
Characteristic Optical Properties of Several Tissues Involved in the Treatment Targeting of Portwine Stain with Dye Laser

Tissue	μ_a (cm^{-1})	μ_s' (cm^{-1})
Blood	20	20
Epidermis	0.1	10
Dermis	0.5	10

b. Is it necessary to know the scattering coefficient and scattering anisotropy factor independently to make an informed decision on the impact of the light–tissue interaction?

4. In the visible and near-infrared region of the optical spectrum between 600 and 1100 nm there are well-documented absorption peaks for oxygenated and deoxygenated hemoglobin. Suppose you want to calculate the following three parameters: the concentrations of oxyhemoglobin and de-oxyhemoglobin, and the relative oxygenation of the blood. At wavelengths 758, 798, and 898 nm, the extinction coefficients for oxygenated hemoglobin are 0.444, 0.702, and 1.123 mm^{-1} and for de-oxygenated hemoglobin 1.533, 0.702, and 0.721 mm^{-1}, respectively. The total absorption coefficients at these three wavelengths in the tissue can be measured using a time-resolved system as 0.293, 0.1704, and 0.1903 cm^{-1}, respectively. Measuring through the finger with a thickness of 1 cm, calculate the relative oxygenation of the tissue, assuming negligible influence of the absorption of the tissue itself at these three wavelengths.
5. What does the Monte Carlo model of light transport in tissue simulate? What two key decisions govern the movement of a photon in tissue? How many photons/s is 1 W of 555 nm radiation?
6. Does the number of photons derived in question 5 have any impact on the choice of the number of photons that should be used in Monte Carlo simulations?

7. What is the condition on the pulse duration τ that must be met for a photothermal interaction to be thermally confined?
8. A 1 mm fiberoptic probe is used to perform a therapeutic procedure on a cancerous lesion in the brain. The radiant energy fluence rate required to excite a particular photosensitizer in the tumor is 0.5 W/cm^2. The laser light is directed out of the tip of a fiber optic ($n_1 = 1.7$) to the tumor, located 2 mm from the fiber tip, and is 1 mm in diameter. The optical properties of the normal brain tissue and tumor are given as $n_{norm} \approx 1.1$, $\mu_{s\_norm} = 5\,cm^{-1}$ and $\mu'_{s\_norm} = 10\,cm^{-1}$, and $n_{tumor} \approx 1.1$, $\mu_{s\_tumor} = 25 cm^{-1}$ and $\mu'_{s\_tumor} = 10\ cm^{-1}$, respectively. You do not want to ablate the tissue for fear of spreading the cancer. So, the ablation threshold of this tissue must not be reached. Instead, since you want to perform selective photodynamic therapy (PDT), you need the photochemical interaction to take place before the photothermal interaction.
   a. Have you been successful?
   b. Calculate the damage function Ω associated with the final tumor tissue temperature? Assume that $A = 10^{81}\,s^{-1}$ and $E = 5 \times 10^5 J/mol$.
   c. What is the critical temperature T_{crit} for the photothermal damage process?
9. What are some of the problems with direct measurements of tissue optical properties? What are some of the limitations with indirect measurements of tissue optical properties?
10. What does the Monte Carlo model of light transport in tissue simulate? Which two key decisions govern the movement of a photon in tissue?
11. You are asked to model an endoscopic laser therapy. The radiant energy fluence rate required to excite a particular photosensitizer in the cancerous lesion is 0.75 W/cm^2. The laser light is diffusely transmitted out of the tip of a second sapphire fiber in a semi-infinite geometry. The tumor is located 1 mm from the fiber tip. The optical properties of the cancer tissue at this wavelength are given as $\mu_a = 3 cm^{-1}$ and $\mu'_s = 15 cm^{-1}$. You do not want to ablate the tissue for fear of spreading the cancer. So, the ablation threshold of this tissue must not be reached. Instead, since you want to perform selective photodynamic therapy (PDT), you want the photochemical interaction to take place before the photothermal interaction.
   a. Calculate the fluence rate Ψ at $z = 1mm$.
   b. For a time duration of 1 s, will the final tissue temperature be above 100°C? Consider the temperature rise to be solely due to conduction.
   c. Calculate the damage function Ω associated with the final tissue temperature? Assume that $A = 10^{83}\,s^{-1}$ and $E = 5 \times 10^5\,J/mol$.
   d. What is the critical temperature T_{crit} for this process?
12. You have the choice of a "top-hat" or a Gaussian beam profile to perform tissue ablation.
   a. Which beam profile is better for tissue ablation, a Tophat or Gaussian profile?
   b. Explain why.
13. A Monte Carlo simulation has as its input light field a "top-hat" spatial beam. Determine the selection rule for the radial position r from the probability density $p(r)$ using a random number for a flat-field beam with radius w, total energy 1 J, and irradiance given by Table 7.2.
14. A simplified model for laser surgery assumes a laser beam whose energy is absorbed in tissue and converted into heat. When the temperature of the heated area is above the vaporization point, the tissue is ablated. What three effects are

TABLE 7.2

Beam Profile for Monte Carlo Simulation

	Value	Location
Irradiance r	$1/\pi w^2$	$r \leq w$
Irradiance r	0	$r > w$

neglected in this model? What are the optimal parameters for laser surgery with little thermal damage?

a. Show that Equation (7.72) can be written as Equation (7.76).
b. Derive Equations (7.77) and (7.78).

15. A 10 mJ laser is absorbed in myocardium with an absorption coefficient of $\mu_a = 140\ \text{cm}^{-1}$. The beam area is $A = 1\ \text{mm}^2$. The initial temperature of the tissue is 37°C.
    a. What is the final temperature after laser exposure?
    b. What process category does this interaction fall into?
    c. What histopathological effect is seen in myocardium?
    d. Calculate the damage function Ω for this interaction. The frequency factor A is $10^{76}\ \text{s}^{-1}$, the time duration is 0.742 s, and the energy per mole E is 5×10^5 J/mol.
16. You are asked to model the thermal profile at an optic–tissue interface during a laser procedure and compare that to the thermal profile produced at an air–tissue interface. Plot the temperature profile as a function of position from the interface at $z=0$ and time t. Assume that the sapphire optic is lossless and that the tissue has an absorption coefficient of $\mu_a=100\text{cm}^{-1}$. The laser radiant exposure is 10 J/cm². Assume the thermal parameters of the tissue to be those of water (appropriate for a water-rich tissue). The thermal diffusivity α_2 and thermal conductivity κ_2 of the sapphire optic are given by $\alpha_2=15\times10^{-1}\text{m}^2/\text{s}$ and $\kappa_2=33\text{W/m°C}$, respectively. Will the tissue be ablated by this laser pulse?

8

Light–Tissue Interaction Mechanisms and Applications: Photophysical

As discussed in the previous chapters, generally three different groups of interaction of light with biological media can be distinguished: photophysical, photochemical, and photobiological. These distinctions are made based on either the physical processes or on the ultimate result. This chapter presents an overview of the photophysical interaction mechanisms, and the research into the underlying dynamics.

All interactions rely on the ease or difficulty with which light can be delivered to a specific location in a biological medium, particularly at greater depths. The optical properties of the target volume may not necessarily be the same as those of the medium the light will need to pass through to effectively deposit the crucial amount of light energy in the target tissue for the desired effect. This will impose certain restrictions on the depth of treatment and diagnosis under the constraint of causing minimal peripheral damage to achieve the desired goal.

8.1 Range of Photophysical Mechanisms

The photophysical interaction of light with biological media for both diagnosis and treatment can be subdivided in the following seven subcategories:

1. Photoablative interaction: vaporization; photodisruption
2. Photoacoustic or electromechanical interaction
3. Spectroscopy
4. Birefringence
5. Interference
6. Polarization
7. Evanescent wave interaction

The first and second mechanisms can be characterized as being photothermal in nature; the remaining five types of interaction are mainly nonthermal, although the evanescent wave principle can be applied with thermal consequences.

8.2 Photoablation

The laser method that is probably ranked highest on the list of practical applications will be photoablation. Some of the advantages of the use of laser for treatment of certain medical conditions are the following:

- Fiberoptic delivery offers minimally invasive microscopic manipulation during arthroscopic and orthoscopic procedures.
- Wound sterilization due to laser heat.
- Tissue cutting while eliminating bleeding and edema due to the cauterizing of high-powered laser light.

The majority of clinical applications of light utilize the conversion of light energy into thermal energy to achieve its goal. The respective procedures can be grouped according to the respective temporal method: continuous wave, pulsed or superpulsed. Another way of listing the applications can be by ranking the actual temperature rise accomplished. Some of the thermal applications can be arranged in the ultimate effect, such as thermal expansion or plasma formation. In the latter case the light energy is ultimately converted into mechanical energy. One distinct application of pulsed laser–tissue interaction used to accomplish tissue vaporization is called photodisruption. Photodisruption will be discussed as well.

Particularly during photoablation the wavelength is usually in a part of the spectrum where absorption is much greater than scattering. This can easily be on the water absorption spectrum or on the absorption peaks of certain prevalent chromophores in the organ that is targeted; proteins or lipids, for example.

The fact that absorption is greater than scattering simplifies the theoretical description of the light–tissue interaction significantly. As a result the attenuation will depend solely on absorption, and there is no concern for the reintroduction of scattered fluence into the path of the beam. The radiance $L(\mathbf{r},t)$ will decay proportionally to the incident radiance and the absorption coefficient:

$$dL \propto \mu_a L_0 \tag{8.1}$$

Additionally, the laser delivery protocol has a one-dimensional boundary condition, yielding a one-dimensional solution:

$$L(z,t) = L_0 e^{-\mu_a z} f(t) \tag{8.2}$$

where $f(t)$ represents the temporal profile of the delivery protocol.

Technically, there will be reflection at the outside interface between the medium the laser beam is delivered from and the tissue that is ablated by absorption, reducing the radiance by a factor $(1-R)$, with R being the reflection coefficient, Fresnel's reflection squared (as described in Section 3.1), for normal incidence:

$$R = \frac{(n_2 - n_1)^2}{(n_1 + n_2)^2} \tag{8.3}$$

Thus,

$$L'(z,t) = (1-R)L(z,t) = (1-R)L_0 e^{-\mu_a z} f(t) \tag{8.4}$$

In addition, the beam profile can be either Gaussian for a free-space beam or top-hat in the near-field of fiberoptic irradiation. This gives for the beam,

$$L'(r,z,t) = (1-R)L_0 f(t) e^{-r^2/2w^2} e^{-\mu_a z} \tag{8.5}$$

where t denotes the time dependence.

For the near-field fiber top-hat irradiation this becomes

$$L(r,z,t) = L_0 f(t) e^{-\mu_a z} \quad \text{for } r < R^* \tag{8.6}$$

and

$$L(r,z) = 0 \qquad \text{for } r > R^* \tag{8.7}$$

where R^* is the radius of the fiber.

At this point fluence and radiance have the same meaning, since there is only a one-dimensional stream of photons, yielding for the fluence Ψ:

$$\Psi(\mathbf{r},t) = \int_{4\pi} L'(\mathbf{r},\mathbf{s},t) d\omega = L'(\mathbf{r},\mathbf{s},t) \tag{8.8}$$

8.2.1 Ablation Threshold

Assuming that for certain threshold fluence Ψ_{th}, enough energy is deposited in the tissue to achieve vaporization, this threshold is given by

$$E_{abl} = \int_{4\pi}\int_t \mu_a L'(\mathbf{r},\mathbf{s},t) dt\, d\omega = \int_t \mu_a \Psi(\mathbf{r},t) dt = m(h_v + c_v \Delta T) \tag{8.9}$$

where ΔT is the temperature rise from steady state to vaporization temperature, c_v the specific heat of the medium, h_v the latent heat of vaporization for the medium, and m the mass that has been evaporated.

8.2.2 Pulsed Laser–Tissue Interaction

Under pulsed light–tissue interaction the equation of radiative transfer can be solved with the following modification: The light source function can be developed in a Fourier series to accommodate the pulse nature of the source. Any periodic function with period 2π can be developed in a series of periodic functions. The Fourier series development is illustrated by

$$Y \rightarrow \sum e^{iny} \tag{8.10}$$

This series is a development in the harmonics of the base function e^{iy}. For a real-valued function this expression can be written as an equivalent sum of sine and cosine functions on the basis of the Euler formula, which states that the complex function is equivalent to

$$e^{iy} = \sin(y) + i\cos(y) \tag{8.11}$$

This is based on the fact that the exponential function e^y can be approximated by the linear function $1+y$ when y is very small.

The Fourier series now becomes a sum of sine and cosine wave functions that, by applying the superposition principle, closely approximate the true duty cycle and delivery protocol of the original light source:

$$f(y) = \frac{a_0}{2} = \sum_{n=1}^{\infty}\left[a_n \cos(ny) + b_n \sin(ny)\right] \tag{8.12}$$

where

$$a_n = \frac{1}{\pi}\int_{-\pi}^{\pi} f(y)\cos(ny)\,dy$$

and

$$b_n = \frac{1}{\pi}\int_{-\pi}^{\pi} f(y)\sin(ny)\,dy$$

Depending on the repetition rate of the pulsed laser and the pulse width, the number of terms in the Fourier series will need to be selected to find the closest match, while maintaining minimal error conditions. For a 50% duty cycle pulsed laser, a monochromatic sine wave term may be sufficient as a source.

However the sine wave will not give a sharp step-function delivery as a pulsed laser would. A pulsed laser will in most cases have a top-hat temporal profile. When moving to very short pulse-duration delivery, a Dirac-delta distribution may be more appropriate as a source. A coarse estimation of which choice is more suited to the substitution can be envisioned by calculating the "photon-cloud" overlap of subsequent pulses in the time-of-flight time frame of the pulse-pause duration.

Depending on the target geometry and the beam diameter, it may be beneficial to develop the source function in spherical, cylindrical, or Cartesian coordinates. The beam profile will also need to be included, most likely the Gaussian spatial distribution, $\exp[-2(r/w)^2]$, where w is the waist of the laser beam.

8.2.3 Pulsed Vaporization

Assume that a laser with irradiance L_0 is incident on a tissue and the tissue is heated to the boiling point to initiate tissue ablation resulting from vaporization, then the ablation energy density w_{abl} becomes

$$w_{abl} = \rho\left[c_v(T_b - T_0) + h_v\right] \tag{8.13}$$

where T_0 is the steady-state tissue temperature, T_b the boiling point, c the specific heat, and ρ the tissue density. The ablation velocity v_{abl} can now be written as

$$v_{abl} = \frac{L_0'}{w_{abl}} \tag{8.14}$$

The ablation depth d is then the integral over the pulse duration of the ablation velocity,

$$d = \int_{t_0}^{t_p} \frac{L'(t)}{w_{abl}} dt \tag{8.15}$$

The time from the initiation of the pulse to the point where ablation starts is the time t_0. At this point the ablation threshold energy per surface area is reached, E_{th}. Combined with the incident total power density $\mu_a L_0'$ (see Equation [8.4]) the first-order approximation of the ablation depth can be written as

$$d = \frac{L_0' - E_{th}}{w_{abl}} \tag{8.16}$$

For a steady-state ablation the vaporized water volume per pulse (V_{PV}) can be written for a laser beam incident on an area A as

$$V_{PV} = \frac{A(L_0' - E_{th})}{w_{abl}} \tag{8.17}$$

The radiant energy deposited at depth d follows the Beer–Lambert law of attenuation and is expressed by Equation (8.4). The ablation depth is reached when the deposited energy at depth d exceeds the threshold; thus,

$$L(d) = E_{\text{th}} = (1 - R)L_0 e^{-\mu_a d} \tag{8.18}$$

When a certain volume of tissue is ablated per unit pulse with pulse duration τ, the ablation depth z_{abl} can be derived from Equation (8.18) and is written as

$$z_{\text{abl}} = \frac{1}{\mu_a} \ln\left(\frac{L'\tau}{E_{\text{abl}}}\right) \tag{8.19}$$

Where E_{abl} represents the ablation power per surface area. For a slow pulse rate the ablation process can be seen as short continuous wave ablation steps. While assuming quasi–steady state for a pulse sequence the ablation velocity v_{pv} can be approximated as

$$v_{\text{PV}} = \frac{V}{H_{\text{abl}}} = \frac{P}{\rho(L_v + c_v \Delta T)} \tag{8.20}$$

where P equals the power density of the laser beam entering the tissue. The expelled volume is now calculated as

$$V_{\text{abl}} = \frac{A}{\mu_a} \ln\left(\frac{\mu_a L'\tau}{E_{\text{th}}}\right) \tag{8.21}$$

Under high pulse repetition rates the solution will need to be submitted to a Fourier expansion. This is particularly true when the thermal relaxation time is in the order of the pulse rate. This will be described in greater detail in the pulse laser treatment sections in Chapter 11, section 11.2.1.

Furthermore, when the pulses follow each other in a time span that is less than the thermal relaxation time the deposited energy will accumulate as

$$E = nE_p \tag{8.22}$$

with n being pulses per second and E_p the energy per pulse.

The first m out of n pulses will be applied to the tissue surface, removing a small layer of tissue, after which the remaining $n - m$ pulses will continue to remove tissue from the developing crater, such that the energy density deposited in the volume during the m pulses $E(z)$ can be written as

$$E(z) = \mu_a m E_p e^{-\mu_a z} = E_{\text{abl}} e^{-\mu_a z} \tag{8.23}$$

After this point in time at the (m+1)th pulse, the depth Δz reached follows from Equation (8.9) as

$$\Delta z = \frac{1}{\mu_a} \ln\left(\frac{(m+1)\mu_a E_p}{E_{abl}}\right) = \frac{1}{\mu_a} \ln\left(\frac{(m+1)}{m}\right) \tag{8.24}$$

After the removal of this volume to the point at depth Δz the remaining laser pulses will remove tissue from the level Δz down in the same manner until all n pulses have been delivered within that second. The total ablation depth is then given by

$$z_{abl} = (n - m)\,\Delta z \tag{8.25}$$

Equation (8.22) can be rewritten as

$$n = E/E_p \tag{8.26}$$

and from the above equation the following expression for m can be derived:

$$m = E_{abl}/\mu_a E_p \tag{8.27}$$

Combining Equations (8.24)–(8.26) gives

$$z_{abl} = \frac{1}{\mu_a}\left[\frac{\mu_a E}{E_{abl}} - 1\right] m \ln\left(\frac{m+1}{m}\right) \tag{8.28}$$

For a Q-switched laser this gives $m \to \infty$ (MHz repetition rate), and the ablation depth can be approximated by

$$z_{abl} = \lim_{m\to\infty} \frac{1}{\mu_a}\left[\frac{\mu_a E}{H_{abl}} - 1\right] m \ln\left(\frac{m+1}{m}\right) = \frac{1}{\mu_a}\left[\frac{\mu_a E}{H_{abl}} - 1\right] \tag{8.29}$$

or

$$z_{abl} = \frac{E}{E_{abl}} - \frac{1}{\mu_a} \tag{8.30}$$

From Equation (8.30) the ablation threshold can be derived by letting Δz approach 0: $z_{abl}\downarrow_0 = (E/E_{abl}) - (1/\mu_a)$ gives $E/E_{abl} = 1/\mu_a$ or for the ablation threshold energy E_{th},

$$E_{th} = \frac{E_{abl}}{\mu_a} \tag{8.31}$$

While for energy delivery below the ablation threshold the energy density in the tissue, $E(z)$, becomes

$$E(z) = \mu_a E e^{-\mu_a z} \tag{8.32}$$

and it can be shown that this holds true for any beam profile as long as there is no heat diffusion.

At any point in time now the ablation depth can be derived using the time-dependent version of Equation (8.32), which gives at time $t = t_0 + \Delta t$ the following expression for the ablation energy:

$$E(z, t_0 + \Delta t) = \left[E_{\mathrm{abl}} + \mu_a P(t_0)\right] e^{-\mu_a z} \tag{8.33}$$

and the depth of ablation after time $t = t_0 + \Delta t$ follows from Equation (8.24) as

$$\Delta z = \frac{1}{\mu_a} \ln\left(\frac{E_{\mathrm{abl}} + \mu_a P(t_0)\Delta t}{H_{\mathrm{abl}}}\right) \tag{8.34}$$

Equation (8.34) can be developed in a Taylor series to represent the periodicity of the ablation process as

$$\Delta z = \frac{1}{\mu_a}\left[\left(\frac{\mu_a P(t_0)\Delta t}{E_{\mathrm{abl}}}\right) - \frac{1}{2}\left(\frac{\mu_a P(t_0)\Delta t}{E_{\mathrm{abl}}}\right)^2 + \frac{1}{3}\left(\frac{\mu_a P(t_0)\Delta t}{E_{\mathrm{abl}}}\right)^3 + \cdots\right] \tag{8.35}$$

The tissue removal per unit time can now be shown to obey the following equation:

$$\frac{\Delta z}{\Delta t} = \frac{P(t_0)}{E_{\mathrm{abl}}} + \text{Rest term} \tag{8.36}$$

By taking the limit of time approaching zero: $\lim_{t \to 0}$, the instantaneous ablation velocity will be found to obey the following expression:

$$v(t_0) = \frac{P(t_0)}{E_{\mathrm{abl}}} \tag{8.37}$$

The coagulation depth z_d from subthreshold ablation can be found by combining Equations (8.20) and (8.23), and assuming that with each pulse the instantaneous coagulation temperature (e.g., T_c=80°C) is reached, from Equations (8.13) and (8.18),

$$\rho c_v (T_c - T_0) = \mu_a E_{\mathrm{th}} \tag{8.38}$$

substituted into

$$\rho c_v (T_c - T_0) = E_{\mathrm{abl}} e^{-\mu_a z_d} \tag{8.39}$$

From Equation (8.39) the coagulation depth z_d can be found.

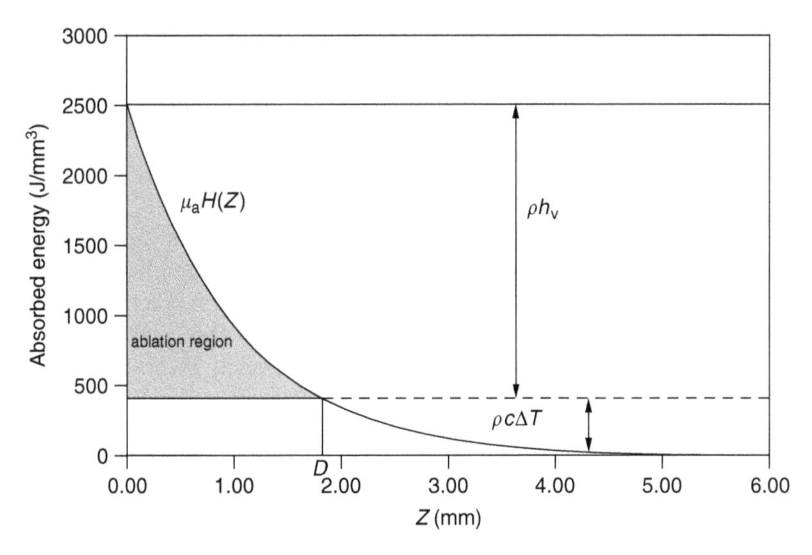

FIGURE 8.1
Schematic representation of the absorbed energy per unit volume for a near-infrared laser pulse as a function of depth.

Figure 8.1 illustrates the absorbed energy per unit volume for a near-infrared laser pulse as a function of depth.

8.2.4 Definition of a Pulsed Laser

During continuous wave or even long pulsed laser irradiation heat conduction will reduce the amount of energy available for tissue vaporization in addition to providing heat for secondary damage to the ambient tissue. A truly pulsed laser can be defined as having a pulse duration shorter than the time required to allow heat to diffuse out of the region directly heated by the laser light. For instance, a definition can be proposed that will restrict the temperature of the laser-heated volume to be greater than 90% of the maximum temperature reached during laser irradiation for the time between laser pulses. The instantaneous temperature distribution can be described as

$$T(z) = T_p e^{-\mu_a z} - T_0 \tag{8.40}$$

where T_p is the arbitrary temperature increase at the tissue surface at the end of the laser pulse and T_0 the equilibrium temperature. From Equation (8.40) it can be derived that there is a time constant $\tau_{\mu a}$ that obeys the following equality:

$$\tau_{\mu_a} = (C_1/\mu_a)^2 \tag{8.41}$$

where C_1 represents a normalization coefficient.

Photothermal laser–tissue interaction is particularly true for ultraviolet and mid-infrared laser irradiation.

8.2.5 Ultraviolet Laser Ablation

Due to the short wavelength and high photon energy of an ultraviolet laser a single photon can disrupt a single molecular bond. This phenomenon can be explained by the close match that may exist between the photon energy ($E=hv$) and the bonding energy in proteins. Some of the lasers operating in this wavelength regime are excimer lasers, chemical lasers, and a select group of frequency-doubled or -tripled lasers.

In pulsed laser ablation the photon energy is not the only factor that will determine the photon interaction; the pulse width will also have a major influence on the probability that a bond may be broken by the photon energy. The pulse width determines the interaction time between the photon and the chemical bond, which is subject to the Heisenberg uncertainty principle.

The generally accepted theory of chemical dissociation states that there needs to be a finite interaction time between the energy source and the atomic interaction. This will rule out any single photon to accomplish a chemical effect, prescribing minimally a two-photon interaction. The two-photon effect is different from that in two-photon microscopy, where the energy of two identical photons adds together to an energy level that matches an excitation band in a chromophore. This is a chemically and physically different situation from bond dissociation, since the excitation process is several orders more likely than the breaking of a chemical bond.

8.2.5.1 Photochemical Breakdown

In photochemical breakdown one photon brings the chemical connection into an excited state, while presumably a second photon provides the bond-breaking instigation.

The minimum number of photons needed to initiate a bond-breaking event is described by the threshold absorbed photons per unit volume and per unit time Π_{th}. While still assuming absorption-dominated light–tissue interaction, on the molecular bonding level, a general description of the number of photons absorbed per unit volume and time during laser–tissue interaction is expressed as

$$\Pi(z,t) = \mu_a \Psi(z,t)\frac{\lambda}{hc} \tag{8.42}$$

The fluence rate in this case is given by the Beer–Lambert law of attenuation:

$$\Psi(z,t) = \Psi_0(t)e^{-\mu_a z} \tag{8.43}$$

The energy of a single photon is given by

$$E_{\text{photon}} = \frac{hc}{\lambda} = h\upsilon \tag{8.44}$$

Bond-breaking will only occur when the following condition is met:

$$\Pi(z,t) \geq \Pi_{\text{th}} \tag{8.45}$$

Neither the molecular concentration nor the molecular bonds themselves affect the photochemical process in this treatment. The initial photon brings a molecular bond in an excited state while the second photon is absorbed with the same efficiency and actually dissociates the bond. The only condition is that a sufficient number of photons interact with one particular bond at any given time. The time-dependent concentration of absorbed photons to initiate a bond-breaking event, ρ_{p}, is defined as

$$\rho_{\text{p}}(z,t) = \int_0^t \left[\Pi(z,t') - \Pi_{\text{th}}\right] \mathrm{d}t' \tag{8.46}$$

under the condition that $\Pi(z,t) \geq \Pi_{\text{th}}$.

A single broken bond will not be noticed, and a critical number of bonds need to be broken to accomplish tissue ablation. The conditions for ablation thus impose a threshold for the critical number of absorbed photons that can be written as

$$\rho_{\text{p}}(z,t) \geq \rho_{\text{th}} \tag{8.47}$$

At this point the expanding volume resulting from breaking bonds is sufficient to expulse the molecular fragments.

The molecular layer closest to the incident light will reach the ablation threshold before the deeper layers. The expanding volume comprising dissociating bonds will at this point expulse the fragmented molecules. Subsequently the ablation front will migrate into the tissue.

An additional complication comes into play regarding the fact that dissociation pressure may build up during a single pulse, but this pressure may not be sufficient to overcome the surface tension of the neighboring molecules covalently bonded to the dissociating molecules. The initial laser pulse will be priming the tissue volume for expulsion, which is accomplished under a second pulse that immediately follows. The power density plays an important role in establishing a critical probability of achieving the threshold number of absorbed photons.

Since the ablation process is quantum-defined, the solution to the ablation depth from an ultraviolet laser pulse train will have a discrete pattern as well. Assuming that during each pulse a thickness Δz is removed with each discrete time interval Δt, the ablation phenomenon will thus have a Dirac-delta distribution. The photoablation bond-breaking process will be described next.

8.2.5.2 Rate of Bond-Breaking

The bond-breaking rate $\dot{n}$ is proportional to the total number of chemical bonds $n(z, t)$ and the local fluence rate $\Psi(z, t)$,

$$\dot{n} = -C_2 \Psi(z,t) n(z,t) \tag{8.48}$$

where C_2 is a rate constant (mm^2/J), which can be seen as the change in local bond density per local fluence. The problem is still one-dimensional only.

Introducing the single molecular bond absorption cross-section as σ_m, the local absorption coefficient for bond-breaking probability is given by multiplying the number of bonds $n(z,t)$ at any given location with the absorption cross-section as follows:

$$\sigma_m n(z,t) \tag{8.49}$$

The locally absorbed fluence rate can now be written as

$$\Psi(z,t) = \Psi_0(t) \left[-\sigma_m \int_0^z n(z',t) \mathrm{d}z' \right] \tag{8.50}$$

which reduces to the standard Beer–Lambert attenuation law with absorption coefficient

$$\mu_{a,m} = \sigma_m n_\infty \tag{8.51}$$

where n_∞ is the limit of $n(z,t)$ under the condition of no ablation, the total bond density.

Combining Equations (8.48) and (8.50) gives for the unbroken bond density,

$$n(z,t) = \frac{n_\infty}{1 + (e^{C_2 \Psi(z,t) t} - 1) e^{-\mu_a z}} \tag{8.52}$$

with $\mu_{a,m}$ defined by Equation (8.51), and $\Psi(z,t) \times t$ is the laser pulse energy E_R during the pulse duration t.

The definition of the pulse energy allows the unbroken bond density to be written as $n(z,E_R)$. The crater depth d can now be defined as the condition when the fraction of the unbroken bonds, F, with respect to the total number of bonds in the target volume has a value between 0 and 1, $0<F<1$, and follows from the following expression:

$$\frac{n(d,E_R)}{n_\infty} = \frac{1}{1 + (e^{C_2 E_R} - 1) e^{-\mu_{a,m} z}} = F \tag{8.53}$$

Equation (8.53) also represents the fraction of the energy that exceeds the ablation threshold.

Solving Equation (8.53) for the crater depth gives

$$d = \frac{1}{\mu_{a,m}} \ln\left(\frac{e^{C_2 E_R} - 1}{1/F - 1}\right) \tag{8.54}$$

with the threshold, $d(E_{th})=0$, this yields

$$E_{th} = \frac{1}{C_2} \ln(1/F) \tag{8.55}$$

Under the condition that the laser energy is much greater than the threshold energy, the ablation depth can be written as

$$d = \frac{k}{\mu_{a,m}} E_R + \frac{1}{\mu_{a,m}} \ln(1/F - 1) \tag{8.56}$$

In the following section the issue of the bond-breaking depth will be addressed in detail.

8.2.5.3 Bond-Breaking Depth

Equation (8.55) provides an expression for the amount of absorbed photons in each molecular layer. The amount of photons absorbed per unit volume, resulting in ablation, is defined as $\rho_p(z,t)$ in Equation (8.46). This allows us to find the absorbed photon energy exceeding the ablation threshold Π_{th} at a time t_k and depth z_j by combining Equations (8.55) and (8.46). Using the fact that the molecular bond-breaking is a discrete phenomenon, the integral in Equation (8.46) can be replaced by a summation over the molecular layers:

$$\rho_p(z_j, t_k) = \sum_{j=1}^{k} \left[\Pi(z, t') - \Pi_{th}\right] \Delta t \tag{8.57}$$

under the condition that $\Pi(z_j,t_k) \geq \Pi_{th}$.

Each time when the absorption of photons exceeds the threshold for ablation a layer of molecules is effectively removed, as expressed in the following equation:

$$\rho_p(z_j, t_k) > \rho_{th}(z, t) \tag{8.58}$$

The ablation threshold $\rho_{th}(z,t)$ is derived from the upper limit of the exponential expression for photon density (Arrhenius value).

Solving Equation (8.54) under the special condition, that is, $\Pi_{th}=0$, which requires $\rho_p(z,t)$ to be equal to the total number of absorbed photons, represents an absolute ablation efficiency. The ablation depth can now be derived as

$$d = \frac{1}{\mu_{a,m}} \ln\left(\frac{\mu_{a,m} E_R}{\rho_{th}}\right) \tag{8.59}$$

which is equivalent to Equation (8.19).

Another interesting case is the quasi-steady-state Q-switched laser ablation, which allows Equation (8.54) to be developed into a Taylor series. Truncating after the first term gives

$$d = \frac{E_R}{\rho_{th}} - \frac{1}{\mu_{a,m}} \tag{8.60}$$

which is equivalent to Equation (8.30).

During the evaporative process of molecular dissociation gas is produced, which is tackled in the next section.

8.2.5.4 Gas Production during Photoablation

The total number of gas molecules produced during photoablation will be proportional to the total number of broken bonds $\rho_p(z,t)$ as described by Equation (8.46) under the assumption that the ablation threshold has been reached.

The total amount of gas molecules released per unit area is now equal to the crater depth times the number of broken bonds.

The vapor volume V_{vapor} can be approximated by using the conversion of liquid into vapor with an expansion factor Ω at atmospheric pressure. The released vapor volume is now defined as

$$V_{vapor} = \Omega V_{liquid} \tag{8.61}$$

The expansion factor can be as large as three orders of magnitude. The ablated liquid volume V_{liquid} is calculated from the ablation depth d. Looking back at Section 8.2.3, the correlation between the ablation depth and the ablation volume can be derived to satisfy the following expression:

$$V_{liquid} = \int_{z=0}^{z=d} \frac{k(Q_0 e^{-\mu_{a,m} z} - Q_{th})\mu_{a,m}}{\rho h_v} dz \tag{8.62}$$

where Q_{th} represents the threshold ablation energy times the laser spot-size area:

$$Q_{th} = AL_{th} \tag{8.63}$$

and analogously the delivered energy Q_0 becomes

$$Q_0 = A\Psi'_0 \tag{8.64}$$

Therefore, the vapor volume is derived as

$$V_{vapor} = \frac{F\Omega}{\rho h_v}\left[Q_0 - Q_{th} - Q_{th}\ln\left(\frac{Q_0}{Q_{th}}\right)\right] = \frac{k\Omega}{\rho h_v}\left[Q_0 - Q_{th}\left\{1 + \ln\left(\frac{Q_0}{Q_{th}}\right)\right\}\right] \tag{8.65}$$

Expressing this vapor volume in terms of laser irradiance gives

$$V_{\text{vapor}} = AL_{\text{th}} \frac{k\Omega}{\rho h_{\text{v}}} \left[\frac{L_0'}{L_{\text{th}}} - 1 - \ln\left(\frac{\Psi_0'}{L_{\text{th}}} \right) \right] \tag{8.66}$$

Proceeding on the pulsed laser–tissue interaction theme the nonablative interaction will next be discussed in the photoacoustic interaction.

8.3 Photoacoustics

Following the description of pulsed laser–tissue interaction the next logical step is the description of the acoustic wave generated by breaking the molecular bonds forming a gas, or just the heating effect from repetitive laser pulses leading to local thermal expansion based on the Boyle–Gay-Lussac law.

8.3.1 History of the Photoacoustic Effect

The photoacoustic effect in both nongaseous and gaseous matter was discovered in the nineteenth century and was first reported in 1880, when the Scottish researcher Alexander Graham Bell (1847–1922), who later moved to the United States, gave an account to the American Association for the Advancement of Science of his work on the photophone. In his paper, he briefly reported the accidental discovery of the photoacoustic effect in solids.

Bell noticed that the mirror being used for carrying the voice signature of his photophone had unexpected vibrations. These vibrations seemed to be related to the magnitude of focused sunlight that was being used. Later experimentation showed that the types of gases present when focused light was used also had an effect on the vibrations detected. What was once seen as a major scientific discovery of the day, photoacoustics had seemingly faded into scientific history. However, discoveries since its resurrection in 1938 have made photoacoustics one of the most exciting new scientific tools available today.

It was not until the 1970s, some 90 years after Bell's original discovery, that the photoacoustic effect in nongaseous matter was "rediscovered." The acoustic signals were initially theorized to be a result of some gas heating interaction on the surface of the material that resulted in a shock wave. It will turn out that the absorptive heating of the medium itself causes expansion, generating an acoustic wave.

Photoacoustics is now being used for everything from nondestructive testing of airplane wing microfractures to imaging techniques for the detection of early-stage cancer.

8.3.2 The Photoacoustic Effect

The photoacoustic effect in its most basic form is an acoustic wave generated by a concentrated light source interacting with a material or gas. The acoustic wave is developed from photonic energy causing the target medium at the material interface to ionize. Ionization is caused by heat generated from a brief state of gas–plasma interaction. Gas expansion and shrinkage causes an acoustic wave to propagate through the material of interest. Figure 8.2 shows a generic example of a laser pulse causing the excitation of an acoustic wave.

Rosencwaig and Gersho were early pioneers in the development of photoacoustic modeling for solids. The Rosencwaig–Gersho model assumes a cell with the diameter of the light beam, which reacts with the material at a light—gas–material interface. This boundary layer of gas, when acted upon by a laser, would be the source for any acoustic wave that would propagate throughout the material after interaction.

The incident periodically fluctuating fluence rate of laser light is given as

$$\Psi = \tfrac{1}{2}\Psi_0(1+\cos\omega t) \tag{8.67}$$

where Ψ_0 is the incident monochromatic light flux (W/cm^2). The light fluence rate determines the illumination of the material and can be responsible for the amount of acoustic wave development for certain materials depending upon their photoabsorption characteristics. The variable ω denotes the chopping frequency of the incident light beam in radians per second.

For a photoacoustic system, the pressure or stress in the material developed from the photoacoustic effect $P_{\mu'_T}$ that is generated within a thermal diffusion length of the sample surface can be approximated by

$$P_{\mu'_T} \approx \beta\alpha'_t(\tfrac{1}{2}\Theta_0) \tag{8.68}$$

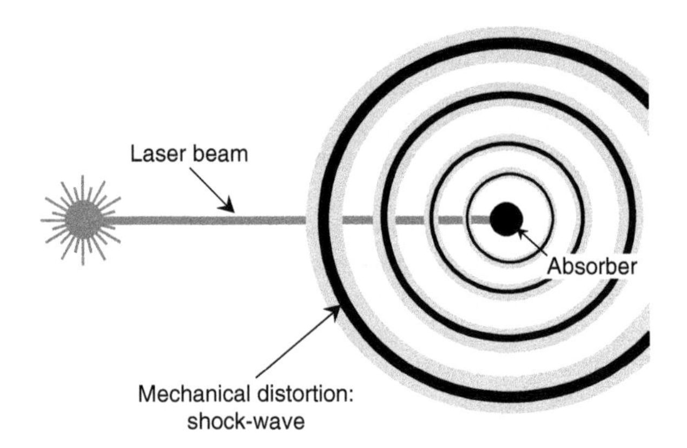

FIGURE 8.2
Generic example of the photoacoustic effect on a material.

where $\frac{1}{2}\Theta_0$ is approximated as the average temperature within this thermal diffusion length if Θ_0 is the temperature at the sample–fluid interface. The pressure at a distance l' away is given by

$$P = P_{\mu'}\left(\frac{\mu'}{l'}\right) = \tfrac{1}{2}B\alpha_t'\Theta_0\left(\frac{\mu'}{l'}\right) \tag{8.69}$$

The following parameters that govern photoacoustics are commonly used in thermodynamics: κ_T defines the thermal conductivity (J/m s K), ρ is the density (kg/m^3), c_v represents the specific heat (J/kg K), $\alpha=\kappa_T/\rho X$ is the thermal diffusivity (m^2/s), with $a=w/2\sqrt{\alpha}$ the thermal diffusivity coefficient (m^{-1}) and $\mu_T=1/a$ the thermal diffusivity length (m).

Substituting the material factors in Equation (8.69) gives

$$P_{\mu_T'} = \frac{\gamma P_0}{2T_0 a' l'}\Theta_0 \tag{8.70}$$

where γ is the ratio of specific heats of gas and solid, P_0 the ambient pressure, T_0 the ambient temperature, a' the thermal diffusion coefficient of the solid, l' the distance from laser to solid, and Θ_0 the temperature at solid–gas interface.

When the gas or fluid at the interface is completely constrained at its borders, then the pressure P is the same everywhere in the cell, as long as the cell dimensions are much smaller than the acoustic wavelength.

The temperature Θ_0 at the sample–gas interface can be approximated by

$$\Theta_0 \approx \frac{H_{abs}}{M_{th}} \tag{8.71}$$

where H_{abs} describes the amount of heat absorbed per unit time within the first thermal diffusion length in the sample, and M_{th} is the thermal mass of this region of the sample. For the case where the thermal diffusion length is smaller than the sample thickness ($\mu_T<l$), and if A is the area illuminated by the light then

$$H_{abs} \approx \frac{\Psi_0 A(1-e^{-\beta\mu_T})}{\omega} \tag{8.72}$$

and

$$M_{th} = \rho c_v \mu A \tag{8.73}$$

Substitution of Equations (8.72) and (8.73) into Equation (8.71) gives temperature Θ_0 as

$$\Theta_0 = \frac{\Psi_0(1-e^{-\beta\mu_T})}{\rho c_v \omega \mu_T} \tag{8.74}$$

Similarly, when $\mu_T > l$, then Equation (8.72) reduces to

$$H_{abs} \approx \frac{\Psi_0 A(1-e^{-\beta\mu_T})}{\omega} \tag{8.75}$$

and Equation (8.73) reduces to

$$M_{th} \approx \rho'' c_v'' \mu_T'' A \tag{8.76}$$

Thus, the expression for temperature becomes

$$\Theta_0 = \frac{\Psi_0(1-e^{-\beta l})}{\rho'' c_v'' \omega \mu_T''} \tag{8.77}$$

where the unprimed symbols represent the sample and the double-primed symbols represent the parameters inside the cell model.

Thermal conduction introduces a phase lag in the photoacoustic signal. There is a $\pi/4$ phase lag due to conduction in the gas and an additional phase lag due to conduction in the sample.

The photoacoustic effect is a function of three equations with a range of unknowns: transport equation (Section 6.1), bioheat equation (Section 7.3), and a displacement description, which is discussed next.

8.3.2.1 *Displacement*

The next declaration describes how the temperature influences the displacement u, and as such creates a pressure:

$$\rho\frac{\partial^2 u}{\partial t^2} - \frac{E}{2(1+\sigma)}\nabla^2 u - \frac{E}{2(1+\sigma)(1-2\sigma)}\nabla(\nabla u) = -\frac{E\beta_c}{3(1-2\sigma)}\nabla\theta \tag{8.78}$$

where σ is the Poisson's ratio, which is defined as the lateral contraction per unit breadth divided by the longitudinal extension per unit length, or the ratio of radial strain to axial strain; β_c is the cubic expansion coefficient ($3\alpha_c$); and E is the Young's modulus, which is defined as the stress over strain of a material. The definition of the Young's modulus is given by

$$E = \frac{F/A}{(L-L_0)/L_0} \tag{8.79}$$

which relates the stress equal to force F over a unit surface area of a bar with length L_0 that is extended to a length L.

Solving this problem relies on clearer choices of approximations when possible.

The correlation between the displacement u and pressure follows from Hooke's law, which describes the tension T in a bar with length L_0 that is extended to a length L as

$$T = E\frac{L - L_0}{L_0} \tag{8.80}$$

or stress equals Young's modulus times strain, expressed in terms of displacement as

$$T = \kappa_s u \tag{8.81}$$

where κ_s equals the spring constant or compression modulus, which is defined as

$$\kappa_s = \rho_0 \left(\frac{dP}{d\rho}\right)_0 \tag{8.82}$$

where ρ is the local density.

In terms of pressure P this translates into

$$P = P_0 - k_s \nabla u \tag{8.83}$$

where the displacement is differentiated in all the directions of propagation of the transverse wave.

The Young's modulus and the spring constant are correlated by the moment of inertia of the medium, I, and the characteristic dimension of the medium in the direction of motion, L, as

$$\kappa_s = c^* \frac{EI}{L^3} \tag{8.84}$$

where c^* is a constant that is dependent on the geometric configuration of the medium.

With the steady-state density ρ_0, the compression modulus is linked to the speed of propagation of the sound wave, v_s, by the following expression:

$$v_s = \sqrt{\frac{\kappa_s}{\rho_0}} \tag{8.85}$$

which technically is the coefficient of the second term in Equation (8.78) when seen as the wave equation:

$$\frac{\partial^2 u}{\partial t^2} = \frac{E}{\rho 2(1+\sigma)}\nabla^2 u + \frac{E}{\rho 2(1+\sigma)(1-2\sigma)}\nabla(\nabla u) - \frac{E\beta(1+\sigma)}{\rho\sigma}\nabla\theta \tag{8.86}$$

compared with

$$\frac{\partial^2 u}{\partial t^2} = v^2 \nabla^2 u \tag{8.87}$$

yielding

$$\kappa_s = \frac{E}{2(1+\sigma)} \tag{8.88}$$

For most tissues the Poisson's ratio will be close to 0.5, which simplifies Equation (8.86) to

$$\rho\frac{\partial^2 u}{\partial t^2} - \frac{E}{3}\nabla^2 u = 3E\beta_c \nabla\Theta \tag{8.89}$$

Using the fact that the motion will be a harmonic oscillation the displacement will be sinusoidal, substitution of a sinusoidal format for the displacement in Equation (8.83) yields the following expression for the correlation between the displacement and pressure wave:

$$P \propto v^2 \rho_0^2 k_s u \tag{8.90}$$

8.3.2.2 Sound Propagation

As an acoustic wave is transmitted through a medium with a reactive nonlinearity, it is distorted along the propagation path. The near-field laser spot incident on the tissue will rapidly disperse to produce a far-field wave. In our theoretical analysis of the ultrasound photoacoustic image formation we will initially make the following assumptions that will be refined later on and expanded to include the observed deviations from these approximations:

- All analyses are made in the Fraunhofer region, i.e., in the far-field.
- The waves will be approximated as plane waves in the far-field.
- The square of the received amplitude is a linear measure of the scattered energy.
- The scattered pressure is weak relative to the incident pressure.

Pulsed laser heating generates a pressure wave. The reverberation produces a sound wave propagating through the biological medium. The resonance frequency ν, created by the periodic local heating and subsequent expansion, depends on the value of Young's modulus of the tissue and the local dimensions. The dependence can be expressed as

$$\nu = A\sqrt{\frac{EI}{m_1 L^4}} \tag{8.91}$$

Here, E is the Young's modulus, I the moment of inertia of the volume where the acoustic stimulus takes place, L the length of the sides of a homogeneous cube of the medium, m_l the mass per unit length, and A a constant that depends on the mode of excitation. This can be rewritten in a more convenient form as

$$v = A\sqrt{\frac{E}{\rho L^2}} \tag{8.92}$$

where ρ is the material density. The strain ε generated as a consequence of the volume expansion is related to the change in volume ΔV with respect to the initial volume V,

$$\varepsilon = \frac{\Delta V}{V} \tag{8.93}$$

The strain is related to the temperature jump, ΔT, as described by the volume expansion coefficient as follows:

$$\Delta V = \beta_c V \, \Delta T \tag{8.94}$$

giving

$$\varepsilon = \beta_c \, \Delta T \tag{8.95}$$

The stress, which is equal to the force per unit area, also describes pressure, generated by this strain and is expressed as

$$P = -K\varepsilon \tag{8.96}$$

Here, K is the bulk modulus that specifies the pressure per unit strain, and T_s is the tension. Combining Equations (8.93), (8.95), and (8.96) results in the following expression for pressure:

$$P = G_\gamma K\beta \frac{\mu_a \psi}{\rho C_p} \tag{8.97}$$

The pressure equation shows the correction factor G_γ, which expresses the fraction of thermal energy from the absorbed light that is converted into mechanical energy. This correction factor is called the Grüneisen coefficient, typically ranging between 0.1 and 1. Assuming a 12% coupling with mechanical energy (a Grüneisen coefficient of 0.12) the relationship between the linear expansion coefficient α_c and the Grüneisen coefficient can be shown to satisfy the following relationship:

$$\alpha_c = \frac{G_\gamma \rho c_v}{3E} \quad \text{or} \quad G_\gamma = \frac{3E}{\rho c_v}\alpha_c \tag{8.98}$$

The maximum pressure (Equation [8.97]), using the volume expansion coefficient can only be reached if the pulse duration t_p does not exceed the mechanical relaxation time of the local medium τ_r. The mechanical relaxation time is directly proportional to the characteristic dimension of the medium volume r, heated by laser light, and is inversely proportional to the speed of sound propagation in this medium v:

$$\tau_r \approx \frac{r}{v} \tag{8.99}$$

Subsequently, the pressure wave will propagate in all directions with equal probability, spreading the mechanical energy over the full solid angle, 4π steradian, of space. In the description of the propagation of this wave, one introduces the velocity potential φ as

$$\mathbf{v} = \nabla\varphi \tag{8.100}$$

or split in three dimensions as

$$u = \frac{\partial\varphi}{\partial x}, \quad v = \frac{\partial\varphi}{\partial y}, \quad w = \frac{\partial\varphi}{\partial z} \tag{8.101}$$

which satisfies the Laplace equation for the velocity potential,

$$\frac{\partial^2\varphi}{\partial x^2} + \frac{\partial^2\varphi}{\partial y^2} + \frac{\partial^2\varphi}{\partial z^2} = 0 \tag{8.102}$$

This velocity potential can be used to derive the particle displacement u and velocity v_p in a spherical polar coordinate system whose origin is at the center of the cavity:

$$u = \frac{\varphi}{R}, \quad v_p = \frac{u}{t} = \frac{\varphi}{tR} \tag{8.103}$$

The pressure can also be computed from the velocity potential as follows:

$$P_\gamma = -\rho\frac{\partial\varphi(r,t)}{\partial t} \tag{8.104}$$

The pressure wave will encompass both compression and expansion at the resonant-frequency wave pattern. The pressure that is proportional to the time derivative of $-\varphi$ will always be initially positive, and then necessarily negative. The acoustic impedance is a measure of how responsive the medium is to sound wave disturbance. The acoustic impedance Z is the ratio of wave pressure to particle velocity and is generally written as

$$Z = \mathrm{p}/v\,\rho v \tag{8.105}$$

At the interface between two separate media with specific acoustical impedance, the acoustic waves will obey Snell's laws of reflection and refraction, and the intensity reflection coefficient R_{acoustic} can be written as

$$R_{\text{acoustic}} = \left(\frac{Z_2 - Z_1}{Z_2 + Z_1} \right)^2 \tag{8.106}$$

Here, Z_1 denotes the acoustical impedance of the medium the wave originates from, and Z_2 denotes the acoustical impedance of the medium the wave enters into, which is on the other side of the interface.

8.3.3 Detection of the Photoacoustic Signal

A pressure wave is generated with the arrival of each laser pulse, and this pressure wave is recorded with acoustic detectors, such as multisensor array quartz piezoelectric detectors, and ultrasound sensors that utilize a thin polymer film as a low Finesse Fabry–Perot interferometer. A fiberoptic device can also be used to deliver the laser pulses. An example of the photoacoustic image formation protocol is shown in Figure 8.3.

8.3.4 Theory of Photoacoustic Wave Propagation

The approach for the theoretical model of photoacoustic wave propagation is summarized here. The acoustic wave equation in the medium that is

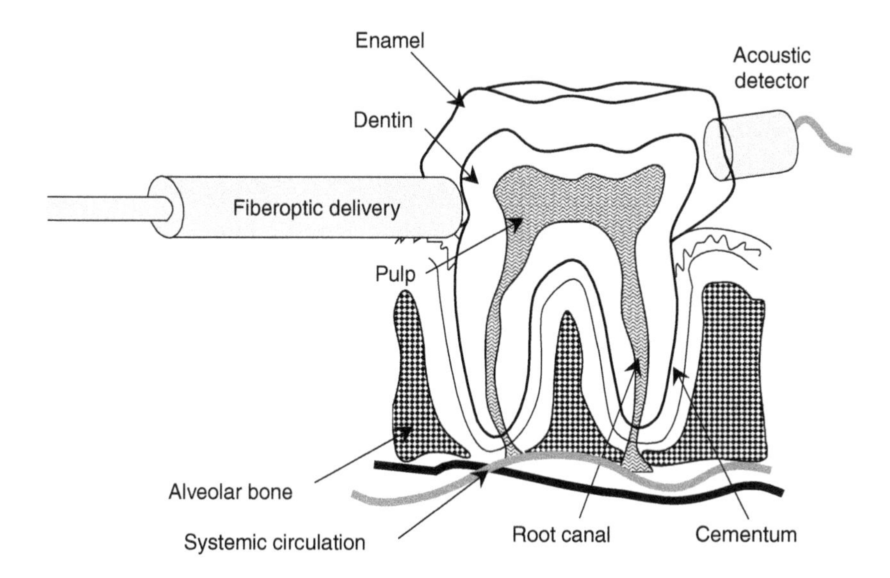

FIGURE 8.3
Schematic for photoacoustic imaging of the tooth. Optical radiation is delivered through a fiber optic, and the acoustic signal is detected by an acoustic sensor.

generated by the light absorption (see Figure 8.2) can be derived from the momentum equation, the equation of continuity, and the equation of state to give the following three conditions, including the initial boundary equation:

$$\frac{\partial \rho(\mathbf{r},t)}{\partial \tau}+\nabla\cdot(\rho v(\mathbf{r},t))=-\int_{t_0}^{t_1}\frac{S(\mathbf{r},t)}{\mathbf{r}\cdot\mathbf{r}'}\delta\tau \tag{8.107}$$

$$\frac{\beta(\mathbf{r})}{\rho_0 C_0^3}\rho\frac{\partial v(\mathbf{r},t)}{\partial \tau}-(v(\mathbf{r},t)\cdot\nabla)v(\mathbf{r},t)+\nabla\cdot P(\mathbf{r},t)=\nabla S(\mathbf{r},t) \tag{8.108}$$

$$P(\mathbf{r},t)=K\rho^{\gamma} \tag{8.109}$$

$$P(\mathbf{r},t)=G_r K\beta_c\frac{\mu_a\psi}{\rho C_p} \tag{8.110}$$

Here, $v(\mathbf{r},t)$ is the displacement velocity, $P(\mathbf{r},t)$ the sound pressure amplitude of the primary wave, $\beta(\mathbf{r})$ a nonlinear parameter distribution, and $\tau=t-\mathbf{r}/C_0$ the propagation time; γ is the ratio of specific heat of tissue and the specific heat of water, and C_0 the speed of sound in the medium. The source function $S(\mathbf{r},t)$ here is proportional to the amount of light absorbed locally ($\mu_a\psi$), divided by the density ρ and the specific heat C_p. The source function is expressed with the same units as for the pressure.

The initial boundary condition is given by Equation (8.108). The integral in Equation (8.107) is evaluated over the duration of the laser pulse, which will experience dispersion during absorption in the medium, and thereby generates the internal source for the acoustic pressure wave. Note, as an acoustic wave is transmitted through a medium with a reactive nonlinearity, it will be distorted along the propagation path.

8.3.5 Electromechanical Effects during Photoacoustic Interaction

When the fluence is high enough, the electric field of the electromagnetic radiation may reach the molecular electric field level. The molecular electric field that holds the individual atoms in the molecule together ranges from 10^7 to 10^{12} V/m depending on the atoms involved and the molecular configuration. When the electromagnetic electric field does exceed the molecular electric field, the light energy can dissociate the chemical bonds and ionize the material. At this point a plasma is formed that can create an expansion and a shockwave. The electric field associated with the light power density $\mathbb{P}$ is given by the equation

$$\mathbb{P}=\frac{P}{2\pi r^2}=\frac{E^2}{\mu_0 c} \tag{8.111}$$

where E equals the electric field of the electromagnetic radiation, P the power of the laser irradiation, r the radius of the laser beam, c the velocity of light, and μ_0 the permeability of vacuum.

One added complication in this scenario is the fact that the formation of the plasma itself may prevent further ionization due to absorption of the laser light in the plasma itself, called plasma shielding.

Plasma formation may occur at laser power densities exceeding 500 MW/cm^2, at this point it can be calculated that the electric field reaches 10^7 V/m.

This phenomenon is applied during laser lithotripsy and photoablative surgery during laser radiative keratotomy. This process is described in greater detail in Section 11.4 under Ophthalmology.

Another phenomenon of electromechanical laser–tissue interaction is called sonoluminescence. Here the production of a flash of light accompanies the bursting of a bubble.

8.4 Birefringence Effects

The principle of more than one optical path leading away from a single source in a medium and resulting from a different index of refraction when traversing the medium in one direction versus another direction is called birefringence. The two paths with different indices are referred to as the ordinary axis and the extraordinary axis, respectively. This is illustrated in Figure 8.4. The ordinary axis associates the index of refraction to the electric field of the incident electromagnetic radiation perpendicular to the optical axis, whereas the

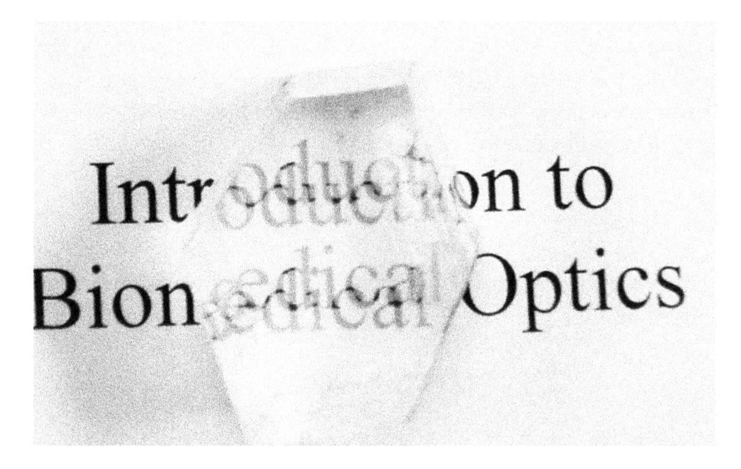

FIGURE 8.4
Birefringence effect showing the ordinary and extraordinary optical axis presenting an image.

extraordinary axis has the electric field parallel to the optical axis to give a different index of refraction. Birefringence is also known as double refraction for this reason. The fact that the index of refraction is different in one direction from another direction inside a medium is most often the result of the molecular orientation when the medium is made up out of a single species of molecules. The molecular matrix can be very regular, and distinctly different in one plane of view with respect to another plane of view.

This phenomenon is observed in solids such as plastics or glasses. The same birefringence can also be observed in nonuniform heated gasses and can show the temperature distribution in the gas. The imaging of temperature effects is based on the fact that the index of refraction is generally a function of temperature as well, as shown in Section 5.2.1.1.

The technique of imaging the birefringence by transillumination is called Schlieren imaging. It is also observed in biological media under nonuniform heating conditions such as laser irradiation; however, to observe the birefringence effects, the tissue will need to be a relatively thin slab to prevent scattering from disrupting the dissociation of optical axis to become visible.

8.4.1 Schlieren Imaging

The Schlieren technique (sometimes also called Töepler imaging) is based on the angular deflection undergone by a light ray when passing through a region characterized by refractive index inhomogeneities.

These optical inhomogeneities are generally caused by density variations in the biological medium, resulting from natural diversity in tissues or resulting from a temperature gradient. The temperature effects are of particular importance in medical imaging. Through this type of imaging detailed local temperature information can be revealed noninvasively.

The index-of-refraction gradients are caused by temperature gradients. The initial assumption can now be made that the temperature gradient profile is made up of infinitesimally thin layers of uniform temperature and uniform index of refraction. At the boundary of each layer Snell's law will apply to the angle of refraction. In other words,

$$n \sin\theta = \text{constant} \tag{8.112}$$

When the index of refraction is a function of temperature, Equation (8.112) can be written as

$$n(T)\sin\theta = \text{constant} \tag{8.113}$$

The temperature in turn is a function of distance to the heat source (point of laser impact), thus making the index of refraction a function of the coordinates of the geometry of the medium. Assuming a one-dimensional axially symmetric configuration, the index of refraction will decrease with lower temperature and with greater distance from the source of heat:

$$n(r)\sin\theta = \text{constant} \tag{8.114}$$

Consider now two successive wavefronts of a probing light source traveling through the medium with the temperature gradient. The distance between the two wavefronts is by definition equal to one wavelength. The wavelength λ is a function of the speed of propagation v, which in turn depends on the density and thus on the local temperature and the frequency v. The frequency remains constant throughout, as dictated by conservation of energy requirements:

$$\lambda = v/\nu = c/\nu n(T) \tag{8.115}$$

The distortion of light passing through a medium that has inhomogeneous temperature distribution can be represented as a medium with changing optical index of refraction due to thermal effects as outlined in Figure 8.5.

The visibility of the Schlieren imaging technique is defined based on the angular deflection resulting from the disparity of the index of refraction due to the temperature effects. The angle of deflection is represented as α in Figure 8.6. Generally, this deflection angle is relatively small and the sine of the angle can be considered equal to the angle in radians. In this case the angle can be found from the displacement of the principle axis on the viewing screen or detector divided by the distance from the reference point in the Schlieren configuration. This reference point will be the location of thermal interaction within the tissue volume.

The distance between the two wavefronts, perpendicular to the respective rays, equals one wavelength, or in terms of angle in the orthogonal coordinate system:

$$\rho_{\mathrm{d}}\sin\theta = \lambda = \rho_{\mathrm{d}}\theta \tag{8.116}$$

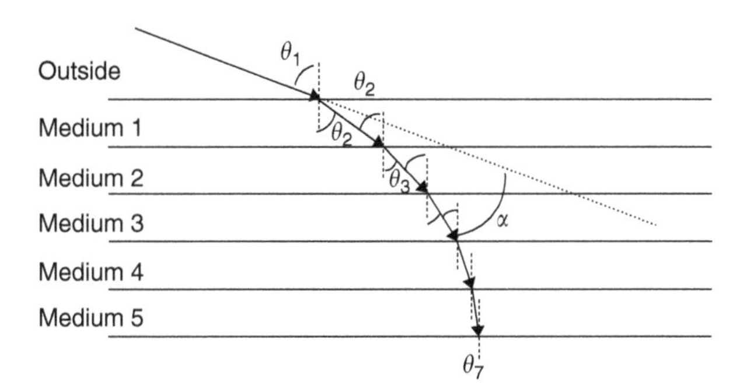

FIGURE 8.5
Ray diagram of propagation in a medium with a disparity in index of refraction, for instance, resulting for a temperature gradient.

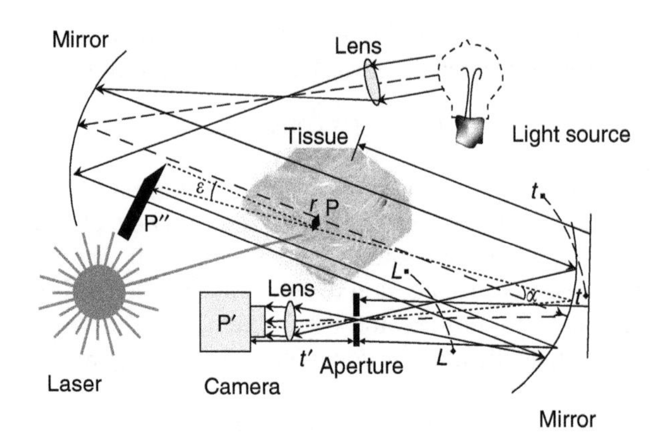

FIGURE 8.6
Configuration of original Schlieren imaging configuration setup. (Courtesy of Giovanni Tanda, DITEC, Universita di Genova, Genova, Italy.)

where ρ_d equals the distance to the conveniently chosen origin, and since the angle θ is small:

$$\sin\theta = \theta \tag{8.117}$$

The second ray is at a distance $d\rho_d$ farther removed from the same reference point, thus giving

$$(\rho_d + d\rho_d)\theta = \lambda' \tag{8.118}$$

Combining Equations (8.116) and (8.118) gives

$$\theta\, d\rho_d = \lambda' - \lambda \tag{8.119}$$

Using

$$\theta = \lambda/\rho_d \tag{8.120}$$

gives

$$\frac{1}{\rho_d} = \frac{1}{\lambda}\frac{d\lambda}{d\rho_d} = \frac{d}{d\rho_d}(\ln\lambda) = -\frac{d}{d\rho_d}(\ln n) \tag{8.121}$$

with

$$\ln\lambda = \ln C^* - \ln f - \ln n \tag{8.122}$$

where C^* is a constant that results from the standard solution to integration of Equation (8.122), and f is by definition constant, the curvature of the ray defined by ρ_d is curved more toward the rising index of refraction.

The angle of deflection $\partial\alpha$ resulting from the gradient in the index of refraction derived from the geometrical analysis of Figure 8.6 is expressed as

$$\partial\alpha = \frac{\partial z}{\partial y} = \frac{(v_y - v_{y+\Delta y})\partial t}{\partial y} = v_0\left(\frac{1}{n_y} - \frac{1}{n_{y+\Delta y}}\right)\frac{\partial t}{\partial y} \tag{8.123}$$

where v_y is the velocity of propagation in point y, and $n=v_y/v_{y0}$. Owing to the infinitesimally small variation in n, the term $v_0\partial t$ can be substituted by $n\partial z$, giving

$$\partial\alpha = -n\partial\left(\frac{1}{n}\right)\frac{\partial z}{\partial y} = \left(\frac{1}{n}\right)\left(\frac{\partial n}{\partial y}\right)\partial z \tag{8.124}$$

The total angle of deflection over a trajectory with length L follows from integration of Equation (8.123) and using as a first-order approximation that $(1/n)$ does not change; thus, substituting the undisturbed average standard index of refraction n_0 of the medium in Equation (8.124), $(1/n_0)$, gives after integration:

$$\alpha = \int_0^L \left(\frac{1}{n}\right)\left(\frac{\partial n}{\partial y}\right)\partial z \approx \left(\frac{1}{n_0}\right)\left(\frac{\partial n}{\partial y}\right)L \tag{8.125}$$

A well-known example of the light diffraction as a result of temperature gradient is seen in high-speed photography of a bullet traveling through air, leaving the Mach trail, shown as refraction discontinuities by back-lit illumination. Another example is the mirroring of objects at a distance seen over an asphalt road surface on a hot day.

The strain e in the medium is a function of the deformation in the respective orthogonal axes, expressed as

$$e_{ij} = \frac{1}{2}\left(\frac{\partial u_i}{\partial x_j} + \frac{\partial u_j}{\partial x_i}\right) \tag{8.126}$$

where i and j represent the respective orthogonal axes for the configuration, giving the two-dimensional strain in Cartesian coordinates as

$$e_{xy} = \frac{1}{2}\left(\frac{\partial u}{\partial x} + \frac{\partial v}{\partial y}\right) \tag{8.127}$$

with u being the displacement in the x-direction and v the motion in the y-direction. Also note that the deformation in the x-direction expressed in the shear angle θ_x can be found from

$$\frac{\theta_x}{2} = \frac{\partial u}{\partial x} \tag{8.128}$$

and so forth.

Another way of expressing the gradient in the index of refraction is in terms of the strain in the medium, by rewriting Equation (8.123) as an expression of the index of refraction, incorporating Equation (8.126), and introducing the elastooptic constant p, giving the following expression without derivation:

$$\Delta n = -\frac{n_0^3}{2} pe \tag{8.129}$$

The elasto-optic constant p needs to be determined experimentally for the various biological tissues. For certain compound and single-component materials there are look-up tables available that list the elastooptic constant values.

This shows that Schlieren imaging can provide temperature distribution information in addition to an impression of the distribution of the material constant n that makes up the organ under examination. Schlieren imaging can thus provide rudimentary optical tissue structure information. Referring to the documented optical characteristics of various tissues the organ can crudely be virtually dissected.

Two types of Schlieren configurations can be distinguished. The first configuration is traditional Töepler construction using spherical mirrors. The second configuration uses Ronchi rulings for focusing.

8.4.2 Conventional Töepler Schlieren Configuration

This type of imaging uses mirrors to collimate light from a light source slit through the test section. As light passes through the test section the light is refracted due to density gradients in the test section. The light leaving the test section is then focused to a point source where a knife-edge is placed to eliminate refracted rays from being focused on the viewing plane. The image is then focused onto the viewing plane where dark regions are formed due to the absence of refracted light and light regions are formed by light rays reaching the plane.

The sensitivity and resolution of Schlieren imaging are limited by the fact that a minimum of 10% change in brightness is needed to produce a detectable difference to draw conclusions on the changes in the index of refraction.

The sensitivity ε'_{min}, which represents the change in angle due to the local change in index of refraction, can be expressed as the goniometric value. Using $\sin(\alpha)=\tan(\alpha)=\alpha$, the thermal dispersion resolution is given by the following equation:

$$\varepsilon'_{min} = 0.1(a/L'_2)\ \text{rad} \tag{8.130}$$

or

$$\varepsilon'_{min} = 20{,}626(a/L'_2)\ \text{arcsec} \tag{8.131}$$

where ε'_{min} is the change in angle resulting in 10% change in brightness in conventional Schlieren imaging, a the image height of the light source above cutoff or knife-edge, and L'_2 the distance from the mirror to the knife-edge location.

The minimum angle change ε'_{min} is directly related to the angle of refraction of the light moving through the test section. The ability to determine the smallest visible refraction angle in the system provides the scenario for Schlieren imaging to detect the smallest density gradients. The resolution of Schlieren imaging is limited by diffraction limits of apertures and lens/mirror geometry.

The spatial resolution d' of the conventional Schlieren technique is defined as the angle of dispersion resulting from thermal effects on the local index of refraction, using the fact that for small angles the angle is equal to the tangent, which is also equal to the sine of the angle,

$$d' = 1.22F'\lambda/A_3 \tag{8.132}$$

Similarly, the lens edge resolution, which is diffraction-limited, is given by the following equation:

$$w' = 1.22(l' - L'_2)\lambda/(m'A_3) \tag{8.133}$$

where w' represents the lens edge diffraction-limited resolution, F' the distance from cutoff knife-edge to the image location, λ the source wavelength, l' the distance from the mirror to the final image location, and A_3 the effective diameter of the focusing lens.

8.4.3 Ronchi Ruling Focusing Schlieren Configuration

This method of Schlieren imaging uses a Ronchi ruling. A Ronchi ruling is a series of thin opaque stripes with an equal width of transparent material between: a grid. The grid is projected on the test object where the light rays are refracted due to density gradients. The light exiting the object under investigation is focused onto a second Ronchi ruling that is one spatial period out of phase with the first Ronchi ruling. The resulting projection is a cancellation image, showing the omission of light due to diffraction resulting from density changes.

The sensitivity and resolution are determined in a similar fashion as for conventional Schlieren imaging.

The sensitivity for the Ronchi ruling type Schlieren imaging is expressed in angular terms as follows:

$$\varepsilon_{min} = 20{,}626aL_g/[L'_1(L_g - l')]\ \text{arcsec} \tag{8.134}$$

where ε'_{min} represents the angle change resulting in a 10% change in brightness, a the light source image height above the cutoff knife-edge, L'_1 the distance from the mirror to the cutoff location, L_g the distance from source grid to imaging lens, and l' the distance from the mirror to the final image location on a screen or on a camera.

The spatial resolution d in contrast is now defined as

$$d = \frac{2(l' - L_g')\lambda}{b} \tag{8.135}$$

The grid diffraction resolution is given by the following equation:

$$w = \frac{2(l' - L_g')\lambda}{mb} \tag{8.136}$$

The distance from the cutoff knife-edge to the image location is given by the following equation:

$$F = (l' - L_g') \tag{8.137}$$

where F is the distance from cutoff (knife-edge) to the image on the plate of camera, w the grid diffraction-limited resolution, m the magnification of the image as described in section 3.1.4.3, and b the distance between cutoff grid lines.

Grids used in focusing Schlieren systems need to be precise and accurate to obtain high-quality images. The following equation calculates the number of line pairs needed to form a quality image on the basis of the type of lens used.

Cutoff grid constraints are defined as

$$\phi = An^* \frac{l' - L_g'}{2l'} \tag{8.138}$$

where Φ represents the pairs of lines involved in forming each point, A the clear aperture of imaging lens, and n^* the grid lines per millimeter at cutoff grid.

One of the advantages of the Ronchi ruling focusing Schlieren system is to adjust the focal plane within the test section. This makes it possible to see certain areas of the test section without interference of density gradients surrounding the area of interest. By acquiring a sequence of images at subsequent depth a three-dimensional image of the sample can be reconstructed.

8.5 Polarization Effects

As in the rest of this book, light is traveling in the z-direction (the z-axis equals the optical axis). Both the electric and the magnetic fields of electromagnetic radiation are always perpendicular to the direction of propagation, which leaves the x-axis and the y-axis as the potential directions to which the electric and the magnetic fields of the electromagnetic radiation can be represented when emitted by a single oscillating charge. Generally the electric field will be taken parallel to the y-axis, and the magnetic field perpendicular to the direction of propagation and to the electric field is then in the direction of the x-axis. Introducing the unit vector of direction as e_i for the x-axis, e_j is pointing in the y-direction, leaving e_k for the z-direction. The direction of polarization is the direction of the electric field by definition. When there are multiple directions of electric field vector present, the combined effect will give a composite polarization. For instance, circularly polarized light will be the vector sum of two perpendicularly oriented electric fields, E_x and E_y, at a constant phase difference of one-quarter wavelength out of phase (assuming that the same amplitude for both the x and y directions will dramatically simplify matters). The expression for circularly polarized light is given by

$$\begin{aligned} E &= E_{x0}\cos(\mathbf{k_k}\cdot\mathbf{z}-\omega t)e_i - E_{y0}\sin(\mathbf{k_k}\cdot\mathbf{z}-\omega t)e_j \\ &= \sqrt{\left(E_{x0}\cos(\mathbf{k_k}\cdot\mathbf{z}-\omega t)\right)^2 + \left(E_{y0}\sin(\mathbf{k_k}\cdot\mathbf{z}-\omega t)\right)^2 - 2E_{x0}\cos(\mathbf{k_k}\cdot\mathbf{z}-\omega t)E_{y0}\sin(\mathbf{k_k}\cdot\mathbf{z}-\omega t)} \\ &= \left(E_{x0}e_i + E_{y0}e_j\sin(-\pi/2)\right)\cos(\mathbf{k_k}\cdot\mathbf{z}-\omega t) \end{aligned} \tag{8.139}$$

or

$$\begin{aligned} E &= E_{x0}\cos(\mathbf{k_k}\cdot\mathbf{z}-\omega t)e_i - E_{y0}\sin(\mathbf{k_k}\cdot\mathbf{z}-\omega t)e_j \\ &= \left[E_{x0}e_i + E_{y0}e_j\sin(-\pi/2)\right]\cos(\mathbf{k_k}\cdot\mathbf{z}-\omega t) \end{aligned} \tag{8.140}$$

which in fact provides clockwise circularly polarized light.
Diagonally linearly polarized light is given by

$$\begin{aligned} E &= E_{x0}\cos(\mathbf{k_k}\cdot\mathbf{z}-\omega t)e_i + E_{y0}\cos(\mathbf{k_k}\cdot\mathbf{z}-\omega t)e_j \\ &= (E_{x0}e_i + E_{y0}e_j)\cos(\mathbf{k_k}\cdot\mathbf{z}-\omega t) \end{aligned} \tag{8.141}$$

When the amplitudes of the x and y fields are not equal, the polarization becomes elliptical.

8.5.1 Polarization in Nature

The scales on the wings of several butterfly species produce interference patterns that enhance the chances of recognition by the opposite sex. These species also have polarized vision. Certain species make use of a prism design in the scales of the wing, which provides a Brewster angle that polarizes the reflected light, while other species use thin-film interference to produce colors as well as polarization.

Another biological application of polarization involves the polarized sky with maximum polarization at 90° with respect to the sun. A large number of animals can detect the polarization under sunlight and use it for navigational

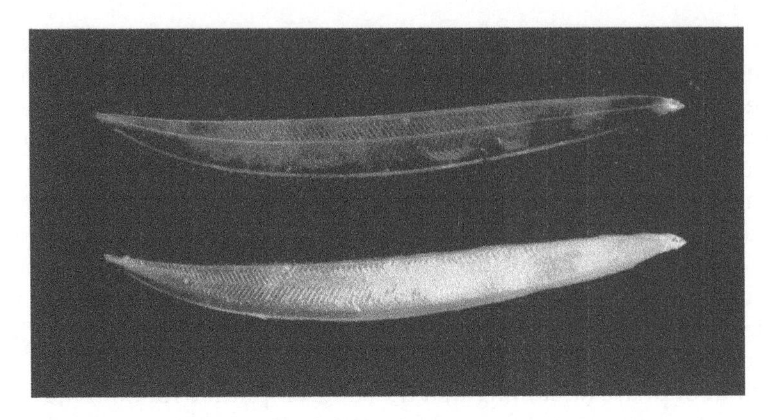

FIGURE 8.7
Two images of the same leptocephalus eel larva. The top image is viewed under unpolarized, transmitted light. The bottom image is viewed under polarized, transmitted light by a camera with a polarizing filter. The increased visibility of the bottom image is due to the presence of birefringent muscle and connective tissue fibers. Birefringent tissues such as these in many transparent organisms allow for a form of camouflage breaking by predators with polarization vision [Courtesy: Edith Widder].

purposes. Yet another animal, the African dung beetle, uses the polarized reflected moonlight for its navigation.

Certain materials, either chemical chains or biological structures, will rotate the angle of polarization of the incident light. This phenomenon is referred to as optical activity of the medium.

In general, the polarization is quickly lost when sending light into a highly scattering medium. This limitation places restrictions on the clinical usefulness of polarized light in diagnostic applications.

Figure 8.7 illustrates the effects of polarized light imaging of a leptocephalus eel larva. The top image is viewed under unpolarized, transmitted light. The bottom image is viewed under polarized, transmitted light by a camera with a polarizing filter. The increased visibility of the bottom image is due to the presence of birefringent muscle and connective tissue fibers.

8.5.2 Polarization in Medical Imaging

Historically, polarized light has been used to distinguish certain tissue types, namely birefringent tissues, because of the tissue's unique structure and the different light–tissue interactions depending on polarization. A birefringent medium has two different refractive indices depending on direction, the ordinary refractive index n_0 and the extraordinary refractive index n_e. Native birefringence is seen in muscle tissue and certain fibrillar collagens, such as the cornea. The birefringence is due to the very regular arrangement of the fibrillar contractile protein macromolecules, actin and myosin, that form the contractile unit of the muscle. In these tissues, unpolarized light can be partially polarized and polarized light can be analyzed to determine things such as thermal damage. Thermal damage is associated with partial and total loss of the native birefringence of these tissues and comes about from dissociation and disruption of the molecules. Reflection is the light–tissue interaction most often used, and photons that interact and scatter in tissue in depth usually lose their polarization state after several scattering events. The amount of reflection depends on the polarization state of the incident light and is described by Fresnel's relations, as discussed Section 3.1.2.4.1 and section 3.1.4.2.

In some cases the polarized light information obtained in histological imaging can provide a direct and irrefutable diagnostic ruling, such as the histology picture of pulmonary silicosis shown in Figure 8.8. Observation under polarization is the criterium for diagnosis.

Polarization effects are also taken advantage of in optical coherence tomography (OCT). Optical coherence tomography is discussed in detail in Chapter 14. Polarization sensitive OCT (PS-OCT) is used to obtain spatially resolved images of polarization changes in muscle, bone, skin, and brain. These fibrous tissues are particularly useful in PS-OCT. Both birefringence and scattering by particles can change the polarization state of light propagating through a turbid media, such as tissue. Except for scatterers arranged in a macroscopic order, scattering would change the polarization state in a random manner. Birefringence is likely to be the dominant factor for producing the images in PS-OCT. This is accomplished by coherently detecting two orthogonal polarization states of a signal formed by interference of light reflected from the tissue sample and the reference arm mirror of a Michelson interferometer. PS-OCT provides high-resolution spatial information on the polarization state of light reflected from tissue, which is not discernible, using existing diagnostic optical methods. An even stronger case can be made for birefringence by measuring the Stokes vector of the reflected light. PS-OCT can be used to completely characterize the polarization state of light backscattered from tissue by determining the depth-resolved Stokes parameters.

Polarization may also play a role in biostimulation. Biostimulation, as described earlier in this chapter, is roughly defined as the potential effects of extremely low levels of light energy on biological tissue. Polarization along with coherence and narrow bandwidth are laser characteristics thought to be of primary importance for understanding the effects of biostimulation. And as such the effect of polarization is one of the open questions yet to be resolved in the mechanism of biostimulation.

8.6 Optical Activity

Optical activity indicates the tendency of a solution or a solid to rotate the direction of linearly polarized light by a certain angle on the basis of the concentration or the composition. The optical activity of the material is

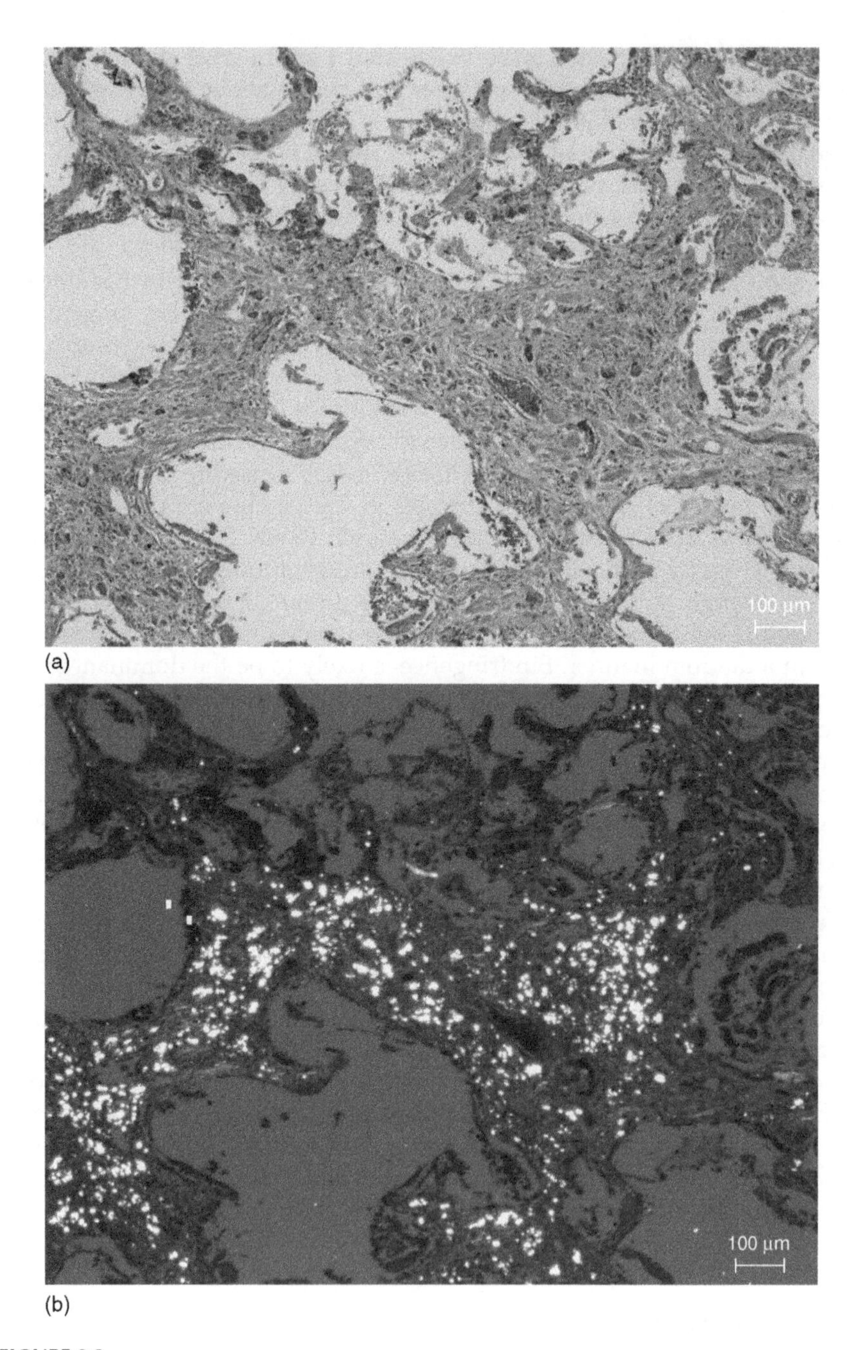

FIGURE 8.8
Pulmonary silicosis stained with hematoxylin and eosin. Image (a) is taken with unpolarized light, and image (b) is taken with polarized light. The striation observed under polarized images provides the crucial clinical information. (From Dr Paulo Sampaio Gutierrez, Heart Institute (Incor) of the University of São Paulo Medical School, São Paulo, Brazil.)

independent of the direction in which the light traverses through the medium, in fact it is a function of the individual molecules that make up the medium or are dissolved in the medium.

The angle of rotation β_r is the result of the different indices of refraction for parallel polarization n_l and for perpendicular polarization n_r:

$$\beta_r = \frac{\pi d}{\lambda_0}(n_l - n_r) \tag{8.142}$$

where d equals the distance the light travels through the optically active medium and λ_0 is the wavelength of the source.

When a medium has an index of refraction for parallel polarization greater than that for perpendicular polarization the rotation will be clockwise or D-rotation (dextrorotary, *dexter*: Latin), whereas in the opposite case the rotation will be counterclockwise (levorotary, *laevus*: Latin). Mixtures of D- and L-isomers are called racemic mixtures.

The specific rotary power of a specimen is defined as the angle of rotation with respect to the optical path length as

$$\text{specific rotary power} = \frac{\beta_r}{d} \tag{8.143}$$

In general, l-active or "left-handed" molecules either are not biologically active or are unhealthy and may even be poisonous. However, predominantly levorotary forms of amino acids are found in the proteins of living organisms.

One example of optically active biological media is the group of amino acids. Amino acids are tetrahedral structures where the central carbon (C_0) is chemically bound to an amino group, a carboxyl group, a hydrogen atom, and a specific variable chain of molecules (R), which gives the amino acid its individual identity. Amino acids are asymmetric molecules or chiral molecules because the four molecular groups attached to each α carbon can be arranged in two possible configurations (except for glycine). These configurations are mirror image isomers or enantiomers. The two species of amino acids can rotate the plane of polarized light either clockwise (dextrorotary) or counterclockwise (levorotary). Living organisms make and use L-forms. Dead protein has a mixture. The longer it has been dead, the closer the D/L ratio may come to 1/1.

This leaves polarization as an option for dating biological tissues.

The basic principles underlying the "biological chronometer" of amino acids are as follows:

1. Proteins are organic macromolecules built up by amino acids
2. Vital amino acids (approximately 20 kinds of amino acids in total) are levorotary, except for glycine, which does not contain any asymmetric carbon atom.
3. Proteins outside vital metabolism (i.e., from dead organisms) degrade into their component amino acids as time goes by.

4. Degradation occurs in such a way that the concentration of levorotary amino acids (L-) decreases with time, while the concentration of dextrorotary amino acids (D-) increases, until an equilibrium point is reached. This process is known as "racemization."
5. The fraction of both L/D in the equilibrium state and the rate of conversion of L-amino acids to D-amino acids depend on how many asymmetric carbon atoms a given amino acid contains.
6. It is the objective to determine the speed of racemization of a given amino acid and construct a temporal relationship, i.e., a functional relation between the rate of conversion of the L- to the D-form, and time.
7. The relative abundances of the two components are determined by standard chromatography, and a timeline for the observed degree of conversion of the L-form to the racemic mixture is also determined.

The process of racemization depends on many environmental factors, and other parameters and variables such as:

- Temperature
- Water concentration in the environment
- pH (acidity/alkalinity) of the environment
- Whether the amino acid is in the bound state or free state
- The size of the protein, if in the bound state
- The precise and specific location of the amino acid in the protein
- Whether the amino acid is in contact with catalytic surfaces, such as clay
- Whether in the presence or absence of aldehydes, and metallic ions
- The concentration of buffer compounds
- Ionic strength of the environment

8.6.1 Glucose Concentration Determination

Several sugars have optical activity. Sucrose is a disaccharide composed of one molecule each of glucose and fructose. The acid-catalyzed hydrolysis of sucrose to glucose and fructose produces two different optically active chemicals. D-Fructose is L-active, whereas D-glucose is D-active. D-glucose is the predominant sugar found in living organisms.

The degree of rotation for a sucrose solution can be found on the basis of the concentration and the specific rotation of sucrose, defined as

$$\beta_r = \frac{\pi}{\lambda_0}(n_l - n_r)d^2[\text{sucrose}]\alpha_D' \tag{8.144}$$

In this equation, β_r is the amount of rotation in degrees, d the path length of the light through the solution in m, [sucrose] the concentration of the solute in kg/L of solution, and $\alpha_D' = 66.48°/\text{m kg}$ is the specific rotation for sucrose at $T = 25.0°\text{C}$.

In general, Equation (8.144) can be simplified for the standard operating procedure in glucose concentration instrumentation using a sodium lamp with the sodium "D" line (an unresolved doublet at 589.0 and 589.6 nm) as light source:

$$\beta_r = d[\text{sucrose}]\alpha'_D \tag{8.145}$$

8.7 Evanescent Wave Interaction in Biomedical Optics

Evanescent waves interact with tissue only within a layer extending a fraction of the wavelength of light away from the wave guide surface. This localization is ideal for biosensing as a thin film of biorecognition molecules can be deposited on the wave guide surface and in the evanescent wave region. This results in the optical interrogation of only the recognition layer and not the bulk of the sample. This has a number of advantages compared to traditional optical methods of analysis.

Evanescent wave diagnostic and therapeutic applications rely on the interaction of light with tissue at the interface between transparent, high-refractive-index optical materials (prisms and fibers) and tissue. When using laser light, the use of evanescent optical waves for precise, controlled tissue ablation (cutting) offers a new approach to examine tissues in the infrared (2–10 μm), where there are substantial differences in the absorption spectra of water-rich and lipid-rich tissues.

8.7.1 Evanescent Optical Waves

Electromagnetic radiation is totally reflected from an interface defined by a high-refractive-index medium n_1 and a low-refractive-index medium n_2 ($n_1 > n_2$) when the angle of incidence exceeds the critical angle (see Figure 8.9). The critical angle θ_c for total internal reflection is defined by Snell's law: $\sin(\theta_c) = n_2/n_1$. The boundary conditions require that the electric field, and hence energy, be present in a layer somewhat less than one wavelength thick, on the n_2 side of the interface. The waves in this layer are called evanescent waves because they decay rapidly to zero away from the interface. The plane wave evanescent electric field amplitude in a transparent medium is given by

$$E_e(x,y,t) = E_e \exp[i(k_x x - \omega t)]\exp[-\gamma_e k_y y] \tag{8.146}$$

where

$$\gamma_e = \sqrt{[n_1^2 \sin^2(\theta_i) - n_2^2]} \tag{8.147}$$

and k_y and k_x are the wave vectors in the planes perpendicular and parallel to the interface. Since the wavefronts or surfaces of constant phase (parallel to the

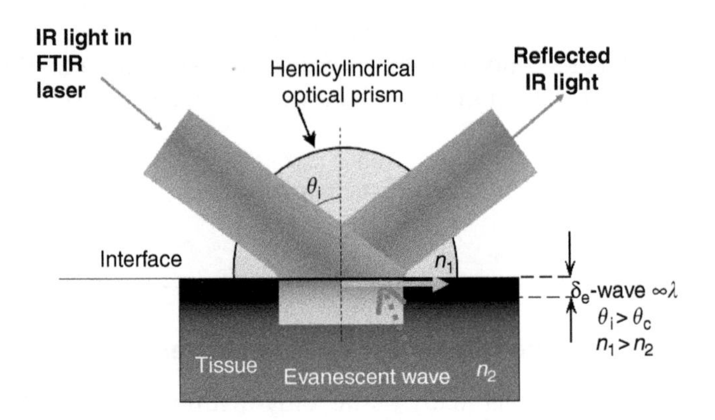

FIGURE 8.9
Schematic representation of the beamline used to launch evanescent optical waves at a prism–tissue interface. The interface is defined by a high-refractive-index medium (prism) in contact with a low-refractive-index medium (tissue); $n_1 > n_2$. The angle of incidence θ_i has to be greater than or equal to the critical angle θ_c for an evanescent wave to be generated.

yz-plane) are perpendicular to the surfaces of constant amplitude (parallel to the *xz*-plane), Equation (8.146) is an inhomogeneous wave. The power (irradiance) is proportional to $E_e^2 e^{-2\gamma k_y y}$.

The penetration depth δ_e of the evanescent wave in a transparent medium is

$$\delta_e = \frac{1}{2\gamma_e k_y} = \frac{1}{4\pi\sqrt{\left[n_1^2 \sin^2(\theta_i) - n_2^2\right]}} \tag{8.148}$$

The amplitude of the evanescent wave decays rapidly in the *y*-direction and becomes negligible at a distance of only a few wavelengths of light.

Now in the case of an absorbing external medium n_2, reflection is *not* total and the reflectance from the interface is $R<1$. When the lower refractive index medium has absorption, the index n_2 is replaced by the magnitude of the complex refractive index, $|n_2| = n_2 - ik_e$, where k_e is the extinction coefficient and is defined as $k_e = \lambda\mu_a/4\pi$. This introduces the absorption coefficient of the external medium (e.g., tissue) into the solution in Equation (8.146) and describes the loss of "total" internal reflection due to absorption within the evanescent wave. The perpendicular and parallel reflectance at the optic–tissue interface, $R_\perp$ and $R_\parallel$, respectively, can be written from Fresnel's relations as

$$R_\perp = \frac{n_1\cos(\theta_1) - n_2\cos(\theta_2)}{n_1\cos(\theta_1) + n_2\cos(\theta_2)} \tag{8.149}$$

and

$$R_\parallel = \frac{n_2\cos(\theta_1) - n_1\cos(\theta_2)}{n_2\cos(\theta_1) + n_1\cos(\theta_2)} \tag{8.150}$$

where $\perp$ and $\parallel$ are the polarizations perpendicular to and parallel to the plane of incidence, respectively.

The fraction of absorbed incident energy in water (mimicking wet tissue absorption in the infrared spectrum) is given by

$$(1-R_{\perp}) = \frac{4\beta_i}{1+2\beta_i+\beta_i^2+\beta_r^2} (1-R_{\perp\parallel}) = \frac{4\beta_i}{1+2\beta_i+\beta_i^2+\beta_r^2} \tag{8.151}$$

where $\beta_{r,c}=n_2\cos(\theta_2)/n_1\cos(\theta_1)=\beta_r+i\beta_i$ is the complex reflection coefficient. The reflection coefficients for the polarization components are related by a ratio of the refractive indices squared, $\beta_{\parallel}=(n_1/n_2)^2\beta_{\perp}$. Because of absorbance in the evanescent wave, the reflectance at the interface is less than 100% and the critical angle loses some of its significance.

The interest in evanescent waves is that they are a novel means for limiting the penetration depth of laser pulse energy independently of the absorbing properties of the external medium (tissue). Fortuitously, the wavelengths of visible and near-infrared light are about right for the evanescent wave field to give the desired 0.1–1 μm penetration depth (δ) analogous to that of the 193 nm excimer lasers now used for extremely precise corneal surgery. In the plane wave approximation for electromagnetic radiation at interfaces between dielectric optical materials and water, we analyze the deposition of energy with depth, predict the requirements for laser pulses to achieve evanescent wave-initiated ablation, and assess the feasibility of ablation within the limits of rugged visible-infrared optical materials such as silica, sapphire, zinc sulfide, silicon, and germanium. This analysis strongly suggests that evanescent wave-initiated ablation of tissue is possible for some tissues in the near-UV and at the water absorption bands of the near- and mid-infrared.

Figure 8.10 shows evanescent optical waves at an interface between sapphire (Al_2O_3) and water and zinc sulfide (ZnS) and water. Shown are (a) the penetration depth of the evanescent wave, (b) the fraction of absorbed incident energy, and (c) the required incident energy for ablation of water (vaporization) by laser energy for $\perp$-polarization as a function of incident angle for wavelengths from 2 to 4 μm. The leftmost part of each curve corresponds to the critical angle for that wavelength, for instance, $\theta_c\approx48°$ at 2.1 μm. Note the order-of-magnitude variation in the evanescent wave penetration depth, except near 3 μm. Near the peak of the absorption curve in water (2.94 μm), the penetration depth versus incident angle is almost constant, varying from about 0.3 to 0.2 μm. Also note that the absorbed fraction of incident energy does not go to zero at the critical angle, because of absorption in the evanescent wave. There is little difference between the absorbed fractions for the two polarizations as a function of wavelength between 1 and 6 μm.

However, the //-polarization always has a slightly larger absorbed fraction at the critical angle. The required incident energy is obtained by multiplying the latent heat of vaporization of water by the penetration depth δ_e and the laser beam area, and dividing by the absorbed fraction $1-R_{\perp}$; (2500 J/cm^3 $\times$ $\delta_e \times (\pi\omega_o^2/2)/(1-R_{\perp})$, and may be a factor of 8 less if the partial vaporization

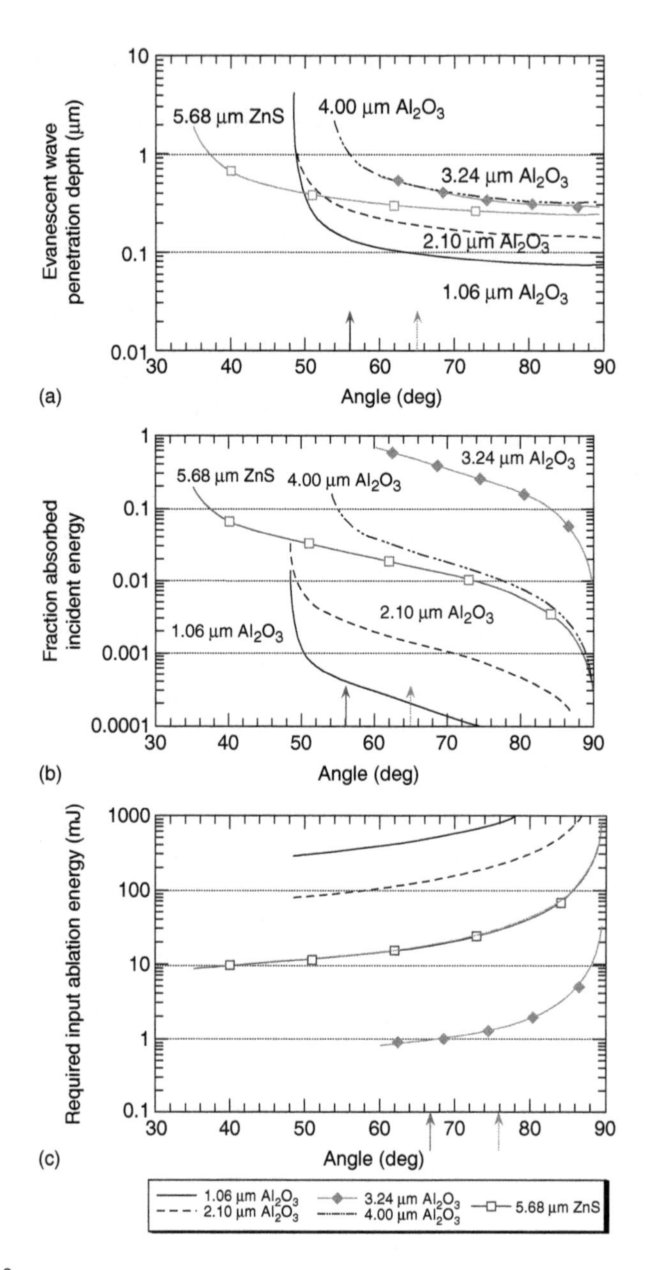

FIGURE 8.10
Evanescent optical waves at an interface between a high-refractive-index optic (sapphire; Al_2O_3, and zinc sulfide ZnS) and water. (a) The penetration depth of the evanescent wave, (b) the fraction of absorbed incident energy, and (c) the required incident energy for ablation of water (vaporization) by laser energy for perpendicular polarization as a function of incident angle for wavelengths from 1 to 6 μm in optical materials such as sapphire (dotted arrow) and zinc sulfide (solid arrow). The required incident energy is obtained by multiplying the latent heat of vaporization of water by the penetration depth δ_e and the laser beam area, and dividing by the absorbed fraction $1 - R_\perp$ it may be a factor of 8 less if the partial vaporization model holds for $E_\sigma \approx 330$ J/cm^3. Beam area calculation assumes a 500 μm Gaussian beam waist, w_o.

model holds for $h_v \approx 3.30 \times 10^8$ J/m^3. The laser beam area calculation assumes a 500 μm Gaussian beam waist, ω_o. The input energy required to induce ablation in water at a sapphire–water interface using a 500 μm Gaussian beam waist is less than about 30 mJ for all wavelengths except 2.1 μm, where the input energy required is roughly 100 mJ. Pulses used for evanescent wave introduction can be delivered at most wavelengths by free electron lasers (FEL), such as the Mark III at Duke University, Durham, NC, USA.

The optical wave penetration depths for several interfaces between a high-index transparent solid and water are compared in Figure 8.11, at a wavelength of 2.1 μm. Figure 8.11a shows the range over which the optical penetration depth can be reduced by using evanescent waves at a wavelength where the normal-incidence penetration depth is substantial (400 μm). The normal-incidence penetration depth is reduced by more than three orders of magnitude by launching evanescent waves. Note that even at a 60° angle of incidence in sapphire, the depth of penetration is reduced by 1000×. Figure 8.11b compares the amount of absorbed incident energy available for ablation. In spite of the small values for the absorbed fraction in Figure 8.11b, the fluence requirements are easily obtained. For instance, in ZnS at 50° where the absorbed fraction is only 0.001, the required fluence is 0.05 J/cm^2 and the energy per pulse is 30 mJ, for a 200 μm Gaussian beam waist. These are modest pulse energy requirements for FELs, and indeed most lasers.

In Table 8.1 the penetration depth for evanescent waves and normally incident waves at 0° are compared for wavelengths ranging from 3 to 6 μm for several optical materials. The critical angle for a high-refractive-index medium–water interface can be quite small (<20°), as shown with Ge in Figure 8.11b. Also shown are the fraction of incident energy absorbed by water from the evanescent wave, the predicted incident fluence and energy required for vaporization in water, and the laser pulse duration for thermal confinement within the evanescent wave depth. Note the order-of-magnitude step in the normal depth of penetration $\delta(0)$ across these wavelengths, with a lower limit of 1 μm. Compare this to the order-of-magnitude "tunability" in δ_e at any one wavelength in Figure 8.11a with a lower limit of 0.15 μm.

The practicality of evanescent-wave-driven tissue ablation at the 1.94 μm water absorption band is apparent from this analysis. In the case of 2.1 μm radiation where the absorbed fraction is small (0.026), the required incident energy at the interface is only 14 mJ with a pulse duration (calculated from Equation [8.19]) of less than 9 μs. Also of interest are wavelengths near 2.7 μm, where transmission through low-hydroxy-fused-silica fibers is high and the refractive index of water is low (n_2=1.18). As shown in Table 8.1, evanescent wave ablation should occur for each wavelength at modest energy densities, well below the intensity threshold for optical breakdown of these optical materials.

8.7.2 Precise, Controlled Light Delivery with Evanescent Optical Waves

Evanescent waves have long been used in attenuated total reflectance spectroscopy, in which the evanescent wave field is used to measure absorption

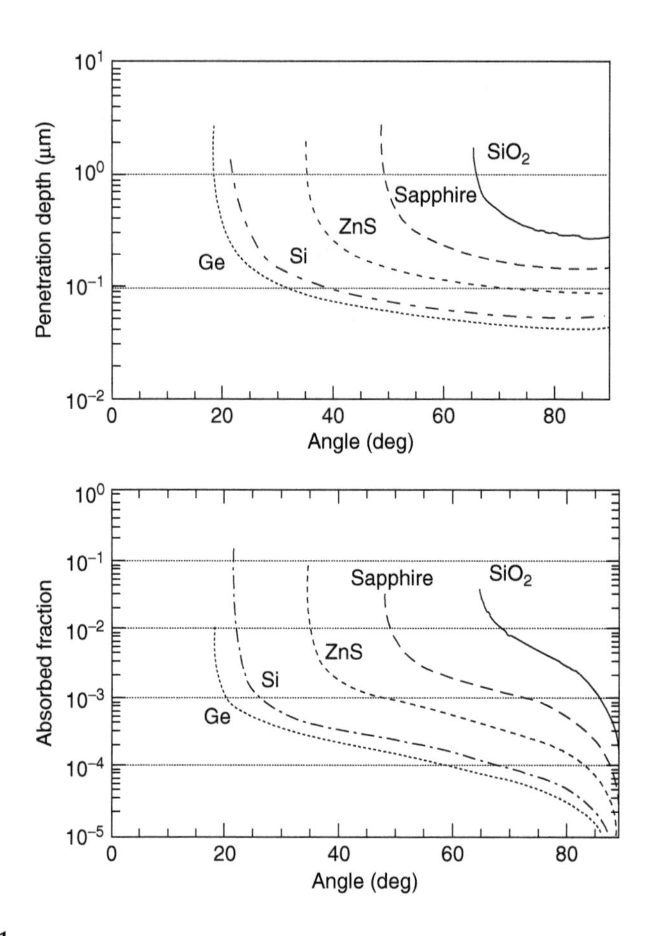

FIGURE 8.11
(a) The penetration depth of evanescent optical waves at interfaces between high-refractive-index transparent solids and water, as a function of incident angle for a wavelength of 2.1 μm. The normal-incidence penetration depth is 400 μm and can be reduced by more than three orders of magnitude by launching evanescent waves. Note that even at a 60° angle of incidence in sapphire, the depth of penetration is reduced by 1000×. (b) A comparison of the absorbed fraction of incident energy available for ablation. Refractive index values for these materials at 2.1 μm are 1.436 (SiO_2), 1.735 (sapphire), 2.262 (ZnS), 3.45 (Si), 4.095 (Ge), and 1.298 (water). (Reprinted with permission from The Optical Society of America: Hooper, B.A., Domankevitz, Y., Linc., and Anderson, R.R., Precise, controlled laser delivery with evanescent optical waves, *Appl. Opt.*, 38, 5511–5517, 1999)

spectra at surfaces in contact with high-refractive-index crystals. Spectroscopic and diagnostic applications at the boundary of fiber optics have been identified using evanescent waves, as well. In the therapeutic realm, evanescent optical waves provide an alternative for confining light energy near the tissue surface. Precise, superficial tissue ablation can also be achieved with evanescent waves generated at a high-refractive-index optic–tissue interface, where the ablation depth may be varied as a function of wavelength, angle of incidence, and refractive indices of the optic–tissue interface. Precise laser surgery traditionally uses laser pulses at wavelengths

TABLE 8.1

Evanescent Wave Ablation Parameters at the Critical Angle

Wavelength (μm)	2.94	3.24	3.50	5.68
Optic	Al_2O_3	Al_2O_3	ZnS	ZnS
Critical angle θ_c	51°	60°	38°	35°
(optic/tissue)	55°	65°	45°	56°
Penetration depth δ_0 (μm)	1	4	27	25
Penetration depth δ_e (μm)	0.26	0.47	0.36	0.33
Absorbed fraction $(1 - R_s)$	0.67	0.5	0.04	0.03
Incident energy for ablation $E_{ablation}$ (mJ)				
(beam diameter ~ 1 mm)	0.8	1	35	14
Ideal pulse duration				
$\tau_{laser} < \tau_r = \delta^2/4\alpha = (4\mu_a{}^2\alpha)^{-1}$ (μs)	~1	~1	~1	~1

that are strongly absorbed at the surface of tissue. However, pulses at these wavelengths (far-UV, far-IR) are not compatible with fiberoptic transmission, making endoscopic surgical procedures inside the body difficult. An alternative for confining light energy near the tissue surface uses evanescent optical waves that are shallow, variable, and controllable at the optic–tissue interface. The optical penetration depth is on the order of microns, is variable as a function of wavelength, angle of incidence, and the refractive indices of the interface, and controllable where the light energy is confined to the surface as a surface wave; hence, there is no free-beam propagation. A new class of precise, controlled laser surgical tools may be achieved in this novel approach for use in both diagnostic and therapeutic endoscopic procedures.

Ideally, lasers can ablate tissue with exquisite precision and almost no damage to the surrounding tissue. This requires extremely shallow optical penetration, such that only a microscopic zone near the tissue surface is affected. Traditionally this has been achieved by using short pulses at wavelengths that are strongly absorbed by proteins (193 nm ultraviolet radiation; e.g., excimer lasers) or water (3 μm infrared radiation; e.g., erbium lasers). Unfortunately, pulses at these wavelengths are poorly compatible with transmission through fiber optics. In this section we present a novel method for achieving precise laser ablation, based on wave optics, in which shallow optical penetration can be achieved, which is not solely dependent on the absorbing properties of the tissue. Using evanescent optical waves, laser energy can be confined to a layer less than one wavelength thick at the surface of high-refractive-index devices. This layer is thin, not because of strong absorption, but because the evanescent optical wave's amplitude decays exponentially as it enters the less dense medium (tissue). Devices can be made in which laser optical energy outside the device is present only within a few micrometers of the device's surface, rather than propagating as a laser "beam" into the tissue.

Maxwell's wave equations were analyzed for all angles of incidence and polarizations to predict the evanescent optical fields arising in water (an analogy for wet tissues) at interfaces with rugged visible-infrared optical materials such as silica, sapphire, zinc sulfide, silicon, and germanium.

Based on this analysis, the useful evanescent field depth is on the order of 0.1–2 μm. For efficient ablation of tissue with minimal thermal damage, a pulse duration achieving thermal confinement on this scale is necessary. For example, for a 1 μm depth of penetration the estimated ideal pulse has a duration of less than or equal to 1 μs and an energy of at least tens of millijoules. A linac-driven FEL is well suited to these requirements. The pulse duration is variable over the desired range and the wavelength can be tuned through the mid-infrared water absorption bands to study deposition of energy, cavitation (vaporization of water), and tissue ablation arising from evanescent wave interactions. The next section presents tissue ablation using evanescent wave–tissue interactions.

8.7.3 Tissue Ablation with FEL-Generated Evanescent Optical Waves

Ablation of biological tissue by lasers occurs predominantly via rapid thermal vaporization of tissue water. Other processes may coexist with thermal vaporization, including photochemistry, explosive mechanical removal, or plasma-initiated shock waves and mechanical fracture. Such ablation by normally incident, well-absorbed short laser pulses is fairly well described by a first-order model that relates optical absorption, depth of ablation and thermal injury, and pulse duration and fluence (energy/area). Since evanescent waves experience strong attenuation, the Beer–Lambert law can be applied and it can be shown that the steady-state fluence rate $\Psi(z)$ inside tissue decreases exponentially with depth z:

$$\Psi(z) = \Psi_0 \exp(-\mu_a z) \tag{8.152}$$

where μ_a is the evanescent absorption coefficient and $\delta = 1/\mu_a$ the characteristic penetration depth.

In this surface layer of thickness δ, the incoming energy is converted into heat, which immediately begins to diffuse into the surroundings in accordance with the thermal relaxation time τ_r of the tissue. Efficient ablation results when the laser pulse duration τ_p is "short," i.e., smaller than τ_r, and when the heat of vaporization for tissue water is delivered to the surface by each pulse. In first approximation, for wavelengths where δ is smaller than the laser spot size, the thermal relaxation time τ_r is found to obey the following equation:

$$\tau_r \approx \frac{\delta^2}{4\alpha} = \frac{1}{(4\mu_a^2 \alpha)} \tag{8.153}$$

where α is the thermal diffusivity of the tissue. Experimental data with a variety of pulsed lasers suggest that tissue ablation requires an energy/volume of $E_V \approx 2500\ \mathrm{J/cm^3}$ (similar to pure water) such that ablation is reliably achieved when the fluence, $F_0 \approx (2500/\mu_a)$ J/cm^2. A partial vaporization model suggests a smaller value, $E_V \approx 330$ J/cm^3, where vaporization takes place at a few nucleation sites and not over the entire surface area of the laser beam.

Each short laser pulse of sufficient incident energy removes a layer of thickness $\sim \delta$ and thermally denatures a layer of several times δ. The optical penetration depth δ is important not only for determining the amount of tissue removed and the residual thermal damage, but also for choosing the optimal laser wavelength and pulse duration. For micron-scale surgical precision using normally incident laser pulses, high tissue absorption coefficients below 220 nm and near 2.94 μm are typically targeted. However, micron-scale precision can be obtained by using evanescent waves, even at wavelengths where optical absorption is weak.

An optical probe, consisting of HeNe laser light reflected from the prism–tissue interface and captured by a fast silicon photodiode, can be used to detect the onset of cavitation in water and the ablation of tissue. At a wavelength of 3.24 μm, ablation of freshly sacrificed porcine aorta and skin performed at a sapphire prism–tissue interface for angles of incidence of 0 and 65° has been investigated. The critical angle for this wavelength, assuming no absorption, is 60°. The 65° angle would then correspond to an evanescent wave at the prism–tissue interface for a transparent tissue. However, at this wavelength the tissue has absorption and at 65° the absorbed fraction $1 - R_\perp$ is 50% (see Figure 8.11c), so the strict definition of the critical angle no longer applies.

Ablation of porcine aorta, cornea, dermis, and lens tissue was studied using the Mark III FEL at Duke University at wavelengths from 2.9 to 3.5 μm. A sapphire hemicylinder prism was used as the dielectric solid to launch evanescent waves at an angle of incidence of 65°. This angle gives an evanescent penetration depth of 0.2–0.5 μm, compared to a normal-incidence penetration depth of 1–30 μm, over this range of wavelengths. In addition, cholesterol-fed rabbit aorta was studied for selective ablation characteristics using evanescent waves delivered through hollow wave guides (HWGs).

The experimentally measured onset for cavitation/ablation using an evanescent wave at 65° is the same for water and porcine aorta (0.33 J/cm^2), where the effective Gaussian beam waist for this incident angle is 200 μm. The ablation threshold energy per pulse is 0.2 mJ and the pulse duration is 1 μs. This is very close to the theoretical value for incident fluence required for reliable cavitation in water, 0.31 J/cm^2 (2500 J/cm^3 $\times$ $\delta_e/(1 - R_\perp)$), which takes into account that tissue is about 70–80% water, $\delta_e = 0.47$ μm and $1 - R_\perp = 0.5$. The ablation mechanism in this case is believed to be the vaporization of tissue water and subsequent explosive mechanical removal of tissue.

Figure 8.12 shows dramatically the range of penetration into tissue for two different light–tissue interactions. Toluidine blue-stained histology sections of ablated porcine aorta are shown in Figure 8.12a and Figure 8.12b. An image is shown for each interaction, a 0° normal-incidence and a 65° evanescent wave. These cuts were made in multiple passes by translating a sapphire prism–tissue interface through the laser beam. Each cut is made with approximately 60 pulses of laser light, with energies per pulse incident on the tissue of 2.5 mJ for the normal-incidence wave and 1.3 mJ for the evanescent wave. These energies per pulse are about an order of magnitude larger than the ablation threshold energy. Figure 8.12 shows dramatically the range of penetration into tissue for these two angles of incidence. At normal incidence, the calculated depth of penetration per pulse is about 4 μm for this wavelength, while the evanescent wave penetration depth is 0.47 μm. For normal incidence (Figure 8.12a), the laser has high transmission through the prism–tissue interface regardless of whether a cavitation bubble (vapor interface) has formed or not, and hence the 60 laser pulses ablate to a depth

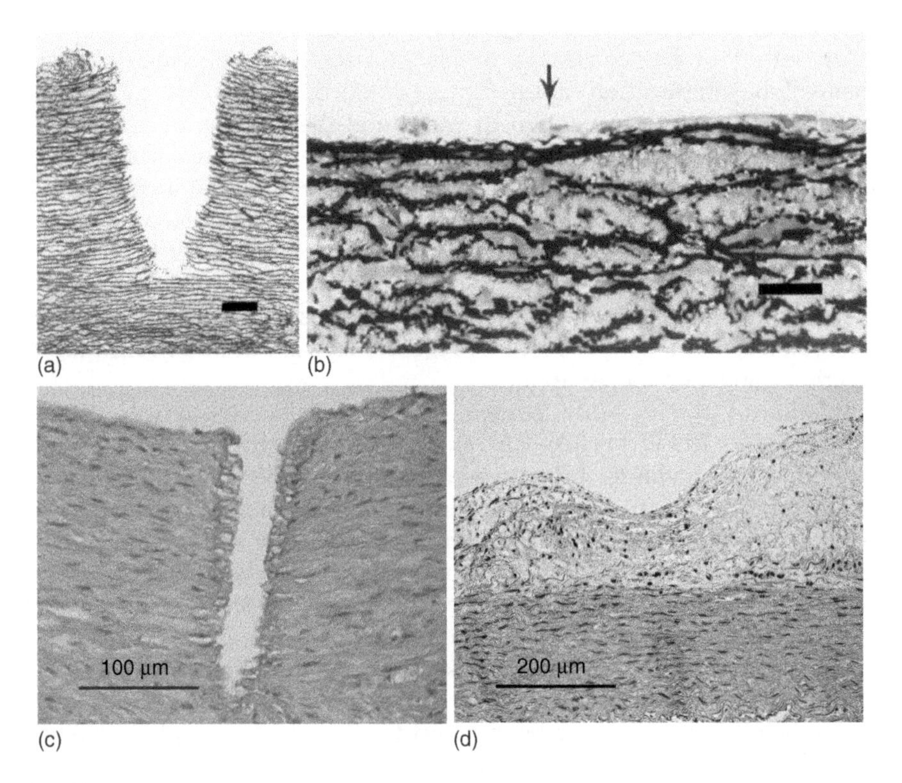

FIGURE 8.12
Illustration of tissue ablation under Mark III free-electron laser irradiation. Irradiation was provided under the following parameters, using (a) and (c) normally incident (0°) and (b) and (d) evanescent optical waves at an optic–aorta tissue interface with free-electron laser (FEL) light. Image (a) is a toluidine blue-stained histology section (bar = 100 μm), (b) is a toluidine blue-stained histology section (bar = 20 μm, arrow is at edge of ablation), (c) is a hemotoxylin and eosin (H&E)-stained histology section (bar = 100 μm), and (d) is a Masson trichrome-stained image (bar = 50 μm) of ablated aorta. Images (a) and (c) show ablation depths of several hundred microns, and images (b) and (d) show evanescent ablation depths of about 4 and 100 μm, respectively. Also visible in (b) are the black elastic layers and the gray smooth muscle cell layers that underlie the endothelial cells at the surface. Note the thick atherosclerotic fatty plaque that covers the vessel wall in (d). (Figure a and b reprinted with permission from The Optical Society of America: Hooper, B.A., Domankevitz, Y., Linc., and Anderson, R.R., Precise, controlled laser delivery with evanescent optical waves, *Appl. Opt.*, 38, 5511–5517, 1999)

of about 500 μm (roughly 60 × 2δ). Figure 8.12b shows a 65° evanescent wave ablation site with the top layer of endothelial and smooth muscle cells removed to the left of the arrow. Also visible are the black elastic layers and the gray smooth muscle cell layers that make up the media (wall) of the aorta. The endothelial cells are several microns thick, indicating that the ablation depth is roughly 4 μm (4 × 2δ_e) and not 500 δm (60 × 2μ), as one sees with normal-incidence irradiation, Figure 8.12a. The ablation depth for the evanescent wave is limited by the penetration depth, independent of the number of laser pulses.

The ablation shown in images in Figure 8.12c and Figure 8.12d was obtained with laser light delivered through a 1 mm inner diameter HWG and presented to a sapphire parallel plate in contact with the tissue in Figure 8.12c, and through a zinc sulfide conic optic tip for the evanescent wave in Figure 8.12d. A hematoxylin and eosin (H&E)-stained histology section of ablated rabbit aorta is shown in Figure 8.12c for 0° normal incidence. The ablated trench is made with 200 pulses of 2.94 μm IR FEL light, with energies per pulse incident on the tissue of 12 mJ. For the evanescent wave ablation of Figure 8.12d, 100 pulses of 3.50 μm FEL light were used at an energy per pulse of 0.8 mJ. This wavelength was chosen to target the fat-rich atherosclerotic plaque that is visible on the top of the vessel wall in this image.

No evidence of optical damage to the sapphire prism due to cavitation was noted, even for fluences much higher than the ablation threshold and after thousands of pulses. In contrast, the ZnS conic optic was damaged in some of the experiments. This may have been due to cavitation of the tissue water at the interface. Damage to the tissue due to cavitation does not appear to be severe in the evanescent cases, based on the histology of the ablated aorta. The dynamics of cavitation and tissue injury at these interfaces need further study.

Extremely precise superficial ablation of tissue using laser-generated evanescent optical waves has been achieved at high-refractive-index optic–tissue interfaces for both a bench-top optic and a catheter configuration. Compared with normal-incidence exposure, the depth of ablation is limited by the evanescent wave penetration depth, independent of the number of laser pulses. For many pulses, on the order of 100, the normal-incidence ablation depth can be several hundred microns, compared to several microns for the evanescent wave interaction. The practical implication is that high-precision, endoscopic laser surgical devices can be produced, with the added advantage of control of the laser energy. Unlike free-beam laser surgery, only tissue in contact with the optical interface of an evanescent wave device is ablated, allowing for safe infrared beam delivery in the operating room. These tools may allow the shape of the interface to be "copied" onto the tissue, and may be ideal for precise resurfacing procedures.

On applying the phase information from the wave phenomenon to biomedical optics the issue of interference comes into play, which is discussed in the next section.

8.8 Phase Interference Effects

Interference uses the phase principle of the wave concept as described in Chapter 2 by allowing waves to interact and add for imaging and therapeutic applications.

The interference radiance pattern generated on the optical axis of an interferometer is given by

$$L = L_1 + L_2 + \sqrt{L_1 L_2}\cos\left(\frac{2\nu d}{c} + \varphi\right) \tag{8.154}$$

where v is the frequency of the light, L_1 and L_2 are the respective radiance collected from the two interfering beams, d the path length difference between the two arms of the interferometer, and φ an additional phase difference between the two interfering arms introduced by either the index of refraction in one of the arms or other measurement artifacts. It is clear from Equation (8.154) that this interference pattern varies as a cosine wave, and that as d increases there will be no change in the variation of the field. The details that can be resolved with interferometry are only a function of the detection accuracy of the light. The signal-to-noise ratio thus plays an important role, since valuable phase information can be lost in phase noise. In general the phase difference will need to be in the order of the same percentage of the wavelength as the signal-to-noise ratio of the measured signal for discriminating any differences. The least significant figure in the detection will stipulate the minimally detectable phase difference.

Some of the interferometers used in medical imaging are Michelson, which uses reflection measurement; Mach–Zehnder, which uses transmission measurement; and Moire, which is a different type of interferometer, using a pattern frequency to extract information instead of the frequency (wavelength) of the light source itself.

Examples of the Michelson, Mach–Zehnder, and Moire interferometer will be described in Chapter 14 in detail.

8.8.1 Interferometry in Medical Imaging

Interferometry is a powerful technique for retrieving the phase of coherent wave fields. The interferometric signal is complex in nature, and can be related to the linear difference between the sample and reference arm phase functions. If both arms are free-space (i.e., the index of refraction $n = 1$), then the difference in phase reduces to $2k\Delta x$, where Δx is the free-space path length difference between the two arms. The interference of these two paths in the interferometer allows one to profile the photons that interact with tissue at depths up to several millimeters. There are several flavors of interferometry used in imaging tissues that have seen considerable work in recent years; light scattering spectroscopy (LSS) and OCT, which take advantage of light scattering, and Fourier transform infrared (FTIR) spectroscopy, which observes absorption in tissue. The scattering interactions take advantage of the fact that tissues in the visible and near-infrared are highly scattering, and light sent into tissue can be recollected in reflection with information about depth, structure, and some function obtained. In the infrared, tissue chromophores are heavily absorptive and absorption dominates scattering. The absorption peaks in the spectrum due to these chromophores are unique for each chromophore, as they are a signature of the vibrational chemical bonds and can provide quantitative functional information about the tissue. A typical configuration is a Michelson interferometer with one of the mirrors replaced by tissue. A scan of the remaining mirror then yields coherent photons that have interfered at a known depth in the tissue, and hence provide depth profile information. Each of the above spectroscopic methods however requires some level of multivariate statistical analysis to extract important diagnostic information from the features of the spectra.

The interferometric implications in medical imaging will be discussed in Chapter 14 in greater detail.

The final interaction mechanism in the photophysical aspects of biomedical light–tissue interaction is the result of nonelastic scattering known as spectroscopy, which will be discussed next.

8.9 Spectroscopy

Light exhibits both wave and particle characteristics when interacting with matter. In general, light is considered a wave when diffraction and interference effects are of interest, and it is considered a particle when energy transfer is considered. Several techniques have been developed over the years that make use of the frequency aspects and as such the spectral analysis of scattered light.

Spectroscopy utilizes the fluorescence caused by irradiation of naturally occurring chromophores as well as fluorescence induced by man-made photochemicals. The primary demonstration of the use of spectroscopy on man-made fluorophores is in the discipline of photodynamic therapy for tumor detection and dosimetry purposes.

Various types of spectroscopy can be recognized based on the direct physical aspects involved in the interaction. The two main types that will be discussed are LSS and FTIR spectroscopy. Specific applications of spectroscopic techniques are ranked under ultrafast and time-resolved spectroscopy and Raman spectroscopy.

8.9.1 Light Scattering Spectroscopy (LSS)

Light scattering spectroscopy has been used to selectively detect the size-dependent scattering characteristics of epithelial cells in vivo. It may be possible to use polarized light for providing quantitative morphological information that could be used for noninvasive detection of neoplastic changes. The fine-structure component in backscattered light from mucosal tissue, which is periodic in wavelength, has been observed. This structure is ordinarily masked by a diffusive background. The origin of this component has been identified as being due to light that is Mie scattered by surface epithelial cell nuclei. By analyzing the amplitude and frequency of the fine structure, the density and size distribution of these nuclei can be extracted. These quantities are important indicators of neoplastic precancerous changes in biological tissue. The ability to measure nuclear size distribution in vivo has valuable applications in clinical medicine. Enlarged nuclei are primary indicators of cancer, dysplasia, and cell regeneration in most human tissues.

In addition, measurement of nuclei of different sizes can provide information about the presence of particular cells, and can thus serve, for example, as an indicator of inflammatory response of biological tissue. This suggests that different morphology/pathology at the mucosal surface will give rise to distinct patterns of nuclear size distributions.

Both LSS and OCT have been used to investigate the case of Barrett's esophagus, a chronic condition in which irritation transforms normal esophagus epithelium into a thin monolayer of columnar cells similar to those used in the cell culture experiments. Such patients have an increased risk of developing dysplastic change, but such change is not visible with an endoscope.

8.9.2 Fourier Transform Infrared (FTIR) Spectroscopy

There is growing evidence to suggest that IR spectroscopy may play a significant role as an adjunct to conventional histopathology in the detection of cancer and other tissue abnormalities. For example, recent research includes studies of skin, oral, mucosa, colon, breast, and cervical tissue and cells. Furthermore, IR spectroscopy has proven to be successful in the ex vivo characterization of human multiple sclerosis plaques, the β-amyloid plaques of Alzheimer disease, and in distinguishing control from Alzheimer's diseased tissue. The IR spectrum reflects the relative amounts of the various classes of biological macromolecules making up the sample of interest: proteins, DNA, lipids, and carbohydrates all contribute unique signatures to the IR absorption profile. Moreover, the sensitivity of the technique extends to the finer variations within these groups of biomolecules. For example, the IR spectra of albumin and globulins differ significantly, thus allowing them to be quantified separately in the IR spectrum of a dried serum film. Lipid spectra differ from one another, depending on the nature of the head group, as well as the length and saturation of the hydrocarbon chain.

In the infrared, FTIR spectroscopy has been used to evaluate glioma brain tumors, which are the most deadly brain tumors.

8.9.3 Ultrafast Spectroscopy

Some photochemical reactions are so fast that they cannot be observed by ordinary spectroscopy: The visual process, photosynthesis, and ion mobility in cellular membranes. These processes need to be studied by pico- and femtosecond pulse lasers.

The theoretical basis for diagnostic applications of spectroscopy will be discussed next.

8.9.4 Time-Resolved Spectroscopy

Assume a two-level system irradiated by an intense short pulse of radiation at $t=0$. This excitation pulse creates a population N_b in the excited state of the medium under investigation. If there is no further introduced radiation, the spontaneous emission rate from level b back to ground state a is

$$\frac{dN_b}{dt} = -N_b A_{ba} \tag{8.155}$$

After integration the following condition is obtained:

$$N_b(t) = N_b(0)\exp(-A_{ba}t) \tag{8.156}$$

The radiance intensity of the emission is written as

$$\begin{aligned} I(\omega)_t &= A_{ba} N_b(t)\hbar\omega \\ &= I_o(\omega)\exp(-A_{ba}t) \end{aligned} \tag{8.157}$$

The exponential term can be expressed as $\exp(-t/\tau_R)$, where τ_R is the radiative lifetime; hence, the lifetime can be written as

$$\tau_R = A_{ba}^{-1} \tag{8.158}$$

The exponentially decreasing radiance can be determined experimentally using a short excitation pulse and detecting the decay of emission radiance; this way the radiative lifetime of the fluorescence can be determined. For an allowed electric dipole transition on an atom the radiation lifetime has a value in the range $\tau_R \approx 10^{-8} - 10^{-9}$ s.

8.9.5 Raman Scattering Spectroscopy

Raman spectroscopy is the detection of inelastically scattered light from molecules shifted in wavelength from the incident light as a result of the interaction with the energies of molecular vibrations.

The simplest model used to describe the Raman effect treats the incident light beam as a stream of particles known as photons that collide with the atoms and molecules within the sample material.

Initially, the molecule may possess thermal vibration (V_I) and rotational (R) energy as shown in Figure 8.13A. The incident photon interacts with the molecule, resulting in the transfer of energy between itself and the molecule as shown in Figure 8.13B. By applying the law of conservation of energy, the photon energy exiting the sample must equal the net photon energy change within the sample plus the initial photon energy as shown in the following equation:

$$E_0 = E_i \pm \Delta E_n \tag{8.159}$$

where E_0 equals the photon energy at output frequency v, E_i the photon energy at incident frequency v_i, $-E_n$ the negative net energy change (Stokes line), and finally $+E_n$ the positive net energy change (anti-Stokes line).

If the photon transfers energy to the molecule within the sample during the collision, the secondary radiation appears in the spectrum at a longer wavelength (i.e., lower frequency) than the incident light wavelength. The secondary radiation spectrum line corresponding to this condition is referred to as a Stokes line. The photon energy given by the following equation is directly related to the light wave frequency:

$$E = hv = \frac{hc_0}{\lambda} \tag{8.160}$$

where h is Planck's constant (6.626×10^{-34} J s), v the light frequency, c_0 the phase velocity of light in free–space, and λ the wavelength of the light.

A decrease in frequency of the emitted photon results in a net energy transfer to the molecule participating in the collision. This energy is stored within the molecule by exciting it to a higher level of rotational or vibrational

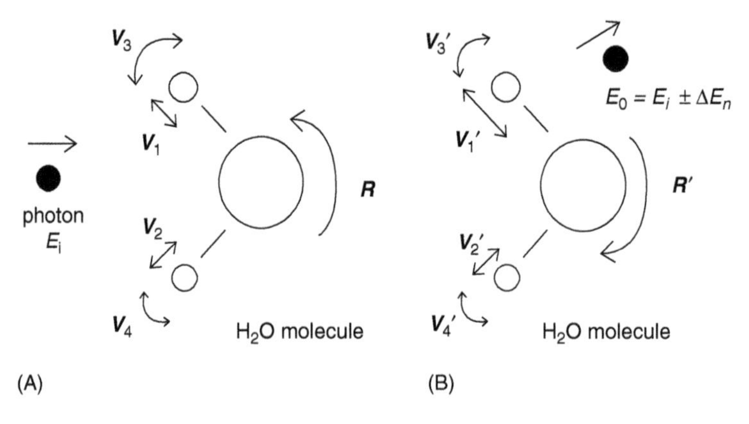

FIGURE 8.13
Possible photon interactions with water molecule. (A) Thermal vibration (V_i) and rotational (R) energy. (B) The incident photon interacts with the molecule resulting in the transfer of energy between itself and the molecule.

energy. If a photon interacts with a molecule possessing stored energy above its equilibrium state, the molecule will transfer energy to the photon, resulting in a net energy increase. The resulting secondary radiation appears at a shorter wavelength (i.e., higher frequency) within the spectrum and is referred to as an anti-Stokes line.

The frequency shifts of the Stokes and anti-Stokes lines (i.e., Raman lines) relative to the incident light spectrum line or band provide a measure of the rotational and vibrational frequencies of the molecule that may be used to determine the composition of the sample material. The position of the lines identifies the type of atomic nuclei and chemical bonds that comprise a molecule, and changes in the position may be used to monitor how the molecule interacts with the environment as in the case of changes in temperature and pressure. As molecules change composition during chemical reactions, their rotational and vibrational frequencies also change. Consequently, monitoring the changes in the Raman line positions provides a means of determining the reactants, intermediate products, and final products present during a chemical process. The frequency shifts are expressed in terms of spectroscopic wave numbers given by the following equation, which differs from the wave numbers used in the wave theory by a multiple of 2π (i.e., $k_w = 2\pi k_s$),

$$k_s = \frac{1}{\lambda} \tag{8.161}$$

where k_s is the spectroscopic wave number (units = cm^{-1}).

The corresponding shift is simply the difference between the wave numbers of the incident light beam and the resulting Raman line. The Raman spectrum ranges from a wave number of a few cm^{-1} to approximately 3800 cm^{-1}. The actual wave frequencies are obtained by simply multiplying the wave numbers by the speed of light as shown in the following equation:

$$\nu = (k_s c_o)\left(\frac{1\ \text{cm}}{10^{-2}\ \text{m}}\right) \tag{8.162}$$

The relationships describing the phase velocity of light in free-space or within some other medium with a given index of refraction are described in Section 2.3.7 in Chapter 2.

The properties of selected example light waves are listed in Table 8.2.

The energy diagram shown in Figure 8.14 illustrates the normal Raman effect and some variants as well as some other types of molecule and photon interactions. The upward pointing arrows correspond to the photon exciting the molecule to a higher energy level for an extremely short period of time. The curved peak connecting each upward and downward transition conveys this short period of time, as opposed to the infrared and fluorescence effects that take considerably longer time for the molecule to

TABLE 8.2
Light Wave Examples

Color	Wavelength (nm)	Frequency (10^{14} Hz)	Wave number (cm^{-1})	Energy (J)
Near–infrared	1000	3.00	10,000	1.98×10^{-19}
Red	700	4.28	14,300	2.84×10^{-19}
Orange	620	4.84	16,100	3.20×10^{-19}
Yellow	580	5.17	17,200	3.42×10^{-19}
Green	530	5.66	18,900	3.75×10^{-19}
Blue	470	6.38	21,300	4.23×10^{-19}
Violet	420	7.14	23,800	4.73×10^{-19}
Near-ultraviolet	300	10.0	33,300	6.62×10^{-19}
Far-ultraviolet	200	15.0	50,000	9.93×10^{-19}

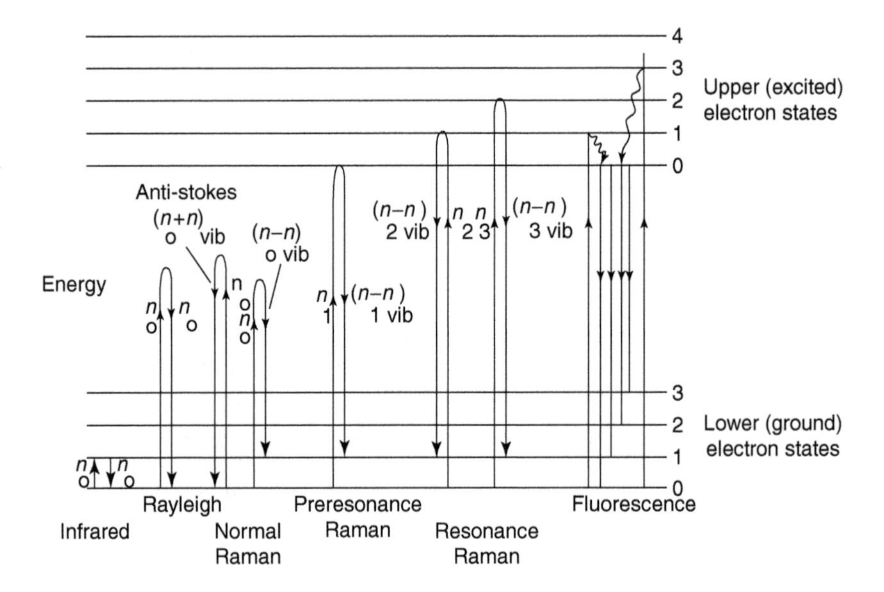

FIGURE 8.14
Energy diagram for the effects of infrared, Rayleigh, Raman scattering (Stokes and Anti-Stokes), normal, pre-resonant, resonant scattering, and fluorescence.

spontaneously return to the original ground-state energy level. The downward transitions correspond to the emission of a photon from the excited molecule during its return to the lower energy state, which may or may not be the original energy state depending on the interaction that has taken place. If the molecule does not return to its original state, then a net energy transfer occurs between the incident photon and the molecule. This results in a frequency shift between the incident photon and the emitted scattered photon. For the normal Raman effect the anti-Stokes process is less probable than the Stokes, since it begins at a higher energy level within the ground-state band. The population of the molecules at energy level 1 (N_1)

relative to the population at the lowest energy level (N_0) follows the Boltzmann distribution described as follows:

$$\frac{N_1}{N_0} = e^{-h\nu_{vib}/kT} = e^{-E_{vib}/kT} \tag{8.163}$$

where k is the Boltzmann's constant (8.617×10^{-5} eV/K), T the temperature (K), ν_{vib} the vibrational frequency, and E_{vib} the vibrational energy.

Therefore, the population N_1 is much smaller than N_0, which results in less anti-Stokes Raman scattering events for a given photon density (i.e., incident light intensity) in the sample. For example, given $T = 300$ K and $\nu_{vib} = 1.44 \times 10^{13}$ Hz then $N_1 = 0.09987$. Consequently, the weak anti-Stokes lines are often ignored, since they may fall below the noise level in the spectrum and only the Stokes line convey any useful information.

For the normal Raman effect shown in Figure 8.14 the energy interaction between the photon and the molecule is purely vibrational and rotational, and no electron transitions occur. In the case of the pre-resonance Raman effect, the incident light frequency is such that almost enough energy is supplied to cause an electron transition within the molecule. While the energy interaction is again only due to the molecular vibration and rotation, this technique has the advantage of producing higher intensity Raman scattering resulting in a better spectrum. The resonance Raman effect is achieved by increasing the incident light frequency above that used for the pre-resonance effect. Now sufficient energy is supplied for an electron transition from the ground state to the excited state within a molecule. However, the electron quickly falls back to the ground state but at a slightly higher energy level, resulting in the emission of a scattered photon at a slightly lower frequency than the incident photon. Resonance Raman scattering has the advantage of producing even higher intensity spectrums. It also has the advantage of introducing the capability of selectively probing a complex system for a desired component. Since each atom and molecule has unique and discrete electron transition states, the resonance Raman effect enhances the ability to uniquely identify a complex molecule or chain. By selecting the correct incident light frequency, only the desired component of the system will absorb the photon energy resulting in an electron transition. The remaining components of the system will not experience electron transitions, since the incident light frequency does not correspond to the possible electron transition levels. However, these components may experience the normal or pre-resonance Raman effect. The resulting spectrum will clearly show the desired component, since the intensity of the resonance Raman effect is much higher than the other effects.

The Raman spectrum plot of water illustrated in Figure 8.15 shows three peaks. The peak furthest to the left corresponds to the incident light wave while the second and third peaks from the left are related to the stretching and bending molecular vibrations. The stretching vibrations are depicted as V_1 and V_2 in Figure 8.13 and the bending vibration is shown as V_3 and V_4, respectively. The large O–H stretch peak indicates that there is a large energy component related

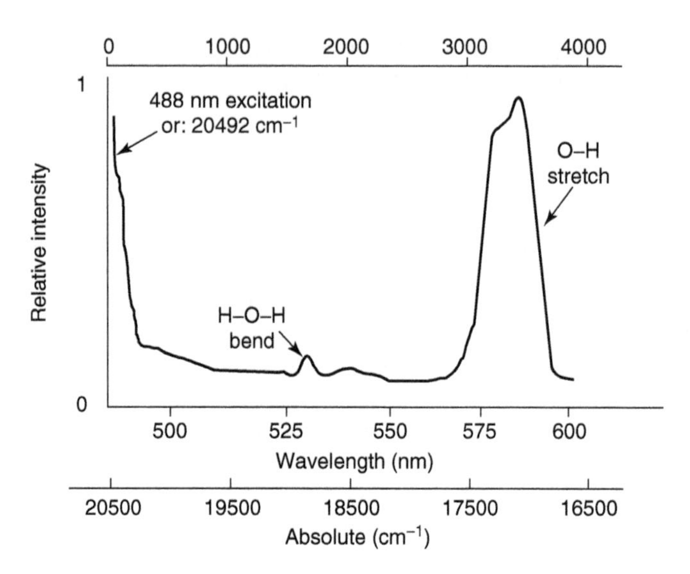

FIGURE 8.15
Raman spectrum of water.

to the interaction between the photon and the spring-like motion of the hydrogen atoms. The much smaller H−O−H bending peak suggests that very little energy is transferred due to the torque-like vibration within the water molecule. It may also be concluded given the magnitudes of the peaks that energy transfer is more probable between the stretching vibration and the photon than between the bending vibration and the photon. The position and magnitude of these peaks uniquely identify the sample as water. The absolute scale given at the bottom of the plot provides the actual wave numbers of the incident and scattered light, while the Raman shift scale provides the relative wave numbers of the scattered light to the incident light. That is, the Raman shift in scale values for the peaks is simply the absolute wave numbers of the peaks minus the wave number of the incident light. The scale at the top of the plot provides the wavelength that corresponds to each of the absolute scale wave numbers. Since the Raman spectrum of water only has one in the visible large peak, it is an excellent medium to suspend biochemicals for observing the chemical interactions. That is, the water component of the solution will not mask or interfere with any of the resulting chemical process Raman spectrum lines.

In addition to the magnitude of the secondary radiation frequency shifts, the state of polarization of light and the intensity of spectral lines are also important characteristics of the scattered light emitted from the sample. The ratio of the horizontal to vertical intensity components of the Raman line is used to define the state of polarization. The degree of polarization is typically measured from the electric field vector (**E**) component of a light wave with respect to some reference plane. The Poynting vector (**S**) describes the relationship between **E**, the magnetic field vector (**H**), and the direction of the wave propagation as outlined in Section 2.3.4 in Chapter 2. The average Poynting vector ($\mathbf{S}_{av}$) given by Equation (5.33) gives the intensity of the radiation.

Let the electric field and the magnetic field be defined as $\mathbf{E}=E_{max} \sin(\omega t-kx)$ and $\mathbf{H}=H_{max} \sin(\omega t-kx)$, respectively. These represent only the field directions of the wave striking the sample. Substituting these into Equation (2.55), we obtain

$$\mathbf{S}=\mathbf{EH}\sin(\alpha)=E_{max}\sin(\omega t-kx)H_{max}\sin(\omega t-kx)\sin(\alpha) \tag{8.164}$$

Since α is the angle between **E** and **H** and is equal to 90°, Equation (8.164) can be rewritten as

$$\begin{aligned}\mathbf{S}&=\mathbf{EH}\sin(\alpha)=E_{max}\sin(\omega t-kx)H_{max}\sin(\omega t-kx)\sin(90^\circ)\\&=E_{max}H_{max}\sin^2(\omega t-kx)\end{aligned} \tag{8.165}$$

Applying the trigonometry identity $\sin(A)=\frac{1}{2}[1-\cos(2A)]$, we obtain

$$\mathbf{S}=\tfrac{1}{2}E_{max}H_{max}[1-\cos(2(\omega t-kx))] \tag{8.166}$$

Taking the average with respect to time, where the cosine wave average is zero, the fluence (Ψ) of the light wave can be expressed as

$$\Psi=\mathbf{S}_{av}=\tfrac{1}{2}E_{max}H_{max} \tag{8.167}$$

As an example, the transverse electromagnetic wave (i.e., light wave) propagating in the positive z-direction as shown in Figure 8.16a will form a polarization angle θ (Figure 8.16b) if the x-axis is chosen as the reference. For a linearly polarized medium this angle will remain constant as the wave propagates through the sample, but the angle will change as a function of the distance traveled through the sample in the case a nonlinear polarized medium (e.g., circular and elliptical polarizations). The direction of **S** in Figure 8.16b is out of the plane of the page. This may be verified by applying the "right-hand rule" that is associated with the cross-product operator given in the equation for the Poynting vector. That is, the direction of propagation may be found by placing the palm of the right hand in the plane of **E** and then folding the fingers into the palm towards **H**. The direction in which the thumb points is the direction of propagation of the wave. As a result of the polarization, a horizontal and vertical intensity component will exist resulting in a state of polarization value for a particular Raman line. The Raman line is considered to be polarized if the state of polarization is less than 0.1, while it is considered to be strongly depolarized if the value is more than 0.5.

The intensity of the Raman lines increase to the fourth power of the incident light frequency unless the frequency falls within the absorption band of the molecules within the sample.

Each atom within a molecule contains electrons that occupy discrete energy levels. Transitions between energy levels are possible when the right amount of energy is introduced. This concept is exemplified in the following equation:

$$\nu_{21}=\frac{(E_2-E_1)}{h} \tag{8.168}$$

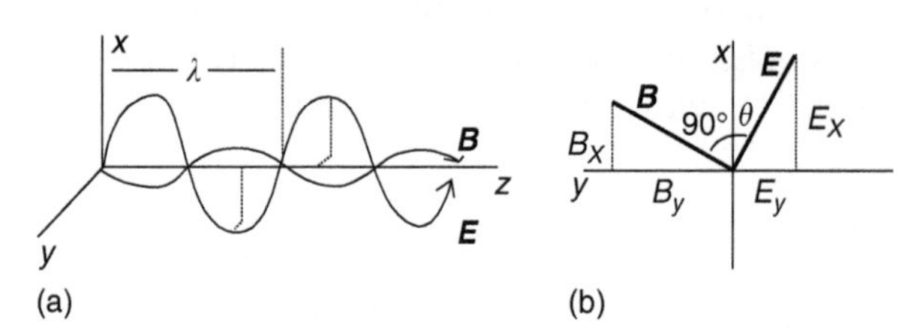

FIGURE 8.16
Illustration of (a) wave nature of light and (b) the energy flow with respect to the field vectors.

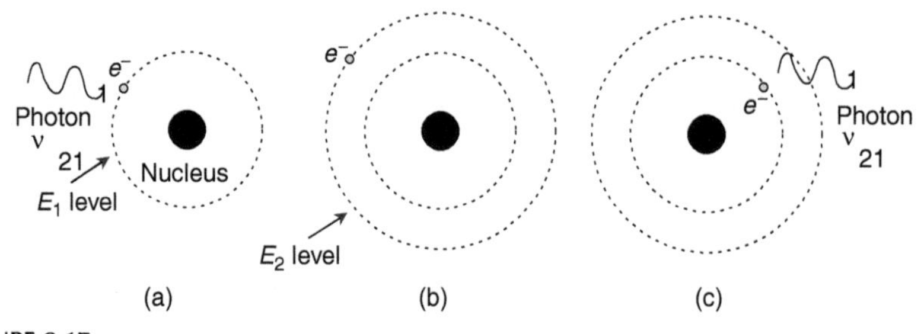

FIGURE 8.17
Diagram of photon absorption and emission: (a) photon absorption, (b) electron in elevated state, and (c) photon emission.

where ν_{21} is the frequency of the photon, which matches the energy difference between level E_2 and level E_1.

When the frequency of the incident light beam satisfies Equation (8.168) for a particular electron, the photon is absorbed by an atom and energy is conserved by elevating the electron to a higher energy state as illustrated in Figure 8.17a and Figure 8.17b. After a period of time, the electron will spontaneously fall back into its original energy level and a photon of the same frequency will be released (Figure 8.17c). In contrast to light absorption and emission, the Raman effect may be viewed as the energy transfer between three discrete states as shown previously in Figure 8.14.

The Raman and infrared spectrums often yield similar results even though the physical processes involved are very different. When possible, both Raman and infrared spectroscopy are used to gain a complete description of the molecular vibrations within a sample. However, Raman spectroscopy is often the only source of information, since many biochemical processes are suspended in water, which has a high-infrared absorbance that masks the actual chemical processes being monitored.

In a Raman spectroscopy configuration a laser provides an intense, coherent, monochromatic incident light beam that is reflected off two intermediate mirrors before striking a focusing lens. Continuous wave lasers are typically used for Raman experiments, but Q-switched and mode-locked lasers are also used. The mirrors not only isolate the laser from any reflected or scattered light from the sample, but also help control the polarization of the light beam striking the surface of the sample. This control is necessary, since if the polarization is along the optical axis, no Raman scattered light will emerge in the direction of the spectrometer. The focusing lens concentrates the photon flow, thus increasing the photon density at the sample surface. This increased density enhances the scattered light created by the Raman effect as the incident light wave passes through the sample. The scattered light emerging from the sample is then focused into the spectrometer chamber by the large collecting lens. The spectrometer then identifies the intensity at each wavelength across the desired bandwidth. The resulting spectrum may then appear as a series of vertical lines. Each line corresponds to a discrete wavelength within the spectrum and their relative height is related to the received light intensity.

8.9.6 Coherent Anti-Stokes Raman Spectroscopy

Coherent anti-stokes Raman spectroscopy (CARS) is used by biochemists to obtain resonant Raman spectrums of highly fluorescent chromophores. The fluorescence present interferes with the collection of the scattered light when the basic Raman technique is used.

In operational setting, for instance, a nitrogen (N_2) laser can be used to pump two tunable dye lasers, which in turn emit light at discrete frequencies ω_1 and ω_2. Each laser beam is then directed to a focusing lens through a series of mirrors. The lens directs the beams to the focal point that lies within the sample. Coherent anti-Stokes emission at frequency $\omega_3 = 2\omega_1 - \omega_2$ occurs when the two beams cross within the sample at the correct phase matching angle. The scattered light reaches peak intensity when the frequency difference between the two dye lasers ($\Delta = \omega_1 - \omega_2$) matches the vibrational frequency of the molecule that is being investigated. Consequently, the frequency for the first dye laser (ω_1) is typically set and the second dye laser is tuned to find the best frequency ω_2. The resulting emitted light at frequency ω_3 is then passed through collecting lens from where it goes to a monochromator, which separates the light beam into its separate frequency components. A photomultiplier can be used to amplify the signal before sending it to the photometer. A chart recorder or computer is attached to the photometer to obtain the spectrum. The CARS setup is illustrated in Figure 8.18.

8.9.7 Time-Resolved Raman Spectroscopy

The time-resolved Raman spectroscopy technique is capable of achieving a complete spectrum in extremely short time spans of the order of nanoseconds or even picoseconds. This allows for the accurate observation of complex biochemical reactions. Not only are the reactants and products identified, but also any transient species or excited electronic states are identified during the chemical reaction.

A short-duration, high-intensity Q-switched or mode-locked laser is focused onto the sample and the Raman scattering events take place. The

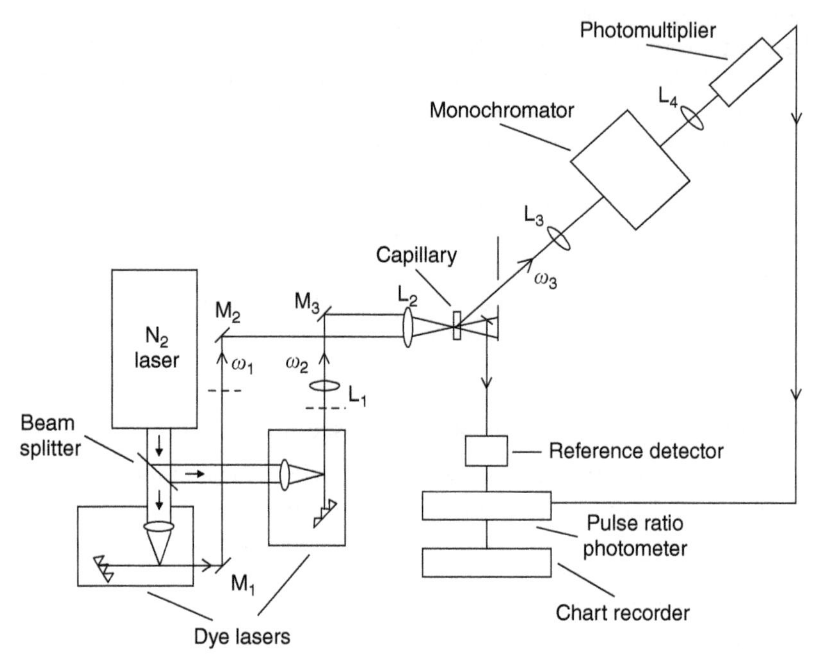

FIGURE 8.18
Schematic representation of the coherent anti-Stokes Raman spectroscopy setup.

scattered light from the sample is then focused into a multichannel spectrograph that breaks the light down into its frequency components. The light is then passed to a multichannel image intensifier whose shutter is synchronized with the laser pulses. This synchronization is made to eliminate light that is not directly related to the Raman scattering effect. For example, Rayleigh and resonance absorption scattering effects also occur, but since the time between photon absorption and emission is different for these effects, it is possible to filter them by controlling the shutter position. The intensified signal is then passed to recording devices.

8.9.8 Raman Spectroscopy Advantages and Disadvantages

Many biological processes may be observed by suspending the biochemical molecules in an aqueous solution. Raman scattering is an advantageous technique for monitoring these processes, since water has a low Raman spectrum compared to other molecules. This allows easy observation of the desired processes, since they appear far above the spectrum of water. In contrast, the infrared absorbance of water is very high making resonance absorption spectroscopy techniques of limited value. Raman spectroscopy also has the advantage of being able to examine a material regardless of whether it is in the solid, liquid, or gas phase. The target material may also be suspended in solutions other than water or may be part of complex molecules or chains that make up films, surfaces, and fibers. The material sample only needs to fill the spot size of the illuminating laser beam to obtain an accurate spectrum. This has the advantage of requiring only a small amount of sample material. However, the typical experimental setup allows the laser beam to move across a larger sample to monitor various regions without interrupting the experiment, and the entire vibrational spectrum (10–4000 cm^{-1}) may be observed by a single scan of a conventional Raman spectrometer. It takes far less than a second to record the complete Raman spectra, and today computers allow continuous data acquisition so that the complete biochemical process may be recorded and reviewed. This is advantageous, since many biochemical processes occur rapidly and have many intermediate states. Also, each species present appears in the Raman spectrum in direct proportion to its concentration, so insight into the stoichiometry of the process may be derived. The Raman scattering technique also gives excellent results when examining homopolar bonds (e.g., C–C and S–S) compared with resonance absorption, which results in extremely weak or undetectable spectral lines.

While Raman spectroscopy has many advantages over other techniques, the basic process also has several disadvantages. The variants of the basic process that are discussed in later sections have been developed to negate or minimize some of these disadvantages. One disadvantage is that a relatively high photon flux is necessary to achieve a measurable Raman spectrum. The high flux may produce unwanted photochemical effects that will disrupt the actual biochemical process, leading to erroneous results. The high photon flux is necessary, since the Raman scattering event has a low probability of occurring and by increasing the flux the resulting Raman spectrum becomes measurable. However, other photon and electron interactions also increase with the increased flux, so steps must be taken to filter these effects from the Raman spectrum. High chemical concentrations that may not represent the actual biochemical process are required to achieve a good Raman spectrum, which is yet another disadvantage of the basic process. In addition, a high level of chemical homogeneity is usually required in a solution.

Several biomedical optics applications are also designed to operate on the basic physical principle of reflection. Examples of reflection methods are described in the next section.

8.10 Endoscopy

The term endoscopy is defined as follows: using an instrument to visually examine the interior of a hollow organ or cavity of the body. Catheters and a rectoscope were used by Hippocrates II (460–377 BC), where inspection of the oral cavity and of the pharynx was routine, including operations on tonsils, uvula, and nasal polyps. This trend continued for the better part of two millennia, until the German physician Phillip Bozzini in 1805 invented what he called the *Lichtleiter.* In the early nineteen hundreds with the advancement of more sophisticated equipment and the discoveries of electrically related technologies, illumination capabilities were greatly improved. New and better quality lenses, prisms, and mirrors dramatically enhanced the potential applications of endoscopy and opened up new avenues of application. In 1910 Victor Elner used a gastroscope to view the stomach, and in 1912 the first partially flexible gastroscope was designed.

The most significant technological development began in the 1950s with the introduction of fiber optics. The use of fiber optics also allowed for the incorporation of ancillary channels for the passage of air, water, suction, and implementation of biopsy forceps or instruments for therapeutic procedures. Obviously, the most significant capabilities fiber optics has brought to endoscopes are their small size, flexibility, and ability to deliver laser light. Lasers are now used in both diagnostic and therapeutic endosopic procedures, including the earlier discussed spectroscopy.

8.10.1 Light Delivery with Optical Fibers

Efficient light delivery is made possible in solid-core optical fibers by use of total internal reflection. Apart from choosing a fiber material that has close to lossless transmission at the wavelength of choice, one needs to focus the light beam so that the diameter of the fiber core is at least three times the radius of the input light beam and the convergence angle of the focused light θ is less than the acceptance angle of the fiber core. The acceptance angle of the fiber is related to the numerical aperture (NA) and the refractive indices of the fiber (n_i) and the medium outside the fiber (n_t) as

$$\mathrm{NA} = \sin(\theta) = \sqrt{(n_i^2 - n_t^2)} \tag{8.169}$$

Fibers are available in diameters from several microns (for single-mode transmission) to about 1 mm.

Wave guides, in contrast, operate by coating the inside of a hollow wave guide (HWG) with a metallic or dielectric coating depending upon the desired wavelength. For HWGs the same rules apply for efficient coupling.

8.10.2 Medical Applications Using Endoscopy

The following are a select group of examples of the use of endoscopy for diagnostic and therapeutic medical applications.

8.10.2.1 Arthroscopy

Arthroscopy was conceived in 1918 by Takagi, who modified a pediatric cystoscope and viewed the interior of the knee of a cadaver. Today arthroscopy is an integral part of orthopedic surgery. One of the main fields of application is examination of the knee joint. Endoscopes have the ability to view off-axis at a range of angles from 0° to 180°, with varying field of view. One drawback, however, is that the larger the field of view the more distorted is the image, similar to the fish-eye lens in photography. A 45° off-axis endoscope, for example, with a 90° field of view can be rotated to visualize the entire forward-looking hemisphere.

The most common procedure in arthroscopy is arthrotomy, which is the exploration of a joint.

8.10.2.2 Bronchoscopy

Bronchoscopy is used to visualize the bronchial tree and lungs. It is widely used in diagnosis and treatment of pulmonary disease.

8.10.2.3 Cardiology

Endoscopy in cardiology is further miniaturized, using intravascular catheters to inspect the inside of blood vessels and more recently the inside of the heart itself. Catheters are thin flexible tubes inserted into a part of the body to inject or drain away fluid, or to keep a passage open.

8.10.2.4 Cystoscopy

Endoscopy is an integral part of the practice of urology. In urology endoscopic procedures include cystoscopy with ureteral catheterization, with fulguration and resection of bladder tumor(s), with direct vision internal urethrotomy, with insertion of stent, removal of bladder stone, and kidney stone fragmentation.

Additionally a wide variety of therapeutic accessories are available for endoscopic treatment, including snares and baskets for stone removal. Other applications include laser lithotripsy of kidney stones. In lithotripsy the photomechanical effect of laser plasma generation is used to shatter hard objects that obstruct the normal passage in cavities.

For instance, the use of Er:YAG laser for the incision of urethral, bladder neck, and urethral strictures, using flexible endoscopes made from germanium fibers to transport laser light, is a major therapeutic advantage.

8.10.2.5 Fetoscopy

Fetoscopy describes the visualization of a fetus in the womb using an endoscopic device without complicating the pregnancy. Fetal surgery this way also is performed with miniaturized surgical tools and laser to perform detailed surgical tasks on a living being of the order of 10 cm in size.

8.10.2.6 Gastrointestinal Endoscopy

In gastrointestinal (GI) endoscopy the entire GI tract can be investigated and treated. Applications are, for instance, in the esophagus, stomach, colon, small bowel and liver, biliary, and pediatric endoscopy.

Laser therapy of GI neoplasms, in addition to management of foreign bodies and bezoars of the upper GI tract, are currently commonplace.

In the colon, nonoperative and interventional management of hemorrhoids is performed with endoscopes. Other procedures include dilatation of colonic strictures, approaches to the difficult polyp and colonic intubation, and clinical approaches of anorectal manometry.

8.10.2.7 Laparoscopy

Laparoscopy or peritoneoscopy is an important diagnostic procedure that allows direct visualization of the surface of many intra-abdominal organs, as well as allowing the performance of guided biopsies and minimally invasive therapy.

8.10.2.8 Neuroendoscopy

In neuroendoscopy laser use is one of the options to perform perforations of obstructions of ventricles in the brain. These perforations allow for the cerebrospinal fluid (CSF) to be diverted so that a permanent communication can be made between the third cerebral ventricle and arachnoid cisterns of the cranial base.

8.10.2.9 Otolaryngology

The use of endoscopes on the interior of the ear is one of the important applications in the field of ear, nose, and throat (ENT) surgery. For instance, the Er:YAG laser with a germanium oxide fiber delivery system is available for tympanoplasty and stapedotomy (middle ear surgery). Additionally, the Er:YAG laser has been used to operate on the eardrum along the ossicles as far as the footplate without carbonization. Micrometer-diameter canals can be drilled under laser vaporization as will be described in Section 11.3 in Chapter 11.

8.11 Summary

This chapter provided a comprehensive review of the photophysical mechanisms available for therapeutic and diagnostic applications of light in medicine. Several specific purely physical applications of light delivery and collection were presented in the discussion of fiber optics and endoscopy.

Further Reading

Boulnois, J.L., Photophysical processes in recent medical laser developments—review, *Lasers Med. Sci.*, 1, 47–66, 1986.

Fitzpatrick, R.E., Comparison of the Q-Switched ruby, Nd:YAG and alexandrite lasers in tattoo removal, *Lasers Surg. Med.*, Suppl. 6, 52, 1994.

Garden, J.M., Bakus, A.D., and Paller, A.S., Treatment of cutaneous haemangiomas by the flashlamp-pulsed dye laser: Prospective analysis, *J.. Pediatr.*, 120, 555−560, 1992.

Garden, J.M., O'Banion, M.K., and Shelnitz, L.S., et al., Papillomavirus in the vapor of carbon dioxide laser-treated verrucae, *J. Am. Med. Assoc.*, 259, 1199−1202, 1988.

Garden, J.M., Polla, L.L., and Tan, O.T., The treatment of port-wine stains by the pulsed dye laser, *Arch. Dermatol.*, 124, 889−896, 1988.

Geronemus, R.G., and Ashinoff, R., Use of the Q-switched ruby laser to treat tattoos and benign pigmented lesions of the skin, *Laser surg. Med.*, suppl. 3, 64–65, 1991.

Goldberg, D.J., and Stampien, T., Q-switched ruby laser treatment of congenital naevi, *Arch. Dermatol.*, 131, 621−623, 1995.

Hecht, E., and Zajac, A., *Optics*, Addison-Wesley Publishing Co., Boston, MA, 1974.

Hooper, B.A. et al., Precise, controlled laser delivery with evanescent optical waves, *Appl. Opt.*, 38, 5511−5517, 1999.

Hooper, B.A., Maheshwari, A., Curry, A.C., and Alter, T.M., Catheter for diagnosis and therapy with infrared evanescent waves, *Appl. Optics*, 42(16): 3205 –3214, 2003.

Hooper, B.A., LaVerde, G.C., and Von Ramm, O.T., Design and construction of an evanescent optical wave device for the recanalization of vessels, *Nucl. Instrum. Methods*, A475, 645−649, 2001.

Jansen, E.D., Van Leeuwen, T.G., Motamedi, M., Borst C., and Welch, A.J., Partial vaporization model for pulsed mid-infrared laser ablation of water, *J. Appl. Phys.* 78(1): 564–571, 1995.

Kasai, K.-I., and Notodihardjo, H.W., Analysis of 200 nevus ota patients who underwent Q-switched Nd:YAG laser treatment, *Lasers Surg. Med.*, Suppl. 6, 50, 1994.

Kilmer, S.L., and Anderson, R.R., Clinical use of the Q-switched ruby and the Q-switched Nd:YAG [1064 nm and 532 nm] lasers for the treatment of tattoos, *J. Dermatol. Surg. Oncol.*, 19, 330−338, 1993.

Kilmer, S.L., Lee, M., Farinelli, W., and Grevelink, J.M., Q-switched Nd:YAG laser [1064 nm] effectively treats Q-switched ruby laser resistant tattoos, *Lasers Surg. Med.*, Suppl. 4, 72, 1992.

Kokhanovsky, A.A., *Polarization Optics of Random Media*, Springer g, Berlin, Heidelberg, 2003.

Lakowicz, J.R., *Principles of Fluorescence Spectroscopy*, Plenum Press, New York, 1998.

Lanigan, S.W., Port wine stains on the lower limb: response to pulsed dye laser therapy, *Clin. Exp. Dermatol.*, 21, 88−92, 1996.

Lanigan, S.W., Sheehan-Dare, R.A., and Cotterill, J.A., The treatment of decorative tattoos with the carbon dioxide laser, *Br. J. Dermatol.*, 120, 819−825, 1989.

Louden, R., *The Quantum Theory of Light*, 2nd ed., Oxford University Press, Oxford, 1983.

Lowe, N.J., Laser therapy of vascular benign pigmented lesions and tattoos, in *Skin Therapy*, Marks, R. and Cunliffe, W.J., Eds., Martin Dunitz, London, 1994.

Mathews, J., and Walker, R.L., *Mathematical Methods of Physics*, Benjamin Cummings, Menlo Park, 1970.

McBurney, E.I., and Rosen, D.A., Carbon dioxide laser treatment of verrucae vulgares, *J. Dermatol. Surg. Oncol.*, 10, 45−48, 1984.

Motley, R.J., Katugampola, G., and Lanigan, S.W., Microvascular abnormalities in port wine stains and response to 585 nm pulsed dye laser treatment, *Br. J. Dermatol.*, 135(Suppl. 47), 13−14, 1996.

Simhi, R., Gotshal, Y., Bunimovich, D., Sela, B.A., and Katzir, A., "Fiber-optic evanscent-wave spectroscopy for fast multicomponent analysis of human blood," *Appl. Opt.*, 35, 19, 3421–3421, 1996.

Swinehart, J., Hypertrophic scarring resulting from flashlamp-pumped pulsed dye laser surgery, *J. Am. Acad. Dermatol.*, 25, 845–846, 1991.
Tanda, G., Application of the Schlieren technique to convective heat transfer measurements, 1999, tutorial posted on the world wide web by the Optical Methods in Heat and Mass Transfer (OMHAT) site: http://dau.ing.univaq.it/omhat.
Taylor, C.R., Flotte, T. J., Gange, W.R., and Anderson, R. R. Treatment of naevus of ota by Q-switched ruby laser, *J. Am. Acad. Dermatol.*, 30, 743–751, 1994.
Taylor, C.R., Gange, R.W., Dover, J.S., Flotte, T.J., Gonzalez, E., Michaud, N., and Anderson, R.R. Treatment of tattoos by Q-switched ruby laser, A dose-response study, *Arch. Dermatol.*, 126 (7): 893–899, 1990.
Thomsen, S., Pearce, J.A., and Cheong, W.-F., Changes in birefringence as markers of thermal damage in tissues, *IEEE Trans. Biomed. Eng.*, 36, 12, 1174–1179, 1989.
Tuchin, V.V., Wang, L., and Zimnyakov, D.A., *Optical Polarization in Biomedical Applications*, Springer Verlag, Berlin, Heidelberg, 2006.
Venugopalan, V., Nishioka, N.S., and Mikic, B.B., The thermodynamic response of soft biological tissues to pulsed infrared-laser irradiation, *Biophys. J.*, 70, 2981–2993, 1996.
Walsh, J.T., and Deutsch, T.F., Pulsed CO_2 laser tissue ablation: measurement of the ablation rate, *Lasers Surg. Med.*, 8, 264–275, 1988.
Walsh, T.J., and Deutsch, T.F., Pulsed CO_2 laser ablation of tissue: effect of mechanical properties, *IEEE Trans. Biomed. Eng.*, 36, 12, 1195–1201, 1989.
Welch, A.J., and Van Gemert, M.J.C., Eds., *Optical-Thermal Response of Laser-Irradiated Tissue*, Chap. 21, Plenum Press, New York, 1995.
Whitaker, S., *Fundamental Principles of Heat Transfer*, Pergamon, New York, 1977, pp. 5–47.

Problems

1. What are the three major mechanisms that can be distinguished in laser–tissue interaction?
2. Calculate the ablation depth of a pulsed Ho:YAG laser operating at 2.9 μm shooting at dental enamel with a repetition rate of 50 Hz, generating 60 mJ per pulse while focused on a 1 mm spot size, assuming 90% debrement.
3. Explain the physical principle of Schlieren imaging.
4. Explain how optical activity can be used to date the age of a cell nucleus, considering the nucleus is made up of amino acids and proteins.
5. Describe the boundary conditions that need to be met to generate an evanescent wave.
6. Explain how time-resolved spectroscopy can improve on the accuracy of the determination whether a cell is malignant or benign.
7. In Raman spectroscopy there are various transitions that can be targeted for detection:
   a. What are some of the complications in Raman spectroscopy technique for sensing purposes?
   b. Are Stokes or Anti-Stokes lines more likely to be used for diagnostic purposes and why?
8. The Morpho butterfly has no pigment in its wings, but it still has a very dark blue iridescent blue wing color. Looking at electron microscopic images of the wing, a

structure can be distinguished that resembles the branches of a pine tree. This structure is covering the wing in a rather dense fashion. The material that makes up this structure is in fact mostly transparent to all visible light, with one distinctive detail: the thickness of, and spacing between the "branches" is extremely regular. Determine what the thickness of the branches and what the spacing between the branches need to be to achieve virtually perfect interference at only the blue light at approximately 400± 20 nm. The complex index of refraction of the scale

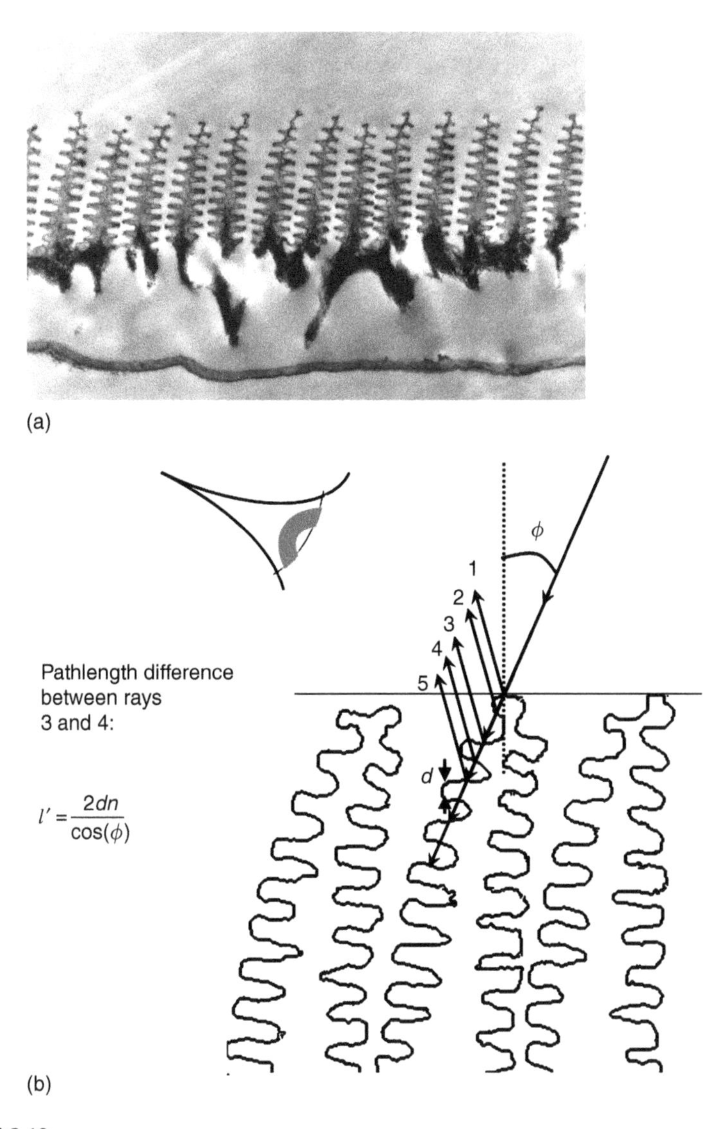

FIGURE 8.19
Morpho Butterfly wings derive their color from the interference pattern resulting from a fixed morphological wing structure. (a) Electron microscope image of the scale of the Morpho Butterfly, (b) interference geometry of the wing structure. (Courtesy of Dr Pete Vukusic from the University of Exeter, Great Britain.)

material mounted on the wing of the butterfly is $n = (1.56 \pm 0.01) + (0.06 \pm 0.01)i$. The complex part is an indication of attenuation. For the calculations use $n = 1.56$. Consider that the light passing through the branches interferes with light reflecting directly off the surface of the branch beneath it in addition to interference between the surface layers of subsequent branches. Structure on top of Morpho butterfly wing (Picture supplied by Dr Pete Vukusic from the University of Exeter Great Britain), Figure 8.19.

9. When monochromatic light passes from air into another material, which of the following does not change, regardless of the material: its speed, frequency, or wavelength?
10. The signal format in an FTIR spectrophotometer is an interferogram. What is so special about an interferogram in a spectrometer?
    a. Give five reasons why you would choose an FTIR spectrophotometer over a dispersive instrument?
    b. State four of the major advantages that an FTIR has over the dispersive technique.
    c. Based on the Nyquist Theorem, what is the short wavelength limit of an FTIR with an internal reference source that emits light at 500 nm?
11. Describe the two types of interaction of light with tissue, fluorescence and Raman scattering, that are currently being used in optical spectroscopy.
    a. How is fluorescence different from phosphorescence?
    b. How is Raman scattering different from Rayleigh scattering?
    c. Biological molecules contain naturally occurring or endogenous fluorophores. What are a few of these fluorophores?
    d. What is the main limitation of the use of fluorescence spectra for diagnosing cancers and precancers?
    e. How might Raman scattering overcome some of the limitations of fluorescence diagnosis of precancers and cancers?
    f. Raman scattering may be particularly well suited to the diagnosis of cancers and precancers. Give reasons.
12. What is the condition on the pulse duration τ that must be met for a photothermal interaction to be thermally confined?
13. Make an argument for why a photodisruptive interaction necessarily needs larger power density and a shorter pulse than a biostimulation interaction.
14. A pulsed laser has 50 μJ of energy in a 1 ps pulse. If the laser operates at a repetition rate of 10,000 Hz, the laser is focused on an area of $A=4\times10^{-8}\text{m}^2$:
    a. What is the time-averaged power?
    b. What is the power in the pulse?
    c. What is the irradiance of the laser beam?
    d. What is the fluence?

9

Light–Tissue Interaction Mechanisms and Applications: Photochemical

This chapter presents an overview of the photochemical interaction mechanisms and the research into the underlying dynamics.

All photochemical interactions as well rely on the ease with which light can be delivered to a specific location in a biological medium and attain the critical fluence rate to induce the chemical reaction, particularly at greater depths. The light will often need to pass through healthy tissues before reaching the target volume, and the optical properties of these tissues may not necessarily be the same at the specifically desired excitation wavelength. This will impose certain restrictions on the maximally allowable depth of treatment.

The photochemical interaction changes the molecular composition of certain chemicals. Some chemicals susceptible to photo-induced changes are naturally occurring in the biological medium (e.g., chlorophyll), while others may be artificially introduced to enhance the effect of the electromagnetic radiation on the tissue in therapeutic or diagnostic aspects. Two examples of natural and artificial enhancement in photochemistry are photosynthesis and photodynamic therapy (PDT), respectively. The former is a response of naturally occurring chemicals to light, and the latter involves man-made interventions.

9.1 Basic Photochemical Principles

A selection of photochemical principles describes the chemical changes induced in biological media under the influence of light. Representative examples are photosynthesis, sun tanning, photochemically induced cell death, and wound healing. The well-known vital natural phenomenon of light used by biological media to enhance the living conditions is photosynthesis. The all-too-familiar practice of sun tanning involves the use of naturally ocurring chromophores to produce vitamins and stimulate certain cellular metabolic processes. A more artificial photochemical principle is photochemical damage or photo-induced cell death, for instance PDT. The relatively unknown photochemical process of wound healing uses light to enhance the cellular metabolism to increase the healing process. Light-induced wound healing is gaining increased acceptance and is becoming increasingly more effective.

Even though the topics of photosynthesis and sun tanning fall outside the general scope of this book, a brief overview will be presented in light of the complex photochemical reaction and its correlation with the topic of PDT.

9.1.1 Photosynthesis

Photosynthesis is the process of converting light energy into chemical energy that can be used by plants, animals, and other organisms. Photosynthesis uses light energy transfer mostly in plants, but also in certain animals, for the chemical production of oxygen that can be used for respiration. Oxygen is used by other organisms, animals, and insects to maintain their metabolism. The oxygen is subsequently used to synthesize glucose, the carbohydrate source of carbon for all the organic molecules that make up all organisms. Photosynthesis can be summarized by the following reaction:

$$6CO_2 + 6H_2O \xrightarrow{\text{light}} C_6H_{12}O_2 + 6O_2$$

Photosynthesis takes place in chlorophyll pigments, which absorb light energy. This energy is used to excite electrons in the chlorophyll molecules, which are then passed on to the electron-transport chain. The cellular energy transfer uses the renewable energy source of adenosine triphosphate (ATP). Adenosine triphosphate consists of three parts: adenine, ribose, and three phosphate groups. Adenine bonded to ribose gives adenosine. Adenosine bonded to two phosphate groups is adenosine diphosphate (ADP). In the energy production cycle in chloroplasts and mitochondria, energy is stored when ATP is produced from ADP and an additional phosphate group "P".

During this process water is split into oxygen and hydrogen ions and electrons. The reaction center of chlorophyll has two sets of molecules involved with a specific response to light. Both sets contain pigment molecules that respond to a single narrow wavelength bandwidth; one absorbs light around 700 nm and is called P700, while the second system absorbs around 680 nm and is called P680.

In general, photosynthesis only has an efficiency of about 40%. This means that about 60% of the incoming light energy is stored as heat, or is re-emitted as light with a lower energy (e.g., fluorescence).

Certain factors that affect the rate of photosynthesis and their interactions are the following:

1. *Temperature*. As with all chemical reactions the higher the temperature, the faster is the reaction. However, there is an upper limit.
2. *Carbon dioxide concentration*. Since this is part of the chemical reaction, it has a crucial limiting factor.
3. *Radiant energy fluence rate*. This is the catalyst and thus needs to be in ample supply.
4. *Water*. It is another component of the chemical reaction.

The photosynthesis process occurs in organelles called chloroplasts on membrane systems known as thylakoids. In the absence of light, respiration reverses the process of photosynthesis, releasing the stored chemical energy. Respiration takes place in cellular organelles called mitochondria.

9.1.2 Sun Tanning

Another example of photochemistry is the effect light has on the metabolic system of certain mammals. Light is used to provide the energy to produce vitamin D.

Most humans are also affected by the psychological impact of light, which apparently is wavelength-dependent. This falls outside the scope of this book. However, the process of getting light to the biological receptors in the tissue medium is covered by the light transport theory.

An additional aspect of exposure of human skin to sunlight results in the conversion of cholesterol from low-density lipids (LDL) into high-density lipids (HDL), which are reportedly better for our general and vascular health.

9.1.3 Light-Reactive Biological Chromophores

A tissue generally contains a number of chromophores that absorb short wavelengths of light. Examples of endogenous chromophores include nucleic acids, with a maximum absorption at a wavelength of 260 nm ($\lambda_{max} = 260$ nm), the aromatic amino acids tyrosine and tryptophan ($\lambda_{max} = 275$ nm), hemoglobin ($\lambda_{max} = 420$ nm), and bilirubin ($\lambda_{max} = 460$ nm). If light incident on the tissue is absorbed by any of these chromophores, it cannot be transmitted to a subcutaneous photosensitizer. Additional information about chromophores is presented in Section 10.3.

It is also known that at wavelengths higher than 1200 nm, light absorption by water molecules is significant. The preliminary conclusion would be that a window of opportunity exists for PDT in the wavelength region ranging from 600 to 1200 nm, where there is maximum penetration into biological tissues. Another fact is that at wavelengths greater than 850–900 nm, the photons may not have sufficient energy to participate in a photochemical reaction.

9.2 Photochemical Effects

The phototoxic effect of hemoglobin derivatives was first described by the German medical student Oscar Raab in the winter of 1897–1898 in Munich, when his Professor Herman von Tappeiner asked him to study the toxicity of aniline dyes in paramecia. Raab found that the time to kill paramecia depended on the intensity of light in the lab. After their examination of many other substances including chlorophyll, they demonstrated that some administered substances could render an organism sensitive to light. They also concluded that molecular oxygen is a necessary ingredient in this phenomenon. They called this photodynamic action (photodynamische erscheinung). Raab also suggested that this could be applied to the treatment of some diseases. Based on the results of the research by Raab and Tappeiner, dermatologist Jesionek was the first to report limited clinical results in 1903 on the treatment of skin cancer with eosin as a photosensitizer. In the past century, numerous researches have dedicated themselves to optimizing this promising therapy on the choices of light sources, the wavelengths of the lights, the photosensitizers and the selective uptake of photosensitizers by the tumor tissue, the dosimetry of the light and the photosensitizers, and the expanding clinical applications of this therapy. The commonly used term to describe the photochemical damage is PDT.

9.2.1 Photodynamic Therapy

Photodynamic therapy or PDT uses light-sensitive drugs to kill cells, in particular cancer cells. Photodynamic therapy is also called phototherapy, photoradiation therapy, or photochemotherapy. This therapy involves the selective uptake and retention of a photosensitizer (compounds that absorb energy from light of specific wavelengths and are capable of using that energy to induce reactions in other nonabsorbing molecules) in a tumor, followed by irradiation with light of a particular wavelength, thereby initiating the destruction of tumor through formation of singlet oxygen.

In 1903, the use of photosentisizers for the treatment of skin cancer was reported for the first time by Professor von Tappeiner who himself applied eosin as a phototoxin. In the period from 1943 through 1961 the attention was focused on the dye derived from hemoglobin, hematoporphyrin IX, for both tumor localization and treatment of certain types of cancer. It eventually turned out that the active ingredient was an impurity in the hematoporphyrin IX, which was further purified and named hematoporphyrin derivative (HpD).

The phototoxicity relies on the activation of photosensitizers by the deposited energy of the incident electromagnetic radiation, which causes the photosensitive drug to form an alternate chemical composition, which is either toxic by itself, or has a by-product that is toxic to certain cellular structures.

The main challenge in PDT is to fabricate photosensitizers that can be excited by light at wavelengths that penetrate deep enough into the biological medium to provide total eradication of the cancer without the need for further surgical intervention.

9.2.1.1 The Photosensitizer Excitation Process

A majority of photochemicals used in PDT are biocompatible chemicals, derived from naturally occurring chemicals either in the human body or in other living organisms, which can be made nontoxic in this altered state, but will form toxins when light energy is applied. A typical PDT session initially involves either intravenous injection (i.v.) or topical application of a photosensitizer such as a porphyrin. Subsequently there is a period of time (usually 48 h) to allow systemic porphyrins (i.v. injection) to be cleared from normal tissues and preferentially be retained by the rapidly growing tissues, blood vessels, and the supporting tissues that grow with malignant tumors, or to allow topical porphyrins to be absorbed by the skin. Ultimately, this sensitization process is followed by light irradiation at a certain wavelength, which activates the photosensitizer.

Normally without light activation, the photosensitizers are in their ground-state energy level, which is a singlet state. Upon absorption of a photon of light of a specific wavelength, a molecule of the photosensitizer is activated to an excited state. This short-lived excited state is also a singlet state, and can go back to the ground state either through "internal conversion" with energy loss in the form of heat or by emitting a photon (fluorescence). The fluorescence has an added advantage that it can be used for diagnostic purposes. In the decay process of the excited porphyrin there is another option: the excited-state photosensitizer can go through the "intersystem crossing" by changing the spin of an electron and converting into a triplet state. The triplet state has lower energy than the singlet state, but a longer lifetime. This increases the probability of collision-induced energy transfer to other molecules such as molecular oxygen, and is the predominant process of PDT. In general, the photosensitizers will produce a reactive oxygen species that in turn forms a chemical bond with the amino acids, unsaturated lipids, and subcellular organelles, amongst a few other cellular building blocks of the medium where it lingers while getting involved in the cellular metabolism.

Figure 9.1 illustrates the chemical transaction of the photosensitizer on the molecular level in a Jablonski diagram. A Jablonski diagram is a schematic portrayal of the relative positions of the electronic and vibrational levels of a molecule. The labels S and T denote the singlet and triplet states of the molecule, respectively. Transitions between subsequent vertical stacks of levels represent radiationless transitions. The "forbidden" transition listed in the illustration refers to the singlet oxygen decay into triplet oxygen since this process technically is a spin-forbidden transition, the excited species has a longer lifetime. In contrast, triplet–triplet energy transfer is forbidden by the dipole–dipole mechanism, since the respective transition moments are extremely small and never take place.

9.2.1.1.1 Role of Cellular Membranes

The primary cellular components involved in the destruction of the cell are cellular membranes. The various membranes that can be distinguished in the cell are the plasma membrane itself, and the membranes of mitochondria, lysosomes, endoplasmatic reticulum, and the cell nucleus. The mitochondria play a crucial role in the PDT targeting process for cytotoxin-mediated cell death. Lysosomal confinement has also been observed, as well as affiliation for the mitochondria due to the membrane potential and the chemical interactions at the membrane in the normal functioning of the mitochondria. The influence of PDT on the nucleus still requires further investigation.

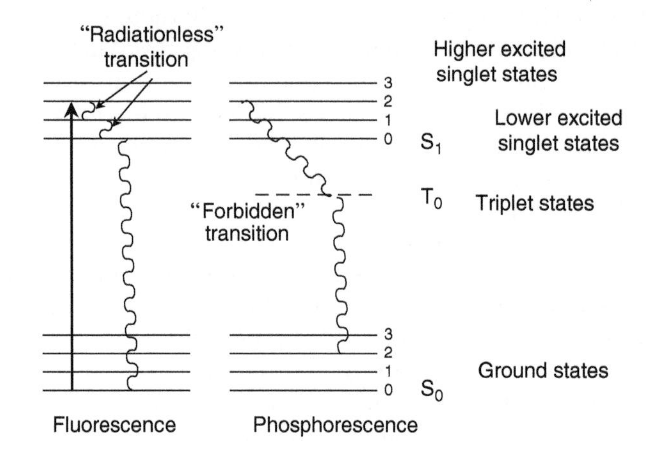

FIGURE 9.1
Jablonski diagram, showing the activation of the photosensitizer. The "forbidden" transition listed in the illustration refers to the singlet oxygen decay into triplet oxygen.

The reactive oxygen species that is formed is the result of two distinguishable processes: Type I and Type II reaction mechanisms.

The Type I reaction mechanism describes the exchange of electrons, electron–proton pairs (hydrogen), with the neighboring molecules. These newly formed chemical components are known as free radicals. Free radicals are formed first by the electron/hydrogen abstraction from a substrate molecule or directly from the photosensitizer. These free radicals in turn interact with smaller molecules, oxygen in particular, to produce an excited state. The free radicals are then ready to react with oxygen to form some highly reactive oxygen species, either peroxide ions, which are positively charged, or superoxide, which can attack many important cellular components.

The Type II reaction mechanism involves the excitation of molecular oxygen to an excited triplet state, which reduces to singlet atomic oxygen ground states.

Based on the published data on optical parameters for various tissues it can be shown that the practical therapeutic window for clinical PDT will need to be in the wavelength range between 600 and 800 nm.

The crucial point remains the photon energy delivered to the entire target volume from the closest proximity to the farthest distance from the point of entry of light from the selected light source. This makes the choice of excitation wavelength just as important as the local optical window for maximum transmission. Ideally, both criteria need to be satisfied simultaneously.

9.2.1.2 The Role of Oxygen

The existence of enough oxygen molecules in the target tissue is critical to the PDT performance. Many tumor tissues can be hypoxic due to either the shortage of blood delivery or the fast consumption of oxygen in the tissue caused by the photodynamic effects. To ensure the necessary oxygen level, some studies used intermitted light delivery or reduced fluence rate of light to allow oxygen diffusion to compete with oxygen consumption and showed increased effects. Some other studies showed that tumor oxygenation can be improved by letting the patient breathe a perfluorochemical emulsion of carbogen (95% O_2, 5% CO_2).

The formation of the "singlet oxygen" designates an excited and highly reactive state of oxygen. Oxygen is among the few molecules that have a triplet ground state. In turn, the two aforementioned reaction mechanisms describing the formation of the mutated oxygen rely on the excitation of the photosensitizer by photon-energy absorption, resulting in a higher energy state of the phototoxin.

Upon reaction with the triplet-state photosensitizer, the oxygen molecules will be converted into their reactive singlet state as a result of the energy transferred from the photosensitizer; subsequently, the photosensitizer itself returns to its ground state. Although in PDT there might be a combination of the two reaction mechanisms, experimental evidence shows the predominance of the Type II mechanism through in vitro studies.

9.2.1.3 Mechanisms in PDT to Induce Cell Death

The generation of these reactive oxygen species can cause damage from several different levels once they attain their critical values of concentration. First it can lead to damage on the cellular and subcellular levels. Four different photochemical mechanisms are described briefly in arbitrary order.

9.2.1.3.1 Apoptosis

Other than direct cell death from necrosis, apoptosis has been found to be very important in PDT. Apoptosis is the programed cell death in which the damaged cells start a sequence of programs to kill themselves without inflammatory responses. Apoptosis in PDT was first described in 1991 and has been observed both *in vitro* and *in vivo*. Photodamage of the mitochondria has been shown to be associated with an apoptotic cell response. In the mitochondria a mechanism called caspase cascade is initiated where aspartate-specific cysteine protease, an enzyme, is used to cleave proteins. The caspase cascade process may result in the formation of pores in the cell membrane, which are partially responsible for the slow but unequivocal cell death.

The administration of exogenous photosensitizing compounds can be substituted by the stimulation of the cellular synthesis of endogenous photosensitizers. For example, 5-aminolaevulinic acid (ALA) is a metabolic precursor in the biosynthesis of heme. Apoptosis is also expected in ALA-PDT, where mitochondria are the production site of the endogenous photosensitizer PpIX induced by ALA. This has also been shown to work both *in vivo* and *in vitro.*

9.2.1.3.2 Photooxidation

The hydrophobic photosensitizers can accumulate in the membranes of the cell and its organelles. Photooxidation of lipids and proteins by the reactive oxygen species in the cell membrane activates some membranous phospholipases, which degrades the phospholipids that are important components of the membrane, causing rapid changes in fluidity, loss of cell integrity, and inactivation of membrane-associated enzyme systems and receptors. Also the cell membrane depolarizes rapidly, leading to an ionic imbalance. In contrast, the hydrophilic photosensitizers, which are less likely to travel through the lipid bilayer of the cell membrane, are more likely to be taken up by the cells through endocytosis and end up localizing in lysosomes. Upon light irradiation, the photosensitizers along with other lysosomal components will be released into the cytosol and bind to the cytoskeleton protein tubulin, where photooxidation causes a depolymerization of the microtubules with accumulation of cells in mitosis, and subsequent cell death.

9.2.1.3.3 *Tumor Vasculature*

Tumor microvasculature is one of the most important targets of phototoxicity, since the endothelial cells of the blood vessels are usually the first place for the accumulation of the photosensitizers. Damage to the endothelial cells leads to thrombosis and vascular stasis. Increased permeability and leakage of albumin from the damaged vessel wall leads to edema in the treated area. Reduced blood perfusion in PDT-treated tissue due to thrombosis formation causes ischemic injury.

9.2.1.3.4 *PDT-Mediated Inflammatory Cell Death*

Photodynamic therapy has been shown to be related to a substantial inflammatory response within the target tissue. The phototoxic damages in the cell membrane lead to the release of metabolites of membrane phospholipids that are strong activators of the inflammatory reaction. Macrophages and T-cell-mediated immune responses toward the tumor tissues are thought to play important roles in PDT.

9.2.1.4 Light Delivery

A laser light delivery system consists of the laser light source, optics, a light guide, and optical target optics. Currently, the coherent monochromatic lasers are coupled to the single-stranded optical fibers to provide the flexibility needed to deliver light to subsurface lesions through surface, interstitial, intracavitary, or endoscopic techniques. Different modifications can be made to adapt the delivery system to specific needs of the target tissue. For example, introducing a lens or diffuser rod to the end of the fiberoptic cable can spread the laser beam uniformly and make the laser more uniform in intensity too.

The curative effect of PDT relies on a threshold amount of radiant energy fluence rate to reach the distal portions of the tumor treated with light-sensitive drug. This is required to excite all photosensitive dye with the threshold amount of energy throughout the entire target volume. The point farthest removed from the entry point of the light is the most obvious concern for the delivered proper dosage.

However, even tumor locations at close proximity may be hidden behind optically dense conglomerations of media, with either high absorption or scattering properties at the therapeutic wavelength, and is thus shielded from receiving the minimum fluence rate for effective destruction of the cells doped with photosensitizer.

The light dosimetry depends on the properties of not only the light used but also the target tissues to which the light is applied. For tissues located on or just under the skin or on the lining of the internal organs, there is a risk of extensive damage with subsequent shrinkage, and also always a small risk of penetration through the organ wall. For light at wavelengths around 630 nm, the penetration depth is usually around 2–5 mm. For shorter wavelengths the penetration depth will start to decline significantly below 540 nm. Longer wavelengths may have deeper penetration but will lack the energy requirements to accomplish the phototoxic effect.

The success of PDT depends upon the total light dose and the fluence rate (the rate of the light delivery). The total light dose, given during PDT, is calculated as the product of the fluence rate and time as

$$\text{Total light dose (J/cm}^2\text{)} = \text{fluence rate (W/cm}^2\text{)} \times \text{treatment time(s)}$$

A wide range of light dose from 30 to 540 J/cm^2 has been used for PDT application using lasers, light-emitting diodes (LEDs), or filtered lamps. Also the fluence rate is usually maintained low in practice, since high fluence rate has been related to significant heating of the tissue. Also, with high fluence rates, it has been shown that the molecular oxygen in the target tissue can be depleted soon, leaving not enough to participate in the photodynamic reaction. Operation at lower fluence rate has been found to increase the treatment effects by allowing time for the reoxygenation in the target tissue. This is what is known as the Arndt–Schultz effects, where a small dose of radiation allows for all processes to finalize within certain boundary conditions, increased dosage will initially have a favorable effect followed immediately by a negative effect, while a very large dose will briefly have a stimulating effect followed by a dramatic and persistent negative effect.

The dosimetry of the photosensitizers itself relies heavily on the balance between the concentrations of the drug in the tumor tissue compared to the concentration build-up in the surrounding healthy tissues at the time of irradiation. Choices of these parameters are based on drug intake in the body including absorption, distribution, transformation, and excretion (pharmacokinetics), in addition to results from drug interaction obtained from animal studies.

9.2.1.5 Photosensitizers

There are numerous factors that play a role in successful PDT treatments. One major factor is the choice of photosensitizer. The photosensitizer must be able to localize to the tumor tissue and possess the photodynamic properties necessary for the generation of reactive chemical species such as singlet oxygen. By choice, photosensitizers are chemically well defined, with high tumor-localizing properties, and inherently nontoxic. Additionally, they should have a strong absorbance in the identified optical window ranging from 600 to 850 nm to maximize tissue penetration.

The photo-reagent drugs are designed to have a stable singlet ground-state electronic configuration. This singlet ground state can be excited to a higher energy level singlet state, which is very unstable in nature. The singlet ground state needs a specific energy to move to the excited singlet state. This excitation energy is provided by the photon packet as $E=hv$. The triplet energy level is lower than the singlet state, but it has a longer lifetime, which provides an opportunity for energy transfer between molecules. The triplet state quantum yield determines the effectiveness of a phototoxin to work as a lethal weapon in the potential to induce cell death. In the triplet-state lifetime the photosensitizer will collide with other molecules, among which is oxygen, and thus produces the reactive oxygen molecules. The energy diagram of the electron bands in this excitation process is illustrated in Figure 9.1. From this diagram it is evident that the excitation process is a renewable process, and the energy transfer can be repeated several times over, thus generating a large amount of reactive oxygen despite the other processes in the degeneration process.

The first photosensitizer used in clinical PDT was HpD and its purified fraction, Photofrin. Since then many more phototoxic chemicals have been developed. Some of the HpD developed are Photofrin I and II patented by T. Dougherty, K. Weishaupt, and U Potter. Other photosensitizers were derived from chlorophyll in the chemical name Pheophorbide A derivatives resulting from the work by E. Snyder and I. Sakata. Additional water-soluble chlorine derivative photosensitive dyes have been developed in Russia under the following names: Photodithazine, Photochlorin, and Radachlorin, among others.

The primary photosensitizer HpD is a mixture containing hematoporphyrin, hydroxyethylvinyldeuteroporphyrin (HVD) and protoporphyrin (Pp), and complex esters, in addition to ether and carbon–carbon linkages of hematoporphyrin. The oligomeric fractions of certain esters are used for the tumor-localizing activity of HpD in vivo. Hematoporphyrin derivitive has further been purified to develop porphyrin, which has a bigger portion of oligomeric material.

In the United States, at the time of publication, the only photosensitizers approved for clinical use are Visudyne (Verteporfin) and Photofrin. Photofrin is a porfimer sodium solution containing a complex mixture of porphyrin esters and ethers. The maximum absorption wavelength of Photofrin molecules is 630 nm.

Photofrin-mediated PDT has proven to be curative for a range of cancers, but there are some drawbacks to this photosensitizer. Since the compound is a complex mixture, the active components are difficult to identify and the synthetic process is hard to reproduce. Furthermore, photofrin is excited at 630 nm; generally this wavelength has a relatively low penetration depth, making Photofrin unsuitable for deeper tumors.

A second generation of photosensitizers has successfully been developed following the success of Photofrin. These include modified porphyrins, chlorins, bacteriochlorins, phthalocyanines, naphthalocyanines, pheophorbides, and purpurins. They show increased efficacy in PDT in many ways, such as improved photophysical properties as demonstrated by their activation wavelengths, hence increasing tissue penetration.

Benzoporphyrin derivative (BPD) is a promising new photosensitizer based on chlorin for use with PDT. It has a peak absorption at 690 nm and reaches peak levels in cancer tissue within 3 h following administration, and reduce to 50–60% within 48 h after injection. The type of prompt proliferation of the photosensitizer within the body is beneficial to the fact that light therapy can be performed the same day as drug administration.

A majority of photosensitizers require at least a 24 h lapse from the time the drug is administered until the target tissue is illuminated with light. Furthermore, rapid clearance of the drug from the body means a decrease in the amount of time the patient will experience photosensitivity. Benzoporphyrin derivative is also excited by a longer wavelength of light than Photofrin. Light of wavelength 690 nm (BPD) will penetrate the skin to a depth of approximately 1 cm compared to a depth of 0.5 cm at a wavelength of 630 nm for Photofrin.

Other examples of photosensitizers being developed include derivatives of napthtalocyanines. Certain groups of researchers synthesized and quantized the ability of zinc(II)- and silicon(IV)-napthtalocyanines to penetrate tissue, preferentially accumulate in cancer cells, and cause cell death upon activation. Studies indicate that hydrophobic molecules tend to accumulate within cancer cells better than in normal cells. The hypothesis was that the hydrophobic nature of napthalocyanines would lead them towards more efficient accumulation, distribution, and retention by tumor cells. To test the hypothesis, the naphthalocyanine complexes were incorporated into liposomes, which resulted in making zinc(II)-naphthalocyanine an excellent photosensitizer due to selective targeting. Furthermore, zinc(II)-naphthalocyanine has a slow tumor clearance, in contrast to fast clearance from skin and has a pronounced phototherapeutic effect on different tumor models. Also, the maximum absorption wavelength for these molecules is 780 nm, which corresponds to relatively deep penetration of tissue by the laser light. Collectively, these qualities make zinc(II)-naphthalocyanine a very succesful molecule for PDT of cancer.

In contrast, the silicon (IV)-napthtalocyanines functioned very poorly as photosensitizers. The hydrophilic nature of the two methoxy (polyetheyleneglycol) groups on silicon (IV) may be one of the reasons why this sensitizer does not accumulate at a high degree in tumor tissue.

The administration of photosensitizing compounds from outside the organism or cell (exogenous) can be substituted by the stimulation of the cellular synthesis of endogenous photosensitizers (from within the cell).

For example, 5-aminolevulinic acid can be administered systemically or topically to induce the immediate precursor of heme, protoporphyrin IX (PpIX), and the relatively slow conversion of PpIX into heme can lead to the build-up of phototoxic levels of PpIX. There are several advantages of ALA-induced PpIX over HpD and photofrin for use in PDT. First, the optimum therapeutic ratio is reached at 2–4 h following ALA administration and the ALA-induced PpIX is cleared from the normal tissues within 24 h. This saves the patients from prolonged protection from light as well as allows repeated treatment as frequently as every 48 h without damage to normal tissues. Second, certain types of tumor tissues exhibit increased accumulation of ALA-induced PpIX. Third, the enzyme ferrochelatase, which catalyzes the incorporation of iron into the porphyrin ring, hence the formation of the heme, has lower activity in some tumors. This lowers the conversion of ALA into heme and results in prolonged elevation of PpIX levels. Furthermore, topically applied ALA cannot readily penetrate the keratinous layer of normal skin but can penetrate malignant lesions. This further enhances the tumor specificity whilst leaving the healthy skin intact. The drawbacks of ALA-induced PpIX photosensitization are the same as that in the case of HpD and Photofrin because it is still mediated by porphrin, and the excitation wavelength is still 630 nm, which is limiting in the penetration depth. 5-Aminolevulinic acid is a precursor for the synthesis of protoporphyrin IX. This compound will readily penetrate through cancerous and other abnormal skin conditions; however, it will not penetrate through normal keratin. One of the primary advantages of ALA over porfimer sodium is the shorter duration of photosensitivity of ALA. In contrast, the low binding efficiency of ALA with respect to porfimer sodium proves to be a disadvantage resulting in the need to administer high doses to achieve clinically relevant levels.

Another noteworthy photosensitizer is chlorophyll *a* derivative, pheophorbide *a* (Ph*a*). Pheophorbide-*a* has maximum absorption at a wavelength of 665 nm. Twenty-four hours following administration of Ph*a* the tumors were surgically exposed and illuminated with argon laser light tuned to a wavelength of 665 nm. The drug Ph*a* also demonstrates the induction of apoptosis upon light activation. Another noteworthy photosensitizer is Foscan (*m*-tetradydroxyphenylchlorin, mTHPC).

9.2.1.6 Dosimetry

In PDT generally the tumor to be treated is rather large compared to the diameter of a laser beam and requires a broad beam illumination protocol. The required low levels of radiance also warrant the use of a broad beam versus a concentrated high-power laser beam. The broad beam irradiation dramatically simplifies the theoretical approach for dosimetry calculations. The treatment wavelengths are specifically chosen to yield a low absorption and moderate scattering, with $\mu_a \ll (1-g)\mu_s$ as the main boundary condition. Combining the conditions of absorption less than scattering and broad beam irradiation encompasses that the diffusion approximation to the equation of radiative transfer is appropriate.

Additionally only the radiant energy fluence rate $\Psi(\mathbf{r},\mathbf{s},t)$ will be of interest instead of the radiance $L(\mathbf{r},\mathbf{s},t)$. Furthermore, it will be shown that both the light delivery and light distribution are acceptably constant over time to apply the steady-state approximation. As shown in Chapter 6, the steady-state diffusion approximation is written as follows:

$$D\nabla^2\Psi(\mathbf{r},\mathbf{s}) - \mu_a\Psi(\mathbf{r},\mathbf{s}) = -q'(\mathbf{r}) \tag{9.1}$$

where

$$D = \frac{1}{3[\mu_a + (1-g)\mu_s]} = \frac{1}{3\mu_{tr}'} \tag{9.2}$$

is the diffusion coefficient and $q'(\mathbf{r})$ the luminous source function resulting from scattering of the incident light within the tissue.

For a broad beam the radiant energy fluence rate will be constant across the width of the beam, removing the radial component $\mathbf{r}$. Furthermore, the light distribution in the tissue will only depend on the distance to the surface z. Additionally the absorption coefficient for the photosensitizer will be combined with the tissue absorption coefficient, where the tissue absorption provides an offset value for a variable absorption due to concentration gradients in the tissue. The photosensitizer absorption will thus be a function of depth as well $\mu_a(z) = \mu_{a,tissue} + \mu_{a,drug}(z)$.

The source function is defined on the basis of the depth dependency of the absorption coefficient, integrated over depth, ζ to the point of interest, as follows:

$$q'(z)\Psi_0[(1-g)\mu_s]\exp\left[-\int_0^z(\mu_a(\zeta)+(1-g)\mu_s)d\zeta\right] \tag{9.3}$$

where Ψ_0 is the incident homogeneous fluence rate from the light source on the tissue.

Implementation of all the discussed conditions and substitution of Equation (9.3) into Equation (9.1) yields

$$\frac{\partial^2}{\partial z^2}\Psi(z) - \mu_{eff}^2(z)\Psi(z) = -3\Psi_0[(1-g)\mu_s](\mu_a(z)+\mu_s')$$
$$\exp\left[-\int_0^z(\mu_a(\zeta)+(1-g)\mu_s)d\zeta\right] \tag{9.4}$$

where μ_s' is the reduced scattering coefficient $(1-g)\mu_s$, and the effective attenuation coefficient is defined this time as $\mu_{eff}(z)=\sqrt{3(\mu_a(z)[\mu_a(z) + (1-g)\mu_s])}$.

Equation (9.4) can be solved using a solution method developed by Wentzel, Kramers, and Brillouin, the WKB method, yielding the following solution:

$$\Psi(z) = \Psi_0 \left\{ \frac{c}{\sqrt{\mu_{\text{eff}}(z)}} \exp\left[-\int_0^z (\mu_{\text{eff}}(\zeta))\mathrm{d}\zeta \right] + c_2(z) \exp\left[-\int_0^z (\mu_a(\zeta) + (1-g)\mu_s)\mathrm{d}\zeta \right] \right\} \tag{9.5}$$

where c_1 and $c_2(z)$ are constants depending on the medium itself and on the boundary conditions, respectively.

Equation (9.5) can subsequently be solved numerically to resolve the depth-dependent radiant energy fluence rate.

9.2.1.6.1 *Photosensitizer Bleaching Rate*

An additional factor that needs to be considered is the depletion of the photosensitizer under the influence of the irradiation itself. The temporal change in photosensitizer absorption is directly proportional to the initial concentration, and thus the initial absorption and fluence rate is expressed as

$$\frac{\mathrm{d}}{\mathrm{d}t} \mu_{\text{a,drug}}(z,t) = -\beta \mu_{\text{a,drug}}(z,t) \Psi(z,t) \tag{9.6}$$

where β is the bleaching rate of the photosensitizer under irradiation.

Equation (9.6) has the following solution for any infinitesimal time period succeeding irradiation $t + \Delta t$ expressed as

$$\mu_{\text{a,drug}}(z, t + \Delta t) = \mu_{\text{a,drug}}(z,t) - \beta \mu_{\text{a,drug}}(z,t) \Psi(z,t) \Delta t \tag{9.7}$$

To justify for the changes resulting from the irradiation process itself, Equation (9.7) will need to be substituted into Equation (9.5), which again will need to be solved, however, this time through an iteration process accounting for continually changing irradiation conditions.

9.2.1.6.2 *Necrosis Threshold Resulting from PDT*

The cell death as a result of drug administration, drug diffusion, drug retention, light delivery, and light absorption is the ultimate goal of PDT. In PDT the efficiency of singlet oxygen production is the ultimate marker for success of the treatment. In addition to the aforementioned parameters, the availability of ground-state oxygen in the treatment volume also needs to be taken into consideration.

The rate changes in the population of singlet and triplet states of the photosensitizer-mediated oxygen as well as the depopulation of the ground-state oxygen can be accounted for on the basis of conservation of energy, including the phosphorescence of the photosensitizer.

For the depopulation of the ground state the expression is as follows:

$$\frac{\partial}{\partial t}[X_0(t)] = -\frac{\Im_D C'}{\Psi_0}\mu_{a,drug}(z)\Psi(z)[X_0(t)] + (A_{10} + k_{10})[X_1(t)] \\ +(A_{20} + k_{20} + \chi[^3O_2(t)])[X_2(t)] \tag{9.8}$$

where $X_0(t)$ is the population of the photosensitizer in the singlet ground state, $X_1(t)$ the population of the photosensitizer in the singlet excited state, $X_2(t)$ the population of the photosensitizer in the triplet ground state, $\Im_D$ a constant derived from the boundary conditions, C' the speed of light in the medium, A_{10} the rate constant of radiative transfer from the singlet excited state to the ground state of the photosensitizer, A_{20} the rate constant of radiative transfer from the triplet excited state of the photosensitizer to the ground state, k_{10} the rate constant of radiationless transfer from the singlet excited state of the photosensitizer to the ground state, k_{20} the rate constant of radiationless transfer from the triplet excited state of the photosensitizer to the ground state, χ the efficiency rate constant for conversion of ground state oxygen into triplet oxygen, and $^3O_2(t)$ the time-dependent triplet oxygen concentration.

Similarly, the situation for the singlet excited state is described as

$$\frac{\partial}{\partial t}[X_1(t)] = \frac{\Im_D C'}{\Psi_0}\mu_{a,drug}(z)\Psi(z)[X_0(t)] - (A_{10} + k_{10} + k_{12})[X_1(t)] \tag{9.9}$$

where k_{21} is the rate constant of radiationless transfer from the triplet ground state of the photosensitizer to the singlet excited state.

And for the triplet ground state it can be found that

$$\frac{\partial}{\partial t}[X_2(t)] = k_{21}[X_1(t)] - \{A_{20} + k_{20} + \chi[^3O_2(t)]\}[X_2(t)] \tag{9.10}$$

The entire photodynamic process is rather slow, which allows us to use a quasi-steady-state approach to solve Equations (9.8)–(9.10) yielding the following solutions:

$$[X_0] = [X_\infty] \tag{9.11}$$

$$[X_1] = \frac{\Im_D C' \mu_{a,drug}(z)\Psi(z)[X_\infty(t)]}{\Psi_0(A_{10} + k_{10} + k_{12})} \tag{9.12}$$

$$[X_2] = \frac{\Im_D k_{21} C' \mu_{a,drug}(z)\Psi(z)[X_\infty(t)]}{\Psi_0(A_{10} + k_{10} + k_{12})(A_{20} + k_{20} + \chi[^3O_2])} \tag{9.13}$$

where $[X_0]$, $[X_1]$, and $[X_2]$ are the respective steady-state concentrations of singlet-ground-state, singlet-excited-state, and triplet-ground-state photosensitizer, and $[X_\infty]$ is the total concentration of the photosensitizer in the tissue derived from the administered amount diluted over the tissue mass.

Similar equations can be derived to account for the transfer of ground state to singlet excited oxygen. Ignoring metabolic depletion of oxygen and oxygen consumption by photobleaching and various cellular oxidation processes provides the following sets of oxygen conversion equations. First, for triplet oxygen,

$$\frac{\partial}{\partial t}\left[{}^3O_2\right] = k_0\left[{}^1O_2\right] - \chi\left[{}^3O_2(t)\right][X_2(t)] + P \tag{9.14}$$

where P represents the oxygen diffusion rate and k_0 the efficiency conversion rate for singlet oxygen. For singlet oxygen,

$$\frac{\partial}{\partial t}\left[{}^1O_2\right] = \chi\left[{}^3O_2(t)\right][X_2] - k_0\left[{}^1O_2\right][X_2] \tag{9.15}$$

Equations (9.13) and (9.15) can be combined under the condition of fluence rate levels below saturation and as such no significant reduction of the ground state population occurs, $[X_1] \ll [X_\infty]$ and $[X_2] \ll [X_\infty]$.

Singlet oxygen production can be written in approximation as

$$\frac{\partial}{\partial t}\left[{}^1O_2\right] = \chi\left[{}^3O_2(t)\right][X_2] \tag{9.16}$$

Additionally the spontaneous decay and phosphorescence will be in direct competition with the creation of singlet oxygen. The rivalry of the triplet state with the singlet state of oxygen as derived from Equations (9.10) and (9.14), under the restrictions mentioned, can be expressed as

$$\Gamma = \frac{\chi\left[{}^3O_2\right]}{A_{20} + k_{20} + \chi\left[{}^3O_2(t)\right]} \tag{9.17}$$

Note that Γ approaches zero when the triplet oxygen levels are low.

These conditions imply that the photochemical process will be effective for high concentrations of oxygen but will be quenched when oxygen levels fall below a critical concentration. This means that the photodynamic process is in constant competition with itself.

The main conclusion from this description is that tissue necrosis will only occur when the fluence rate reaches a minimum value, which will be location-dependent. Additionally a minimum level of oxygenation is also required. In contrast, too high levels of radiant energy fluence rate will quench the photochemical destruction process.

9.2.1.6.3 *Fluorescence in PDT*

Under the assumption that the fluorescence source function is proportional to the light distribution inside the tissue with the factor γ_{FL}, the quantum yield for fluorescence times the absorbed fluence rate, gives the following expression:

$$FL(\mathbf{r},\mathbf{z}) = \gamma_{FL}\Psi(\mathbf{r},\mathbf{z})\mu_{a,drug}(\mathbf{z}) \tag{9.18}$$

Equation (9.18) reduces to a one-dimensional expression under broad beam illumination. Further analysis will show that the influence of the concentration of photosensitizer closest to the surface will have the greatest contribution to the fluorescence signal.

The steady-state fluorescence and phosphorescence can be derived from Equations (9.19) and (9.20), respectively. Equation (9.19) will provide the fluorescence as

$$\mathrm{FL} = A_{10}[\mathrm{X}_1] = \frac{\Im_D A_{10} C' \mu_{a,drug}(z)\Psi(z)[X_\infty(t)]}{\Psi_0(A_{10}+k_{10}+k_{12})} \tag{9.19}$$

whereas the phosphorescence *PH* follows from Equation (9.20) as

$$\mathrm{PH} = A_{20}[\mathrm{X}_2] = \frac{\Im_D A_{20} k_{21} C' \mu_{a,drug}(z)\Psi(z)[X_\infty(t)]}{\Psi_0(A_{10}+k_{10}+k_{12})(A_{20}+k_{20}+\chi[{}^3\mathrm{O}_2])} \tag{9.20}$$

The quantum yield can now be derived to be

$$\gamma_{FL} = \frac{k_{21}A_{20}}{(A_{20}+k_{20})A_{10}} \tag{9.21}$$

The fluorescence and phosphorescence can be measured and used for diagnostic purposes to determine drug-delivery efficiency and tumor localization. The constants A_{10} and A_{20} and k_{10}, k_{12} and k_{20} are the same as defined in section 9.2.1.6.2.

9.2.1.7 Clinical Applications

Photodynamic therapy is now a rising modality for cancer treatment. Because of its limitation of penetration depth, it is currently widely used to treat tumors in or just under the skin area or on the lining of some internal organs. Also, in ophthalmology, it can be applied to the age-related macular degeneration (AMD) with discharge.

Photodynamic therapy targets the neovascularization by sealing the leaky vessels and causes vascular endothelial damage with vessel occlusion and destruction, preventing further leakage from the vessels. Another very promising application of PDT is as adjuvant therapy for surgery to treat brain tumors, malignant mesothelioma, and intraperitoneal tumors.

There are numerous other clinical applications in various countries and selected clinical trials underway. Examples of the use of PDT include the treatment of skin cancers, bladder tumors, macular degeneration, gynecological neoplasms, breast cancer, atherosclerosis, rheumatoid arthritis, pancreas cancers, ovarian tumors, nasopharyngeal tumors, atherosclerosis, and intimal hyperplasia.

Multiple studies report the successful use of PDT to treat superficial tumors. Although most superficial tumors can be treated with surgery and laser ablation, the cosmetic outcome is often undesirable. Surgery and laser ablation may cause significant scarring. However, depending on the application, several cosmetic laser applications have been shown to be quite effective. In addition, PDT causes minimal damage to the connective tissue, specifically collagen and elastin. Therefore, healing occurs with more regeneration and less scarring.

On average, 3–4 days following administration of the photosensitizer, skin tumors are illuminated with light, for instance an argon laser tuned to a wavelength of 652 nm to match the excitation of the photosensitizer. Generally lesions treated with low dosages of light heal within 4–6 weeks and show little if any signs of scarring. In contrast, lesions treated with high dosages of light may take up to 8 weeks to heal and additionally may show significant scarring. Therefore, the dosage of light administered is critical to the treatment outcome. By using low dosages of light, PDT proves to be an effective alternative to surgery or laser ablation and has a more desired cosmetic outcome.

Several examples of the use of PDT in clinical setting are described next.

9.2.1.7.1 *Inhibition of Intimal Hyperplasia*

One example includes the use of PDT to inhibit intimal hyperplasia, or in other words, restenosis. A common result of angioplasty or vascular intervention therapy is the proliferative and migratory response of smooth muscle cells as a result of vascular injury at the site of the vascular injury; this process is called restenosis. The ability of PDT to inhibit restenosis has been shown to inhibit cell growth in the carotid artery after inflation stretching by means of balloon inflation. The site treated with a photosensitizer and subsequent exposure to laser light revealed no intimal hyperplasia.

With respect to non-cancer applications of PDT the operator must be extremely careful regarding the amount of tissue that is exposed to the laser light. Unlike cancer cells, the therapeutic index between smooth muscle cells at the location of vascular injury does not differ greatly from that of adjacent tissues. Since these smooth muscle cells are not cancerous, there is no preferential accumulation of photosensitizer within these cells. This problem can be reduced by limiting the exposure of tissue to BPD. By sealing a compartment in the carotid artery, the blood is evacuated, while subsequently introducing BPD into the area with hyperbaric pressure.

Other tumor treatments are available for the head and neck, the bladder, and the skin. Recent new developments in nasal tumors will also be discussed.

9.2.1.7.2 *Treatment of Head and Neck Tumors*

One of the world's major cancer concerns is the development of squamous cell carcinoma (SCC) of the head and neck. Squamous cell carcinoma is the second most common skin cancer. The first most common skin cancer is basal cell carcinoma. Each year, more than 200,000 Americans succumb to SCC. The SCC surfaces from the epidermis and resembles the squamous cells that comprise most of the upper layers of skin. Squamous cell carcinomas are most common in areas exposed to direct sun, but may occur on all areas of the body including the mucous membranes. Usually SCCs remain confined to the epidermis. However, eventually they will penetrate into the underlying tissues if not treated on time. When the carcinoma penetrates the dermis and underlying tissue, it can be disfiguring. In isolated cases, they can even metastasize to distant tissues and organs and become fatal.

An illustration of a rather large tumor growth on the neck and the treatment protol was illustrated earlier in Figure 1.2.

9.2.1.7.3 *Treatment of Urinary Bladder Tumors*

The urinary bladder offers a unique geometric design that lends itself to particular treatment protocols. When treating cancerous growths in the lining of the urinary bladder, a large area can and will be irradiated at once, thus yielding a broad beam irradiation protocol. In addition, the location is relatively awkward and cannot be irradiated selectively. In the treatment of bladder tumors the selection criteria will be left to the localization of the photochemical, while using diffuse wide-field irradiation.

The treatment protocol for the transitional cell carcinoma of the urinary bladder also takes into consideration that the disease pattern involves that the tumor often sprouts up from the flat lining into the interior of the bladder cavity. With this in mind, the protruding polyps and lining all need to be treated simultaneously. For this purpose the irradiation mechanism usually involves the placement of an isotropically diffusive light source held in place by a multistring nylon positioning umbrella that is flexed against the bladder wall in at least four positions. This situation is illustrated in Figure 9.2. The diffusing section at the tip of a fiber radiates laser light in a broad field covering a virtually spherical geometry. In addition to the broad diffuse irradiation from the specially designed fiberoptic delivery device, the bladder is filled with a purely scattering solution such as soymilk solution (Intralipid® and Nutralipid®, to name some, but a few trademark sterile solutions are also used). The treatment will thus take place under diffuse irradiation preferably from the center of the bladder, which in turn is filled with a scattering solution. In this way the resulting irradiation is relatively homogeneous and diffuse across the entire bladder wall. The selective

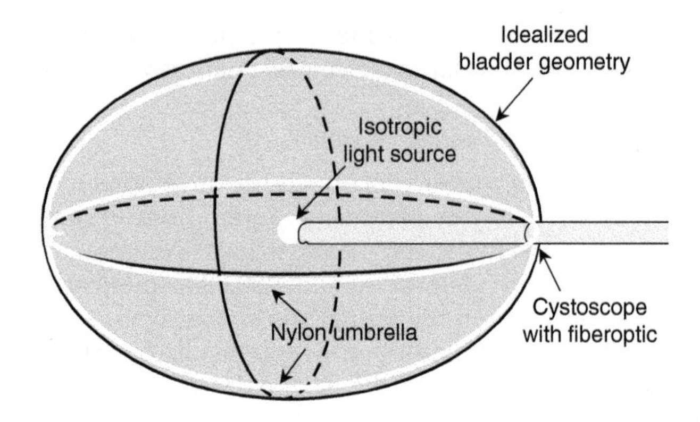

FIGURE 9.2
Illustration of a diffusive, isotropic light source held in place by a multistring nylon positioning umbrella flexed against the bladder wall for the irradiation of bladder tumor. The diffusing section at the tip of a fiber radiates laser light in a broad field covering a virtually spherical geometry. (From the work by W.M. Star and J.P.A. Marijnissen, Daniel Den Hoed Clinic, Rotterdam, The Netherlands.)

uptake of photochemical will need to be relied on for complete treatment efficacy.

In terms of dosimetry, the calculations for the radiance will be considerably different for this type of bladder irradiation compared to external collimated broad beam irradiation of the skin, for instance.

Additional new work is being done on the treatment of nasal tumors.

9.2.1.7.4 *Treatment of Nasal Tumors*

As mentioned earlier the treatment objective in clinical biomedical optics is delivering the proper light dose to the appropriate locations. The dosimetry requirements place major constraints on both the light-delivery mechanism as well as the wavelength selection. The treatment of nasopharynx carcinoma (NPC) is one of those particular applications that has required the development of an appropriate light applicator that will provide maximum coverage and subsequently high cure rate.

Delivering light to the entire nasal anatomy accounting for the soft palate as well as the fixed anatomy (bone and cartilage) is just as important as the actual choice of wavelength.

The nasopharyngeal cavity can be accessed from both the nose and the mouth, providing various access pathways for treatment options. An illustration of the nasal cavity is shown in Figure 9.3. Additionally, deep-seated recurring tumors may still require interstitial irradiation by insertion of fiberoptic delivery devices. Figure 9.4 shows the patient treatment protocol of the PDT group of the Erasmus University in Rotterdam during patient trial studies in Indonesia with a specially designed light diffuser inserted into the nasal cavity.

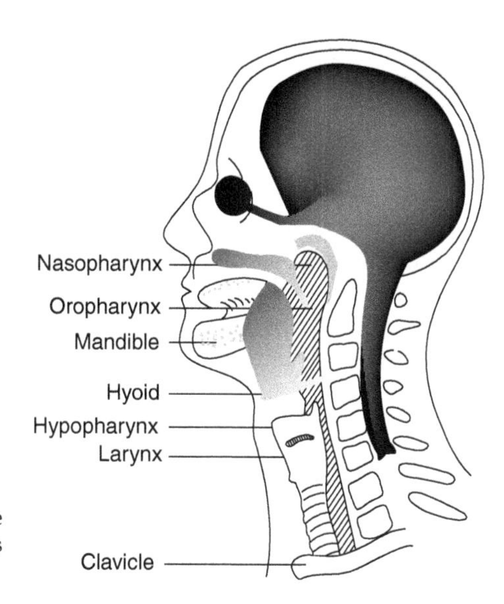

FIGURE 9.3
The nasal cavity and its location in the head. (From Sterenborg H.J.M., Erasmus University, Rotterdam, The Netherlands.

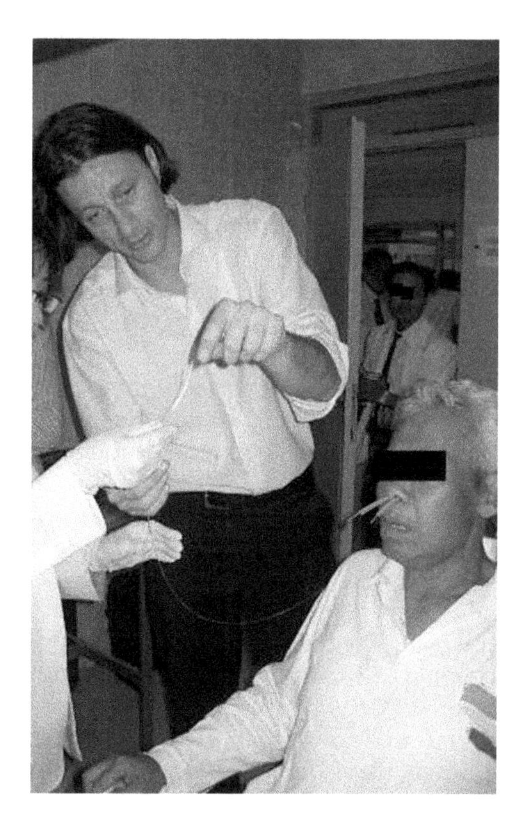

FIGURE 9.4
Patient treatment protocol with a specially designed light diffuser inserted into the nasal cavity. (From Sterenborg H.J.M., Erasmus University, Rotterdam, The Netherlands.)

In addition to PDT for cancer treatment there are several trial studies on noncarcinogenic applications of PDT currently in progress.

9.2.1.7.5 Other Experimental and Controversial PDT Research

Other non-cancer applications of PDT currently being investigated include control of oral biofilms for the reduction of dental caries and periodontal disease, synovectomy for the treatment of rheumatoid arthritis, menorrhagia, and female sterilization. Overall, these studies demonstrate the potential diversity of PDT and its versatility in clinical applications.

The choice of light source for the treatment options in PDT is offering a wide selection as will be discussed in the next section.

9.2.1.8 Light Sources

Any light source can be used with PDT assuming it contains the wavelength of light necessary to activate the photosensitizer. For the therapeutic PDT irradiation, both coherent and noncoherent light sources can be utilized. Initially, broad-spectrum light sources such as xenon arc lamps or slide projectors equipped with red filters to get rid of the short-wavelength lights have been applied to PDT. They are still in use for *in vitro* and preclinical *in vivo* studies of tumors in or under the skin. However, the unique properties of lasers make them the most efficient light source for PDT. As the coherent light source, laser has become the current standard for PDT application, although LEDs as noncoherent light source are also used in some cases.

The advantages of lasers over other options of light sources are that they are monochromatic, of high-power output, and coherent. The monochromatic nature of the light with respect to PDT is important for two reasons.

Primarily, the monochromatic character allows light to match exactly the absorption wavelength of the photosensitizer in the area where the penetration in tissue is the largest, hence avoiding heating the tissue with unnecessary light that does not participate in the photochemical reaction. The coherence makes it easy to be coupled to the fiber optics for endoscopic light delivery system, which is used for tumors on the linings of body cavities or travel through the lumen of needles into tissues for interstitial implants.

Photodynamic therapy using the filtered lamps as light sources is limited because they are mostly applicable to the directly accessible areas such as the skin or just under the skin. Although they are relatively simple, reliable, and of low cost, they cannot be used for optic fiber delivery due to their poor coupling efficiency into single fibers. Also, the light intensities are low with these devices.

Some other noncoherent light sources such as LEDs are also used for PDT. Figure 9.5 illustrates the use of an LED lamp for PDT treatment developed by Dr. Latchezar Avramov at the Institute of Electronics of the Bulgarian Academy of Sciences in Sofia, Bulgaria. Although LEDs emit in wavelength

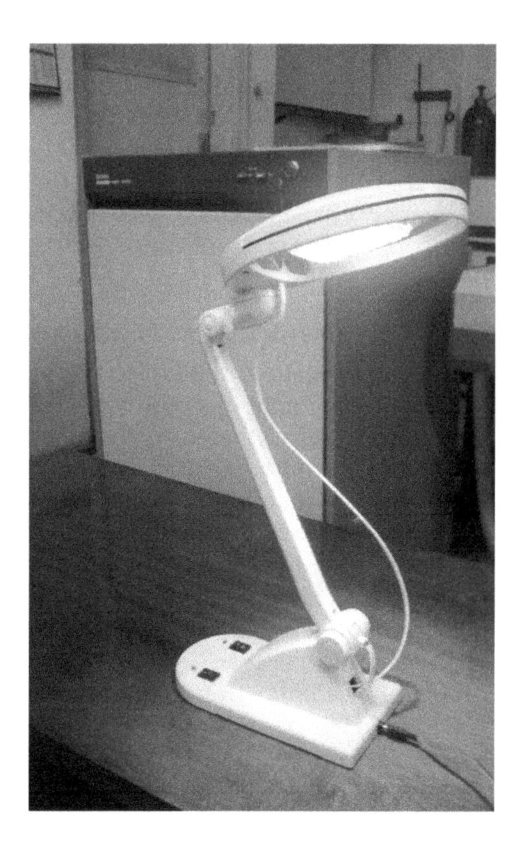

FIGURE 9.5
LED light source for PDT. (Developed by Latchezar Avramov, Institute of Electronics, Bulgarian Academy of Sciences, Sofia, Bulgaria.)

bands that are generally much wider than those from a laser, typically around 25 nm, the wide bandwidth sometimes interferes with the application due to other photochemical as well as photothermal interactions; this makes LED light not always the ideal light source for every application.

The second advantage of the use of laser light is the selective choice of wavelengths to maximize transmission of light through the tissue, and enhance treatment efficiency.

Optimal results from PDT require light to be transmitted through biological tissue and then maximally absorbed by the photosensitizer. Even though this seems relatively simple and straightforward, there are a few inherent problems. In general, biological tissues are highly scattering, and therefore, our bodies are practically impenetrable by light. This limitation can be overcome to a limited degree by using longer wavelengths of light. The longer the wavelength the deeper the light will penetrate the tissue. This works up to a point in the mid-infrared around 2 μm, where absorption due to water limits efficient light penetration.

9.2.1.9 Advantages of PDT

Photodynamic therapy does not cause systemic toxicity as seen with traditional chemotherapy. In general, the photochemical application of PDT is less toxic than chemotherapy. The advantages associated with PDT include selective targeting and destruction of cancer cells. Photodynamic therapy causes minimal, if any, damage to connective tissue when appropriate light dosages are employed. It also eliminates the use of ionizing radiation as used in the alternative radiation therapy.

Finally, PDT exhibits the potential for healing due to the decreased damage to connective tissue and results in the regeneration of tissue as opposed to scarring, yielding a more desirable cosmetic outcome.

9.2.1.10 Disadvantages and Limitations

The major limitation of PDT is the inability to treat systemic diseases. Photodynamic therapy is only effective if light can be applied directly to the photosensitizer; therefore, the technician must know where the lesion is located. Two other significant limitations of PDT are the ability of light to penetrate tissue and the retention of photosensitizers within the body over an extended period of time. Although much research has been concentrated on designing photosensitizers that are excited by infrared light, the current technology only allows for penetration depths of about 3 cm in the brain for excitation in the near-infrared, and often less than 1 cm for other tissues.

Also, after administration of photosensitizers the patients become photosensitive. The photosensitizer is not easily removed from the biological system, leaving the patient subject to several additional weeks of solar photosensitivity. Depending on the location of the tumor, treatment of sensitive areas can be painful due to the fact that the phototoxins are not selectively targeting tumor cells only, but attack every cell that has a concentration of the photosensitive dye present and is subjected to light. The overall healing process may take several weeks. Brief exposure to sunlight will result in severe burning and blister formation. Therefore, depending on the half-life of the photosensitizer employed, patients are advised to stay out of sunlight for at least 1 month and may be asked to do so up to 6 months following treatment. This photosensitivity cannot be overcome by the use of sunscreens, since sunscreens block UVB rays. These UVB rays are of much shorter wavelengths than the light that activates the photosensitizers.

Other photochemical applications are found in the realm of biostimulation and wound healing. Some of the implications of wound healing will be discussed next. Biostimulation will be discussed in greater detail in Chapter 10, under photobiological aspects.

9.2.2 Wound Healing

Acute and chronic wounds can be problematic in many different populations of patients, and wound management is an area where light therapy may be quite beneficial. Diabetic patients often have complications with wounds due to impaired sensation and blood flow, decreased collagen synthesis, and a malfunctioning immune system. The National Institute of Diabetes and Digestive and Kidney Diseases estimates that 15% of diabetes patients will develop a foot ulcer and 12–24% of these patients may require amputation of the limb.

To understand how light affects wound healing, a brief overview of the wound healing process will be discussed. Within seconds of wound infliction, the body begins the process of hemostasis. This first step reduces blood loss through vasoconstriction, formation of a platelet plug, and the initiation of the coagulation cascade. Activated platelets release agents involved in the next stage of wound healing: the inflammatory response. Chemokines and cytokines, and cell-signaling molecules, initiate the extravasation of white blood cells such as neutrophils and macrophages. Macrophages are critical for the phagocytosis of wound debris and possible contaminating bacteria, for uptake of necrotic material, and finally for secretion of chemokines or growth factors. Chemokines are attractant cell-signaling molecules responsible for recruiting cells such as fibroblasts to the wound site.

The incoming fibroblasts initiate the next stage of wound repair, or the tissue-remodeling phase. The fibroblasts are responsible for producing and depositing large amounts of collagen, the matrix upon which new endothelial cells will create tissue. Tissue remodeling also involves a balanced combination of synthesis and degradation of matrix proteins, which leads to the final stage of scar formation. As collagen is deposited and forms tight fibrils that become cross-linked, scar tissue is formed. These later stages of inflammation, cell proliferation, and tissue remodeling have been shown to be affected by light.

Visible light in the range of red has long been known to have healing effects. It was not until a monochromatic, coherent light source was introduced, the He–Ne laser, that the medical community revisited the area of photobiomodulation in a more scientific manner. Growing evidence supports the theory that coherence is not a vital property of the light used for wound healing. Other light sources such as LEDs have become commonplace in the limited but expanding field of biostimulation.

Light emitted from semiconductors is noncoherent and current research suggests that deep penetration may not be possible with LEDs. However, the requirement for coherence is still under debate.

The healing effect of light on burn wounds showed no significantly increased healing effects. Light-emitting diodes can increase collagen production, promote angiogenesis, and increase expression of growth factors. Examples of the effect of irradiation with various light sources are:

- Fibroblast proliferation (e.g., under LED irradiation)
- Increase in cellular adhesion due to NIR radiation mediated by melatonin
- Induction of reactive oxygen species.

9.3 Summary

This chapter presented the various photochemical processes that can be recognized in therapeutic, diagnostic, and laser applications of light to biological specimens. The description of the photodynamic process is the most important photochemical process in biomedical optics to date. The description of wound healing was only a brief overview; the predominantly chemical effects can be found in various specialized journals and books.

Further Reading

Anderson, R.R., and Parrish, J.A., The optics of human skin, *J. Invest. Dermatol.*, 77, 13–19, 1981.

Eyring, R.E., Rate equation and the temperature dependence of solvolytic reactions, *J. Chem. Phys*, 25, 2, 375–376, 1956.

Glassberg, E., Lask, G., Rabinowitz, L.G., Tunnessen, W.W., Capillary haemangiomas: case study of a novel laser treatment and a review of therapeutic options. *J. Dermatol. Surg. Oncol.*, 15, 1214–1223, 1989.

Jablonski, A., Efficiency of anti-stokes fluorescence in dyes, *Nature*, 131, 839–840, 1933.

Jablonski, A., Zur Theorie der Polarisation der Photolumineszenz von Farbstofflosungen, *Z. Physik*, 96, 236–246, 1935.

Karu, T.I., Low power laser therapy, in *Biomedical Photonics Handbook*, Tuan Vo-Dinh, Ed., CRC Press, Boca Raton, FL, 2003, chap. 48, pp. 48–1 to 48–25.

Kelly J.F., Snell M.E., and Berenbaum, M.C., Photodynamic destruction of human bladder carcinoma, *Br. J. Cancer*, 31, 2, 237–244, 1975.

Lanigan, S.W., and Cotterill, J.A., Use of a lignocaine-prilocaine cream as an analgesic in dye laser treatment of port wine stains, *Lasers Med. Sci.* 2, 87–89, 1987.

Reid, R., Physical and surgical principles governing carbon dioxide laser surgery on the skin, *Dermatol. Clin.*, 9, 297–316, 1991.

Sterenborg, H.J.C.M. et al., Phosphorescence-fluorescence ratio imaging for monitoring the oxygen status during photodynamic therapy, *Opt. Express*, 12, 1873–1878, 2004.

Street, M.L., and Roenigk, R.K., Recalcitrant periungual verrucae; the role of carbon dioxide laser vaporization, *J. Am. Acad. Dermatol.*, 23, 115–20, 1990.

Tunér, J., and Hode, L., *Low Level Laser Therapy—Clinical Practise and Scientific Background*, Prima Books, Spjutvägen, 1999.

Wheeland, R.G., Clinical uses of lasers in dermatology, *Lasers Surg. Med.*, 16(1), 2–23, 1995.

Problems

1. Give three examples of photochemical light–tissue interaction.
2. Give a brief description of the photodynamic mechanism.
3. What is the main optical difference between PDT and cutting?
4. Calculate the total light dose at a depth of 3 mm from a broad beam He-Ne laser (3 cm beam diameter) irradiating the skin after 5 min of irradiation at a power of 50 mW in the center of the beam?
5. What do photosynthesis and photodynamic therapy (PDT) have in common?
6. What is produced in the PDT photochemical interaction that kills cells? What primary cellular component is involved in the destruction of the cell in PDT?
7. List some of the clinical applications of PDT.
8. In the treatment of cancer, photodynamic therapy (PDT) involves the accumulation of photosensitizers within the malignant tissue and their subsequent excitation, destroying the cancer tissue. One such agent is hematoporphyrin derivative (HpD). Taking into account light–tissue interaction, what is the best wavelength for PDT with this photosensitizer?
9. How can the irradiance just below an air–tissue interface be larger than the incident irradiance? Assume a tissue that has scattering.
10. Wound healing has a direct implication on what cell types?
11. What are the two types of photomechanical phenomena that have been used clinically in surgery and therapy?
12. What has been the predominant photochemical procedure to date?

10

Light–Tissue Interaction Mechanisms and Applications: Photobiological

This chapter presents an overview of the photobiological interaction mechanisms, and the research into the underlying dynamics.

The photobiological concept describes biological effects or aspects of light–tissue interaction. The topics described in this chapter are biostimulation, as well as photocoagulation, port-wine stain treatment, and tissue welding. The last three procedures fall under the umbrella of photothermal effects. One particularly interesting topic in the nonthermal applications of light for biological monitoring is the use of quantum dots, which is described last in this chapter.

10.1 Photobiological Biostimulation

It has been demonstrated that irradiated cells receiving less than optimal nutrient conditions grow just as well or better than control cells. In these research protocols the control cells are not exposed to light and are maintained under optimal nutrient conditions. It has been shown that low-level light therapy (LLLT) has direct implications for at least the following different cell types:

- Keratinocytes
- Fibroblasts
- Macrophages/mast cells

As part of the photobiostimulation process, basement membrane proteins, integrins, and calcium-binding proteins were shown to exhibit an increase in the number of receptors for a chemical or drug. As a result the cells are more reactive to the effects of the agent. Additionally, gene expression analysis revealed that many genes were upgraded in their response, making them more reactive to the effects of the agent under irradiation by light.

Additionally, certain families of genes involved in muscle differentiation have been shown to experience an altered expression under the influence of light. The upregulation of these types of genes supports the theory that light treatment is a valid technique to be used for biostimulation.

Red to near-infrared irradiation can increase DNA synthesis and cellular adhesion. Additionally, melatonin has been shown to act as a modulator of cell adhesion upon illumination. Melatonin acts as a free radical scavenger and an antioxidant, but is not involved in redox reactions. It was found that melatonin negatively affected cell adhesion under irradiation with light. Because levels of melatonin differ significantly between day and night in humans, the study suggests the importance of using LLLT during daytime hours, when physiological levels of melatonin are lowest in the human body.

Different cell lines and cell culture systems have been used, and a broad range of wavelengths have been investigated for their potential stimulatory effects. The majority of research since the 1980s, however, has focused on red, far-red, and near-infrared radiation. One of the first glimpses into the healing effects of red radiation came from Boulton and Marshall, who investigated He–Ne laser stimulation of cultured human fibroblasts. They found that the growth rate of these cells was increased on exposure to low-level laser light, and that attachment of the cells to a substrate was enhanced. The birth of photobiomodulation in the late 1970s and early 1980s leads to research concerning the molecular mechanisms of light's stimulatory effects.

In 1960, Gordon and Surrey discovered that light from the red and far-red regions of the visible light spectrum encouraged oxidative phosphorylation in the mitochondria. The mitochondrion is the organelle of the cell responsible for energy production through aerobic respiration. It is composed of two membranes: the inner and outer mitochondrial membranes. It is here that pyruvate is oxidized to carbon dioxide and the energy released from these reactions is utilized for ATP synthesis. As enzymes are reduced during the oxidation processes, reoxidation must occur through the transfer of electrons to the final electron acceptor, oxygen. The terminal enzyme in the electron transport chain is cytochrome *c* oxidase, and it is this particular molecule that is thought to be the main component in photoactivated biostimulation.

Isolated mitochondria display increased ATP synthesis, membrane potential changes, and increased phosphate exchange rates between ADP and ATP when exposed to red and near-infrared light.

There is theoretical evidence that cytochrome *c* oxidase is the primary photoacceptor. The main theory of how cytochrome *c* oxidase may be involved in the absorption of light and stimulation of ATP synthesis goes as follows. The electron transfer in ATP is accelerated as a result of change in the redox properties of the enzyme. There is also the influence of a very small but significant rise in temperature upon absorption of light, which indicates a structural change to the enzyme. During the four-electron reduction of oxygen to water, free radicals (hydroxyl and superoxide) are produced and the mitochondria may be able to reabsorb these radicals and use them for the oxidative phosphorylation of ADP. Additionally, singlet oxygen plays a minor role in the process.

The fact that ATP synthesis is increased during light irradiation of cell cultures led to further research in the area of photomediated fibroblast proliferation. As shown in Table 10.1, a variety of light sources with different properties were used to support the hypothesis suggesting that light can accelerate fibroblast proliferation.

Studies also suggest that low-energy visible light can stimulate the production of reactive oxygen species. It is thought that the production of H_2O_2 may mediate, directly or indirectly, the phosphorylation of calcium transporters. These results can be directly applied to the field of cardiology. For example, light-activated improved preservation of transplant hearts by means of biostimulation, while concurrently reducing infarct size and speeding recovery after heart damage shows great promise.

Other applications of biostimulation are, for instance, the specific illumination of biological photoreceptors identified at various anatomical locations to provide psychological benefits. This effect is shown when fibroblasts are irradiated with a visible light source. The fibroblasts showed evidence of production of reactive oxygen species after 10 min of irradiation. The reactive oxygen has been documented in psychology journals as an antidepressant, especially when inducing the production of reactive oxygen species in skin. Such oxygen species have been shown in nuclear magnetic imaging to participate in signal transduction pathways in the brain leading to mood changes.

TABLE 10.1

Comparison of Various Experimental Designs of Wound-Healing Mechanism

Study	Cells of Light	Source(s) (nm)	Wavelength (mW)	Power (fluence)	Radiant Exposure (J/cm^2)
Vinck et al.	Chicken embryonic fibroblasts	LED:			
		Green	570	0.2–10	0.1 (3 min)
		Red	660	15–80	0.53 (2 min)
		Infrared	950	80–160	0.53 (1 min)
		Laser (GaAlAs)	830	1–400	1 (5 s)
Webb et al.	Scar-derived human fibroblasts	Laser diode	660	17	2.4 (31 s) 4 (52 s)
Almeida-Lopes et al.	Human gingival fibroblasts	Laser diodes	670 780 692 786	10 50 30 30	2 (fluence set at hand piece, time of exposure determined to allow homogenous fluence for all diodes)

A chain of molecular events starts with the initial absorption of light by a photoreceptor, which leads to signal transduction and amplification, and finally results in a photo response. The absorbed light activates the respiratory chain and subsequently the oxidation of the NAD pool. This oxidation changes the oxygen status of both the mitochondria and the cytoplasm. As a result the membrane permeability changes, which in turn has an effect on the Ca^+ flux. The Ca^+ ions in turn influence the levels of cyclic nucleotides. The cyclic nucleotides modulate the synthesis of both DNA and RNA. This process is the basis of cell proliferation and as such an evidence of photobiostimulation. This process has an upper limit, which is the natural homeostasis of the cell and means that only poorly performing cells can be stimulated.

The stimulation effect of light in biological tissues depends on four parameters: the wavelength of the light targeting specific receptors, the light fluence rate $\Psi_{source}(z)$, the total irradiation time Δt_{irr}, and the energy density required for activation $(E/A)_{act}$, where A represents the cross-sectional area of the light source or the targeted area itself. Illumination will generally take place under a broad uniform beam, allowing this type of simplification. The stimulation conditions can thus be expressed as

$$(E/A)_{act} = \Psi_{source}(z)\,\Delta t_{irr} \tag{10.1}$$

where the fluence rates for stimulation minimally need to exceed the threshold fluence rate, $\Psi_{source}(z) \geq \Psi_{th}$, keeping in mind that the source fluence rate is the light distribution inside the tissue, and decreases with the distance to the light source. In contrast, too high fluence rate will negatively affect the biostimulation process, as described by the Arndt–Schultz law. Equation (10.1) seems rather simple, which it is, but the components are quite complex themselves. The conditions for biostimulation will still need to obey the diffusion equation in addition to the Arndt–Schultz law. Additionally specific wavelengths have been identified for biosensors such as HeLa cells, which are activated over a range of wavelengths from 330 to 860 nm, while ATP can be stimulated under irradiation at 632.8 nm (the He–Ne laser wavelength).

10.2 Photobiological Effects

The photobiological interactions can be divided into photothermal and nonthermal photo interactions. The most crucial and wide use of the photobiological interaction in the thermal subdivision is photocoagulation: the destruction of tissue for treatment by means of targeted thermal destruction. An area of special interest is the photothermal treatment of port-wine stain by selective ablation of blood vessels underneath the skin surface. Another thermal interaction mechanism is the process of tissue welding, where mostly collagen is used as a thermally activated glue. The process of wound healing in photobiology is still a highly debatable process, but has certain very important indicators that support the potential usage in medicine.

10.2.1 Photothermal

Laser irradiation deposits energy in the medium under irradiation, the radiant energy flux times the exposure time, as described in Chapter 5.

Before the photothermal concepts can be described two definitions need to be introduced. These concepts are enthalpy and entropy.

Enthalpy H is defined as the energy of a system stored in the chemical bonds between the atoms that make up the molecules in the system.

The next issue of entropy is related to the internal energy of a molecular system.

The work done on a system combined with the administered heat results in a change in internal energy. This is known as the First Law of Thermodynamics:

$$dU = dQ - dW \tag{10.2}$$

The First Law of Thermodynamics states that energy cannot be destroyed or generated.

The change in internal energy divided by the temperature of the system at the time is defined as the entropy of the system:

$$dS = dQ/T \tag{10.3}$$

The energy deposited in the tissue when absorbed will raise the local temperature, and can induce a chemical reaction by changing the enthalpy of a system. The system contains the atoms forming the complex molecules that make up the carbohydrates, amino acids, fats, and proteins of a biological medium. The deposited light energy is considered the reaction enthalpy, where the change in enthalpy is the final enthalpy minus the initial enthalpy.

10.2.1.1 Photocoagulation

Photocoagulation can be reversible or irreversible, depending on the total amount of energy delivered and the time frame over which this energy is delivered. In general a reversible chemical reaction takes place in infinitesimally small steps in time and the chemical composition changes take place gradually. In the reversible process the system is always at equilibrium, and reverts back to the starting conditions with no additional energy deposited. As a matter of fact the energy initially deposited will be released in the process of reversal to the initial state.

10.2.1.2 Reversible Photocoagulation

In a reversible process of photocoagulation the changes in the molecular bond can revert to the original configuration without a change in entropy. This can be the case at the edge of a biological volume under laser irradiation under in vivo circumstances. Histological examination may reveal changes that are not classified as denaturation; cell membranes are still intact and the nucleus has survived as well. Under normal conditions this region will resemble normal healthy tissue after a period of healing. Typical proteins have a range of energy where they are stable. For instance, the range of stability between folded and unfolded protein strands is between 7 and 15 kcal/mol at 37°C. This stability energy level is the free energy of the protein.

10.2.1.3 Irreversible Photocoagulation

In the process of irreversible photocoagulation the cell membrane and nucleus will be destroyed. The Second Law of Thermodynamics states that for an irreversible process in an isolated system the entropy always increases:

$$\Delta S = S_{\text{final}} - S_{\text{initial}} > 0 \tag{10.4}$$

The disorder associated with the increase in entropy has been defined by Boltzmann as the natural log of all the number of configurations possible in a system ω_c times the Boltzmann constant:

$$S = k \ln \omega_c \tag{10.5}$$

Equilibrium corresponds to $\partial S = 0$, where S is the maximum entropy attainable.

To describe the process of irreversible damage resulting from photocoagulation, the change in internal energy of the molecules that make up the tissue is considered.

The change in internal energy under external influences is now defined as the change in enthalpy of the system:

$$\Delta H = U + \Delta(PV) \tag{10.6}$$

where $\Delta(PV)$ is the work done by changing pressure and volume. Under constant pressure this becomes the volume labor $P\Delta V$.

Using the ideal gas law ($PV = nRT$), Equation (10.6) transforms into an equation that can be applied to a chemical reaction:

$$\Delta H = U + \Delta nRT \tag{10.7}$$

where Δn equals the change in molecules due to a chemical reaction, R is the universal gas constant, and the temperature of the system is measured in K. The enthalpy change of a system is considered to be negative for an exothermic reaction, a chemical reaction that produces heat. During an endothermic reaction heat is absorbed, rendering the enthalpy change positive.

Combining the definitions of enthalpy and entropy provides the change in free energy of a system. The Gibb's free energy G expresses the energy of the system. The change in free energy is now given by

$$\Delta G = \Delta U - T\,\Delta S - S\,\Delta T + \Delta nRT + nR\,\Delta T \tag{10.8}$$

The denaturation (coagulation) process can now be written as

$$\Delta G_{\mathrm{den}} = \Delta H_{\mathrm{den}} - T\,\Delta S_{\mathrm{den}} \tag{10.9}$$

or in terms of the chemical reaction,

$$a\mathrm{A} + b\mathrm{B} = c\mathrm{C} + d\mathrm{D} \tag{10.10}$$

which yields $\Delta G = xJ$ and

$$\Delta G_{\mathrm{den}} = \Delta G_{\mathrm{native}} + RT\ln\left[\frac{[\mathrm{C}]^c[\mathrm{D}]^d}{[\mathrm{A}]^a[\mathrm{B}]^b}\right] \tag{10.11}$$

where [A] and [B] are the respective concentrations of the native molecules, and [C] and [D] are the concentrations of the denatured states of the respective molecules.

In case of denaturization the change in entropy δS_{den} will per definition be positive, which is described as follows:

$$\Delta S_{\mathrm{den}} = k\ln(\omega_{\mathrm{c,denatured}}/\omega_{\mathrm{c,native}}) \tag{10.12}$$

This positive entropy change is the result of the fact that denatured proteins will have more configurations available than the native proteins rendering the change in entropy positive, since the ratio of denatured configurations over native protein configurations obeys the following expression:

$$\omega_{\mathrm{c,denatured}}/\omega_{\mathrm{c,native}} \gg 1 \tag{10.13}$$

This states that the change in free energy will be negative only when the temperature is high enough; thus, the native state is stable at low temperature and the denatured state is stable at high temperature.

Combining this with the process of photocoagulation, the energy deposited by the laser light is the activation energy of the denaturation process. An example of the reaction process with hypothetical energy levels is presented in Figure 10.1.

The reversible damage is illustrated on the left up-slope of the energy graph, while the irreversible process takes place as soon as the crest of the energy curve is reached, and enough energy has been deposited to initiate the total chemical modification of the protein molecules.

The standard Gibbs energy difference between the transition state of a reaction and the ground state of the reactants is calculated from the experimental rate constant k_r via the conventional form of the Eyring absolute rate equation using the fact that the energy for each reaction obeys quantum theory due to the discrete nature of the chemical reactions. The Eyring rate process description is shown in the following equation:

$$\Delta G = RT\left[\ln\left(\frac{k}{h}\right) - \ln\left(\frac{k_r}{T}\right)\right] \tag{10.14}$$

where k is the Boltzmann constant and h the Planck constant ($k/h = 2.08358 \times 10^{10}\ \mathrm{K}^{-1}\,\mathrm{s}^{-1}$). The values of the rate constants, and hence Gibbs energies of activation, depend upon the choice of concentration units (or on the choice of thermodynamic ground state).

The standard enthalpy of activation ΔH° is the enthalpy change that appears in the thermodynamic form of the rate equation obtained from conventional transition state theory. The quantity ΔS° is the standard entropy of activation. Solving Equations (10.7) and (10.8) for Δn gives an expression for

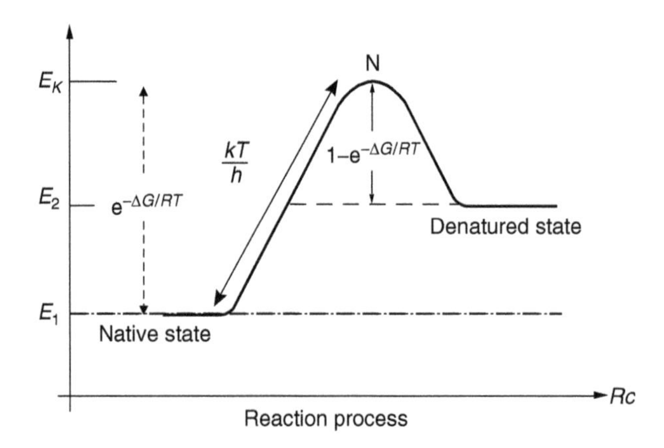

FIGURE 10.1
Illustration of the process of development of reaction enthalpy in relation to thermal change.

the probability of conversion between the native and the denatured state k_r, as expressed in the following equation:

$$k_r = \frac{kT}{h} e^{\Delta S^\circ/R} e^{-\Delta H^\circ/(RT)} \tag{10.15}$$

or rewritten as

$$k_r = Ae^{-E/(RT)} \tag{10.16}$$

Using the fact that the transition from the native to the denatured state takes place at a rate expressed by

$$kT/h \tag{10.17}$$

where h is Planck's constant, and the empirical reaction constants A and E are defined as

$$A = \frac{kT}{h} e^{\Delta S/R} \tag{10.18}$$

and, respectively,

$$E = \Delta H \tag{10.19}$$

These constants need to be obtained experimentally, since there is not just a single molecule involved in the denaturation of a biological specimen.

The physical expression of the denaturization will be a tissue volume that has been destroyed. The volume of denaturization can be derived from the so-called damage integral, which is described next, and is identified by histological markers of nuclear destruction.

10.2.1.4 Damage Integral

When there are initially N_0 molecules of one species in a volume of tissue, the number of surviving molecules is directly proportional to the native number, and the number of denatured molecules dN as a function of time can be shown to obey the following equation:

$$\mathrm{d}N = k_r N \,\mathrm{d}t \tag{10.20}$$

The number of surviving native molecules as a function of time can be derived as

$$N(t) = N_0 e^{-\Omega(t)} \tag{10.21}$$

where

$$\Omega(t) = \int_0^t k_r(t')\,dt' = \int_0^t \frac{kT(t)}{h} e^{\Delta S/R} e^{-\Delta H/[RT(t)]}\,dt' \quad (10.22)$$

The expression $\Omega(t)$ is called the *Damage Integral*, which extends from 0 to infinity, such that $N(t)$ extends from N_0 to zero. At any time, the number of denatured molecules, $N_{\text{denatured}}(t)$, equals $N_0 - N(t)$. In general, the damage integral incorporates the average values of all molecules involved in the denaturation process of a section of tissue.

The value of k_r depends on the temperature of exposure and the values of ΔS and ΔH, which are the characteristics for some measurable transitions in the tissue. There are many such transitions. For example, the irreversible loss of enzymatic activity involves denaturation of labile proteins, the coagulation of egg white and the whitening of soft tissues such as muscle or liver during cooking involves denaturation and aggregation of proteins, and the coagulation of collagenous tissue to yield gelatin involves the loss of collagen fiber bundle structure. The rate constant k_r at a given temperature can be different for different transitions. For example, second-degree burns can be addressed with $\Omega > 10$, and for third-degree burns it holds that $\Omega > 10{,}000$.

Typical values of ΔS and ΔH for examples of thermal coagulation are listed in Table 10.2. The rate constant for each example varies as a function of temperature. It is common to consider the exposure time (t_Ω in s) required to achieve $\Omega = 1$, which corresponds to $1/k_r$, where $k_r = 1/t_\Omega$ or k_r, as a function of temperature. The more complex the molecular structures being denatured, the steeper are the rate constant changes with temperature, since the change in entropy (ΔS) is greater as well as the number of bonds to be broken concurrently (ΔH) are larger in number. The less structured the chemical matrix that needs to be denatured, the more gradual is the change in rate constant because the changes in ΔS and ΔH are not as aggressive with respect to temperature.

TABLE 10.2

Thermal Coagulation Parameters

Tissue	ΔS (J/[mol K])	$\Delta H \times 10^3$ (J/mol)
Rat liver necrosis	392	221
Dog heart whitening	510	259
Pig liver whitening	553	276
Mouse dermal contraction	996	425
Rat tail collagen birefringence loss	831	367
Collagen	825	368

The entropy ΔS released by denaturation of a molecular structure involved in an irreversible thermal transition is to the first approximation linearly related to the enthalpy ΔH of the transition. In general, the choice of ΔH as the independent variable is conceptually connected to the energy of all bonds that cooperatively hold a molecular structure together. The entropy increase is more often considered to be the dependent variable that describes the entropic energy change involved with breaking of the bonds.

Equation (10.22) can be rewritten with the experimentally obtained damage constants A' and E as

$$\Omega(t) = A' \int_0^t e^{-E/[RT(t)]} dt' \tag{10.23}$$

Published data list the following average empirical values for skin tissue: $E = 627$ kJ/mol and $A' = 3.10 \times 10^{98}\ \mathrm{s}^{-1}$; for collagen the published values are, respectively, $E = 89$ kJ/mol and $A' = e^{130}\ \mathrm{s}^{-1}$.

The damage region $0 \leq \Omega(t) \leq 0.53$ is generally considered not to cause any permanent damage to the tissue, while for $0.53 < \Omega(t) < 5$ tissue will be coagulated. Generally $\Omega(t) = 1$ is associated with reversible tissue damage. The value $\Omega(t) = 1$ is the point of onset of tissue coagulation, with a 63% probability of tissue denaturization. A value of $\Omega(t) = 4.6$ corresponds with a 99% probability of tissue denaturization. Carbonization can be considered to occur under the condition that the damage integral satisfies the condition $\Omega(t) \geq 5$.

Using these numbers, for instance, to determine the time required for a second-degree burn on skin exposure to 135°C yields a value of approximately 2.5 minutes, and a third-degree burn can be expected after an order of magnitude of 1 day and 18 hours of total exposure. Figure 10.2 illustrates the time–temperature correlation to achieve irreversible damage.

10.2.2 Medical Applications of the Photothermal Effect

Thermal coagulation of tissue was one of the earliest uses of lasers in medicine. The argon ion laser (488 nm wavelength) is used to coagulate the retina of the eye to cause a local scarring that tacks down a retina that is at risk for detachment. Coagulation of small tumors is a common application and can be achieved with various lasers, although an Nd:YAG laser (1064 nm wavelength) will better penetrate a tissue and coagulate a larger volume than most other lasers. The body generally resorbs the coagulated tissue as part of the wound-healing process.

10.2.2.1 Thermal Effects in Treatment of Port-Wine Stains

A dermatologic application of thermal injury is the treatment of vascular lesions such as port-wine stains by a pulsed laser (1–10 ms pulse duration

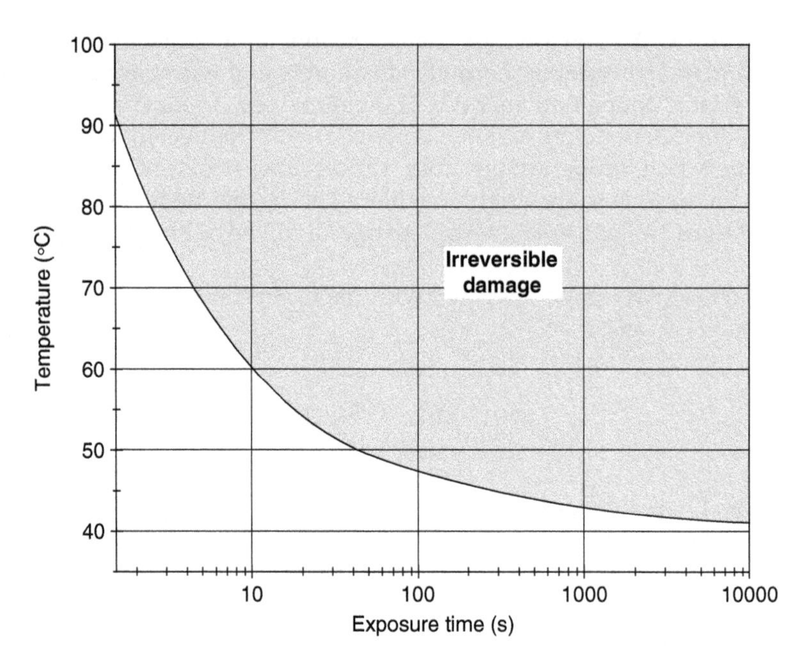

FIGURE 10.2
Graphical display of the scale of the exponential relationship for damage resulting from the time–temperature combinations that may result under laser irradiation.

in the 532–585 nm wavelength range). Port-wine stain is a pronounced birthmark that affects 3 in 1000 people, and is often located on or near the face. The pulsed laser deposits energy selectively in the blood vessels due to the strong absorption of light by hemoglobin. The high temperatures transiently achieved in the blood vessels cause thermal damage to the vessel wall and shrink the vessel. The body's wound-healing response then removes the affected vessel, hence removing the port-wine stain lesion.

Irreversible tissue damage occurs at about 58 or 60°C. However, the instantaneous coagulation of blood does not occur until a temperature of 90–100°C is reached. This makes avoiding unwanted thermal damage difficult, even using the selective absorption of blood. One answer to this problem has been to cool the skin cryogenically while lasing the blood vessels. The human aorta is highly scattering at 632.8 nm, and a good candidate for using the scattering approximation. The same would not be true for the human liver near 515 nm, but would be at 635 nm. Human whole blood has a dramatic change in scattering and absorbing characteristics between 577 and 685 nm. This property is exploited in the treatment of port-wine stains. A dye laser is often used near 577 nm to coagulate the blood because of its strong absorption peak there, while the dermis and epidermis theoretically remain unharmed because of their much lower absorption coefficients (even though they are near to the surface of the blood vessel's interior wall). Table 10.3 lists various optical properties of tissues at wavelengths relative

TABLE 10.3

Selected Optical Properties of Tissues Involved in Port-Wine Stain Treatment

Tissue	λ (nm)	μ_a (cm^{-1})	μ_s (cm^{-1})	G
Human aorta	632.8	0.52	316	0.87
Human aorta	1060	2.0		
Human whole blood	577	376	9.6	
Human whole blood	685	2.65	1413	0.99
Human liver	635	2.3	313	0.68
Human liver	515	18.9	285	

to port-wine stain treatment. It has become a standard procedure to treat port-wine stains using a 585 nm pulsed-dye laser. The 585 nm wavelength is actually closer to an absorption peak for blood than 577 nm. In addition, absorption coefficients for the dermal and epidermal layers are slightly lower at longer wavelengths, which makes 585 nm the optimum wavelength, though both wavelengths are used. In order to avoid thermal damage lower powers are used than are necessary to properly coagulate the blood, so that 100% removal is rare (especially in darker colored stains). By cooling the surface higher power over longer periods of time can be used than the 7–10 J/cm^2 typically used. Under cryogenic cooling energy densities up to 10 J/cm^2 have been reported compared to 8–9 J/cm^2 used on those receiving standard treatment. A typical treatment protocol uses a 20 ms pulse followed by a 30 ms delay to keep the skin surface at a temperature elevation between 5 and 10°C. In addition to facilitating faster healing, pain sensation and visible damage due to the laser were also reduced, indicating that higher powers can be used with skin cooling while actually reducing thermal damage. Tissue components such as proteins and nucleic acids absorb ultraviolet wavelengths most effectively, water absorbs infrared most efficiently, and within the visible wavelengths proteins and pigments (such as melanin and oxygenated hemoglobin protein in blood) absorb strongly in certain wavelength regions. These characteristics can be exploited in the treatment of port-wine stain to selectively deposit heat in blood vessels while leaving the dermis and epidermis relatively unharmed.

The diffusion approximation, as referred to here, is a simplification and uses only the first two terms of the sum of the Taylor series described in Section 6.1. It is represented by the differential equation:

$$\nabla^2 \Psi_s(x,y,z) - \kappa^2 \Psi_s(x,y,z) = -Q(x,y,z) \tag{10.24}$$

which can be rewritten in cylindrical coordinates as

$$\nabla^2 \Psi(r,z) - 3\mu_a(\mu_a + (1-g)\mu_s)\Psi = 3\mu_s(\mu_s + (1+g)\mu_a)\frac{\Psi_0}{4\pi} e^{-(\mu_a+\mu_s)z} \tag{10.25}$$

where $Q(x, y, z)$ represents the light source and g is an anisotropic factor that determines how much more light will scatter in the forward direction than in other directions.

Commercially available software packages, such as Maple (Maple Soft, Waterloo, Ontario, Canada), can be used to find an appropriate solution to Equation (10.25). One possible solution that can be found for the fluence rate of scattered light is as follows:

$$\Psi_s = \frac{\Psi_{\text{beam}}}{4\pi} \frac{A_2}{[(\mu_a + \mu_s)^2 - A_1]} e^{-(\mu_a + \mu_s)z} \tag{10.26}$$

The coefficients A_1 and A_2 are complex constants. The total light intensity due to both the unscattered components as well as the scattered part is a Beer–Lambert law solution combined with the solution in Equation (10.26). For a Gaussian beam profile this will give

$$\Psi_s = \left[\Psi_0(\beta r) e^{-\sqrt{A_1 + \beta^2} z} + \Psi_0 \left(1 - \frac{1}{4\pi} \frac{A_2}{[(\mu_a + \mu_s)^2 - A_1]}\right) e^{-(\mu_a + \mu_s)z}\right] e^{-2(r/w)^2} \tag{10.27}$$

where β is a constant that depends on boundary conditions and Ψ_0 represents the maximum incident energy fluence rate. The solution expressed in Equation (10.27) is a relatively precise analytical approximation, however, with very strict boundary conditions. This, in fact, is an analytical approximation that will work for a broad beam. In case a pencil beam is used as a source the solution will require a different approach, in fact the numerical solution described in Chapter 7. The wide beam approximation is only allowed due to the relatively high absorption involved, which significantly reduces the optical mean free path length. As such the beam to optical mean free path length ratio comes in the appropriate range.

In the treatment of port-wine stains the laser is pulsed to deliver a specific dose of energy. Typical incident energy densities used are in the range 3–8 J/cm^2. If the pulse rate is 0.5 ms and the energy density is 3.45 J/cm^2, the laser power would be 6900 W/cm^2. This value will be incorporated into Equation (10.27) as Ψ_0 to find the heat generation at the level of the blood vessel and then to examine the transient temperature of the blood vessel under the simplifying case where spatial variations within the tiny vessel are ignored. The standard time-dependent heat conduction equation for the blood vessel is now formulated as

$$g(z) = \rho C_v \frac{\partial T_b}{\partial t} \tag{10.28}$$

which has a solution

$$T_b = T_{b0} + \frac{g(z)}{\rho C_v} t \tag{10.29}$$

Next the solution of transient temperature distribution in dermis with constant temperature boundary conditions can be formulated.

The standard heat conduction equation (bioheat equation without blood flow) in one dimension, still under the assumption of broad beam irradiation, can be written as

$$\frac{\partial^2 T}{\partial z^2} + \frac{g(z)}{\kappa} = \frac{1}{\alpha}\frac{\partial T}{\partial t} \tag{10.30}$$

With the heat generation function, $g(z)$, derived from Equations (10.29) and (10.30), this nonhomogenous equation is solved by splitting into two equations: a steady-state problem:

$$\frac{\partial^2 T_s}{\partial z^2} + \frac{g(z)}{\kappa} = 0 \tag{10.31}$$

and a homogenous problem:

$$\frac{\partial^2 T_h}{\partial z^2} = \frac{1}{\alpha}\frac{\partial T_h}{\partial t} \tag{10.32}$$

The condition for using this approach is that the initial temperature for Equation (10.31) must be written as the actual initial temperature (T_0) minus the solution to Equation (10.31); that is,

$$T_h(t = 0) = T_0 - T_s \tag{10.33}$$

The solution to Equation (10.31) can now be derived as

$$T_s = T_0 - \frac{\mu_a \Psi_0}{\mu_t^2 \kappa}\left(1 - \frac{A_2}{4\pi(\mu_t^2 - A_1)}\right)e^{-\mu_t z} - \frac{\mu_a I_0}{\kappa(A_1 + \beta^2)}e^{-\sqrt{A_1+\beta^2}z} \tag{10.34}$$

where the total coefficient of attenuation for the light, μ_t, is defined as

$$\mu_t = \mu_s + \mu_a \tag{10.35}$$

The solution for Equation (10.31) can now be rewritten as

$$T_{\mathrm{h}}(z,t)=\frac{2}{\pi}\int_{\phi=0}^{\infty}\mathrm{e}^{-\alpha\phi^{2}t}\int_{z'=0}^{\infty}\left[T_{0}-T_{\mathrm{s}}(z')\right]\sin\phi z\sin\phi z'\,\mathrm{d}z'\,\mathrm{d}\phi \tag{10.36}$$

where $T_s(z')$ is defined in Equation (10.34), which can be solved to give

$$\begin{aligned}T_{\mathrm{h}}(z,t)=\frac{I_0\mu_{\mathrm{a}}}{2k}\mathrm{e}^{x^2/4at}&\left\{\frac{1}{\mu_{\mathrm{t}}^2}\left(1-\frac{A_2}{4\pi(\mu_{\mathrm{t}}^2-A_1)}\right)\right.\\&\left[\mathrm{e}^{\mu_{\mathrm{t}}}-\frac{x}{2at}\left(1-\mathrm{erf}\left[\frac{1}{2}\left(\mu_{\mathrm{t}}-\frac{x}{2at}\right)\sqrt{4at}\right]\right)\right.\\&\left.-\mathrm{e}^{(\mu_{\mathrm{t}}+x/2at)^2at}\left(1-\mathrm{erf}\left[\frac{1}{2}\left(\mu_{\mathrm{t}}+\frac{x}{2at}\right)\sqrt{4at}\right]\right)\right]\\&+\frac{1}{A_1+\beta^2}\left[\mathrm{e}^{\left(\sqrt{A_1+\beta^2}-x/2at\right)^2at}\right.\\&-\left(1-\mathrm{erf}\left[\frac{1}{2}\left(\sqrt{A_1+\beta^2}-\frac{x}{at}\mathrm{e}^{\left(\sqrt{A_1+\beta^2}+\frac{x}{2at}\right)^2at}\right.\right.\right.\\&\left.\left.\left.2at\left(1-\mathrm{erf}\left[\frac{1}{2}\left(\sqrt{A_1+\beta^2}+x/2at\right)\sqrt{4at}\right]\right)\right)\sqrt{4at}\right]\right\}\end{aligned} \tag{10.37}$$

After the time interval of 0.5 ms specified for the power level of the incident beam, Equation (10.37) can be solved exactly by numerical means to show the temperature at the proximal side of the blood vessel as well as for the distal side. The solution under the aforementioned irradiation conditions gives for the vessel wall closest to the light source a temperature of 130°C, while near the back of the vessel (the distal side) the temperature barely ascents above the ambient 37°C starting value. This wide range is due to the high absorption coefficient for blood. As it has been shown, it is desirable to cool the skin surface, which would allow for higher laser energies or longer pulse times, which, as was seen above, would elevate the temperature at the distal portions of the blood vessel to higher levels.

In the extended example of port-wine stain treatment the optical and thermal characteristics determine how the laser light propagates. These optical and thermal characteristics are determined by the biological structure of the material in question. For example, in the case of highly scattering tissue such as the dermis, the scattering approximation provides more realistic values for light intensity distribution than a simple exponentially decreasing estimation. In addition, the absorption response of different biological media, such as proteins, nucleic acids, water, and pigments, to different wavelengths of laser light must also be considered. In conclusion, treatment of port-wine stain requires careful consideration of the interplay between biological and physical factors.

Yet another area of research in the realm of nonthermal effects of electromagnetic radiation causes proteins to unfold. The energy that would be absorbed would be processed and used for different functions depending on a state of vibration. Thus, energy from an electric field may put energy into this state and trigger significant biological changes. However, it may take several minutes from the moment electromagnetic exposure occurs for the irradiated parts of the tissue to reach their final equilibrium temperatures. Additional consideration needs to be given for modeling the thermodynamic and heat transfer properties due to blood flow effects. A variety of theoretical models have been used to predict the blood flow effects; however, the Pennes heat sink model is the most commonly used and is believed to be the most accurate.

Figure 10.3 illustrates the different biological effects created by the various types and operating modes of lasers.

10.2.2.2 Tissue Closure and Welding

Tissue welding has seen a revival in recent years, as more is learned about the biochemistry of wound healing. There are a number of ways that one can weld tissue: through photothermal heating of the tissue, through use of a solder or adhesive activated by light to produce a tissue weld; and by combination of solid and liquid solder.

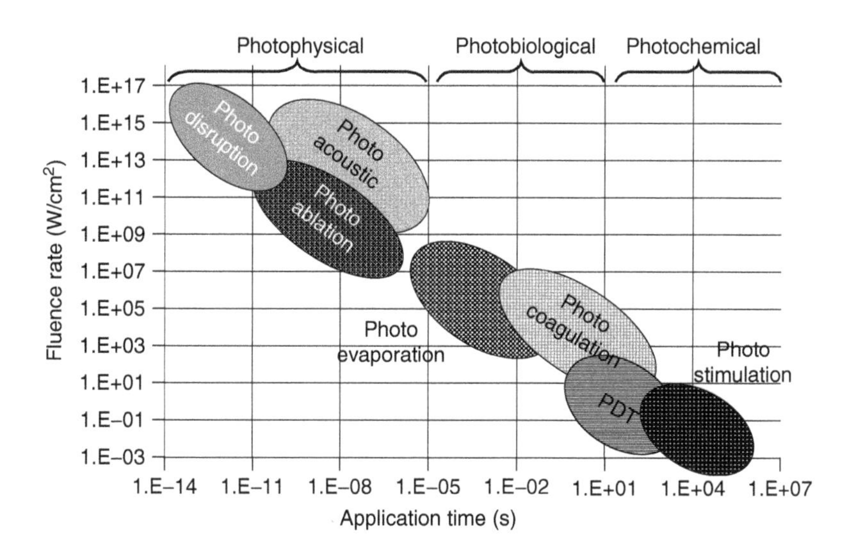

FIGURE 10.3
Overview of photochemical, photobiological, and photophysical effects under various irradiation conditions. (Modified from Boulnois, J.L., *Lasers Med. Sci.*,1, 47–66, 1986).

Common solder agents are blood and albumin. When necessary dyes are added for targeting wavelength selective absorption. Solid solders include biodegradable polymer thin films.

Direct laser welding uses lasers to heat endogenous components, such as water, to coagulate tissue proteins with direct laser–tissue interaction. The advantages are that the tissue repair is fast, with little or no foreign body reaction, and watertight. Disadvantages include low acute strength and thermal damage to the tissue. Laser soldering adds to the repertoire, with chromophore-enhanced protein "solders" that augment the welding process. Advantages versus direct welding are that acute wound strength is improved, and the use of solders allows for differential absorption or the use of different laser wavelengths tuned to a particular protein/tissue combination, making laser tissue repair with solders more flexible. Solid solder strips were introduced next in concert with solder in an attempt to strengthen the weld, with the advantage that the combination is stronger than liquid solder acutely and after hydration. Apparently the bounded water provides a vehicle for proteins to coagulate and form an adhesion. The disadvantages of solder strips are that they are brittle, are difficult to make in large dimensions, and have poor shelf-life. Other solders include the biodegradable polymer polylactic-co-glycolic acid thin film, which improves acute weld strength with liquid albumin solder. The polymer film serves to reinforce the cohesive strength of the liquid albumin solder. And even more recently, a protein cross-linker genipin has been added to albumin solder in an attempt to make a stronger tissue weld. Applied laser power densities are usually in the order of several watts per square meter.

Some of the early tissues that were welded include skin and blood vessels. Vessel closure, called anastomosis, was an early target because of the desire to stop bleeding induced by a cut or during surgery.

The effectiveness ε_w of laser tissue welding depends on the feeding speed, on the motion of the laser beam over the tissues that need to be joined, v, and on the potential for welding those tissues. Often a welding agent such as indocyanine green-doped albumin (derived from bovine serum colored green to enhance absorption at the near-infrared wavelengths used) is a collagen that is applied to facilitate the welding process. Especially in vascular welding there is a need for solder. Other factors involved in the welding efficiency are the thickness of the media, d, and the absorbed laser energy over the duration of exposure, t. The absorbed energy can be expressed as

$$\mu_a \Psi'(r,t) \times t \tag{10.38}$$

Using the latent heat of the tissue h_t to find the effectiveness ε_w, we obtain

$$\varepsilon_W \propto \frac{v d \mu_a \Psi'(r,t) h_t t}{(1-R)\Psi_0} \tag{10.39}$$

where reflection from the surface of the interface is taken into account as $(1-R)\Psi_0$.

Histological examination of tissue welds shows negligible evidence of collateral thermal damage to the underlying tissue. Microstructural study often reveals albumin intertwining within the tissue collagen matrix. In general laser–tissue soldering is considered to be an effective alternative for tissue repair with improved tensile strength and minimal collateral thermal damage over conventional tissue suturing.

Additionally, tissue welding has the potential for endoscopic surgical applications.

10.2.2.3 Current Tissue Closure Methods

The current methods for tissue closure include sutures, staples and clips, and tissue adhesives (glues). Each of these modalities has advantages, but each is also fraught with disadvantages that make pursuing alternative methods, such as laser-assisted tissue closure, worthwhile. Advantages of sutures are that they are inexpensive, provide a strong closure, and are suitable for most tissue types. Their disadvantages are that they have risk of inflammation and they are not suitable for endoscopic use. The advantages of staples and clips are their uniform results, fast application, and suitability for endoscopic use. Their disadvantages are a "one size fits all" approach, large mechanical forces required for implementation, and their bulkiness, which makes microsurgery implementation difficult. Some advantages of tissue adhesives are that they are inexpensive, fast to apply, watertight, and amenable to endoscopic use. Their disadvantages are low strength and tissue toxicity. With these methods in mind, the options of tissue closure with the use of light become more appealing.

10.2.3 Photobiological Nonthermal Interaction

Two examples of nonthermal photobiological interactions are vision and quantum dot labeling. The vision process will be described first, followed by quantum dot labeling. Other nonthermal photobiological interactions is decribed in Chapter 9, sections 9.1.1, 9.1.2 and 9.2.2 respectively: photochemical interaction.

10.3 Excitation of Chromophores

Chromophores are clusters of atoms that have the capability of existing in at least two energy states: a ground state and an excited state. The chromophore may be raised to the excited state by energy absorbed from incident electromagnetic radiation. In most cases the specific chromophores can only be raised to a higher energy level by photons within a relatively narrow energy (wavelength) bandwidth.

When the energy carried by the incoming photon $h\nu$ exactly matches the energy gap between two energy levels in a molecule, it can effectively be absorbed by this molecule. Hemoglobin and melanin are the most important chromophores in human bodies. Hemoglobin has significant absorption in the blue and green spectra, thus leaving the red light to be backscattered, giving blood a red color.

A different kind of chromophore is rhodonine. Rhodonine brings us into the area of mammalian vision. The rhodonines are a homologous chemical family that can be derived from retinenes. Retinenes are oxidized vitamin A variations. Vitamin A is a retinol, an alcohol in chemical base composition. The retinol in turn is oxidized to a retinal. Retinenes are based on the carboxyl-ion group. Rhodopsin is a retinal photoreceptor consisting of a carotenoid protein linked with a chromophore (retinal).

There are four rhodonine structures, each with a specific maximum absorption wavelength. The four rhodonines are rhodonine(5) (5-*cis*-retinal basis), most sensitive at 625 nm; rhodonine(7) (7-*cis*-retinal basis) with an absorption peak at 532 nm; rhodonine(9) with its peak response at 437 nm; and finally rhodonine(11) in the ultraviolet at 342 nm. The difference between each of these rhodonine molecules is the location in the molecular chain of the attachment of oxygen atoms to the respective sites on the molecule. In dilute solution all absorb near 493 nm; however, in liquid crystalline state they exhibit their specific absorption spectra.

Bacteriorhodopsin is a protein that is part of the purple membrane of the retina. Next to the photosynthetic proteins found in plants and bacteria, bacteriorhodopsin is one of the few molecular pigment compounds that have the capability to directly convert light energy into different forms of energy, electrical energy in this case since it acts as a proton pump. It converts light into an electrochemical gradient across a bacterial membrane, where in turn the proton gradient synthesizes ATP.

10.4 Optic Nerve-Cell Depolarization under the Influence of Light (Vision)

The process of vision is described as follows. Rhodonine coats a protein substrate attached to the dendrites of the optic nerve. This protein substrate is called opsin, which is suspended between multiple dendrites. The opsin substrate is coated with rhodonine. In fact rhodopsin is rhodonine linked with opsin. This structure is the fundamental building block of the cones that provide color vision. Under the influence of photons rhodopsin undergoes isomerization ultimately leading to the activation of a protein called transducin. The protein transducin in turn activates CGMP phosphodiesterase, which cleaves cGMP, a relatively small messenger molecule. The function of cGMP in the cone cells is to keep the Na^{+} channels open, thus creating a transmembrane

TABLE 10.4
Various Photoreceptors Occurring in Nature and their Function

Photoreceptor (chemical composition)	Function
Bacteriorhodopsin (opsin protein)	Proton pump in bacteria
Chlorophyll/carotenoids (reaction center polypeptide)	Photosynthesis in plants
Cryptochrome (protein)	Regulation of circadian rhythm (day–night) in living organisms
Photolyase (protein, with enzyme activity)	DNA repair resulting from ultraviolet light damage
Phototropin (protein, with kinase activity)	Regulation of phototropism in plants in response to blue light
Phytochrome (protein, with kinase activity)	Photomorphogenesis in plants and cyanobacteria
Rhodopsin (opsin)	Black and white vision and color vision in higher animals

potential. With the drop in cGMP as a result of the cleaving, the transmembrane potential drops as well. Following isomerization, the retinal is cleaved from the protein. As a result an opsin is left without a chromophore. The reconstruction of the cis configuration requires a certain amount of time, which explains the time delay in vision adaptation going from bright light levels to dim light. Table 10.4 lists some of the naturally occurring photoreceptors.

10.4.1 Quantum Photon Dots as Biological Fluorescent Markers

Quantum photon dots (QDs) are unique semiconductors that behave like single atoms and bear nanomolecular sizes, making them attractive as biological fluorescent agents. Whereas the gold standard of biological fluorescent markers, the organic dyes and fluorescent proteins, has numerous disadvantages to its both short-term and long-term usage, the unique properties of QDs point to a growing potential as the biofluorescent agent of choice in the near future.

10.4.2 Optical Properties of Quantum Dots

Quantum dots are synthesized primarily using elements from groups II—to VI (such as CdSe, CdTe, CdS, and ZnSe) or groups III—to V (such as InP and InAs). Crystals of diameters 2–6 nm are considered suitable for biological usage because of their small size and unique optical properties. Their small size leads to what is referred to as a quantum confinement effect, meaning particles that have smaller physical dimensions than their exciton Bohr radius. What results is a nanoscale molecule that behaves very similarly to a single atom. Excitation of an electron to the conductance band (higher energy state) and subsequent return to the valence band results in the release of a photon. It has been noted experimentally that QDs can be excited along a wide range of wavelengths. However, Gaussian emission results in a single symmetric peak, with typical widths of 20–30 nm (full-width at half-maximum, or FWHM) in the visible region, with some as narrow as 13 nm. This excitation and emission profile is in stark contrast with the organic dye, rhodamine. Organic dyes often have a narrow range of excitation (~520–530 nm for rhodamine) and their emission spectra are asymmetrical, with the peak tailing off in the red region leading to the problem of overlap with other organic dyes. Consequently, the use of organic dyes results in the need for several different sources of energy to generate different colors in the same cell/tissue. This also raises the concerns of the cost associated with the need for multiple sources of excitation when using organic dyes for labeling. Quantum dots, in contrast, can be excited along a broad range of wavelengths, reducing the need down to one source of energy.

Another interesting quality of QDs that is of great benefit to the biologist is the diversity of color QDs can produce. It has been noted that as the size of QDs of the same preparation decreases, the absorbance onset and emission maximum increase in energy, resulting in size-tunable fluorophores. To the operator, this means many QDs can be excited with a single source of light, allowing detection of these QDs to occur simultaneously. For instance, ZnS-capped CdSe quantum dots excited with a near-UV lamp have 10 distinguishable emission colors. The respective emission maxima are located at 443, 473, 481, 500, 518, 543, 565, 587, 610, and 655 nm. Other quantum dots are, for instance, made with CdSe and other materials.

If one were to use fluorophores using organic dyes, this would require excitation with one wavelength of energy, collection of data, switching to another source of energy to excite the second organic dye, and finally the use of computer software to build composite images to view where the organic dyes have accumulated in a particular specimen. This step is eliminated when using QDs.

The advantage that QDs have over their organic dye companions is their ability to fluoresce for long periods of time, and furthermore to resist photobleaching. Organic dyes have been shown to have decay rates on the order of ~1–5 ns. Quantum dots have been shown to have decay rates in the hundreds of nanoseconds, which is ~100-fold greater than those of the organic dyes.

The emission of these nanocrystals is due to a transition that would be spin-forbidden according to Hund's law in a system of light atoms. Because of strong spin-orbit coupling and quantum confinement effects, the emission has a radiative rate that is on the order of $10^7\ s^{-1}$.

An important challenge to overcome has been solubility and delivery of QDs to their target cells and tissues. Traditional QD synthesis occurs in organic solvents resulting in a nanocrystal that is insoluble in the aqueous environment of cells. As a result, several methods have been employed to eliminate this limitation. One method has been to coat the surface of the CdSe/ZnS QD with a silica/siloxane material. This forms a shell on the ZnS-capped QD making it soluble in intermediate solutions (partially aqueous solutions). Further modification with bifunctional methoxy compounds (amphipathic compounds, meaning molecules that have both nonpolar and polar components), such as aminopropyl trimethoxysilane or trimethoxysilyl propylurea, makes the nanocrystal soluble in an aqueous buffer.

Some other methods of rendering QDs soluble are electrostatic attraction to biomolecules and hydrophobic attraction to molecules like tri-*n*-octylphosphine oxide.

Certainly one consideration for QDs is the safety in organisms. Heavy metals are an independent cause of cell death and damage, and in particular cadmium is a known cause of metabolic dysfunction through its ability to uncouple ATP production from proton transport in mitochondria.

Another important consideration when using QDs is the targeting of specific proteins in cells and tissues. This is commonly accomplished through immunodetection. Traditionally, an organic dye is coupled to an antibody that recognizes a specific antigen on the target protein of interest. Thus, it is intuitive that coupling QDs to the same molecules would be an attractive application.

10.5 Summary

This chapter described the main subdivision of light–tissue interaction that has biological implications. The photothermal processes of coagulation and the damage integral were discussed in addition to the specific application of the photothermal effect to remove shallow blood vessels in port-wine stain treatment. The biological process of vision was also briefly elaborated on. The relatively new topic of quantum dots in biological use was also discussed.

Further Reading

Almeida-Lopes, L., Rigau, J., Zangaro, R.A., Guidugli-Neto, J., and Jaeger, M.M.M., Comparison of the low level laser therapy effects on cultured human gingival fibroblasts proliferation using different irradiance and same fluence, *Lasers Surg. Med.*, 29(2), 179–184, 2001

Anderson, R.R., and Parrish, J.A., Microvasculature can be selectively damaged using dye lasers: a basic theory and experimental evidence in human skin, *Lasers Surg. Med.*, 1, 263–276, 1981.

Apfelberg, D.B., Maser, M.R., Lash, H., and Rivers, J., The argon laser for cutaneous lesions. *J. Am. Med. Assoc.*, 245, 2073 – 2075, 1981.

Apfelberg, D.B., Maser, M.R., Lash, H., and White, D.N., Treatment of xanthelasma palpebrarum with carbon dioxide laser, *J. Dermatol. Surg. Oncol.*, 13, 149 – 51, 1987.

Apfelberg, D.B., Maser, M.R., Lash, H., and Rivers, J., Analysis of complications of argon laser treatment of port wine haemangiomas with reference to striped technique, *Lasers Surg. Med.*, 2, 357–371, 1983.

Boulnois, J.L., *Lasers Med. Sci.*,1, 47–66, 1986.
Dubertret, B., Skourides, P., Norris, D.J., Noireaux, V., Brivanlou, A.H., and Libchaber, A., In vivo imaging of quantum dots encapsulated in phospholipid micelles, *Science*, 298(5599), 1759–1762, 2002.
Gardner, E.S., Reinisch, L., Stricklin, G.P., and Ellis, D.L., In vitro changes in non-facial human skin following CO2 laser resurfacing, *Lasers Surg. Med.*, 19, 379 – 387, 1996.
Greenwald, J., Rosen, S., Anderson, R.R., Harrist, T., MacFarland, F., Noe, J., and Parrish, J.A., Comparative histological studies of the tunable dye (at 577 nm) laser and argon laser. The specific vascular effects of the dye laser, *J. Invest. Dermatol.*, 77, 305 – 310, 1981.
Henriques, F.C., and Moritz, A.R., Studies of thermal injury I, *Am. J. Pathol.*, 23, 531–549, 1947.
Hruza, G.J., and Dover, J.S., Laser skin resurfacing, *Arch. Dermatol.*, 32, 451–458, 1996.
Lanigan, S.W., and Cotterill, J.A., Psychological disabilities amongst patients with port wine stains, *Br. J. Dermatol.*, 121, 209–215, 1989.
Morelli, J.G., Tan, O.T., Garden, J., Margolis, R., Seki, Y., Boll, J., Carney, J.M., Anderson, R.R., Furumoto, H., and Parrish, J.A., Tunable dye laser (577 nm) treatment of port wine stains., *Lasers Surg.* Med., 6, 94 – 99, 1986.
Motley, R.J., Katugampola, G., and Lanigan, S.W., Microvascular abnormalities in port wine stains and response to 585 nm pulsed dye laser treatment, *Br. J. Dermatol.*, 135 Suppl. 47, 13–14, 1996.
Motley, R.J., Katugampola, G., and Lanigan, S.W., Videomicroscopy of vascular patterns in port wine stains predicts outcome, *Lasers Surg. Med.*, (Suppl. 8), 94–99, 1996.
Noe, J.M., Barsky, S.H., Geer, D.E., and Rosen, S., Port wine stains and the response to argon laser therapy: successful treatment and the predictive role of colour, age and biopsy, *Plast. Reconstr. Surg.*, 65, 130 – 36, 1980.
Polla, L.L., Tan, O.T., Garden, J.M., and Parrish, J.A., Tunable pulsed dye laser for the treatment of benign cutaneous vascular ectasia, *Dermatologica*, 174, 11 – 17, 1987.
Reid, W.H., Miller, I.D., Murphy, M.J., Paul, J.P., and Evans, J.H., Q-Switched ruby laser treatment of tattoos, a 9-year experience, *Br. J. Plast. Surg.*, 43, 663 – 669, 1990.
Renfro, L., and Geronemus, R.G., Anatomical differences of port-wine stains in response to treatment with the pulsed dye laser, *Arch. Dermatol.*, 129, 182–188, 1993.
Ruiz-Esparza, J., Fitzpatrick, R.E., and Goldman, M.P., Selective melanothermolysis: a histological study of the Candela 510 nm pulsed dye laser for pigmented lesions, *Lasers Surg. Med.*, Suppl. 4, 73, 1992.
Tan, O.T., Carney, M., Margolis, R., Seki, Y., Boll, J., Anderson, R.R., and Parrish, J.A., Histologic responses of port-wine stains treated by argon, carbon dioxide and tunable dye lasers, A preliminary report, *Arch. Dermatol.*, 122(9), 1016 – 1022, 1986.
Tan, O.T., Murray, S., and Kurban, A.K., Action spectrum of vascular specific injury using pulsed irradiation, *J. Invest. Dermatol.*, 92, 868–871, 1989.
Tan, O.T., Sherwood, K., and Gilchrest, B.A., Treatment of children with port wine stains using the flash lamp-pulsed tunable dye laser, *New Engl. J. Med.*, 320, 416–421, 1989.
Taylor, C.R., Gange, R.W., Dover, J.S., Flotte, T.J., Gonzalez, E., Michaud, N., and Anderson, R.R., Treatment of tattoos by Q-Switched ruby laser, *Arch. Dermatol.*, 126, 893–899, 1990.

Torres, J.H., Motamedi, M., and Pearce, J.A., Experimental evaluation of mathematical models for predicting the thermal response of tissue to laser irradiation, *Appl. Opt.*, 32, 597–606, 1993.

Van Gemert, M.J.C., and Welch, A.J., Treatment of port-wine stains: analysis, *Med. Instrum.*, 21, 213–217, 1987.

Varma, S., and Lanigan, S.W., The psychological, social and financial burden of tattoos, *Br. J. Dermatol.*, 135 Suppl. 47, 37, 1996.

Vinck, E.M., Cagnie, B.J., Cornelissen, M.J., Declercq, H.A., and Cambier, D.C., Green light emitting diode irradiation enhances fibroblast growth impaired by high glucose level, *Photomed. Laser Surg.*, 23(2), 167–71, 2005.

Webb, C., and Dyson, M., The effect of 880 nm low level laser energy on human fibroblast cell numbers: a possible role in hypertrophic wound healing, *J. Photochem. Photobiol. B*, 70(1), 39–44, 2003.

Problems

1. Give an example of photobiological interaction?
2. Calculate the time duration required for cell denaturization of skin up to a depth of 2 mm, assuming irradiation by a titanium:sapphire laser tuned to 790 nm and operating at 3 W with a beam diameter of 2 cm at the center of the beam.
3. The blood vessel is filled with blood and has a diameter of 2 mm. The diode laser is operating in continuous wave mode at a wavelength of 585 nm, with an output power of 15 W. The tissue optical properties are listed in Table 10.3. Calculate the temperature at the proximal and distal edge of a blood vessel that runs parallel with the surface of the skin at a depth of 2 mm.
4. Give the definition of quantum dots.
5. Calculate the coagulation depth of tissue under arbitrary laser irradiation. Assume that tissue is mostly water, and has all the physical parameters associated with water (the ablation energy of tissue is approximately 2.6 J/mm^3), the coagulation temperature is 80°C. (Hint: the absorption coefficient at the arbitrary laser wavelength will be unknown.)
6. Which beam profile is better for tissue ablation, a top-hat or Gaussian profile? Explain why?
7. An Ho:YAG laser with 10 mJ per pulse is absorbed in myocardium with an absorption coefficient of $\mu_a = 165\,cm^{-1}$. The beam area is $A = 1\ mm^2$. The initial temperature of the tissue is 37°C. The frequency factor has the following value: $A = 10^{76}\ s^{-1}$, the time duration is 0.742 s, and the energy per mole is $E = 5\times10^5\,J/mol$.
   What is the final temperature after one laser exposure?
   What physical process category does this interaction fall into?
   What histopathological effect is seen in myocardium?
   What is the damage function Ω for this interaction?
8. List some of the advantages and disadvantages of laser-assisted tissue welding.

Section II

Therapeutic Applications of Light

11

Therapeutic Applications of Light: Photophysical

With the growing number of laser wavelengths available and the constant search for better treatment modalities, therapeutic laser applications are continuously added to the list of choices in the physicians' toolbox. It will be virtually impossible to cover all available clinical or experimental treatment methods that use light in either collimated or diffuse format, and in multi- or single-wavelength operation.

In this section an overview is presented of several representative therapeutic applications of light, or in particular, when laser light is used for major medical procedures with a distinct advantage over mechanical or chemical treatment.

Selected advantages of the use of light in clinical treatment setting are as follows:

- Cauterizing action of high-powered laser light reduces or even eliminates bleeding and edema.
- Laser light used for treatment combined with optical vision and miniaturized tools combined in a single application, e.g., endoscope.
- Minimized surgical intervention
    - Fiberoptic delivery.
- Minimized risk of cross-contamination
    - Wound sterilization due to laser heat.
- Lower risk of scar formation and increased wound-healing potential.
- Elimination of electrical interference with monitoring equipment in contrast to high-frequency electrical cauterizing tool frequently called by the generic name "bovie" (from the inventor William T. Bovie).
- Fiberoptic or light guide assisted vision.

In most clinical applications of light the conversion of light energy into thermal energy is the primary mode of operation. The major exception to this rule is photodynamic therapy, but there are others as well, as discussed in Chapters 9 and 10. In case the operative mode of light–tissue interaction is thermal, the respective procedures can be grouped according to the method of heat generation, continuous wave pulsed or super pulsed, or the actual temperature rise accomplished. Some of the thermal applications can be arranged in the ultimate effect, such as thermal expansion or plasma formation. In the latter situation the light energy is ultimately converted into mechanical energy. A lesser but significant field of therapeutic applications of light in medicine uses the conversion of light energy into chemical energy. In selected light applications the light–tissue interaction is aimed at achieving a specific biological effect, and thus qualifies as photobiological. The photochemical and photobiological therapeutic applications of light are described in Chapters 12 and 13, respectively.

As described in Chapters 8–10, the use of light in medicine can be subdivided as follows:

- Photophysical
  - Photoablative/vaporization/photodisruption
    - Photoacoustic
    - Electromechanical
- Photochemical
  - Photodynamic Therapy
  - Wound healing
- Photobiological
  - Photothermal
    - Coagulation
      - Tissue welding
    - Port-wine stain treatment
- Photo-nonthermal
    - Quantum dots

One item of interest in all therapeutic irradiation applications is the fact that within one optical free path length a buildup of light will occur due to the inherent tissue scattering. The next section illustrates the effect of light buildup under broad beam irradiation, which means that the light delivery conditions are several times wider than the tissue optical mean free path at the respective irradiation wavelength.

11.1 Delivery Considerations

In many therapeutic biomedical optics applications the conditions for broad beam irradiation can be found to be an appropriate fit. These conditions will combine the effective attenuation coefficient at the irradiation wavelength and the spot size to determine whether a broad beam approximation is warranted. Especially in oncology the use of light arrays are customary and the conditions for a broad beam are definitely satisfied.

Under the broad beam the boundary conditions justify the diffusion approximation of the equation of radiative transfer. In particular, the Kubelka–Munk three-flux photon stream from Section 6.1.3.2 will describe the light interaction with a first-order approximation homogeneous medium quite well.

We will now show that under broad beam irradiation a phenomenon of photon accumulation will occur, which can be considered as an amplification of the fluence rate compared to the incident light power density.

Let us assume a homogeneous medium being irradiated with a beam with finite diameter. The beam diameter is greater than 10 times the optical mean free path length (travel between interactions: absorption or scattering). It can be shown that the diffusion equation satisfies these conditions. The diffusion equation provides a solution for the diffuse forward radiance and diffuse backward radiance. The highest total radiance at a point directly underneath the interface between two media is under collimated irradiation. The three-flux approximation can now be written as follows:

$$I_{\text{tot}} = F_{\text{c}} + 2[F_{d+} + F_{d-}] \tag{11.1}$$

Under diffuse irradiance the total radiance changes to

$$I_{\text{tot}} = 2F_d + 2[F_{d+} + F_{d-}] \tag{11.2}$$

where F_d and F_{d+} need to be identical, thus giving

$$I_{\text{tot}} = 4F_{d+} + 2F_{d-} \tag{11.3}$$

Under Kubelka–Munk theory the diffuse backward flux was the result of backscattered forward flux, with the proportionality factor R_∞ being the whole body bulk reflection coefficient. In this way a thin layer just below the surface area of the interface between the two media can be represented by an infinitely thin layer with reflection coefficient r_2 at the distal side, and at the proximal side the light follows Fresnel's law of reflection giving a reflection coefficient r_1. The backward flux defined by Kubelka–Munk is the bulk reflection coefficient $r_2 = R_\infty$ times the forward flux, and the diffuse forward flux is the reflection coefficient at the proximal surface times the backward diffuse flux. Theoretically, both the diffuse forward and the diffuse backward flux will experience the Kubelka–Munk reflection an infinite number of times. Remember, in the Kubelka–Munk theory the reflection coefficient incorporates the combined effects of absorption and scattering, so losses are permitted. However, the diffusion equation approximation is only valid when absorption is much smaller than scattering.

The relative radiance immediately below the boundary between medium 1 and medium 2 for a layer approaching zero thickness can be expressed for the collimated incident radiance from Equation (11.1) as

$$m_s = \left.\frac{I_{\text{internal}}}{I_{\text{incident}}}\right|_{\delta\to 0} = 1 + 2\left[r_2 + r_2 r_1 + r_2^2 r_1 + r_2^2 r_1^2 + \cdots\right] \tag{11.4}$$

The ratio of the incident radiant fluence rate on the internal radiant fluence rate is the amplification factor m_s resulting from internal scattering.
The recurrent sums in Equation (11.4) can be replaced by summations that can be solved by applying tables of summation rules found in most mathematical reference books. The summation in Equation (11.4) appears to be the sum of two recurrent semi-infinite summation series as follows:

$$\begin{aligned} m_{s,c} &= \left.\frac{I_{\text{internal}}}{I_{\text{incident}}}\right|_{\delta\to 0} = 1 + 2\left[r_2 + r_2 r_1 + r_2^2 r_1 + r_2^2 r_1^2 + \cdots\right] \\ &= 1 + 2\left[\sum_{n=1}^{\infty} r_2^n r_1^n + \sum_{n=1}^{\infty} r_2^n r_1^{n-1}\right] \\ &= 1 + 2\left[\frac{r_2 r_1}{1 - r_2 r_1} + \frac{1}{r_1}\frac{r_2 r_1}{1 - r_2 r_1}\right] = 1 + 2\frac{r_2(r_1 + 1)}{1 - r_2 r_1} \end{aligned} \tag{11.5}$$

while the solution for the diffuse irradiance in Equation (11.6) provides for the amplification due to the combined effect of internal backscattering and boundary discontinuity reflection as

$$\begin{aligned} m_{s,d} &= \left.\frac{I_{\text{internal}}}{I_{\text{incident}}}\right|_{\delta\to 0} = 2 + 2\left[r_2 + r_2 r_1 + r_2^2 r_1 + r_2^2 r_1^2 + \cdots\right] \\ &= 2 + 2\left[\sum_{n=1}^{\infty} r_2^n r_1^n + \sum_{n=1}^{\infty} r_2^n r_1^{n-1}\right] \\ &= 2 + 2\left[\frac{r_2 r_1}{1 - r_2 r_1} + \frac{1}{r_1}\frac{r_2 r_1}{1 - r_2 r_1}\right] = 2 + 2\frac{r_2(r_1 + 1)}{1 - r_2 r_1} \end{aligned} \tag{11.6}$$

It is to be noted that in case light is incident on tissue with index of refraction $n_{\text{tissue}} = 1.5$ from air with $n_{\text{air}} = 1.0$, there will be loss of the incident light due to specular or diffuse reflection entering the second medium, i.e., $r_{\text{external, collimated}} = 0.04$ and $r_{\text{external, diffuse}} = 0.0969$. The surface reflectivity here is derived from Fresnel's law as described in Chapter 3 (Equation [3.27]).

The theoretical derivation of the internal diffuse reflection from the tissue at the air interface and the whole body reflectivity can be found in

Section 3.4.3. The internal diffuse reflection from tissue to air at 623 nm equals $r_{\text{external, diffuse}} = 0.56$, while the whole body reflectivity is $R_\infty = 0.591$ in case of, for instance, skin irradiated with a helium–neon (He–Ne) laser.

The following situation will arise for various laser-irradiation protocols. For instance, under collimated broad beam illumination of the skin during photodynamic therapy (PDT) with a He–Ne laser, the radiance directly below the surface of the epidermis becomes 3.76 times greater than the incident radiance. Under diffuse Light Emitting Diode (LED) illumination the diffuse radiance in the epidermis is 4.76 times greater than the incident radiance.

These are serious considerations to be taken into account when designing an irradiation protocol, since this may result in subcutaneous burns, while trying to achieve the threshold radiance inside the tumor located below the epidermis.

A selected parade of clinical applications relying on the physical aspects of light–tissue interaction is discussed in the following sections. The list starts off with pulsed laser use in cardiology followed by the field of laser dentistry and the photophysical therapy in ophthalmology. The final physical application of light in medicine is the description of the working of optical tweezers.

11.2 Pulsed Laser Use in Cardiology

In cardiology the use of pulsed laser application attempts to vaporize thrombus formation. Thrombus vaporization is still being evaluated on a limited scale, since there are numerous antithrombotic drugs available with a high degree of efficiency. Additional pulsed laser applications are found in the transmyocardial revascularization attempts.

To describe the pulsed photophysical effect the principles of pulsed laser–tissue interaction will be discussed.

11.2.1 Pulsed Laser–Tissue Interaction

Pulsed laser operation provides high peak power per pulse, but low mean power when averaged over the duty cycle of the pulse delivery protocol. A laser delivers its energy in one or more short pulses, as distinguished from a continuous wave (CW) laser. For instance, the laser emits bursts of light energy over periods of less than 0.25 s. Generally, the pulses are separated by equal or longer pause time between pulses. The pulses may be repeated and comprise a train of pulses, or the laser delivers a single shot pulse only when this is required for the desired effect. Pulse durations can vary from milliseconds to microseconds, nanoseconds, or even femtoseconds, as defined by half-peak power points on the leading and trailing edges of the pulse.

The source of the energy converted into work is the temperature reached during the absorption of the laser light. The work will be provided by the increasing temperature of the tissue volume. This is why the timing of the process is so crucial to maximize the energy transfer.

As noted earlier, pulsed laser–tissue interaction consists of three stages: light absorption, acceleration of the ejected tissue followed by stagnation. With each pulse this process is repeated when the pulse length is shorter than the thermal relaxation time of the tissue. This is especially true for pulsed laser ablation of hard tissues such as plaque, bone, and mineral aggregates such as kidney stones.

The tissue expansion during the ablation of ventricular muscle is considered to be adiabatic. By definition, during adiabatic expansion no heat goes in or out of the system. The ablation energy is delivered and the next step is the expansion itself. However, there will be work performed on the surroundings as the tissue expands. The work W is described by the First Law of Thermodynamics and is known as the work of expansion, and it is the change in pressure ΔP multiplied by the change in volume ΔV. The work of expansion is expressed in the following equation:

$$W = \Delta P \times \Delta V = nRT \tag{11.7}$$

where n is the molecular count and R the universal gas constant, and the process is assumed to proceed isothermally since no additional heat is delivered after the laser pulse has ceased.

The basic thermodynamic equations of interest are the following conservation laws: conversion of mass, momentum, and energy.

Conservation of mass is given as

$$\frac{\partial \rho(r,\theta,z,t)}{\partial t} + \nabla \cdot \left(\rho(r,\theta,z,t)\mathbf{u}(r,\theta,z,t)\right) = 0 \tag{11.8}$$

Conservation of momentum is expressed by

$$\rho(\mathbf{r},\theta,z,t)\left(\frac{\partial \mathbf{u}(r,\theta,z,t)}{\partial t} + \mathbf{u}(r,\theta,z,t) \cdot \nabla \mathbf{u}(r,\theta,z,t)\right) + \nabla P(r,\theta,z,t) = 0 \tag{11.9}$$

And finally, the conservation of energy (density) is shown in the following equation:

$$\frac{\partial P(r,\theta,z,t)}{\partial t} + \nabla \cdot \left(P(r,\theta,z,t)\mathbf{u}(r,\theta,z,t) + \frac{\kappa_{\mathrm{T}}}{\tau_{\mathrm{pulse}}} \nabla T(r,\theta,z,t) \right) = \frac{\mu_{\mathrm{a}} \Psi_0 z_{\mathrm{abl}}}{\tau_{\mathrm{pulse}} V} \tag{11.10}$$

where $\rho(r,\theta,z,t)$ is the tissue density, $\mathbf{u}(r,\theta,z,t)$ the velocity of tissue vaporization or expulsion rate, $P(r,\theta,z,t)$ the pressure in the tissue, κ_{T} the thermal conductivity, and V the absorption volume, which is roughly equal to the laser spot size times the absorption depth $V = A_{\mathrm{laser}} Z_{\mathrm{abl}}$, and the ablation energy density threshold can be found from Equation (8.13). The equation of conservation of momentum (Equation [11.9]) will reduce to the Bernoulli equation when the ejection of tissue vapor is in equilibrium, immediately after the initial onset of the tissue explosion. At the time of onset of the explosion there will be acceleration, but when the tissue is being expelled, there is no more acceleration.

The acceleration process can be solved from Equations (11.8)–(11.10) in many different ways, mostly by making several assumptions on boundary condition. One assumption is to treat the process after the initial light absorption as isobaric, meaning under constant pressure. This is not a very meaningful approximation, since it will only describe the process of the buildup leading to the tissue expulsion, while the ablation process puts more weight on the actual tissue removal aspects. The tissue removal is described by the Rayleigh–Taylor instability of ablation, which is outlined next.

11.2.1.1 Ablation Rate

The thermodynamics of the pulsed energy deposition in the tissue and the resulting vaporization can be used to find the ablation velocity, or the ablation rate.

We also assume that the tissue is expelled out in the opposite direction of the incident laser beam. In this case the expulsion rate becomes a one-dimensional problem $\mathbf{u} = v_a$.

For soft tissues such as muscle the expulsion rate can be approximated as if the process takes place under isobaric conditions when the timeframe is small enough. The solution for the ablation velocity v_a can be shown to satisfy the following equation for soft tissue ablation:

$$v_a = \frac{f \mu_a m_s \Psi_0 z_{abl}}{h_v} \tag{11.11}$$

where m_s is the buildup of fluence directly underneath the surface as given by Equation (11.5) and f is the efficiency of the tissue escape from the ablation trough. The remaining parameters are the latent heat of vaporization h_v, and the incident fluence rate time the tissue absorption.

During the expansion following the initial light absorption, Equation (11.10) can be integrated over the ablation volume V_a. For this purpose and due to the symmetry of the system it is convenient to write Equations (11.8)–(11.10) in spherical coordinates instead of cylindrical coordinates. This yields the following set of equations:

Conservation of mass is expressed as

$$\frac{\partial \rho(\mathbf{r},\theta,t)}{\partial t} + \nabla\cdot(\rho(\mathbf{r},\theta,t)\mathbf{u}(\mathbf{r},\theta,t)) = 0 \tag{11.12}$$

The conservation of momentum is expressed as

$$\rho(\mathbf{r},\theta,t)\left(\frac{\partial\mathbf{u}(\mathbf{r},\theta,t)}{\partial t}+\mathbf{u}(\mathbf{r},\theta,t)\cdot\nabla\mathbf{u}(\mathbf{r},\theta,t)\right)+\nabla P(\mathbf{r},\theta,t)=0 \tag{11.13}$$

While the conservation of energy density is written as

$$\frac{\partial P(\mathbf{r},\theta,t)}{\partial t}+\nabla\cdot\left(P(\mathbf{r},\theta,t)\mathbf{u}(\mathbf{r},\theta,t)+\frac{\kappa_T}{\tau_{\text{pulse}}}\nabla T(\mathbf{r},\theta,t)\right)=\frac{\mu_a\Psi_0 z_{\text{abl}}}{\tau_{\text{pulse}}V} \tag{11.14}$$

Integration of Equation (11.14) after dividing by the product of the pressure and volume, which is the work done by the tissue explosion, yields an expression for the ablation velocity as

$$\begin{aligned}\mathbf{u}(\mathbf{r},\theta,t)\propto&\frac{\mu_a\Psi_0 z_{\text{abl}}}{\tau_{\text{pulse}}P(\mathbf{r},\theta,t)V}-\frac{\partial P(\mathbf{r},\theta,t)/\partial t}{P(\mathbf{r},\theta,t)}\mathbf{r}\\&+\frac{\kappa_T}{\tau_{\text{pulse}}}\frac{\nabla T(\mathbf{r},\theta,t)}{P(\mathbf{r},\theta,t)}+\mathbf{u}_{\text{rot}}(\mathbf{r},\theta,t)\end{aligned} \tag{11.15}$$

where $\mathbf{u}_{\text{rot}}(\text{r},\theta, t)$ is the rotational part of the velocity, a mathematical expression signifying the integrand constant in spherical geometry. The rotational velocity follows from $\mathbf{u}_{\text{rot}}(\text{r},\theta,t) = \nabla\times\mathbf{u}(\mathbf{r},\theta,t)$ or $\mathbf{u}_{\text{rot}}(\text{r},\theta,t) = \nabla P(\text{r},\theta,t) \times \nabla P(\text{r},\theta,t)$. The solution methods for the ablation velocity in Equation (11.15) and the ejected mass in Equation (11.12) are generally dependent on the Péclet number. The Péclet number is the ratio of convective to conductive heat transfer and will significantly alter the ablation process when convection or conduction is dominant. The Péclet number is defined as

$$Pe=\frac{Vz_{\text{abl}}}{\alpha_T} \tag{11.16}$$

where α_T is the thermal diffusion coefficient $\alpha_T =(\kappa_T / \rho c_p)$ with c_p the specific heat. In laser vaporization the Péclet number will generally be small since there is negligible convection in the tissue, and will approach zero for high-absorption situations such as excimer and erbium:YAG laser ablation.

All the assumptions made so far indicate a steady-state solution to the conservation law propositions, as long as the initial artifacts of the tissue explosion are ignored. It will also be evident that the vaporized tissue leaves the biological medium in one direction only in a first-order approximation due to the mechanical confinement. Under the assumption that the

Péclet number is zero and the expulsion phenomenon is quasi-steady state, Equation (11.15) can be used to find an expression for the ablation velocity in a one-dimensional format as shown below:

$$v_{\mathrm{a}} \propto \frac{\mu_{\mathrm{a}} \Psi_0 z_{\mathrm{abl}}}{\tau_{\mathrm{pulse}} P(\mathbf{r},\theta,t) V} + \frac{\kappa_{\mathrm{T}}}{\tau_{\mathrm{pulse}}} \frac{\nabla T(\mathbf{r},\theta,t)}{P(\mathbf{r},\theta,t)} \tag{11.17}$$

This one-dimensional approach is not entirely justified for soft tissue ablation, and the three-dimensional expansion is presumably the foundation for the neovascularization in the ventricular wall as observed in transmyocardial laser revascularization (TMLR).

It can be shown that Equation (11.17) reduces to Equation (11.11) for a thermally confined ablation, which can be expected in tissue with high water content. For most pulsed laser ablation interactions the thermal relaxation time of the tissue, which is in the order of microseconds, satisfies the conditions for thermal confinement under the applied laser pulse widths.

In reference, the Bernoulli equation (Equations [11.9] and [11.13]) gives for the ablation velocity the following relationship:

$$\rho(\mathbf{r},\theta,t) v_{\mathrm{a}} \propto \sqrt{\frac{\Delta P(\mathbf{r},\theta,t)}{\Delta V(\mathbf{r},\theta,t)}} \tag{11.18}$$

Table 11.1 gives a list of the ablation fluence at various wavelengths for plaque compared to healthy tissue at the same wavelength as well as the relative efficacy of ablation.

TABLE 11.1

Ablation Fluence for Plaque and Normal Tissue at Various Ultraviolet, Visible, and Near-Infrared Wavelengths

	Plaque		Healthy Tissue	
Wavelength (nm)	***I* (J/cm²)**	**Efficacy (mg/J)[a]**	***I* (J/cm²)**	**Efficacy (mg/J)**
266	0.8	1.2	0.8	1.2
290	3.91		3.85	
355	4	0.21	4	0.21
482	42	0.21	101	1.05
532	24	0.13	34	0.10
658	127		>295	
1064	45	6	42	7

[a] The efficacy is the ablation mass per deposited laser energy.

11.2.2 Laser Plaque Molding (Angioplasty)

Earlier in the use of angioplasty the traditional balloon was combined with a laser diffuser to apply heat to mold plaque. The combination of balloon angioplasty with laser thermal molding of the plaque was discontinued by the sole producer of this technique. In contrast, different laser balloon-assisted applications are still being investigated. One example is the thermal release of pharmacologic agents. The light delivery for both plaque molding and drug release will adhere to the light distribution protocol outlined in Chapter 6.

In analogy with the technique of PDT discussed in Section 9.2.1.3, the release of photochemicals to inhibit proliferation of smooth muscle cell growth and resulting intimal hyperplasia is one laser application still open in the treatment of atherosclerotic plaque.

Another application for vascular diseases is implantation of temperature-responsive memory stents. The stent is wrapped around a balloon, which can expand the stent to widen an occluded vessel. The latest development uses thermally stressed stents that will assume a predetermined shape and remain at this shape when the temperature during the setting stage reached a set level. The thermal memory of the stent provides a more stable setting for the stent in addition to increased success in maintaining a patent lumen for an extended period of time.

11.2.3 Laser Thrombolysis

Laser thrombolysis is a procedure that is being developed to treat cardiovascular disease and stroke by removing clots occluding arteries of the heart and the brain. Laser thrombolysis is an interventional procedure that removes clot by delivering microsecond pulses via a fluid-core catheter. The removal of the clot results in a restoration of blood flow while maintaining vascular integrity. Ablation was initiated at surface temperatures just above 100°C. A vapor bubble was formed during ablation. The vapor bubbles expanded and collapsed within 500 μs after the laser pulse. Less than 5% of the total pulse energy is coupled into the bubble energy. A large part of the delivered energy is unaccounted for and is likely released partly as acoustic transients from the vapor expansion and partly wasted as heat.

Other tissue-removal techniques are valvulotomy and valve debridement.

11.2.4 Laser Valvulotomy and Valve Debridement

Conventional bypass surgery is not always an option to remedy poor perfusion to the lower extremities. In situ removal of venous valves combined with a proximal arteriovenous fistula to perfuse the foot retrograde is an alternative. The use of laser for valvulotomy shows promise for in situ laser ablation of the venous valves.

For instance, using an excimer laser operating at 308 nm required an ablation energy of less than 20 mJ per pulse at 20 Hz to advance the catheter at each valve in the vein of the leg. In contrast, calcified valves and vessel wall required energies in excess of 45 mJ per pulse to penetrate. Combined with the appropriate catheter selection using the proper diameter to fit the vein lumen resulted in ample clearing of the valves. The larger diameter catheters have in excess of 100 fibers as shown in Figure 11.1.

Another potential laser application is in valve debridement. Two valves in the heart, the mitral valve, which is the inflow valve of the left ventricle, and the aortic valve are both susceptible to calcification. Calcified valves will be less pliable and may not close properly. The mitral valve is susceptible to stenosis resulting from rheumatic fever. Rheumatic fever often results from certain infections, creating an attack on the valve leaflets by the immune system. As a result the leaflets become stiff and thickened, and will not open properly.

Other complications are, for instance, calcification of the ascending aorta and aortic valve. The aortic valve allows outflow of the left ventricle into the aorta, supplying the rest of the body with oxygen and nutrients. The most common cause of aortic stenosis is in patients of the age of 65 years and older and is called "senile calcific aortic stenosis." With growing age the protein collagen of the leaflets of the valve is destroyed and calcium is deposited on the valves. Resulting from the reduced mobility of the leaflets by calcification the normally present turbulence across the valve increases. The increased turbulence causes scarring, thickening, and eventually stenosis of the valve. The common method of repair of the valve is replacement. A better option would be in situ valve repair by means of laser debridement, removal of the calcified layer by laser vaporization, and plasma formation.

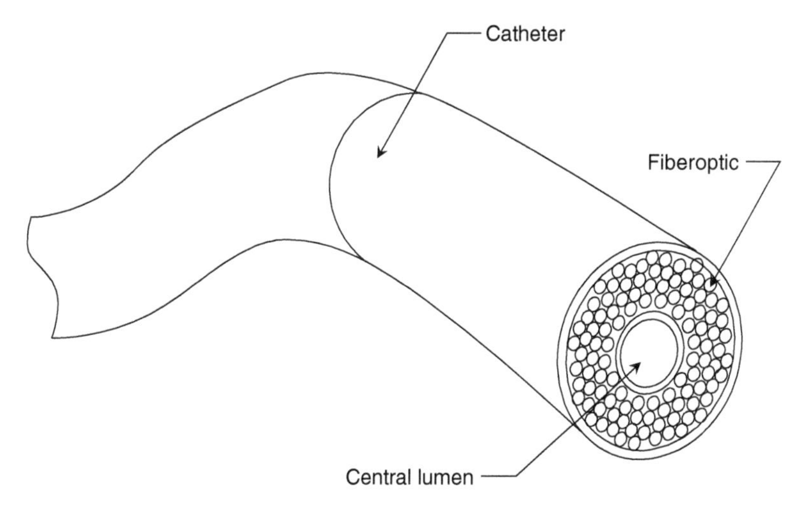

FIGURE 11.1
Multifiber catheter configuration.

The next section describes the relatively new field of laser transmyocardial revascularization.

11.2.5 Transmyocardial Revascularization

Transmyocardial laser revascularization is an option for patients whose coronary arteries are badly damaged or are in generally bad health to prevent any vascular treatments as described in the previous section, to increase blood flow. There are other solutions available to increase blood perfusion of the heart. It is also suggested that inflicting perforations can increase blood supply to the heart muscles, as in a reptilian heart. This process is described under neovascularization, which is the formation of new blood vessels. Neovascularization of the ischemic myocardium has been attempted for many years. Some suggestions used mechanical perforation and hot-tip transmyocardial revascularization. Later these attempts were improved upon by excimer, and pulsed near- or mid-infrared laser vaporization to achieve the same goal.

Laser can be used to drill tiny holes in the heart muscle itself, under the expectation that new blood channels will develop. In the process of pulsed laser irradiation one might also say that the heart, which is a soft tissue, is tenderized by the mechanical effects of the explosion. The laser energy can be introduced into the ventricular cavity through a catheter equipped with a fiber optic or directly through surgical intervention. The transcatheter approach yields the most appealing solution, since it is minimally invasive. The process of tissue vaporization is described in Section 8.2. Generally, the channels drilled by delivery of laser energy are on the order of 1–1.5 mm in diameter.

The channel generated by the vaporization of the heart muscle through the delivered laser energy eventually stimulates the heart muscle to sprout new blood vessels that supply blood to deprived tissue. The new growth is called angiogenesis and may require the stimulation of certain body proteins as well.

Some of the laser techniques applied use CO_2 laser (10.6 μm), XeCl laser (306 nm), pulsed Nd:YAG laser (1440 nm), pulsed Er:YAG laser (2.94 μm), or pulsed Ho:YAG laser operating at 2.1 μm. The CO_2 laser does not lend itself to fiberoptic delivery and is restricted to open chest application. The remaining laser wavelengths can be applied by fiber optic. The interaction mechanism however differs significantly between the four laser choices. The difference lies in both the absorption rate and the pulse duration for the laser operating at a specific wavelength.

Generally the vaporization energy density for tissue is in the order of $2\ J/cm^3$. The ablation values for all lasers range from 2 to $5\ J/cm^3$. Table 11.2 lists selected laser and ablation values.

A CO_2 laser generates pulses of 100 ns to 1 ms duration. The CO_2 lasers used for TMLR can be up to 800 W, operating at 200 Hz, with ablation fluence of $5 \times 10^6\ J/m^2$. In Q-switched mode the CO_2 laser can run up to 100 kHz, with pulse durations ranging from 100- to 500-ns and pulse energies between 0.02

TABLE 11.2

Selection of Lasers Available for Transmyocardial Revascularization with their Respective Ablation Thresholds

Laser	Mode of Operation	Wavelength (nm)	Pulse Duration	Frequency (Hz)	Energy Density Per Pulse (J/cm²)	Ablation Threshold (J/mm³)
CTE:YAG		2670	200 μs		0.83	2–7
CTH:YAG		2100	250 μs		5.31	10–15
CO_2	Quasi-CW		20–50 ms	Single-shot	2000–5000	2–5
	Ultrapulsed	10640	1.4 ms	200	250	5
	Q-switched		250–350 ns	200–20000		
Ho:YAG	Various	2100	100–500 μs	5–20	130–250	2–5
Er:YAG		2940	100–800 μs	100	0.2	1–4
Nd:YAG		1440	600 μs	5–20	130–250	5
Tm:YAG		2010	175 μs		18	3–8
XeCl	Multifiber	308	10–200 ns	40	5–10	2–4

and 0.05 J. In contrast, a XeCl excimer laser only produces 10–200 ns pulse width and 30 mJ per pulse operating at less than 100 Hz for this purpose. The excimer laser reportedly has an ablation energy density ranging from 2 to 4 J/mm^3, using pulse energy densities in the range of 5–10 J/mm^2. An erbium:YAG laser operates at approximately 100 Hz in standard pulsed mode, with pulse durations ranging from 100 to 800 μs and several hundred millijoules of energy per pulse.

Owing to the short pulse duration and fast pulse rates the ablation process is difficult to be described theoretically. Not all the tissue that is vaporized in one pulse leaves the volume of the medium and blocks the next laser pulse from reaching the tissue beyond the temporary solid surface location in the crater.

The ablation energy density $w_{\rm abl}$ can be approximated from the heat needed to raise the tissue water temperature from 37 to 100°C.

$$w_{\rm abl} = \rho(c_{\rm v}\, \Delta T + h_{\rm v}) \tag{11.19}$$

where ρ is the tissue density, $c_{\rm v}$ the heat capacity of muscle tissue, $h_{\rm v}$ the latent heat of vaporization, and ΔT the temperature rise.

For muscle these values are approximately as follows: $\rho = 1.06 - 1.09$ kg/m^3, $c_{\rm v} = 3551 - 3722$ J/kg K with a latent heat of $h_{\rm v} = 1.802 \times 10^6$ J/kg K. For a temperature rise of 63 K this amounts to an ablation energy density ranging from 2.15×10^6 to 2.20×10^6 J/m^3. A more detailed analysis can be made using the description in Section 8.2.3.

The full effect of the absorbed laser energy from each pulse is inflicted when the pulse duration is less than the thermal relaxation time of the tissue and the delivered energy exceeds the ablation threshold. The TMLR thermal relaxation time $\tau_{TMLR,r}$ is defined as the thickness of the absorption layer squared divided by the thermal conductivity κ_T,

$$\tau_{TMLR,r} = \frac{d_a^2}{4\kappa_T} \tag{11.20}$$

where the absorption layer is equal to the reciprocal effective attenuation coefficient or the reciprocal absorption coefficient, depending on the albedo.

The characteristic ablation time is the pulse width or the laser energy density divided by the power density Ψ_0:

$$\tau_{abl} = \frac{\Phi}{\Psi_0} \tag{11.21}$$

provided the minimal energy density exceeds the threshold energy requirement for ablation as given by Equation (11.19) or

$$\Phi_{th} = w_{abl} d_a \tag{11.22}$$

Combining Equations (11.20) and (11.22) will give the minimum requirements for laser pulse parameters to achieve ablation, where the pulse width needs to be less than the thermal relaxation time. The ablation depth is a critical component that will rely heavily on the absorption coefficient, and as such will be wavelength-dependent:

$$\tau_{pulse} \ll \frac{\Phi_{laser}^2}{4w_{abl}^2 \kappa_T} \tag{11.23}$$

Pulses that do not fulfill the requirement of Equation (11.23) still produce damage when the pulse energy is high enough, but because of the thermal diffusion little or no tissue removal will take place; however, it will generate collateral damage.

In the case the time duration for a shock stress wave to migrate out of the pulse-irradiated tissue volume (τ_Σ) is shorter than the laser pulse duration, the pulse is said to be inertially confined. This condition is represented as follows:

$$\tau_{TMLR,r} \leq \tau_\Sigma = \frac{d_a}{v_s} \tag{11.24}$$

where v_s is the speed of sound in tissue, approximately 1500 m/s.

The ablation depth was described earlier in Section 8.2.3 (Equation [8.30]), and reads as follows:

$$z_{abl} = \frac{E}{E_{abl}} - \frac{1}{\mu_a} \tag{11.25}$$

Note that with increasing absorption, as would occur due to carbonization, the vaporization plume will absorb a significant amount of the incident laser energy. As a result of this plume absorption the effective ablation depth will have a plateau under continuous ablation.

Additionally, the repetition rate will also need to be in line with the thermal relaxation time to maximize ablation. When all conditions are satisfied, the rate of ablation can be found.

Additional hard tissue laser interaction can be found in dental applications.

11.3 Dentistry and Oral Surgery

Laser was used by Stern and Sognnaes in 1964 to investigate the acid demineralization reduction of tooth enamel. However, it was soon discovered that the ruby laser caused damage to the pulp of the tooth, due to absorption. The use of the Ho:YAG laser was also investigated with mixed results. As a result of this revelation, laser research in dentistry, was predominantly aimed at the soft tissues in the oral cavity, and particular care was taken to find ways to protect the teeth from the laser-related artifacts, heat stress, and shock waves.

11.3.1 Photo Curing

Other applications include the use of lasers to cure light-sensitive composite polymers used in dental restoration; however, other ultraviolet light sources may be just as effective since only low-power irradiance is required.

11.3.2 Dental Drill

Since the initial investigation of Stern over 40 years ago, several unique laser dental applications have evolved for restorative dentistry, namely laser ablation of dental hard tissue, caries inhibition treatments by localized surface heating, and surface conditioning for bonding. The laser was also seen as an alternative for mechanical drills in cavity treatment in the teeth. The effects of the Er:YAG laser were especially encouraging, and showed resemblance to the interaction of a mechanical drill with dentin.

At the center of the tooth is the dental pulp, which is a confined connective tissue that is highly susceptible to temperature excursions in the tooth. Heat may accumulate in the tooth to dangerous levels during multiple pulse laser irradiations. It is generally accepted by the dental community that a temperature rise of as little as 5.5°C in the pulp may cause irreparable pulpitis and require root canal therapy. Therefore, any viable laser dental procedure has to minimize the accumulation of heat in the tooth. This can be achieved by using a laser wavelength tuned to the maximum absorption coefficient of the tissue irradiated, and by judicious selection of the laser pulse duration to be on the order of the thermal relaxation time of the deposited laser energy.

The best results for laser drilling applications have been obtained when a water jet is applied simultaneously with the laser irradiation. The beneficial effects of water are attributed to the cooling effects on the dentin, and subsequently the reduction in the shock wave effect generated by rapid heating resulting from the pulsed Er:YAG ablation, and a marked reduction in charring as can also be expected. An added benefit is the reduction of airborne particles resulting from the pulsed laser ablation under continuous water rinse. Airborne particles are always a great concern during explosive laser ablation because of the increased risk of contamination and the spread of infections.

The laser offers several potential advantages over the high-speed drill for the removal of dental hard tissue. The laser procedure is well tolerated, and there is reduced pain due to diminished noise and vibration. Moreover, carious tissue can be preferentially removed because of the higher volatility of water and protein under laser irradiation, since these materials have a higher rate of occurrence in carious tissue and have selectively higher absorption.

Lasers can be tightly focused to drill holes for micropreparations with very high aspect ratio (depth/width), well beyond those attainable by dental drills that are limited by the size of the dental bur. Therefore, lasers have the potential of substantially reducing the amount of tissue that needs to be removed for cavity preparations.

Dental hard tissue ablation is a water-mediated explosive process at 2.94 and 2.79 μm. During rapid heating, the initially confined water can create enormous subsurface pressures that can lead to explosive removal of the surrounding mineral matrix. Studies of hard tissue ablation in the $\lambda = 3$ μm region indicate that large intact particles are ejected with high velocity from the irradiated tissue. Moreover, the inherent highly ordered structure of dentin and enamel is conserved during the ablation process. In contrast, the mechanism of ablation of dental hard tissue at the highly absorbed CO_2 wavelength of 9.3 and 9.6 μm is apparently mineral mediated.

The strongest absorption bands of dental hard tissue that are readily accessible to conventional laser systems are in the near-infrared regions of 2.7–3.0 μm and 9–11 μm. Solid-state erbium lasers ($\lambda = 2.94$ and 2.79 μm) can be used to ablate bone, dentin, and enamel due to the strong water and OH– absorption of apatite. Er:YSGG laser emission is coincident with the narrow (OH–) absorption band at $\lambda = 2.8\ \mu m$ while Er:YAG ($\lambda = 2.94\ \mu m$) emission overlaps the broad water absorption centered at $\lambda = 3\ \mu m$. Studies of ablation using the free-running Er:YAG and Er:YSGG ($\lambda = 2.79\ \mu s$) laser systems (pulse duration 190–250 μs) reveal efficient ablation of enamel and dentin while minimizing heat deposition.

The erbium lasers were the first laser wavelengths to ablate hard tissue at a rate comparable to the high-speed dental drill. Carbon dioxide lasers with pulse duration between 5 and 100 μs are also very promising for dental hard tissue applications, particularly those operating at 9.3 and 9.6 μm wavelengths. Carbon dioxide laser radiation at 9.6 μm is most highly absorbed by the mineral phase. The absorption coefficient is 10 times higher at 9.6 μs than at 10.6 μm, the wavelength currently used by medical CO_2 lasers. CO_2 lasers can be used to ablate enamel with a greater efficiency than the erbium dental laser system.

For pulsed laser drilling, a laser beam is focused to a fine spot at or just below the material surface. Usually the plume from the ablated tissue will need to be evacuated by forced air flow or by suction. This assisting gas is usually an inert gas such as CO_2. The evacuation of the debridement also prevents tissue from splattering on the focusing lens or fiber optic, thus keeping a clear field for the laser delivery. Laser-drilled holes are often high in quality, with little risk of fraying edges or breakout. On the other hand, deep holes can result in taper effects, depending on laser wavelength and the corresponding absorption coefficient, and the pulse mechanism (frequency, duty cycle, etc.).

Different pulse mechanisms are described in the theory of laser operation in Chapter 5. The mechanism of pulsed laser–tissue interaction is described in detail in Section 8.2.

11.3.3 Etching

Acid etching is widely used in clinical dentistry to facilitate the mechanical retention of resin-based materials in the enamel surface of the teeth or in the dentine.

To increase the bonding of dental prosthesis, laser etching was investigated as an alternative to acid etching. A marked threefold increase in bond strength was observed using a CO_2 laser for etched dentin with respect to acid etching, whereas Nd:YAG laser etching was significantly less successful. No benefits were seen for laser over acid etching of enamel.

Several commercial organizations have developed laser systems with the aim of modifying dental hard tissues such as enamel. One of those lasers is the Er:YAG laser operating at the wavelength of 2.94 μm, which is used as a potential alternative to the acid-etching technique.

Lasers are also widely used in the shaping of dental prosthesis from silicon blocks and other available mineral deposit materials that can be sculptured by properly selected laser wavelengths and operating modes.

The lasers used for etching are standard pulsed lasers in most cases; however, femtosecond laser pulses have been shown to work very effectively, without some of the risks of cracking the enamel due to the heat latency seen in microsecond pulse duration. Pulsed laser operation with microsecond pulse duration with fluences above 25 J/cm^2 have been shown to produce macroscopic pitting of the enamel surface, while a fluence below 12 J/cm^2 seemed to have little effect on the enamel surface when examined both macro- or microscopically. Thus, fluences of 15–24 J/cm^2 are considered most appropriate for effective etching. In contrast, femtosecond pulses can be obtained by Q-switched laser operation.

11.3.4 Tooth Hardening

Lasers can be used to effectively modify the chemical composition of the remaining mineral phase of the enamel and the dentin. This occurs because the mineral hydroxiapatite, found in bone and teeth, contains carbonate inclusions that make it highly susceptible to acid dissolution by acids generated by bacteria in dental plaque. Upon heating to temperatures in excess of 400K, the mineral decomposes to form a new mineral phase that has increased resistance to acid dissolution. Thus, surfaces can be rendered resistant to acid dissolution via low-energy irradiation, and, as a side effect of laser ablation, the walls around the periphery of a cavity preparation will also have an enhanced resistance to further decay.

11.3.5 Scaling

More recently, the laser is being investigated again for scaling and cleaning purposes; however, with the increased knowledge of optical and thermal properties of dental hard and soft tissues, there is a better chance for success than in the inaugural trials. The increased awareness of the basic light interaction with the oral tissues, especially the dentin and enamel, leads to tooth resurfacing and hardening procedures, which are rapidly gaining widespread acceptance.

The use of lasers in jaw surgery will fall under laser surgery, which will be discussed in Section 13.6.

The next clinical application that predominantly relies on physical aspects of light–tissue interaction is in ophthalmology, which will be described in the next section.

11.4 Ophthalmology

Immediately with the discovery of laser, incandescent and gas-discharge lamp retinal photocoagulation was substituted by laser photocoagulation. It still took nearly 20 years for the introduction of photodisruption for the non-invasive treatment of secondary cataracts. Other laser applications are diabetic retinopathy and glaucoma, and recently laser reshaping of the cornea for vision correction purposes. Some of the laser wavelengths used over time for various opthalmologic coagulation applications are the blue-green and green argon laser lines, frequency doubled Nd:YAG, yellow and red lines of the krypton laser, tunable dye lasers in the 570–630 nm range, diode lasers in the 780–850 nm spectrum, and the Nd:YAG laser. The excimer laser and CO_2 laser are being used for corneal etching, refractive keratoplasty, photorefractive radial keratectomy, and radial keratotomy.

In refractive keratoplasty, radial incisions are made in the cornea to reduce the curvature to rectify myopia. The laser has an advantage over the widely used surgical approach, since the depth of incision can be controlled, hence providing increased refractive accuracy. In photorefractive radial keratectomy, radial keratotomy, and laser-assisted in situ keratomileusis (LASIK), a fresnel-like lens is carved in the cornea to adjust the focal length of the eye. In LASIK, the lens behind cornea is changed by nibbling with an excimer laser. Basically, a computer-programmed refractive pattern is written on the surface of the lens of the eye, taking the place of a contact lens. The lens pattern is carved by a controllable rotating iris diaphragm positioned in the path of an excimer laser beam. By the application of the ablation pattern myopia, hyperopia, and astigmatism can all be treated. The ablation mechanism for the cornea is similar to that described in Section 11.2.1.1 for laser ablation of soft tissue.

In the treatment of diabetic retinopathy, the argon laser is still the preferred and virtually single treatment option to weld the retina to the choroid. The laser treatment of diabetic retinopathy relies on photocoagulation of pervasive neovascularization. This process is similar to the port-wine stain treatment, which will be described in the clinical applications in Section 13.3.

In the removal of donor tissue from eye bank eyes the excimer laser is used for perforation, with minimal trauma, and better transplant success.

Also, the removal of cataract by laser therapy is a major improvement over surgical dissection.

Additional examples of the use of laser in ophthalmology are laser peripheral iridotomy and argon laser trabeculoplasty.

Peripheral iridotomy refers to the clearing of an obstruction that prevents the flow of aqueous fluid to the drainage area where the iris meets the cornea. Argon laser trabeculoplasty is another procedure in the treatment of glaucoma to help reduce pressure. Generally in the laser treatment of glaucoma a fistula is created between the episcleral space and the anterior chamber of the eye, thus reducing the interocular pressure.

Ophthalmologic diagnostic laser applications include the scanning laser ophthalmoscope, invented by Robert Webb in 1979, the laser interferometer in optical coherence tomography, and the laser flare and cell meter, next to the laser doppler velocimeter. The latest diagnostic application in ophthalmology is optical coherence tomography, which will be discussed along with other diagnostic applications in Chapter 14.

One particular use of light is the ability to exert a force resulting from a sharp change in the momentum of the delivered photons. This application is described in the next section.

11.5 Optical Tweezers

Using laser wavelengths in the optical window roughly ranging from 780 to 960 nm generally allows for deep penetration of light into the biological medium with low loss due to absorption and relatively low scattering depending on the biological medium.

As the light encounters a transition to a medium with a different index of refraction, the light as an energy carrier will apply radiation pressure P as described briefly in Section 5.3.4.5 in Chapter 5. The pressure results from the changing momentum p when scattering or refracting from an interface, or upon absorption. Especially in the event of absorption it is clear that there will be a change in momentum, since at the time of absorption all photon momentum will dissipate.

Pressure is the force per unit area and the force F is equal to the change in momentum per unit time as defined by the following equation:

$$F = \frac{\partial p}{\partial t} \tag{11.26}$$

The momentum resulting from the electric field E of a single photon is expressed as

$$p = \frac{E}{c} \tag{11.27}$$

where c is the speed of light.

As previously discussed the energy flow of electromagnetic radiation through a unit surface area is described by the Poynting vector, which has the units of power per unit area as shown in the following equation:

$$\mathbf{S} = \frac{1}{\mu_0}\mathbf{E} \times \mathbf{B} \tag{11.28}$$

where μ_0 is the dielectric permeability of vacuum, and $\mathbf{E}$ and $\mathbf{B}$ are the electric field vector and the magnetic field vector of the electromagnetic radiation, respectively.

The magnitude of the Poynting vector can be rewritten due to the fact that $E = cB$ as expressed by the following equation:

$$S = \frac{EB}{\mu_0} \tag{11.29}$$

The intensity of the sinusoidal electromagnetic fields is the average magnitude of the Pointing vector, which relies on the fact that $\langle \sin \theta \rangle = 1/2$, yielding the following equation:

$$I = S_{\text{average}} = \frac{E_{\max} B_{\max}}{2\mu_0} = \frac{E_{\max}^2}{2\mu_0 c} = \frac{c B_{\max}^2}{2\mu_0} \tag{11.30}$$

Subsequently the energy density for the electric and magnetic fields is given by the next equation in terms of the electric field intensity u_{E} as

$$u_{\mathrm{E}} = \frac{1}{2} \varepsilon_0 E^2 \tag{11.31}$$

where ε_0 is the dielectric permittivity of vacuum.

The energy density of the electromagnetic radiation written in terms of the magnetic field intensity u_{B} is given by

$$u_{\mathrm{B}} = \frac{B^2}{2\mu_0} \tag{11.32}$$

Combining Equations (11.31) and (11.32) yields the total energy density of the combined effect of the electric and magnetic fields as

$$u = u_{\mathrm{E}} + u_{\mathrm{B}} = \varepsilon_0 E^2 = \frac{B^2}{\mu_0} \tag{11.33}$$

Rewriting Equation (11.33) for the average energy density is presented as

$$u_{\text{average}} = \frac{1}{2} \varepsilon_0 E_{\max}^2 = \frac{B_{\max}^2}{2\mu_0} \tag{11.34}$$

Now the energy density can be reapplied to the intensity in Equation (11.30) to yield

$$I = S_{\text{average}} = cu_{\text{average}} \tag{11.35}$$

With respect to the interaction at the event of absorption, the momentum of the incident electromagnetic radiation is the total energy U of the electromagnetic wave absorbed on a surface area A per unit time interval Δt. The total energy of the photon is the average energy density multiplied by the volume that contains the photons, as provided by the next equation , taking into account the integral over all angles for the Poynting vector

$$U = Vu_{\text{average}} \tag{11.36}$$

In case there is total absorption this yields the following expression for the momentum:

$$p = \frac{U}{c} \tag{11.37}$$

Using the fact that the intensity is related to the average energy, it gives the expression for the radiation pressure as

$$P = \frac{I}{c} \tag{11.38}$$

In case there is no absorption but perfect reflection the change in momentum will be double that of absorption and the pressure on the reflecting surface becomes as given by the following equation:

$$P = \frac{2I}{c} \tag{11.39}$$

or expressed in terms of the total power of the light source P (collimated beam) as

$$P = \frac{2I}{cA} \tag{11.40}$$

Substitution of Equation (11.27) into Equation (11.26) yields the following expression for the total force under total absorption of a photon:

$$F_{\text{tot,abs}} = \frac{1}{c}\frac{\partial E}{\partial t} \tag{11.41}$$

And, respectively, for total reflection the change will be twice that of absorption, since the direction is totally reversed. The force resulting from a reflection event is given by

$$F_{\text{tot,scatt}} = \frac{2}{c}\frac{\partial E}{\partial t} \tag{11.42}$$

Even though Equation (11.42) shows the force resulting from total perpendicular reflection, during diffraction the directional change is only minor and as such the resulting force will be constructed from the perpendicular and parallel components of the diverted ray with respect to the original direction.

Even though the force generated by a low-power and moderately powerful focused laser beam is only of the order of pN, it can be seen that this force and resulting pressure can be applied to manipulate very small objects such as single-cell organisms, bacteria, and DNA strands. It is also feasible to visualize that a focused laser beam may be able to dissociate molecules and cut DNA during meiosis. Note that high-power laser may produce more pressure and force, but it will also generate enough heat to destroy the biological entity.

To simplify the theoretical derivation of the forces exerted by light under tight focusing on cellular components we assume a spherical geometry. During the interaction of light with microscopic objects we can assume a laser beam focused on a spherical object of micrometer or nanometer size. B.A. Brown and P.R. Brown gave a detailed synopsis of the optical tweezers principle in *American Laboratory*, Vol. 33(22), 2001.

Two situations can be distinguished here. One situation is the case where the particle size is greater than the wavelength; the other entails the condition that the particle size is smaller than the wavelength. In case the particle diameter is smaller than the wavelength the particles fall in the Rayleigh regime, and a model representing the formation of a dipole in the particle under the influence of radiation force is introduced. When the particle has a diameter greater than the wavelength, the particle falls in the Mie regime and the forces can be illustrated by geometrical optics. To envision the forces acting on a biological particle under Mie conditions, ray optics will provide an insight.

11.5.1 Rayleigh Regime Particle Force

In the case of Rayleigh regime particles suspended in a medium with index of refraction n_{medium} the scattering force is expressed in terms of the scattering cross-section σ_{scatt} of the particle with radius r as shown in the following equation:

$$F_{\text{scatt}} = n_{\text{medium}} \frac{S_{\text{average}}\sigma_{\text{scatt}}}{c} \tag{11.43}$$

with the scattering cross-section defined as

$$\sigma = \frac{8}{3}k^4 r^6 \left(\frac{\left(n_{\text{particle}}/n_{\text{medium}}\right)^2 - 1}{\left(n_{\text{particle}}/n_{\text{medium}}\right)^2 + 1} \right)^2 \tag{11.44}$$

where n_{particle} is the index of refraction of the biological particle (e.g., protein strand of DNA) and k is the wave number expressed as

$$k = \frac{2\pi n_{\text{particle}}}{\lambda} \tag{11.45}$$

Under the influence of the force from Equation (11.43) the particle is polarized and hence experiences a force in the direction of the gradient of the electric field strength. The ultimate force is called the gradient force and is given by

$$F_{\text{grad}} = \frac{\alpha}{2} \nabla \langle E^2 \rangle \tag{11.46}$$

where

$$\alpha = n_{\text{medium}}^2 r^3 \left(\frac{\left(n_{\text{particle}}/n_{\text{medium}}\right)^2 - 1}{\left(n_{\text{particle}}/n_{\text{medium}}\right)^2 + 1} \right) \tag{11.47}$$

From Equation (11.46) it is clear that the direction of force is towards the brighter light or towards the center of the Gaussian profile. For a particle with a high relative dielectric permittivity, the force will maintain the particle in the focal point of the beam. With the particle trapped in the focal point, it can be moved with the movement of the position of the focal point.

11.5.2 Mie Regime Particle Force

When the wavelength is smaller than the particle size, the Mie regime applies and the force scheme can be found from geometrical optics. The ray diagram for a focused beam on a relatively large particle is shown in Figure 11.2. Figure 11.2 illustrates the conditions where a laser beam is diffraction-limited when focused inside the spherical particle. On the surface of the sphere the rays obey the Snell's law of refraction to form a trapezoid of rays inside the sphere in a cross-sectional view. The refraction of light causes a change in the momentum of electromagnetic radiation. When the

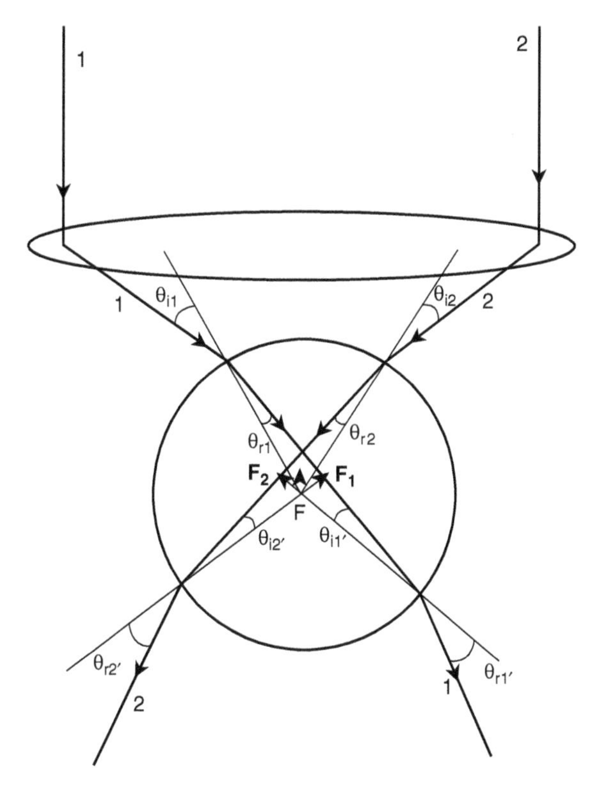

FIGURE 11.2
Ray diagram of a beam focused on a spherical particle above the center, yielding an upward force. (Courtesy Paul Lloyd Larsen, University of Texas Austin, TX.)

particle is illuminated straight from above, the change in momentum in a horizontal plane parallel to the focal plane will cancel out while the change in momentum perpendicular to the focal plane will provide a net force. For the geometrical approximation the angles of the rays interacting with the spherical specimen are taken with respect to either the normal to the surface of the sphere or the radial line from the center of the sphere. In Figure 11.2, the incident angles are represented as $\theta_{i\#}$, where # stands for the respective ray (two rays are used for convenience, each on either side of the central axis through the sphere). The situation is assumed to be symmetric around the central axis, which is plausible since there are an infinite number of rays that can be combined to form mirror pairs. The refracted rays are similarly indicated as $\theta_{r\#}$. The fact that the configuration has symmetry is expressed in the equation indicating the summation of the components. In the use of congruent but mirrored rays the numbering has been dropped in the theoretical description; the ray numbering is mostly used for clarification in the illustrations. The primed parameters are for the rays leaving the sphere.

Owing to conservation of momentum the momentum of the laser beam is transferred to the particle. Using trigonometry it can be shown that the rays will have a resultant as illustrated in Figure 11.2. Note that the rays only represent a cross-sectional view, and the total effect is obtained by integration over the circumference of the sphere.

As the ray refracts at the surface of the sphere it will obey Snell's law of refraction (see Chapter 2). The new direction will be the vector described by the parallel and perpendicular components of the refracted vector with respect to the coordinate system of the incoming photon. The factor of the magnitude of change in the direction perpendicular to the original path is given by the expression

$$\sin(\theta_i - \theta_r) \tag{11.48}$$

While for the factor of the magnitude of the vector component parallel to the original path of the photon, the value is multiplied by the following equation:

$$\cos(\theta_i - \theta_r) \tag{11.49}$$

The force resulting from the directional change of the path of photon is now described in terms of parallel and perpendicular changes with respect to the incident ray. It is clear from Figure 11.2 that both the perpendicular and parallel components need to be transformed to a Cartesian coordinate system that has the axes running along the edges of the page to find the direction of final force, especially since the incoming rays are parallel to the page edge and define the original coordinate system. Using the optical axis of the lens through the center of the "scattering" sphere as the reference the directional changes can be put in perspective. The tangent line with the circle at the point of intersection of either ray forms an isosceles with the optical axis. The two equal angles of the isosceles are $(\pi/2)-\theta_i$, which leaves for the remaining angle, which is the angle of the extended incident ray with the optical axis $\pi-2((\pi/2)-\theta_i)= 2\theta_i$. From this point the angle that the refracted ray makes with the optical axis can be found from the triangle formed by the refracted ray, tangent line, and optical axis. The angle between the refracted ray and the tangent line is $(\pi/2)-\theta_r$; the other angle of the triangle was $(\pi/2)-\theta_i$, leaving for the angle of the refracted ray with the optical axis $\theta_i+\theta_r$. Thus, the change in angle of the incident to the refracted ray at entry into the sphere is from $2\theta_i$ to $\theta_i+\theta_r$ with respect to the optical axis. A similar event takes place when the refracted ray exits out of the sphere again, once more being refracted according to Snell's law of refraction, contributing to the changing impulse as well.

Due to the symmetry of this particular setup the horizontal components cancel out. The resulting vertical force on the spherical particle can be described by Equation (11.50) using the trigonometric analysis of the rays in

Figure 11.2, after combining the vertical components of both rays 1 and 2 at entry and exit of the sphere. After applying the trigonometric relations for the vertical and horizontal components Equation (11.44) yields

$$F_{\text{scatt}} = \frac{n_1 \mathcal{P}}{c}\left\{1 + R\cos 2\theta_{\text{i}} - \frac{T^2[\cos(2\theta_{\text{i}} - 2\theta_{\text{r}}) + R\cos(2\theta_{\text{r}})]}{1 + R^2 + 2R\cos(2\theta_{\text{r}})}\right\} \quad (11.50)$$

where the factors R and T are the Fresnel coefficients of reflection and transmission, respectively, as discussed in Section 3.1, and are defined as follows:

The overall reflection coefficient is given by

$$R = \left(\frac{E_{\text{r}}}{E_{\text{i}}}\right)^2 = \left(\frac{n_2 - n_1}{n_2 + n_1}\right)^2 \quad (11.51)$$

and the overall transmission coefficient is given by

$$T = \left(\frac{E_{\text{t}}}{E_{\text{i}}}\right)^2 = \frac{4n_1^2}{(n_1 + n_2)^2} \quad (11.52)$$

The resulting force on the particle is expressed as the gradient force, with the force acting in the direction of the gradient of the focal point. This means that the force will be towards the focusing lens when the beam is focused above the center of the sphere and away from the focusing lens when focused below the center of the sphere.

Similarly, when the laser beam is focused on either of the side from the center of the sphere, the resultant force will act in the direction of the off-center side as illustrated in Figure 11.3. The expression of a lateral force resulting from lateral off-center focusing is given by the goniometric interpretation of the resultant momentum as illustrated in Figure 11.3. The lateral force is given by the following equation:

$$F_{\text{gradient}} = \frac{n_1 \mathcal{P}}{c}\left\{R\cos 2\theta_{\text{i}} - \frac{T^2[\cos(2\theta_{\text{i}} - 2\theta_{\text{r}}) + R\cos(2\theta_{\text{r}})]}{1 + R^2 + 2R\cos(2\theta_{\text{r}})}\right\} \quad (11.53)$$

11.5.3 Size Region between the Rayleigh and Mie Regime

For laser interaction with particle size in the order of the wavelength the electric field can still be considered relatively constant over the diffraction-limited spot

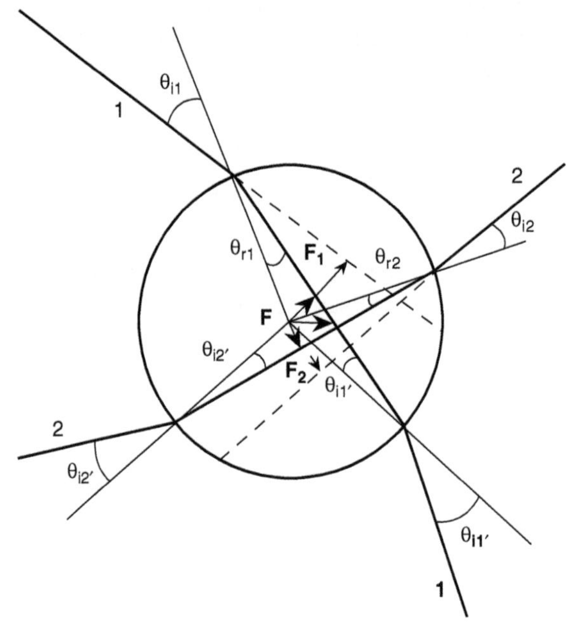

FIGURE 11.3
Ray diagram for a laser beam focused on a spherical particle with off-center focusing, yielding a lateral force. (Courtesy Paul Lloyd Larsen, University of Texas Austin, Austin, TX.)

size. This means that the dipole approximation can be used, with a slight correction applied. The dipole energy of the particle in the electromagnetic field of the laser focus is equal to the laser energy density integrated over the volume of the spherical particle, as expressed by

$$W = -\alpha \int I \,\mathrm{d}V \tag{11.54}$$

Both the electric and magnetic fields can be written as a three-dimensional Gaussian beam with a waist that envelops the sphere. For convenience the frame of reference is located in the center of the particle. The use of cylindrical coordinates simplifies the theoretical description due to cylindrical symmetry. The Gaussian beam profile in cylindrical coordinates will look like the following expression:

$$I(\rho, z) = I_0 \exp\left(-\frac{\rho^2}{2w^2} - \frac{z^2}{2w^2\varepsilon^2}\right) \tag{11.55}$$

where w is the waist of the beam in the radial direction, ρ the radial distance, and ε the focal point depth of the beam in the axial direction. The dipole polarization forms a potential well, which exerts a force that pulls the particle to the center of the focal point of the beam. The equation for the force is presented without derivation to be formulated as

$$F(r) = \alpha 4\pi I_0 w^2 e^{-(\alpha^2+r^2)/2}[\mathrm{Rr}\cosh(\mathrm{Rr}) - \sinh(\mathrm{Rr})] \tag{11.56}$$

where R is the normalized radius with respect to the radius R of the assumed spherical particle $\mathrm{R}=(R/w)$, and r is the normalized location in the radial direction $\mathrm{r}=(\rho/w)$.

11.5.4 Applications of Optical Tweezers

Some applications of optical tweezers are, for instance, in vitro fertilization, genetic manipulation, and study of the mechanical properties of biological entities such as erythrocytes and DNA. One interesting application is the use of optical tweezers to encourage neuron growth.

Another use of the electromagnetic force is to separate structures. This application is more appropriately referred to as an optical scissor. The optical scissors can cut DNA and, when switched back to tweezers, recombine DNA.

Target heating is a major concern while operating laser tweezers, especially with continuous wave laser irradiation. Adding the required light focusing will only increase the risk for complications. The deposited laser energy under continuous wave irradiation for an optical tweezer can result in a temperature rise of 1.15–1.45°C per 100 mW of laser power. Generally the target is contained in a liquid medium that will provide relatively inefficient heat dissipation. As a result, in common operational mode, the deposited laser power can result in 10 times the temperature rise mentioned before. The use of pulsed laser irradiation in optical tweezers has significantly reduced the heat accumulation, but the peak power is often considerably greater.

11.6 Summary

The therapeutic applications of light described in this chapter can be placed under the photophysical concept due to the particular nature of the light–tissue interaction. The therapeutic use of lasers in cardiology and dentistry has been described and another relatively new concept of optical tweezers introduced.

Further Reading

Bell, G.I., Taylor instability on cylinders and spheres in the small amplitude approximation. *Technical Report* LA-1321, Los Alamos Scientific Laboratory, 1951.

Brown, B.A., and Brown, P.R., Optical Tweezers: Theory and Current Applications, *American Laboratory*, 33(22), 13–20, 2001.

Choy, D.S., Stertzer, S.H., Myler, R.K., Marco, J., and Fournial, G., Human coronary laser recanalization, *Clin. Cardiol.*, 7(7), 377–381, 1984.

Copeland, M., Cram, G.P., Edwards, G., Ernst, D., Gabella, W., and Jansen, E.D., First human surgery with a Free-electron laser, in: *Biomedical Applications of Free Electron Lasers. SPIE Biomedical Optics Symposium*, Edwards, G., Sutherland, R., Eds., SPIE Bellingham, 2000.

Duck, F.A., Thermal properties of tissue, in *Physical Properties of Tissue*, Vol. 335, Duck, F.A., Ed., Academic Press, London, 1990.

Jelínková, H., Nemec, M., Kubecek, V., Šulc, J., Dvorácek, D., Kvasnicka, J., Kokta, M., Miyagi, M., Shi, Y., and Matsuura, Y. Ablation effect comparison of solid-state mid-infrared lasers applied for TMLR, *Proceedings of SPIE Vol.4244: Lasers in Surgery: Advanced Characterization, Therapeutics, and Systems XI*, SPIE, Bellingham, 461–467, 2001.

Kimel, S., Svaasand, L.O., Cao, D., Hammer-Wilson, M.J., and Nelson, J.S., Vascular response to laser photothermolysis as a function of pulse duration, vessel type, and diameter: Implications for port wine stain laser therapy, *Lasers Surg. Med.*, 30, 160–169, 2002.

McGuff, P.E., Bushnell, D., Soroff, H.S., and Deterling, R.A. Jr., Studies of the surgical applications of laser light (light amplification by stimulated emission of radiation), *Surg. Forum*, 14, 143–145, 1963.

Plesset, M.S., On the stability of fluid flows with spherical symmetry, *J. Appl. Phys.*, 25(1), 96, 1954.

Rode, A., Gamaly, E., Luther-Davis, B., Chan, A., Lowe, R., and Hannaford, P., Subpicosecond laser ablation of dental enamel, *J. Appl. Phys.*, 92, 4, 2153–2158, 2002.

Ross, E.V., Ladin, Z., Kreindel, M., and Dierickx, C., Theoretical considerations in laser hair removal, *Dermatol. Clin.*, 17(2), 333–55, 1999.

Venugopalan, V., Nishioka, N.S., and Mikie, B.B., Modeling of thermal injury produced by laser ablation of biological tissue, *Proc. Soc. Photo-Opt. Instrum. Eng.*, 1886, 102–111, 1993.

Science Reference Services, The Library of Congress, http://www.loc.gov/rr/scitech/tracer-bullets/laserstb.html.

Problems

1. What are some of the advantages of laser treatment over conventional (surgical) approaches?
2. Give an example of photomechanical use of light.

3. What are the advantages of using pulsed laser light in laser angioplasty over balloon angioplasty and over continuous wave laser ablation?
4. Explain the mechanism involved in laser lithotripsy and name two applications.
5. Name minimum two criteria that need to be satisfied to provide the basis for effective photoablation.
6. Calculate the force applied on a DNA strand of 50 nm under focused laser irradiation with 1064 nm at 150 mW under a 2 μm spot size.
7. What are the conditions that must be satisfied for a photothermal interaction to be thermally confined?
8. What are the conditions that need to be satisfied for a photomechanical interaction to be stress-confined?
9. Derive Equation (11.38).
10. Consider a sucrose solution in a vial of 0.1 mm thickness. The polarized light of sodium D line near 589 nm is used to find the optical rotation of this solution. The optical activity of sucrose at 589 nm of a solution with a sample path (l) of 0.1 m and ([C]) of 1 g/mm is +66.47°.
    a. Calculate the optical activity of a 10% solution.
    b. Consider that an average person has a blood sugar level of 100 mg/dL or 5.55 mmol/dL. What will the rotation be over a 3 mm thickness (e.g., thickness of an earlobe), assuming the earlobe consists of only blood?
    c. What would the rotation be if only 10% of the tissue of the earlobe is blood?
    d. Show how the measurement can be improved by applying a differential measurement, $(I_1+I_2)/(I_1-I_2)$

12

Therapeutic Applications of Light: Photochemical

The primary photochemical therapeutic applications of the use of light in medicine are photodynamic therapy (PDT) and vascular welding. Other clinical applications that may rely on light as the tool of choice are various soft tissue treatments.

The photochemical interaction in vascular welding will be described first.

12.1 Vascular Welding

To speed up the recovery process, tissue closure can often be performed faster, and can be less traumatic by actually welding the fissure together instead of using sutures, or staples, that may need to be removed later on, or with the use of clips. All of the latter mechanical methods of closure make use of foreign bodies, with chances of infections, and rejection, and additionally scarring. Also the sutures may need to be applied in a confined space, leaving much room for extended surgery time, with added risk factors.

Tissue welding was first used in vascular applications by Jain and Gorich in 1979. Early trials for arterial anastomosis were unsuccessful. But major improvements are still being made. The more attractive applications are in venus welding. Additionally, successful tissue welds are made in microvascular arterial conduits.

Laser-assisted vascular anastomosis was at first studied with neodyminium:yttrium-aluminum-garnet (Nd:YAG) laser and later replaced by CO_2 laser and argon laser. The CO_2 laser was especially popular in microvascular anastomosis due to its shallow absorption and the highly maneuverable devices that had been developed. Later on argon tissue welding replaced CO_2 due to its ability to produce stronger bonds, as well as for Nd:YAG. For microvessel surgery, the THC:YAG (2.15 μm) and Raman-shifted Nd:YAG (1.9 μm) were found to be more effective because of the matching penetration depth to the vessel wall thickness. Other pulsed laser systems were also introduced in this discipline. Often adhesive materials that react well to laser photocoagulation, for instance in the form of collagen, have been introduced to enhance the welding bond strength.

Other laser welding or soldering practices are applied in many fields of surgery for sealing air leaks after lung biopsy, for skin closure in plastic surgery, for fallopian tube repairs, for peripheral nerve welding in the field of neurosurgery, and for sealing corneal incisions after surgical treatment of cataract. These are a few minimally invasive applications of laser.

Tissue welding is still an active area of research, and it is likely that a combination of tissue, laser, and solder/adhesive will be found that successfully joins tissue in a permanent fashion.

In cosmetic surgery the laser is used to remodel soft tissue and has the highest success rate of scar prevention. In cosmetic soft tissue treatment currently alternative light sources are found in the titanium:sapphire laser and the new growing selection of high-power diode lasers. The photochemical effects of laser therapy in cosmetic surgery are discussed next.

12.2 Cosmetic Surgery

Plastic surgery is most often used to describe the surgical interventions that do not have a direct medical indication. In the field of cosmetic surgery, laser has provided several unique solutions to highly desirable reconstructive surgical demands. There is skin resurfacing, using laser coagulation of subcutaneous collagen, and the ever-popular hair removal with laser ablation, which is even more effective and longer lasting than conventional techniques. Tattoo and port-wine stain removal are also part of this category.

Blepharoplasty or eyelid surgery is a procedure to remove fat and excess skin from the upper and lower eyelids. Eyelid surgery can be used to correct drooping upper lids and puffy bags below your eyes. These are specific and limited procedures and do not include the removal of crow's feet or other wrinkles.

The main advantage of laser over traditional stainless steel scalpels and scissors is that laser incisions tend to bleed less and thus the time it takes to perform most operations may be decreased slightly (because there is far more to most operations than simply controlling bleeding). With laser blepharoplasty, eyelid fat can be removed without the need for clamping to prevent bleeding. The first laser to be used in blepharoplasty was the Neodymium: Yttrium-Aluminum-Gamet (Nd: YAG) laser. The next-generation laser was the carbon dioxide (CO_2) laser.

In hair removal the laser light is used to destroy the hair follicles by heating and cavitation from shock waves. Both in the photothermal and photomechanical destruction of hair follicles the melanin in the skin is the target chromophore for the wavelength selection. The photomechanical effect is most dramatically achieved by the use of Q-switched laser irradiation. Additionally, lasers are used to produce a photochemical effect to induce hair removal by the photodynamic production of free radicals and singlet oxygen. The photosensitizer for photochemical hair removal is aminolevulinic acid. Aminolevulinic acid is converted into protoporphyrin IX, which in turn generates singlet oxygen.

In scar revision as well as in wrinkle reduction the use of laser is in the collagen heating and restructuring. Additionally, in skin resurfacing it was initially believed that the laser tissue evaporation was the main property to reach successful skin rejuvenation, but as it turned out the ablation depth was significantly less than the topographical features that needed treatment, even after several passes. Apparently the success rate was improved by using lasers that have a much slower thermal relaxation time during light–tissue interaction, pointing to the conductive heat effects as the most significant contribution. The use of CO_2 lasers has the advantage of short pulse length, which is less than the thermal diffusion time. This fact ensures that the laser energy is delivered over a time frame less than the thermal relaxation time, which is in the order of 20–40 ms. During these dermal ablation procedures the fluence is just barely above threshold for ablation, removing only a thin surface layer. The ablation threshold for the millisecond pulses of the CO_2 laser has been established to be around 5 J/cm^2 for skin. The overall power density of the CO_2 laser is around 10,000−50,000 W/cm^2, with a typical fluence of around 40 J/cm^2.

In the case Er:YAG laser is used for skin resurfacing the interaction mechanism changes considerably. The absorption for the Er:YAG wavelength is roughly 13,000 cm^{-1} compared to an absorption coefficient of 800 cm^{-1} at 10.6 μm. Owing to the higher absorption the interaction is much more volatile under typical fluence of 5–8 J/cm^2 for the Er:YAG laser. Some of this is the effect of the 250 μs pulse shape of the ER:YAG, which comprises a train of 1–2 μs pulses, acting similar to a jackhammer on the skin surface. This effect is enhanced by the fact that the thermal relaxation time for Er:YAG ablation is roughly 1 μs. The overall result is a greater ablation volume for the Er:YAG laser than for the CO_2 laser.

A large portion of wrinkle removal is focused around the restructuring of the collagen underneath the dermis. The heat deposited by the laser will denature the collagen, which involves thickening and shortening of the collagen fibrils. Additionally, the chemical structure of the collagen changes with some optical activity involved. Normal collagen is in a crystalline form with right-turning properties, while the heat-treated collagen becomes a gelatinous substance that has lost most of its birefringence.

In tattoo removal it has been shown that the presence of titanium dye in the tattoo significantly reduces the efficiency of specific diode laser removal procedures. Furthermore, both wavelength and pulse duration have been shown to influence the removal efficacy. Pulse durations greater than 1 ms were generally found to be less effective.

Cosmetic surgery is sometimes controversial; however, there are numerous cases of traffic victims, and war casualties in addition to the often necessary palliative treatments following other medical treatments leading to scarring and disfiguring diseases, that require some form of cosmetic surgery.

Nonetheless, certain cosmetic features may have psychological considerations that require serious medical attention.

Table 12.1 lists various laser applications for cosmetic purposes and the most frequently used laser devices.

Additional photochemical use of laser is found in soft tissue treatment of inflammatory diseases, which is described in the following section.

12.2.1 Inflammatory Disease Lesion Treatment

One of the most common inflammatory skin lesions is acne vulgaris, which affects at least 85% of adolescents and young adults in various forms and severity.

The treatment of acne vulgaris has successfully been attempted through near-infrared laser irradiation at 805 nm in combination with topical indocyanine green application. Other treatments involve the ALA/protoporphyrin IX PDT protocol. Even regular sunlight exposure has been known to improve the skin conditions. Other acne treatments have been with ultraviolet-A irradiation. The treatment of acne vulgaris can be based on either thermotherapeutic interaction or photochemical interaction. The photochemical interaction relies on the gram-positive microaerophilic skin bacterium *P. acnes*. The bacterium *P. acnes* in the skin produces a selection of porphyrins, such as protoporphyrin, uroporphyrin, and coproporphyrin III. All the aforementioned porphyrins absorb light in the ultraviolet and blue spectrum with peaks in the Soret band (400–420 nm) and Q-band (500–700 nm), with selective peak absorption wavelengths in the latter range. The Soret band has by far the highest absorption, several orders of magnitude greater than the Q-band absorption peaks. The porphyrins produce free radicals under absorption of light, which are highly reactive and cause bacterial destruction.

The fluorophore activation in acne makes identification with the Woods' lamp an added convenience.

The efficacy of the light treatment will depend on the type of acne lesion, the amount of radiant energy fluence rate delivered at the site, and whether the delivered light matches the excitation wavelengths. The absorption in the Soret band is especially favorable over the Q-band in efficiency.

One complication in the treatment of acne with ultraviolet light is the fact that for best clinical results the light should be applied at the site of the infections, the sebaceous follicles. To reach the sebaceous follicles the light will need to penetrate the epidermis, which is rather difficult at the ultraviolet wavelengths, or even blue light.

Irradiation of acne sites in the green or red spectrum has shown good results, provided enough fluence is delivered. The required higher fluence and fluence rate at longer wavelengths contribute a risk of thermal damage. In contrast, it has been long known that chemical reactions proceed faster at higher temperatures, and the longer wavelength irradiation seems to rely on

TABLE 12.1

Laser Wavelength for Cosmetic Applications

Condition\Wavelength	Alexandrite 720–860 nm	Argon 488, 514 nm	Copper Vapor 578 nm	Diode 790–830 nm	Dye 360–950 nm	Er:YAG 2.94 μm	Frequency Doubled Nd:YAG 532 nm	Nd:YAG 1064 nm	Ruby 694 nm	CO_2 10.6 μm
Blepharoplasty								** ++		*** +++
Hair removal	** +			*** ++ Dark skin				**&(Q) + Very dark	*** ++ Light skin	
Nonablative skin rejuvenation/skin resurfacing						** ++				** ++
Scar revisions						** ++				** ++
Tattoo removal	*** +++			** ++	** ++	** ++		***(Q) +++		
Wrinkle removal/reduction						** ++				** ++

Note: Frequency: *, low; ***, moderate; ****, high. Efficacy: +, low; +++, moderate; ++++, high. (Q), Q-switched; &(Q), long pulse and Q-switched.

that. Better results have been reported with combination red or even infrared irradiation combined with blue light. Even near-infrared irradiation at the 1450 and 1540 nm wavelengths have been known to produce positive results, predominantly due to the thermal injury process quenching the sebum production.

Warts, in contrast, are hard rough lumps on the skin of the hands and feet, usually caused by infection with certain viruses. Warts in general and specifically plantar verruca, a wart located on the soles of the foot can be treated by CO_2 laser vaporization and flash lamp pulsed dye laser.

One additional phototherapeutic application is in psoriasis, a skin disorder with an inflamed appearance often combined with edematous skin lesions covered with a silvery white scale. Psoriasis can very successfully be treated by repeated ultraviolet-B irradiation. This type of treatment frequently takes place on what appears to be a solarium, also known as tanning bed.

Table 12.2 shows a few laser applications for inflammatory cosmetic treatments.

12.2.2 Pigmented Lesion Treatment

In the treatment of pigmented lesion of the skin the main goal is to remove the discoloration and make the end product a relatively good blend with the natural skin tone. The choice of ablation wavelength is directly based on the color of the pigmented lesion, while the application protocol depends on minimizing peripheral thermal damage. The following pigmented lesions can be distinguished: Café-au-Lait, freckles, lentigos, nevi, melasma, and rosasea. Each condition will be described briefly in the following paragraphs.

Café-au-lait spots on the skin are frequently associated with neurofibromatosis type 1 and can be indicative of a type of cancer. The café-au-lait phenomenon itself is a form of hyperpigmentation combined with raised skin.

Freckles is a form of hyperpigmentation where the skin will not be raised. The severity of the occurrence of freckles can be raised by excessive sun exposure, and has a hereditary component as well; some of this is associated with the development of melanoma.

Lentigos, also known as age or liver spots, are hyperpigmented areas of the skin that has been frequently exposed to sunlight. They are however not related to the liver or liver function as the name sometimes may seem to implement.

Nevi are benign cell-growth regions (neoplasms), mostly composed of melanocytes. Melanocytes are pigment-producing cells in the epidermis.

Melasma is a benign dark skin discoloration of the face resulting from sun exposure, often symmetrical.

Rosacea is a relatively common skin condition of uncertain etiology with the potential for significant facial deformity and ocular impediment. In rosacea four developmental stages can be recognized: (1) facial reddening, (2) erythema sometimes combined with edema with additional ocular symptoms, (3) papules and pustules, and finally (4) rhinophyma. Some of the causes are attributed to excessive sun exposure, alcohol consumption,

TABLE 12.2

Laser Wavelength for Inflammatory Treatments

Condition\Wavelength	Alexandrite 720–860 nm	Argon 488, 514 nm	Copper Vapor 578 nm	Diode 790–830 nm	Dye 360–950 nm	Er:YAG 2.94 μm	Frequency Doubled Nd:YAG 532 nm	Nd:YAG 1064 nm	Ruby 694 nm	CO_2 10.6 μm
Acne vulgaris				***	*			**		
				+++	++			++		
Warts/verruca					***					***
					++					++

Note: Frequency: *, low; ***, moderate; ****, high. Efficacy: +, low; +++, moderate; ++++, high. (Q), Q-switched; &(Q), long pulse and Q-switched.

prolonged exposure to cold weather, stress, consumption of hot beverages, and certain other food items. Generally the occurrence of rosacea is about two to three times more likely in women than in men.

Several of the aforementioned pigmentation disorders may mask the diagnosis of certain types of skin cancer and usually require frequent inspection for changes. Changing shape, outline, and color are often indications of a deviating pathological indication.

In Table 12.3 the combinations for various laser wavelengths pertaining to the treatment of pigment lesions are listed.

The most well-known clinical application of light is probably in oncology for the photochemical dye-assisted necrosis of cancerous tissues. The photodynamic applications for tumor therapy are outlined in the next section. An in-depth discussion of the PDT process was described in Chapter 9.

12.3 Oncology

Examples of the use of laser in oncology are the treatment of locally invasive malignant tumors such as gliomas in the field of neurosurgery, endobronchial tumors in ear-nose, and throat surgery, skin cancer, colorectal cancer, and urinary bladder cancer. The treatment protocol for the urinary bladder is illustrated in Figure 9.2.

There are three major laser types used for clinical ablative applications in oncology: the carbon dioxide laser, the Nd:YAG laser, and the argon laser.

Other laser wavelengths are available to activate photosensitizing drugs in PDT while causing minimal activation of endogenous molecules within the surrounding tissue as described in Section 9.2.1. A brief review of PDT is presented next.

12.3.1 Photodynamic Therapy

Even though the history of this type of treatment dates back to the beginning of the nineteenth century, applications are still limited. The main advancement in PDT is the wavelength selectivity of new dyes, and the knowledge of penetration depth of preferred wavelengths. In this modality the treatment relies on the light activation of a photochemical drug, a photosensitizer, that is selectively absorbed and retained in the tumor. All photosensitizers rely on the same principle; light is absorbed in the photochemical, inducing a chemical reaction where singlet oxygen is produced. Singlet oxygen in high dosage is toxic, and causes cell death. The yield of singlet oxygen in turn is directly proportional to the amount of light fluence incident on the cells containing the dye.

The photosensitizer must however adhere to several criteria: selective uptake in the tumor, activation by wavelengths that can penetrate deep into the tissue, nontoxic behavior, and ability to destroy malignant tissue growths. The effectiveness of the photosensitizer depends largely on the

TABLE 12.3

Laser Wavelength for Pigment Lesion Treatment Applications

Condition\Wavelength	Alexandrite 720–860 nm	Argon 488, 514 nm	Copper Vapor 578 nm	Diode 790–830 nm	Dye 360–950 nm	Er:YAG 2.94 μm	Frequency Doubled Nd:YAG 532 nm	Nd:YAG 1064 nm	Ruby 694 nm	Co_2 10.6 μm
Cafe-au-lait								***(Q) +		
Freckles							***(Q) +		***(Q) +++	
Lentigos						*** +++	*** +++	***(Q) +++	***(Q) +++	*** +
Melasma				*** ++	*** ++			***(Q) +		
Nevi	*** +++							***(Q) +++	*** +++	
Rosacea							*** +++			

Note: Frequency: *, low; ***, moderate ****, high. Efficacy: +, low; +++, moderate; ++++, high. (Q), Q-switched; &(Q), long pulse and Q-switched.

localization of the drug in the tumor, and the light transmittance of the tissue that surrounds the tumor, at the activation wavelength. This puts a lot more weight on provisions about the knowledge of the optical properties of the tissues that encapsulate the tumor, and of the tumor itself. The light needs to be dispensed preferentially from an epithelial surface. Any avoidance of mechanical disturbance of the tumor, with the chance of spreading the neoplastic cells, is preferred. Mechanical interaction may actually result in the spread of tumor cells in the surrounding tissues.

Certain photoporphyrins selectively bind to brain tumors since the blood–brain barrier is altered due to the tumor presence, while under normal conditions the photochemical will not pass the blood–brain barrier. Figure 12.1 illustrates the use of light for the treatment of skin cancer. Figure 12.1a shows the tissue before injection with ALA and subsequent irradiation with LED light. The LED light source is shown in Figure 12.2. An example of an invasive treatment of a large tumor on the neck with fiberoptic laser light delivery is shown in Figure 12.3.

Owing to its light specificity, PDT has been the driving force in light–tissue interaction research.

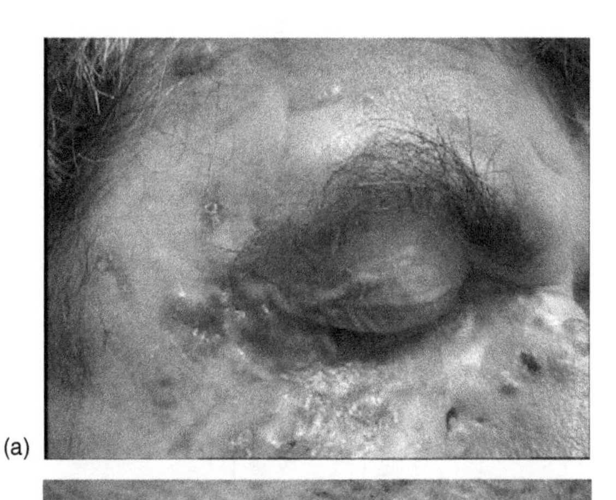

(a)

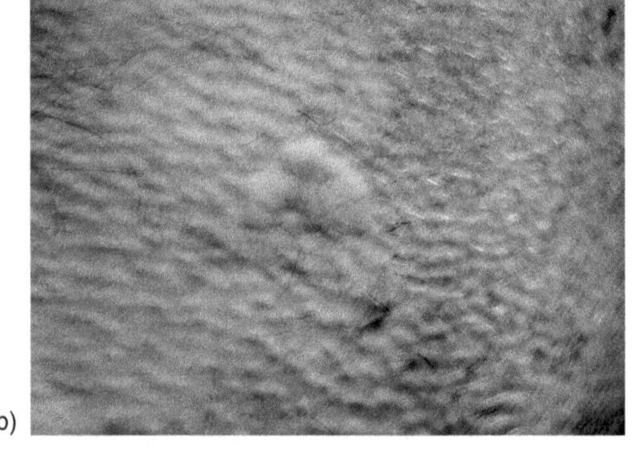

(b)

FIGURE 12.1
Skin cancer treatment in Sofia, Bulgaria. (a) Pathologic skin cancer condition of the face around the eye before treatment, (b) skin appearance 4 months after PDT treatment of basal cell carcinoma. (Courtesy of OPTELLA Ltd., Sofia, Bulgaria.)

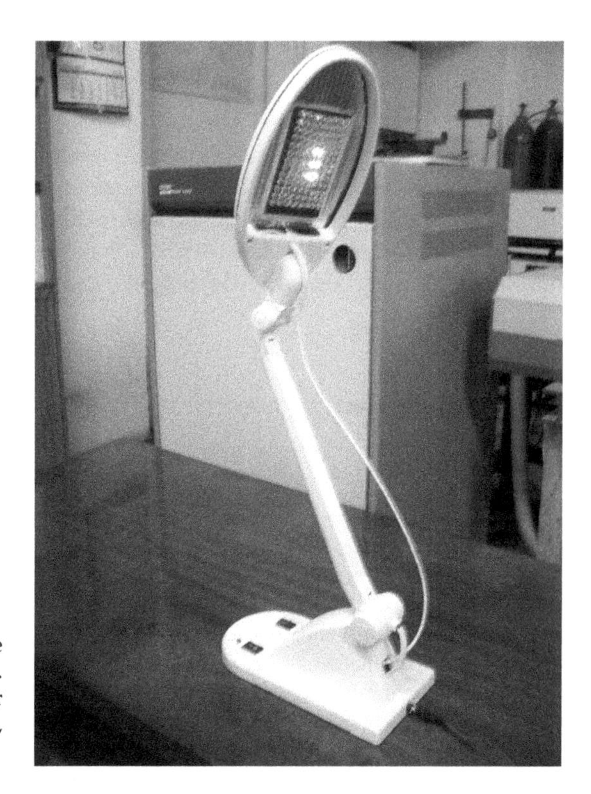

FIGURE 12.2
LED light source used to treat the skin cancer patient in Figure 12.1. (Courtesy of Dr Latchezar Avramov, OPTELLA Ltd., Sofia, Bulgaria.)

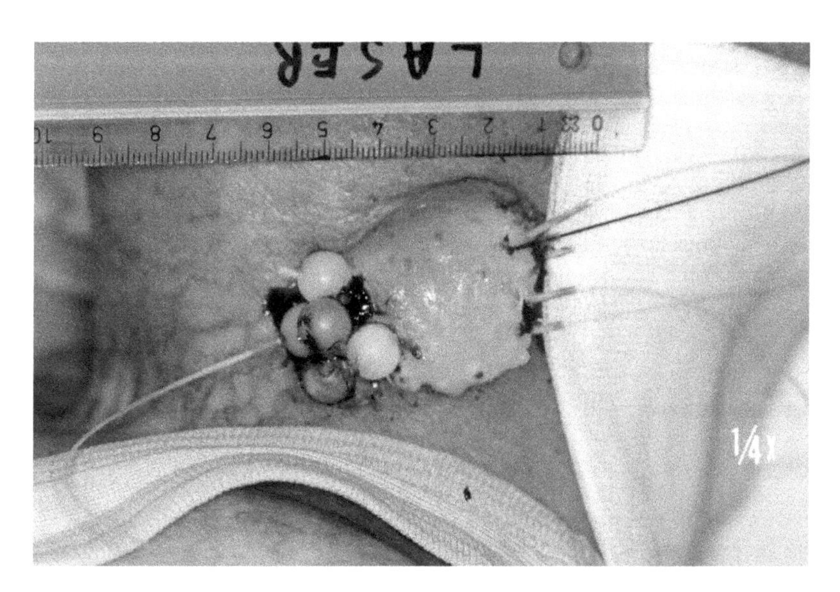

FIGURE 12.3
Invasive fiberoptic laser PDT treatment of a neck tumor. (Courtesy of Dr J.P.A. Marijnissen, Daniel Den Hoed Clinic, Rotterdam, The Netherlands).

12.4 Summary

This chapter provides an overview of what may be considered the use of light in photochemical treatment. The use of light to weld tissue and remove infectious residue can be seen as chemical applications. Other chemically related afflictions in skin treatments for cosmetic use or involving pigmentation are also discussed. Additionally the pinnacle photochemical use in PDT is revisited.

Further Reading

Aksan, A., McGrath, J.J., and Nielubowicz, D.S., Thermal damage prediction for collagenous tissues, Part I: a clinically relevant numerical simulation incorporating heating rate dependent denaturation, *ASME J. Biomed. Eng.*, 127, 85–97, 2005.

Finley, J.L. Arndt, K.A., Noe, J., and Rosen, S., Argon laser–port-wine stain interaction. Immediate effects, *Arch. Dermatol.*, 120, 5, 613–619, 1984.

Fitzpatrick, R.E., Goldman, M.P., and Ruiz-Esparza, J., Treatment of benign cutaneous pigmented lesions with the Candela 510 nm pulsed laser, *Lasers Surg. Med.*, Suppl. 4, 73, 1992.

Jacques, S.L., The role of skin optics in diagnostic and therapeutic uses of lasers, in *Lasers in Dermatology*, Steiner, R., Kaufmann, R., Landthaler, M., and Braun-Falco, O., Eds., Springer, Berlin, 1991.

Jacques, S.L., Laser–tissue interactions: photochemical, photothermal and photomechanical, *Surg. Clin. North Am.*, 72(3), 531–558, 1992.

Mariwalla, K. and Rohrer, T.E., Use of lasers and light-based therapies for treatment of acne vulgaris, *Lasers Surg. Med.*, 37(5), 333–342, 2005.

Mirabella, F., *Internal Reflection Spectroscopy*, Practical Spectroscopy Series, Vol. 15, Marcel Dekker, New York, 1992.

Nanni, C.A. and Alster, T.S., Long-pulsed Alexandrite laser-assisted hair removal at 5, 10, and 20 ms durations, *Lasers Surg. Med.*, 24(5), 332–337, 1999.

Rol, P., Fankhauser, F., Giger, H., Dürr, U., and Kwaniewska, S. Transpupillar laser phototherapy for retinal and choroidal tumors: a rational approach, *Graefe's Archive Clin. Exp. Ophthalmol.*, 238(3), 249–272, 2000.

Problems

1. Give an example of photochemical use of light.
2. Give an example of photothermal use of light.
3. Calculate the relative increase in energy fluence rate at one optical free path length depth, for a titanium:sapphire laser tuned to 790 nm, irradiating a skin tumor with the following optical properties: μ_a=0.067cm^{-1}, μ_s=82.9cm^{-1}, and g=0.895.
4. What are the four primary differences between laser cutting and laser welding?

13

Therapeutic Applications of Light: Photobiological

The photobiological therapeutic applications of light described in this chapter outline the biological implication of the use of light as the treatment of choice for certain diseases and conditions that require some form of medical intervention. The photobiological effects are mostly thermal in nature and often rely on some form of coagulation to achieve a cure.

A selected parade of clinical applications using light is presented next in alphabetical order. The list starts off with laser use in cardiology, followed by selected dermatological applications. It proceeds with light use in the gastrointestinal (GI) tract, general surgery, gynecology, neurosurgery, orthopedics, pulmonary, otolaryngology, podiatry, and finally urology.

13.1 Cardiology and Cardiovascular Surgery

In modern cardiology, lasers have been rigorously studied and successfully used in the areas of angioplasty, arrhythmia ablation, and plaque removal, to name but a few applications.

The following examples of laser applications in cardiology give a total view of the wide range of capabilities of laser treatment. A relatively new treatment option is transmyocardial revascularization, which is described in Chapter 11, along with coronary and peripheral laser angioplasty, due to the pulsed light source nature of the treatment. In the same realm of photophysical pulsed laser ablation fall laser thrombolysis and laser valvulotomy, which are described in Chapter 11 as well. On a more biological note we mention the use of laser arrhythmic node ablation and ventricular and atrial coagulation. The photobiological use of lasers in cardiology starts out with a description of cardiovascular photocoagulation in the next section.

13.1.1 Arrhythmogenic Laser Applications

The occurrence of atrial and ventricular arrhythmias tends to increase with age. Certain hereditary conditions may make a person prone to arrhythmias.

However, acquired heart disease is still the most important factor predisposing a person to arrhythmias.

Cardiovascular diseases cause 12 million deaths in the world each year, according to the third monitoring report of the World Health Organization, 1991–1993. They cause half of all deaths in several developed countries and are one of the main causes of death in many developing countries. In the United States, all cardiovascular diseases combined claim the lives of about 505,000 females annually while all forms of cancer combine to kill about 257,000 females. For men the mortality for cardiovascular disease was 455,152 according to 1995 statistics. Cardiovascular diseases claimed 954,720 lives in 1994 (41.8% of all deaths). More than one sixth of all people killed by cardiovascular diseases are under age 65.

About half of all deaths from heart disease are sudden and unexpected, regardless of the underlying disease. Sudden cardiac deaths (SCDs) are the result of an unresuscitated cardiac arrest, which may be caused by almost all known heart diseases. Most cardiac arrests are due to rapid or chaotic activity of the heart (ventricular tachycardia or fibrillation) and some are due to extreme slowing of the heart. These events are called life-threatening arrhythmias and are responsible for an estimated 350,000 SCDs in the United States each year.

A heart that is scarred or enlarged as a result of diseases (e.g., atherosclerotic heart disease) may eventually develop life-threatening arrhythmias. Arrhythmias are predominantly caused by myocardial infarction, creating scar tissue mixed with normal muscle tissue. Figure 13.1 illustrates the histology of an aneurysm, a diseased heart where collagen has replaced deceased muscle.

Atrial fibrillation (AF) is the most common sustained cardiac arrhythmia encountered in clinical practice and is more often seen in association with

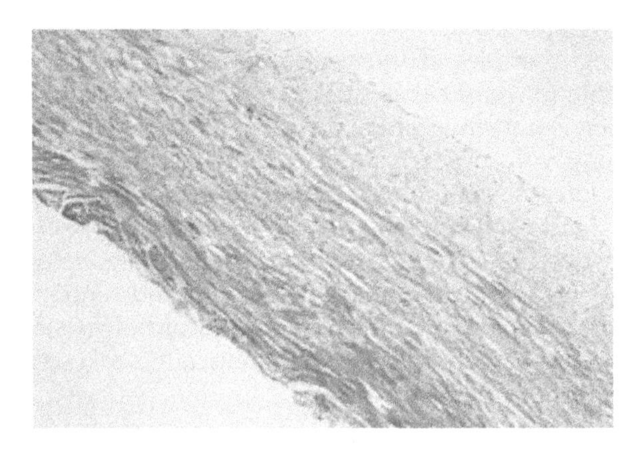

FIGURE 13.1
Representative histology slide of aneurysm (Masson triChrome stain) where collagen has replaced deceased muscle, consisting of surviving myocardial muscle cells (red), collagen (blue), fat (clear), and blood vessels.

hypertension or coronary artery disease in the setting of myocardial infarction. Atrial fibrillation is also the most powerful and treatable cardiac precursor of stroke. The risk for AF doubles each decade after the age of 55. As with mitral stenosis, the presence of AF and mitral annular calcification results in an amplification of risk for stroke. It has been estimated that 2.2 million Americans have intermittent or sustained AF. Occurrence of AF above age 65 is estimated at 5.9% of the population.

Several techniques have evolved for the treatment of life-threatening, drug-resistant recurrent postinfarction ventricular tachycardia. To offer cure instead of control of arrhythmias, alternatives to implantable defibrillators were sought.

Among the competing treatment alternatives for heat coagulation and ablation of biological tissue are the following methods currently at the disposal of clinicians: chemical treatment, cryogenic ablation (cryo), direct current resistive heating, laser photocoagulation, microwave ablation, radio frequency ablation (RF), and ultrasound coagulation.

The use of lasers in cardiology, i.e., the use of photocoagulation to cure patients with ventricular and supraventricular arrhythmias and atrial arrhythmias, is presented as an alternative to other coagulation sources such as radio frequency and cryogenic freeze.

Alternatively, several surgical techniques have evolved over the years for the treatment of life-threatening, drug-resistant recurrent postinfarction ventricular tachycardia. The most widely used procedures are endocardial resection, and implantation of an automatic cardioverter/defibrillator. However, surgical intervention involves an elevated risk with increased morbidity and mortality. Moreover, not all heart disease patients are candidates for surgery. More recently, open and closed chest ablation by a variety of energy sources have opened new opportunities for more efficacious and less invasive treatment of drug-resistant ventricular tachycardia.

13.1.2 Laser Photocoagulation

In laser photocoagulation of the diseased heart muscle the aim is to destroy the surviving muscle cells that are mixed with collagen tissue resulting from oxygen deprivation in case of a heart attack. The surviving muscle cells are still capable of depolarization, and the heart depolarizes from cell to cell; thus, every cardiac muscle cell is involved in the depolarization and contraction process. The electrical transmission between healthy myocardial muscle cells and surviving muscle cells in collagen areas is not very effective and causes a delay in depolarization. When the depolarization front exits out of the aneurysm area, the healthy muscle may be repolarized and ready for the next contraction cycle. This phenomenon results in two or three heartbeats resulting from a single pacemaker pulse from the atrioventricular node. All curative methods are designed to eliminate electrical conduction through scarred tissue, thus preventing additional depolarization waves from stimulating subsequent contractions.

Laser photocoagulation relies on hyperthermia to cause irreversible cell damage, leading to cell death. The temperature profile and temperature history will determine cell death, transforming native molecules into a denatured state by breaking hydrogen bonds and altering ionic interactions. Generally, it is accepted that the denaturization effects start at temperatures ranging from 44 to 47°C. Following cell death enzymes are released that will further lead to tissue necrosis. The thermal damage effect is described in detail in Section 10.2.1.

A method of looking at the time requirement to reach cell death is an Arrhenius relationship describing the correlation to the exposure time at a given temperature in equivalence to the time of tissue exposure to a temperature of 44°C. Looking at the damage curve in Figure 13.2 the temperature–coagulation relationship can be shown to correspond to the following expression:

$$t_{44} = t_T \times A_r^{(T-44)} \tag{13.1}$$

where t_{44} is the equivalent time at 44°C exposure, t_T the treatment time at temperature T, and A_r the reaction activation constant for the denaturization process. The reaction constant A_r is dependent on the temperature itself, and ranges from approximately 2, at temperatures above 44°C, to 6 at 39°C. This first-order approximation will provide an indication of the exposure time

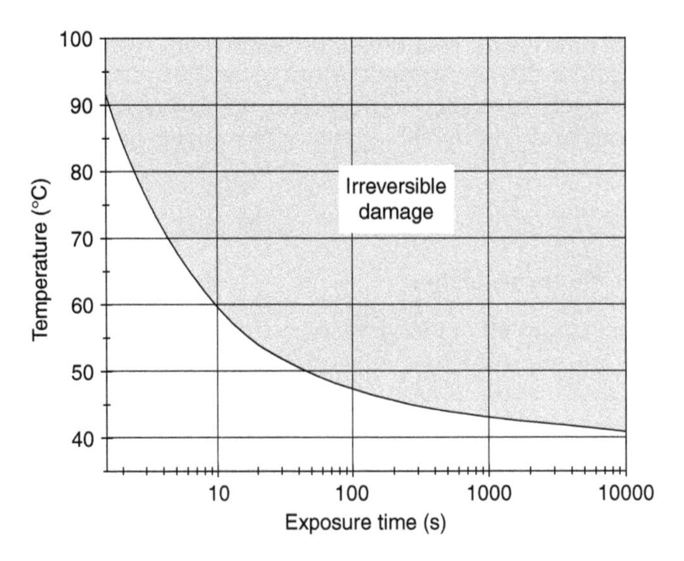

FIGURE 13.2
Graphic representation of the temperature–coagulation relationship of biological tissues as a function of exposure time at any particular temperature. Higher temperatures will result in coagulation much faster than temperatures only several degrees above the body temperature. Note that exposure to 42°C will after prolonged exposure also cause coagulation.

needed to create a lesion under known light absorption conditions, yielding the temperature rise that can be derived from Equation (13.1).

The zero-order steady-state temperature can be approximated from the absorbed light energy as shown in the following equation:

$$T = \int_t \frac{\mu_a \Psi}{\rho c_v} dt + T_0 \tag{13.2}$$

which is the absorbed energy over the duration of exposure converted into temperature rise, and added to the temperature of the biological medium T_0. The tissue density is given by ρ and the specific heat of the tissue is C_v, which can be approximated by the specific heat of water in the case of soft tissue irradiation. One important issue has been ignored in Equation (13.2): The light distribution is not uniform, $\Psi(x, y, z, t)$. The nonuniformity becomes less of a concern when an "infinitely" wide beam is applied, however still resulting in an axial dependence.

Another way of looking at the coagulation process is to consider the specific absorption rate (SAR). In other words, the rate of absorbed light energy (SAR) needs to equal the specific heat of the tissue times the temperature rise, divided by the irradiation time, plus the coagulation rate (rate of mass coagulation per unit time) times the latent heat of coagulation (h_c), as expressed in the following equation:

$$\text{SAR} = \mu_a \Psi = c_v \frac{\Delta T}{\Delta t} + \frac{\Delta m_{\text{coag}}}{\Delta t} h_c \tag{13.3}$$

Additionally laser photocoagulation otherwise operates under normothermic conditions, in contrast to cryogenic ablation. The heart slows down its depolarization under lower temperatures, which makes it imperative to monitor the success of the procedure under normal operating conditions.

To maintain a high tissue temperature and ensure continuous photocoagulation without mechanical side effects, laser photocoagulation is performed under continuous wave (CW) irradiation.

Figure 13.3 shows a schematic representation of the laser delivery to the heart by means of a fiberoptic catheter device.

During laser photocoagulation of the ventricular muscle the boundary conditions prescribe that the energy transfer on the surface of the coagulated volume has to be continuous. This means that the following holds true at the surface:

$$\kappa_{\text{T,air}} \frac{\partial T}{\partial z} = \kappa_{\text{T,tissue}} \frac{\partial T}{\partial z} \tag{13.4}$$

assuming laser delivery in the z-direction, where $K_{T,i}$ is the thermal conductivity of the respective environment.

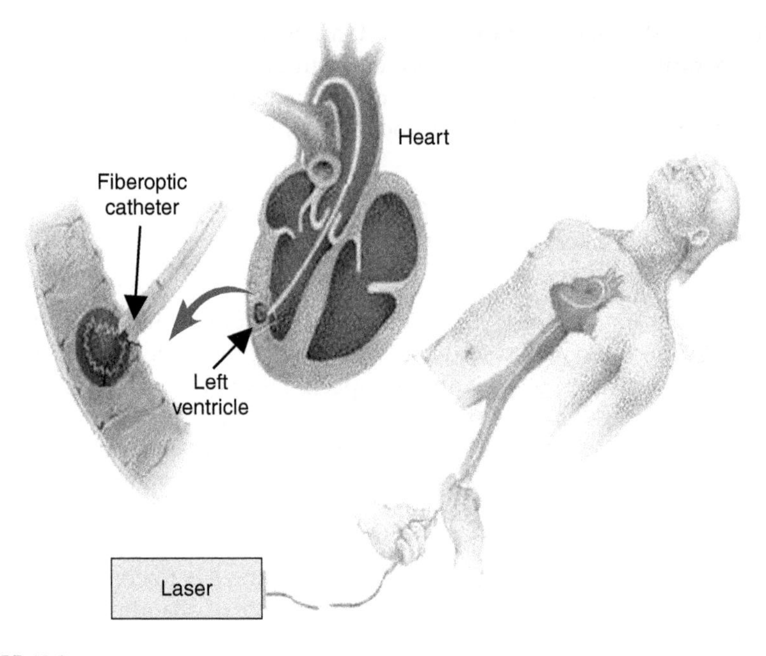

FIGURE 13.3
Schematic representation of the laser procedure to treat drug-resistant postinfarction ventricular tachycardia by means of fiberoptic catheter delivery laser photocoagulation.

Additionally, the heat flux out of the tissue being coagulated needs to equal the heat flux into the healthy surrounding normal tissue. During the coagulation process itself the tissue does not change significantly in thermodynamic sense, yielding equal heat diffusion coefficients for the normal and the coagulated tissue. Mathematically this is expressed as follows:

$$\kappa_{\mathrm{T,coagulated\ tissue}} \frac{\partial T}{\partial r} = \kappa_{\mathrm{T,normal\ tissue}} \frac{\partial T}{\partial r} \tag{13.5}$$

This assumption is however no longer valid when dehydration occurs.

The process of photocoagulation follows the light absorption theory described in Section 3.2 in Chapter 3. To target the exact location of the arrhythmogenic focus the coagulation may need to reach the opposite side of the ventricular wall.

Under any condition the location of the source of the electrical discontinuity caused by a heart attack can be endocardial, epicardial, and even in the septum of the ventricles. Ablation of persistent arrhythmogenic foci in hearts frequently demands the creation of deep transmural lesions, similar to the situation for infarct tissue. To reach these regions for effective coagulation the laser energy may need to penetrate more than 10 mm of "healthy" muscle before the diseased section can finally be ablated. This situation is illustrated in Figure 13.4.

Nonlinear laser photocoagulation may sometimes offer solutions. Collagen strands have been proven to act as light guides, and can effectively route the laser light to the perimeter of the fibrotic area. The treatment options for ventricular arrhythmias were even more appealing when nonlinear optical interaction became evident under high-power irradiance, creating narrow cylindrical lesions up to 18 mm deep. Contact fiber laser delivery is favorable over far-field irradiation due to the high-power density and resulting nonlinear optical effects. An example of a standard lesion under 20 W Nd:YAG laser photocoagulation is shown in Figure 13.5. The matching light distribution and temperature distribution are shown in Figure 13.6a and Figure 13.6b, respectively.

Red and near-infrared laser photocoagulation currently offers distinct advantages over other surgical, electrical, and thermal ablation methods for the treatment of cardiac arrhythmias. Lasing can be performed in the normothermic beating heart, which allows immediate verification of treatment efficacy and instantaneous termination when the arrhythmia has been removed. From light propagation computer simulations, a first-order prediction of the approximate lesion dimensions can be made. The laser-catheter

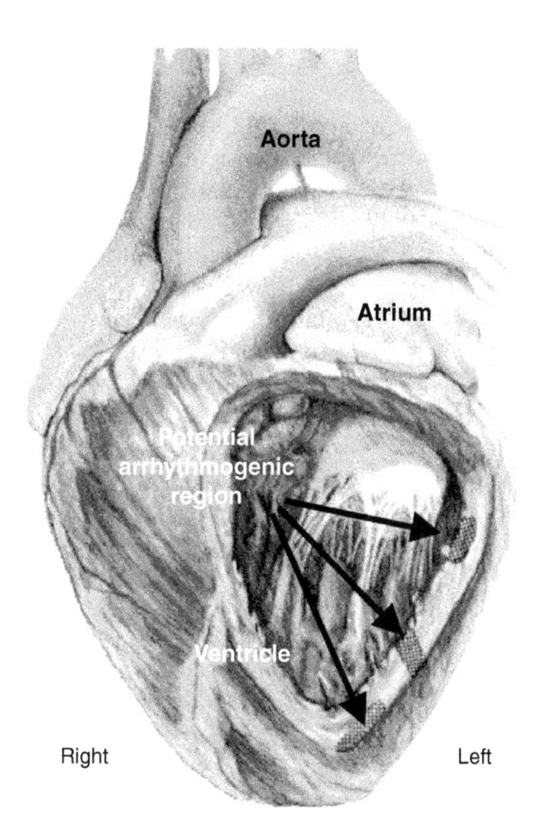

FIGURE 13.4
Illustration of potential arrhythmogenic locations resulting from infarct and inflammation.

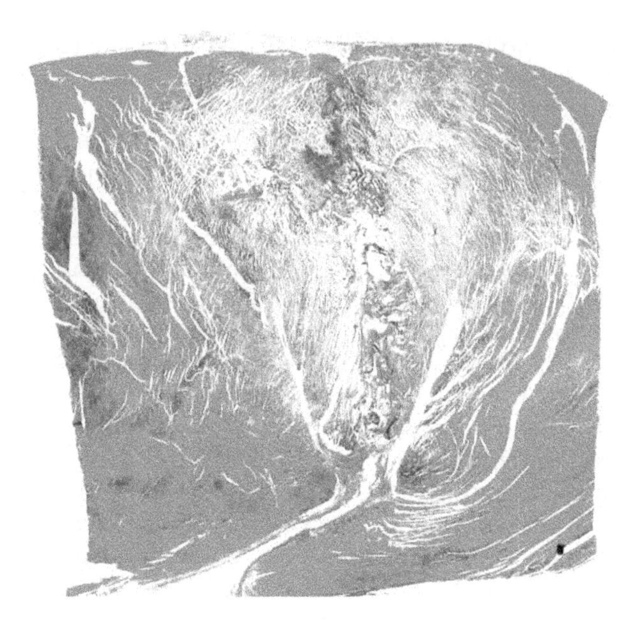

FIGURE 13.5
Lesion created by 20 W of Nd:YAG irradiation at 1064 nm, from a 400 mm fiber at a distance of 0.3 mm with the ventricular wall for an exposure time of 20 s.

technique can be employed to treat heart disease in virtually all patients suffering from arrhythmias, whereas many other surgical procedures require patients to be in otherwise generally good health. This, in turn, will reduce the number of hospitalizations connected to cardiovascular complications in advanced stages of heart and related disorders. With the introduction of a complete transcatheter approach, the need for open chest surgery can be eliminated.

Despite the preliminary advantageous findings in regard to the use of lasers to treat cardiac arrhythmias, it is still highly unpredictable in determining the success rate for each individual patient. The main blame for this rests on the lack of understanding of the light–tissue interaction under high-power density during in vivo application. Sometimes the effort to generate a transmural lesion will result in a much broader lesion than initially desired, with excessive peripheral damage. However, the peripheral damage caused in these cases is still less than the peripheral damage caused by cryo and RF in the respective efforts to create transmural lesions. To generate reproducible results in lesion formation, more research is needed in solving the underlying mechanisms involved during light–tissue interaction. These research efforts also need to address the three-dimensional matrix of tissue nonuniformity in diseased hearts (Figure 13.7).

Figure 13.5 shows a lesion created by 20 W of Nd:YAG irradiation at 1064 nm from a 400 mm fiber at a distance of 0.3 mm with the ventricular wall after 20 s of laser exposure. In contrast, Figure 13.8 shows a lesion created under the same laser parameters, however, with the fiber optic in direct contact with the

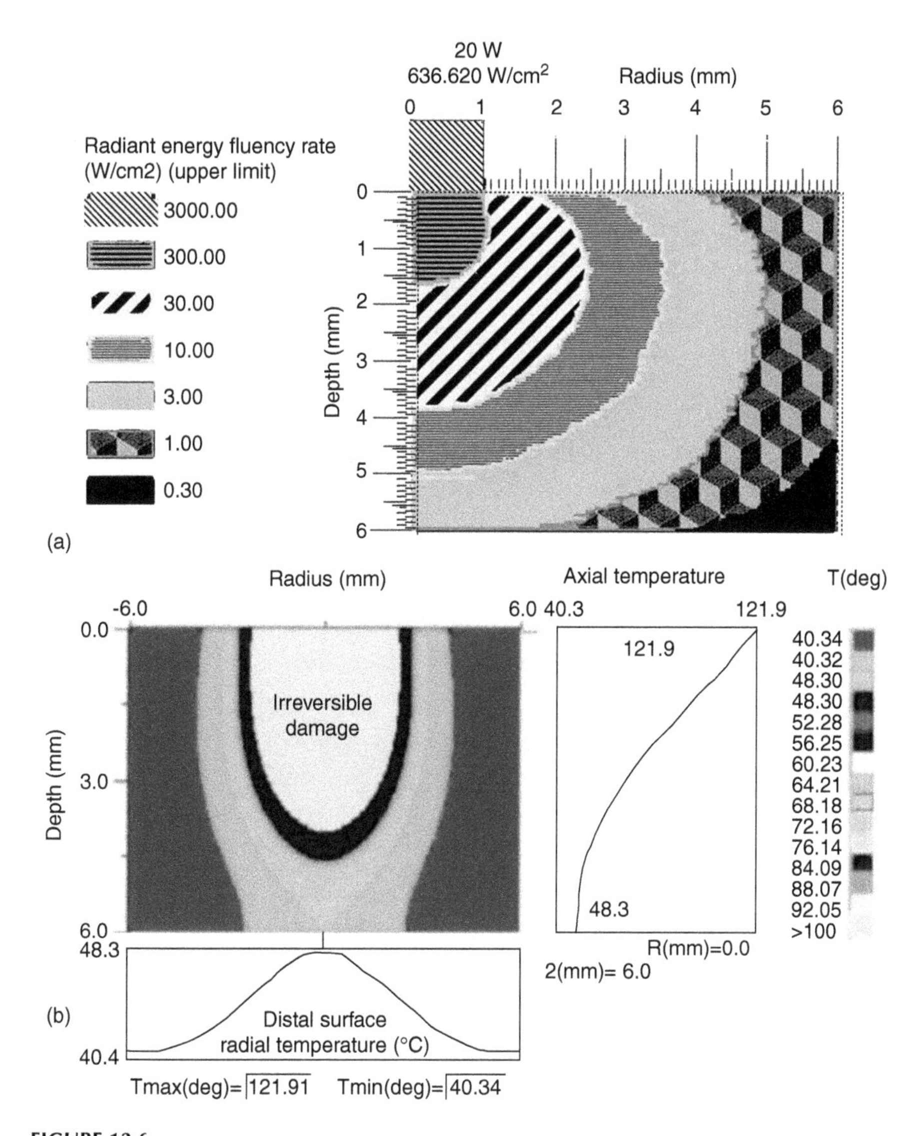

FIGURE 13.6
Computer-generated data pertaining to the lesion shown in Figure 13.5. (a) The matching light distribution for the lesion created in Figure 13.5, (b) the calculated temperature distribution at the end of 20 s of laser exposure for the lesion created in Figure 13.5.

endocardial side of the myocardium, for a duration of 5 s. Clearly, nonlinear optical effects are in effect under such high-power density irradiation. The light and temperature distributions for the fiberoptic contact photocoagulation protocol are shown in Figure 13.9a and Figure 13.9b, respectively. The optical properties for these simulations were found through curve-fit procedures after making some theoretical approximations for the first-order optical properties.

All methods except laser photocoagulation plateau for lesion volume with increasing energy delivery because they rely predominantly on passive heat

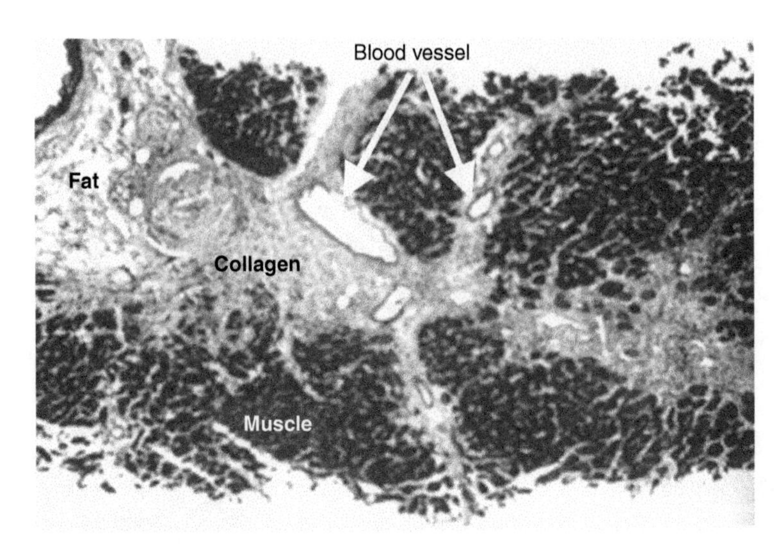

FIGURE 13.7
Chagas disease histology slide, illustrating the complex three-dimensional histological and optical nonuniformity.

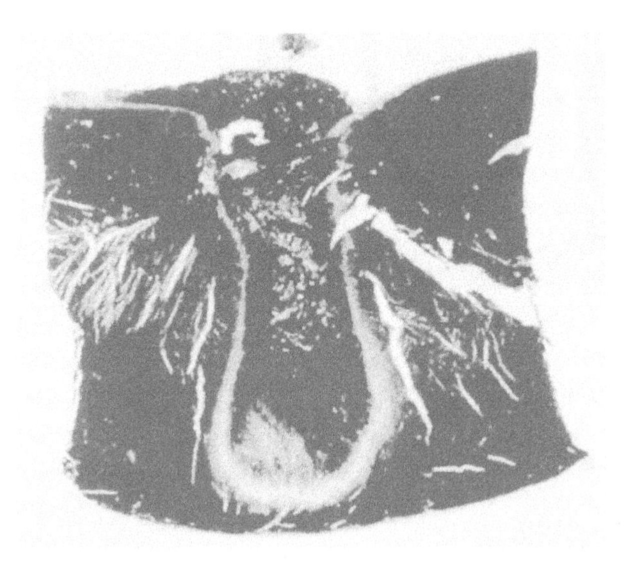

FIGURE 13.8
Representative cylindrical lesions created in myocardial tissue with 20 W of neodymium:YAG laser photocoagulation at 1064 nm under contact 400 μm fiberoptic delivery over 5-s duration. The high-power density during contact fiberoptic laser delivery apparently creates saturation effects, resulting in nonlinear optical behavior of the myocardial muscle tissue.

transfer. Furthermore, the potential of creating cylindrical lesions under high irradiance offers treatment options for deep-seated arrhythmogenic foci, without the risk of excessive peripheral damage to the healthy surrounding tissue.

The ablation conditions are found by entering the light distribution in the bioheat transfer equation, and subsequently calculating the damage

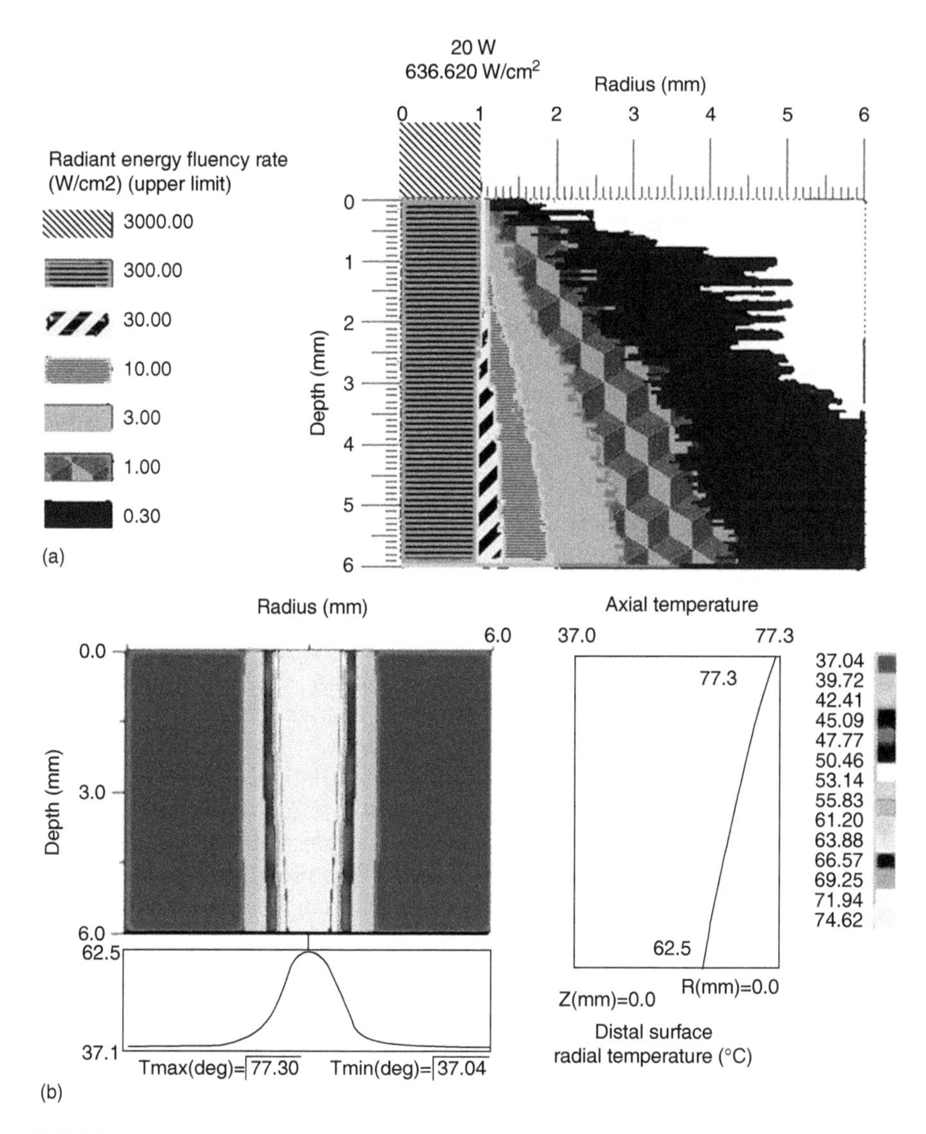

FIGURE 13.9
Computer-generated data pertaining to the lesion shown in Figure 13.8. (a) The matching light distribution and (b) matching temperature distribution. The nonlinear optical effects under this kind of high-power density irradiation can be expressed in the optical properties that are entered in the Monte Carlo simulation to match this type of lesion. The optical properties are a scattering coefficient μ_s that has a reduced value of 125 cm^{-1}, the absorption coefficient that has been left unchanged at 0.4 cm^{-1}, and the scattering anisotropy factor that has been set to extremely forward scattering at $g = 0.997$.

integral. The temperature distribution inside the scattering medium can be determined by calculating the light distribution times the local absorption coefficient, which yields the heat source function q, and entering this into the bioheat transfer equation with blood flow. Next by applying the damage integral the potential total volume of coagulation can be estimated. Since the light distribution is in the regime of low absorption and scattering dominates, the light distribution can only be solved numerically by, for instance, Monte Carlo simulation. As a result the temperature distribution will need to be solved numerically as well, followed by the damage integral for each volume element in the irradiated site.

The choice of wavelength for treatment will largely determine the extent of curative photocoagulation and peripheral damage. It is therefore important to select a laser wavelength that will have low absorption, preferentially combined with a low degree of scattering to avoid lateral damage when attempting to reach a deep location. The most likely wavelengths to generate deep lesions can be seen from the penetration depth measurements presented in Figure 13.10, and seem to be in the 800–840 and 1050–1080 nm range, which suggests the use of diode or Nd:YAG laser for treatment. Additional experimental observations have shown optical anomalies under high-power density irradiation, resulting in saturation of scattering, and showing predominantly net forward light propagation. These observations were derived from histological observations of lesions generated by contact fiberoptic laser light delivery, yielding cylindrical lesions (Figure 13.8).

One notable phenomenon in photocoagulation is the changes in optical properties associated with the coagulation process. The red muscle turns gray as a result of denaturization. This visible change indicates an increase in scattering over the visible spectrum. Even at near-infrared wavelengths the optical properties will change under coagulation. Another marked example of changing optical parameters is the denaturation of albumin (the pure form of protein in egg, egg white), which goes from transparent to white under the influence of coagulation. This phenomenon is of great importance when considering the photocoagulation process, since not only does the temperature change and is heat dissipated by the phase transition

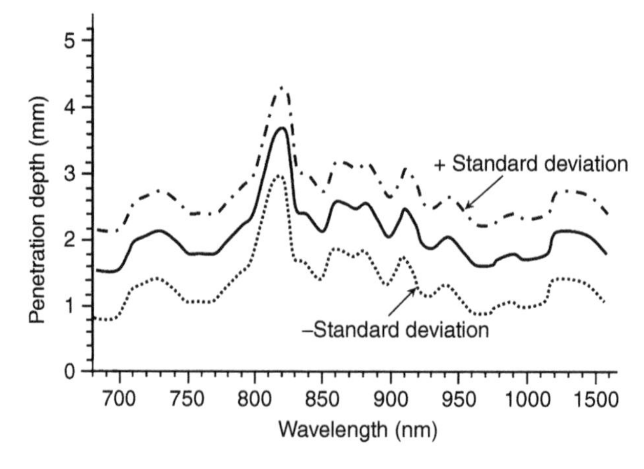

FIGURE 13.10
Penetration depth as a function of wavelength for myocardial tissue.

of normal to denatured state, but also the amount of light energy penetrating to a certain depth will change as photocoagulation starts to occur.

13.1.3 Arrhythmic Node Ablation

The use of laser to slow down the fast heart rate in patients by modification of the sinus rhythm has also shown great potential for a treatment option. Photocoagulation is used to destroy the Sino-Atrial (SA) node cells that have a natural ability to produce a higher depolarization sequence rate. The SA node has the cells organized from fast depolarization to slow depolarization, making it relatively easy to selectively eliminate the occurrence of fast heart rates, thus reducing the potential for arrhythmias.

13.1.4 Atrial Ablation

Several other applications are geared to the use of laser for the treatment of atrial arrhythmias. One of the most accepted methods in atrial arrhythmia treatment is the production of linear coagulation lesions similar to the surgical maze procedure that is the current state of the art in the treatment of atrial rhythmic dysfunction. Figure 13.11 shows the drawn lesion pattern in "white" as seen from the endocardial side of a heart resulting from irradiation by 815-nm diode laser at 20 W from the epicardial side.

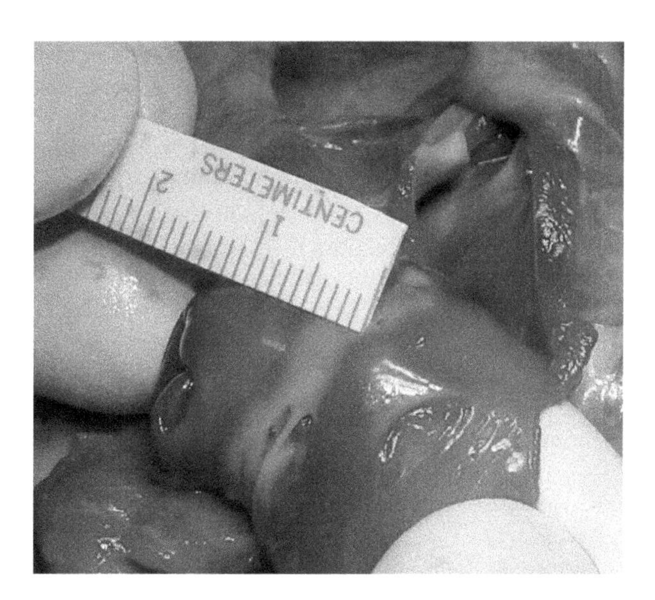

FIGURE 13.11
Linear lesions created in the atrial wall of the heart by dragging a fiber optic over the epicardial wall under continuous laser irradiation from a neodymium:YAG laser operating at 20 W. This image shows the photocoagulation lesion seen from the endocardial side, which results from close-contact epicardial fiberoptic laser photocoagulation.

Other examples of laser photocoagulation are also in GI medicine applied to surgery in the GI tract, which will be described later in this chapter.

13.2 Soft Tissue Treatment

The main biological advantage of laser therapy is the fact that laser ablation minimizes bleeding. Moreover, swelling is controlled and there is usually little if any postoperative discomfort. For maximum comfort though, most of the laser procedures do, however, require a local anesthetic.

The main selection of soft tissue treatments is found in the medical discipline of dermatology. Specific dermatological laser treatments will be discussed in the following section.

13.3 Dermatology

The carbon dioxide laser appears to be the most preferred laser in dermatology. Mostly owing to its short penetration depth, high degree of accuracy can be obtained as a cutting tool in the ablation of superficial tissue layers and for cutting surgery. Owing to its shallow penetration depth the CO_2 laser can also be used to vaporize thin layers of soft tissue. The shallow penetration depth is primarily due to the water absorption in the intra- and extracellular fluid. The laser as a cutting tool provides the opportunity to make bloodless incisions, since blood vessels up to 0.5 mm can immediately be sealed off.

In all dermatological applications the incident laser radiation will have a portion reflected off the skin surface due to the mismatch in index of refraction going from air to skin. The incident radiance needs to be multiplied by the loss factor of the specular reflection, $(1-r_s)$, to obtain the actual light radiance entering the tissue. Furthermore when the surface is wrinkled and uneven or even trabeculated, the angle of incidence of the light will not always be perpendicular, leading to varying reflection percentages as a function of angle of incidence depending on the location.

Examples of laser-assisted treatment protocols for dermatological applications are for instance:

- Cosmetic
  - Blepharoplasty
  - Hair removal
  - Nonablative skin rejuvenation/skin resurfacing
  - Scar revisions
  - Tattoo removal
  - Wrinkle removal/reduction

- Inflammatory
  - Acne vulgaris
  - Warts and verruca

- Pigmented lesion
  - Cafe-au-lait
  - Freckles
  - Lentigos
  - Melasma
  - Nevi
    - Blue nevi
    - Nevus of ota and ito
  - Rosacea

- Vascular lesions
  - Hemangioma
  - Port-wine stain removal
  - Telangiectasia
  - Actinic cheilitis

The frequency-doubled Nd:YAG laser is often used for the treatment of vascular lesions and benign pigmented lesions. Dye lasers, which are tunable in the green and yellow, and argon lasers are also used in erasing tattoos and the removal of port-wine stain; this is all due to the high degree of oxyhemoglobin and hemoglobin absorption. Another alternative laser source in dermatology is the krypton laser operating at the 568 nm wavelength, and also the treatment of pigmented lesions such as melanomas. The treatment of the proliferating vascular cancer, Kaposi's sarcoma, is often trusted to the flashlamp-pumped dye lasers or the argon laser. Other vascular malignancies are treated with the same modality due to the high blood absorption. Scar formation remains a problem with these laser wavelengths. To prevent scar formation currently alternatives are being sought in the titanium:sapphire laser and the new growing selection of high-power diode lasers.

The therapeutic applications of laser light use their thermal effect to quickly coagulate tissues, with blood being one of those tissues. Thus, lasers lend themselves ideally for the treatment of vascular lesions in the discipline of dermatology.

13.3.1 Vascular Lesion Treatment

Actinic cheilitis or actinic cheilosis is a degenerative change of the lower lip primarily in men resulting from sun damage. It manifests itself as chronic ulcers, puffiness, and leukoplakia (white plaques). It usually presents itself after the age of 50 and is the result of excessive exposure to sunlight in the adolescent years. Increased sun exposure, mainly due to occupational conditions, increases the chances of developing actinic cheilosis. There is a 6% chance of developing cancer from this condition. Actinic cheilitis primarily affects people with a light complexion, who are prone to sunburn. Laser ablation can be used to remove and congeal the ulcerous areas.

Superficial vascular lesions such as hemangiomas, port-wine stain, and facial telangiectasias can successfully be removed with the help of laser irradiation.

Hemangiomas are vascular lesions combined with abnormal cell growth that can take on a significant size. One form is infantile hemangioma, which is a benign cell growth at an abnormal rate that usually forms in newborns during the first few months. Girls are known to be three to five times more likely to develop hemangioma than boys. This type of birthmark is more common in Caucasians. Most infantile hemangiomas are subject to rapid initial growth, forming a massive blister-type expulsion. Other hemangiomas will have the neoplasm develop underneath the skin and are noticed by only a slight pink discoloration, but may still be quite large in size. When the hemangioma forms close to bone tissue, bone erosion may result.

Capillary malfor, or port-wine stain, is a birthmark formation made up of a bed of microvascular extensions just below the epidermis. Since these blood vessels are much closer to the skin surface than regular vessels, the skin will show the red blood color. The capillaries can be removed by photocoagulation without interfering with the regular blood supply to the area. In port-wine stain removal it has been shown that when operating with pulse durations less than the thermal diffusion time the temperature rise achieved at the site of the blood vessel is most effective, while reducing the peripheral thermal damage.

When the laser pulse duration is less than 0.5 ms, the irradiation process can be considered thermally confined, since this is less than the thermal relaxation time of the skin tissue. The local temperature rise $T(r, \theta, z)$ reached after exposure to a pulse with duration t can be approximated as follows for a pencil beam hitting a blood vessel with a mean free path length on the order of the beam width:

$$T(r,\theta,z) = \frac{\mu_a \Psi(r,\theta,z)}{\rho c_v} t = \frac{\mu_a \Psi(r,\theta,z)}{\kappa_T/\alpha_T} t = \frac{\mu_a \Psi(r,\theta,z)\alpha_T}{\kappa_T} t \tag{13.6}$$

where $\Psi(r, \theta, z)$ is the local fluence rate, μ_a the absorption coefficient, ρ the tissue density, C_v the specific heat, α_T the thermal diffusivity, and K_T the thermal conductivity. The fluence rate can be found from the diffusion approximation, or from Monte Carlo simulations, depending on the local light–tissue interaction boundary conditions.

Whereas when the light reaches the blood vessel, the absorption in the blood will be much greater than that for the tissue and the thermal relaxation time for the contents of the blood vessel will be greater than that for the tissue and will also be greater than the pulse duration. The blood vessel will not satisfy the thermal confinement conditions and heat will leach into the vessel wall during the laser pulse. The blood vessel with diameter D_b exposed to a laser pulse with duration τ_p will heat up, with diffusion taken into consideration.

The light from a parallel laser beam will see the blood vessel with different thicknesses depending on the radial location with respect to the center of the vessel. The schematic representation of the location of the blood vessel and the geometry involved is shown in Figure 7.11. In the dead center of the vessel the thickness will be D_b; however, towards the edge the thickness will recede to zero corresponding to the angular position of the location on the wall with respect to the center of the vessel, which is represented by

$$D_b' = D_b \cos(\alpha) \tag{13.7}$$

where α is the angle with the normal to the incident wavefront.

Inside the blood vessel absorption and scattering are roughly equal, and in first order, the attenuation will be in accordance with the Beer–Lambert law. The attenuation through the thickness of the blood vessel can now be described as a function of the location with respect to the centerline as follows:

$$\Psi(r) = \Psi_0 (1 - e^{-D_b \cos(\alpha) \mu_{eff}}) \tag{13.8}$$

Integrating Equation (13.8) over the width of the blood vessel, or over α from $-\pi/2$ to $+\pi/2$ yields for the overall thickness $D_b/4$. Applying this finite vessel geometry to a broad beam irradiation, the irradiation per unit length of the blood vessel sees a cross-section of $A=(\pi D_b^2/4)\mu_{eff}$. The effective cross-section results from integrating Equation (13.8) over the vessel volume, taken per unit length, assuming the mean free optical path length is greater than the vessel diameter as shown in the following equation:

$$\text{Absorbed fluence in blood vessel} = \Psi_0 D_b \int_{\alpha=0}^{\pi/2} \cos(\alpha)(1 - e^{-D_b \cos(\alpha)\mu_{eff}})\, d\alpha = \Psi_0 A l \tag{13.9}$$

where l is the irradiated vessel length.

Including conduction in the derivation of the bioheat transport equation, we obtain:

$$\rho c_p \frac{\partial T}{\partial t} = Q + \nabla \cdot (\kappa_T \nabla T) \tag{13.10}$$

where C_p is the specific heat and Q the heat source resulting from absorbed laser light.

After rewriting Equation (13.10) with the conduction term included yields for the temperature profile dT as a function of time τ in the following expression:

$$\begin{aligned} dT(r,\theta,z) &= \frac{\mu_a \Psi(r,\theta,z)\pi D_b^2}{(\kappa_T/\alpha_T)\pi D_b\left(D_b + 4\sqrt{(\alpha_T\tau)}\right)} dt \\ &= \frac{\mu_a \Psi(r,\theta,z)\alpha_T}{\kappa_T + 4\kappa_T D_b\sqrt{(\alpha_T\tau)}} dt \end{aligned} \tag{13.11}$$

where the thermal diffusion length is less than the vessel diameter ($\sqrt{\alpha_T\tau} \ll D_b$).

Substitute $\tau = \tau_p - t$ in Equation (13.11) and use the laser fluence in the tissue defined as $I(r, \theta, z) = \Psi(r, \theta, z)\tau_p$. Now integration of Equation (13.11) over the pulse duration yields an expression for the temperature rise ΔT as shown in the following equation:

$$\begin{aligned} \Delta T(r,\theta,z) &= \int_0^{\tau_p} \frac{\mu_a I(r,\theta,z)\alpha_T}{\tau_p\left[\kappa_T + 4\kappa_T D_b\sqrt{\alpha_T(\tau_p - t)}\right]} dt \\ &= \frac{\mu_a I(r,\theta,z) D_b^2}{16\tau_p\kappa_T}\left\{\ln\left[\frac{D_b^2\left(D_b - 4\sqrt{(\alpha_T\tau_p)}\right)}{\left(D_b^2 - 16\alpha_T\tau_p\right)\left(D_b + 4\sqrt{(\alpha_T\tau_p)}\right)}\right] + \frac{8\sqrt{(\alpha_T\tau_p)}}{D_b}\right\} \end{aligned} \tag{13.12}$$

where it is assumed that the vessel diameter is less than the mean free optical path length. In case the vessel diameter would be too large, the absorption of laser light inside the vessel will be nonuniform and a temperature gradient will be created from the proximal to the distal portion of the vessel.

The ultimate temperature after several laser pulses will be the summation of the effect of each individual pulse.

The percentage of coagulation of blood inside the vessel can be found by applying the damage integral as the natural log of the ratio of concentration of healthy red blood cells at the beginning of the pulse, $[C_0]$, divided by the remaining healthy red blood cells at the end of the laser pulse, $[C_0]-[C_{denatured}]$. This condition is similar to the expression derived in Section 7.3.3, and is shown in the following equation:

$$\Omega(t) = \ln\left(\frac{[C_0]}{[C_0]-[C_{denatured}]}\right) = A\int_0^{\tau_p} \exp\left(\frac{-E}{RT(r,\theta,z)}\right) dt \tag{13.13}$$

The percentage of denatured blood, $[C_{\text{denatured}}]$, at the end of the laser pulse, $\prod_b(r, \theta, z, \tau_p)$, can be found after rewriting Equation (13.9) to give the following expression:

$$\prod\nolimits_b(r,\theta,z,\tau_p) = 1 - \frac{1}{e^{\Omega(\tau_p)}} \qquad (13.14)$$

Telangiectasia is the widening of superficial blood vessels. One form is hereditary hemorrhagic telangiectasia, which affects both males and females in equal proportions and is not race-specific. It is a genetic disorder of the blood vessels that affects about one in every five thousand people.

In the laser treatment of leg veins (varicose veins), it is important to reach deep below the skin surface. The deeper light penetration requirements warrant longer wavelengths and longer pulse duration. The uniform heat production at longer wavelengths results in optimal clearance of most leg telangiectasias, compared with volumetric heating at shorter wavelength, e.g., argon, which results in additional peripheral damage.

Table 13.1 shows selected laser wavelengths for vascular applications and the popularity in medical practices.

There are certain phototherapeutic applications in medicine that can only be applied through either a catheter or an endoscope. The catheter applications fall mainly under the cardiovascular use of lasers in medicine; however, on another microscopic scale is the treatment of fetuses either in or out of the womb. The delicate surgical application options for laser therapy on fetuses are described in the next section.

13.4 Fetal Surgery

Laser photocoagulation is the main prerogative of laser light that is used in closing connecting blood vessels on the surface of the placenta to prevent sharing of multiple babies' blood with each other. Another fetal laser surgery is the removal of embryonic tumors, called sacrococcygeal teratoma (SCT).

One particular example of laser surgery in fetal applications is in the treatment of the twin-to-twin-transfusion syndrome. Twin-to-twin-transfusion syndrome (TTTS) is a serious complication when babies share the same placenta during twin and triplet pregnancies that occurs in 10–15% of twin pregnancies. Normally, the blood supply for each baby is separate even though the placenta may be shared. In TTTS, blood from one baby can mix with the blood of the other baby through small connecting blood vessels in the placenta. In case the blood is not shared equally between the babies, one of the babies may suffer over time from complications of too much circulating blood. In contrast, the other fetus may suffer from problems associated with not having enough blood. For instance, the baby with excessive blood

TABLE 13.1

Laser Wavelength for Vascular Applications

Condition\ Wavelength	Alexandrite 720–860 nm	Argon 488, 514 nm	Copper Vapor 578 nm	Diode 790–830 nm	Dye 360–950 nm	Er:YAG 2.94 μm	Frequency-Doubled–Nd:YAG 532 nm	Nd:YAG 1064 nm	Ruby 694 nm	CO_2 10.6 m
Actinic cheilitis										
Hemangioma					**					
					++					
Port-wine stain removal	**	*	***	**	****					
	++	+	+++	++	++++					
Telangiectasia	***			**	**			***&(Q)	***(Q)	
	+++			++	++			+++	+++	

Note. Frequency: *, low; ***, moderate; ****, high. Efficacy: +, low; +++, moderate; ++++, high. (Q), Q-switched; &(Q), long pulse and Q-switched.

can develop heart failure (cardiomyopathy) and swelling (edema or fetal hydrops), in addition to developing excess amniotic fluid (polyhydramnios); whereas, the baby with the lack of blood can develop poor growth (intrauterine growth restriction), a lack of amniotic fluid (oligohydramnios) connected to particular problems, in addition to problems known to be associated with severe and prolonged blood loss (anemia).

A fetal endoscope called a "fetoscope" is used to apply laser surgery to correct the circulatory imbalance underlying the disproportionate distribution of blood between the siblings. Laser energy is used to ablate the placental anastomoses, thus interrupting fetal blood-flow transfusion and restoring the circulatory balance. In nonselective laser photocoagulation, all anastomosed vessels that cross the intertwin septum are ablated, thereby creating a dichorionic placenta. Otherwise, in a selective approach, the ablation is limited to the participating vessels. Fetal and neonatal survival rates following selective ablation are higher than those following nonselective ablation, with a lower rate of spontaneous abortion.

The standard surgical treatment for embryonic tumors (SCT) is complete excision after birth in case it is not detected prenatally. Sacrococcygeal teratoma is a tumor growth originating in more than one embryonic germ layer. Most of these tumors are benign. However, the chances of malignancy increase with increasing age and require immediate attention when diagnosed. The predominant indication is by ultrasound detection showing abnormal size of the uterus. When SCT is treated prenatally, it will prevent the development of fetal hydrops. Fetal hydrops are extremely vascular tumors that frequently develop as the result of vascular shunting between low-pressure vessels within the tumor. In case of large lesions this type of fetal hydrops may lead to cardiovascular collapse resulting in 100% fetal mortality. The laser photocoagulation offers the ultimate tool to perform the ablation of the vessels leading to the tumor in the womb through fiberoptic delivery.

The next section will briefly describe the clinical applications of light in the discipline of gastroenterology.

13.5 Gastroenterology

The therapeutic use of lasers today in gastroenterology can be classified under two major methods of interaction:

- Thermal vaporization of tumors
- Coagulative debridement

The thermal vaporization of tumors can be done with various near-infrared lasers and is frequently applied in the large intestine and during colorectal cancer treatment.

Coagulative debridement of cancer with the CW Nd:YAG is the second most popular technique. A desirable treatment of various esophageal cancers is applied by laser photocoagulation, for the relief of malignant dysphasia.

Various other GI laser applications are found in the large intestine, the stomach, and the biliary tree. In the majority of laser applications in the GI tract the procedures can and will be performed under endoscopic observation. Section 8.10.2 in Chapter 8 presented a brief description of endoscopy as a surgical tool.

In the intestines the laser is frequently used to palliate tumors by vaporization. In contrast photodynamic therapy (PDT) is gradually finding more recognition here as well.

The stomach is one of the main areas of therapeutic photodynamic applications for gastric cancer. A detailed description of PDT is found in Section 9.2.1 in Chapter 9.

Other laser applications involve the suppress GI hemorrhage and coagulative ablation of peptic ulcers, mainly by Nd:YAG irradiation. The main advantage of laser surgery in the GI tract is the coagulation of vascular malformations. Examples of pathological vascular conditions in the GI tract are angiodysplasia and hereditary hemorrhage telangiectasia. The coagulative laser application is favored based on the high morbidity and mortality rate a result of excessive bleeding following conventional surgical procedures.

In line with the alphabetic listing of clinical laser applications general surgery will be discussed in the next section.

13.6 General Surgery

Two issues of concern in general surgery are cutting and welding.

The primary advantage of laser cutting is the cauterization effect. Q-switched laser ablation of bone tissue has been shown to be able to produce a narrow, 2–6 μm thick thermally altered layer on the edges of 1–5 mm deep laser cuts. With increasing cut depth no significant accumulation of the thermal damage was observed. Laser incisions in the trabecular tissue of bone marrow resulted in a layer of thermal damage ranging from 100 to 200 μm. For clinical purposes these thermal side effects are considered minor and most Q-switched and other short-pulsed laser systems such as the CO_2 laser are generally acceptable for hard tissue surgery.

In general an acceptable combination of thermal relaxation time and pulse duration needs to be found to prevent the thermal damage from interfering with the healing process.

The interaction of femto- and picosecond laser pulses with brain tissue has shown a high degree of accuracy for cutting without thermal or structural changes to adjacent tissue. The ablation efficiency of brain tissue by visible and near-infrared femtosecond pulses from the tunable titanium:sapphire has been shown to exceed that of the picosecond pulses of the Nd:YLF operating at identical pulse energy.

Tissue welding is a developing art and depends on the introduction of welding agents in cases where the tissue itself has no potential for thermal attachment. Collagen has good welding potential, but muscle does not. The heat generated in muscle tissue may cause too much peripheral damage, causing excessive cell death and preventing any chances for adhesion. Tissue welding is discussed in detail in Section 10.2.2.2.

The gynecological relevance of light therapy will be discussed next.

13.7 Gynecology

As in gastroenterology, in gynecological applications the laser light is most often delivered through fiberoptic devices, with sapphire tips attached. The fiber optic is positioned through a laparoscope, or a suction irrigator.

An example of bare fiber laser photoablation in the genital tract is selective photocoagulation of erythema in idiopathic vulvodynia for example, with either flashlamp-excited dye laser, argon laser, or frequency-doubled Nd:YAG laser irradiation. Since all of these lasers can target the chronically inflamed blood vessels at the surface, with minimal scarring and appropriate desensitization, they are the most used laser wavelengths in gynecology.

The CO_2 laser again has extensive surgical applications in cutting and resurfacing, while genital tumor treatment follows the same approach as in gastroenterology.

It was a long time after the introduction of laparoscopic surgical approach for the abdominal cavity in 1901 that laser laparoscopic surgery for female sterilization was introduced in 1967. Other endoscopic laser applications are the treatment of endometriosis and intraperitoneal adhesions. In general the laser has to compete with the laparoscopic use of the electrocautery by means of Bovie devices.

Descending down the alphabetic clinical disciplines we will describe neurosurgical photobiologically relevant treatment applications in the next section.

13.8 Neurosurgery

In neurosurgery the coagulation modality of lasers operating in the blue–green wavelength band is often the main reason for the use of laser irradiation. Surgically altered areas can be cauterized and coagulated by sweeping of a bare laser beam. Additionally, lasers have been used in endoscopic and microscopic surgery to vaporize tumors that may be difficult to reach with conventional surgical tools. For these procedures either a Nd:YAG or CO_2 laser is often used or a combination of both laser systems.

Tumors regression can be accomplished by interstitial laser heating. Interstitial heating can be accomplished by insertion of a fiberoptic probe inside a tumor. In this case the probe will be some kind of diffuser that has been designed to provide homogeneous illumination of the surrounding tissue. Under low-level irradiance and long exposure time slow bulk uniform heating can be accomplished. This type of laser hyperthermia is carefully maintained at temperatures between 42 and 43°C, selectively generating irreversible damage in neoplastic cells only. The duration of the exposure time is directly correlated to the established tissue temperature as described by the irreversible damage curve in Figure 10.2.

Other neurosurgical applications are laser discectomy and laser-assisted cranial drilling and cutting.

Additional work is being done in the laser-assisted endoscopic bypassing of the obstructed third ventricle of the brain, to relieve the pressure leading to hydrocephaly.

The next field of medical practice has probably the longest history of the use of light for both treatment and diagnosis. The clinical applications of light in ophthalmology will be shared next.

13.9 Ophthalmology

Immediately with the discovery of the laser, incandescent and gas-discharge lamp retinal photocoagulation was substituted by laser photocoagulation. It still took nearly 20 years for the introduction of photodisruption for the noninvasive treatment of secondary cataracts. Other laser applications are diabetic retinopathy and glaucoma, and recently laser reshaping of the cornea for vision correction purposes. Some of the laser wavelengths used over time for various ophthalmologic coagulation applications are the blue–green and green argon laser lines, frequency-doubled Nd:YAG, the yellow and red lines of the krypton laser, tunable dye lasers in the 570–630 nm range, diode lasers in the 780–850 nm spectrum, and the Nd:YAG laser. The excimer laser and CO_2 laser are being used for corneal etching, refractive keratoplasty, photorefractive radial keratectomy, and radial keratotomy.

In refractive keratoplasty, radial incisions are made in the cornea to reduce the curvature to rectify myopia. The laser has an advantage over the widely used surgical approach, since the depth of incision can be controlled, hence providing increased refractive accuracy. In photorefractive radial keratectomy, radial keratotomy, and laser-assisted in situ keratomileusis (LASIK), a Fresnel-like lens is carved in the cornea to adjust the focal length of the eye. In LASIK the lens behind the cornea is changed by nibbling with an excimer laser. Basically, a computer-programmed refractive pattern is written on the surface of the lens of the eye, taking the place of a contact lens. The lens pattern is carved by a controllable rotating iris diaphragm positioned in the path of an excimer laser beam. By the application of the ablation pattern myopia, hyperopia and astigmatism can all be treated. The ablation mechanism for the cornea is similar to the laser ablation mechanism of soft tissue as described in Section 11.2.

In the treatment of diabetic retinopathy, argon laser is still the preferred and virtually single treatment option to weld the retina to the choroid. The laser treatment of diabetic retinopathy relies on photocoagulation of pervasive neovascularization. This process is similar to the port-wine stain treatment described in Section 10.2.2.

In the removal of donor tissue from eye bank eyes the excimer laser is used for perforation, with minimal trauma, and better transplant success.

Also, the removal of cataract by laser therapy is a major improvement over surgical dissection.

Additional examples of the use of laser in ophthalmology are laser peripheral iridotomy and argon laser trabeculoplasty.

Peripheral iridotomy refers to the clearing of an obstruction that prevents the flow of aqueous fluid to the drainage area where the iris meets the cornea. Argon laser trabeculoplasty is another procedure in the treatment of glaucoma to help reduce pressure. Generally in the laser treatment of glaucoma a fistula is created between the episcleral space and the anterior chamber of the eye, thus reducing the interocular pressure.

Ophthalmologic diagnostic laser applications include the scanning laser ophthalmoscope, invented by Robert Webb in 1979, the laser interferometer in optical coherence tomography, and the laser flare and cell meter, next to the laser Doppler velocimeter. The latest diagnostic application in ophthalmology is optical coherence tomography, which will be discussed along with other diagnostic applications in Chapter 14.

Additional applications for chest, throat, head and neck will be discussed in the following sections.

13.10 Pulmonology and Otorhinolaryngology

After the initial trials with a neodymium:glass laser on the vocal folds by Geza Jako in the late 1960s, and further CO_2 experiments, various other otolaryngologic laser surgery applications were soon developed. Owing to the anatomical layout treatment in this particular discipline, a large number of laser applications is administered through microscopic instrumentation or endoscopes.

The removal of recurrent respiratory papillomatosis in the treatment of laryngeal diseases has advantages over conventional surgical techniques due to the reduced postoperative edema. The accuracy of laser delivery through micromanipulators also provides lasers an edge over conventional surgery.

Another frequent use of the laser is in the reduction of laryngotracheal stenosis, in combination with the placement of a stent to permanently dilate the larynx.

Other applications further down the air pipe found their way, one of them being the management of tracheal stenosis.

Owing to the coagulative effects of the laser it was also seen as a viable technique to perform uvulopalatoplasty, which reduces the size of the uvula to minimize sleep apnea.

Special pediatric laser microinstrumentation for otolaryngologic surgical interventions has also been developed over the years.

Other microsurgical laser applications have found their niche in the reshaping of the stapes in the inner ear.

One major concern in the laser delivery for otolaryngological procedures is the fact that laser irradiation produces heat while vaporization is desirable. Considering that the patient will require anesthesia, ventilation with oxygen poses a significant risk factor due to the explosion hazard. The instrumentation will need to be protected appropriately, plus the anesthesia will need to be adjusted to minimize the risks for explosion. One solution involves gating the laser application with the respiratory rhythm, thus avoiding the pure oxygen in the field of high temperatures during laser vaporization.

13.11 Otolaryngology, Ear, Nose and Throat (ENT), and Maxillofacial Surgery

Lasers are used in otolaryngology for variety of reasons. One characteristic is the improved precision of the surgeon with the narrowly focused light. As explained in previous sections, the main advantage of laser is the fact that they can reduce bleeding by coagulating blood vessels when the tissue is cut. The other characteristic is the ability to aim narrow passages or manipulate fiber optics in endoscopes to reach areas that are otherwise inaccessible.

The ability of lasers lies in the fact that the special properties of specific wavelengths can selectively treat different problems without peripheral damage, in contrast to thermal devices that do not discriminate in the passive heat transfer.

Both the coagulative and ablative laser properties can be used to their full extent in various microsurgical ENT applications. Otolaryngologists use laser systems for various disorders of the head and neck, in particular management of hearing loss and balance complaints and treatment of laryngeal, sinus, and esophageal disorders.

Some examples of laser applications in otolaryngology and maxillofacial surgery are listed in Table 13.2.

TABLE 13.2

Examples of Laser Applications in Otolaryngology, ENT, and Maxillofacial Surgery

Allergic Rhinitis	Granulomas	TMJ Laser Arthroscopies
Laser arytenoidectomy	Laryngeal/subglottic stenosis	Tongue surgery
Blepharoplasty	Leukoplakia	Tonsillectomy
Endoscopic sinus surgery	Nodules	Turbinectomy
Laser bronchoscopies (rigid and flexible)	Papillomatosis	Laser-assisted uvulopalatoplasty (LAUP)
Excision of carcinomas	Laser stapedectomy	

Specific laser treatments are available for the voice box, throat, mouth, nose, and ear. In particular, nodules or polyps on the larynx and blood vessel defects in the upper airway can effectively be treated with by laser. Additionally, hard tissue laser applications are found in the removal of stapes from the middle ear for the treatment of otosclerosis. Otosclerosis is a fixation of the stapes footplate. Optical drilling is used to precisely remove bone to improve hearing. The optical drilling technique uses the hard tissue removal technique described in Section 11.1.

Other hard tissue applications are, for instance, modification of the temporomandibular joint, or the jaw joint (TMJ). This condition is usually associated with clicking or popping of TMJ, headaches, ear pain, ringing in the ears and neck, and shoulder and upper back pain. The treatment involves reconstructive jaw surgery, electively assisted by laser.

Papillomatosis refers to the endoscopic removal of laryngeal and pharyngeal cancers by CO_2 laser. Frequency-doubled Nd:YAG (KTP) and Ho:YAG lasers are used to manage diseases in the nose, paranasal sinuses, larynx, and trachea. Examples are laryngeal/subglottic stenosis, tonsillectomy, the excision of an arytenoid cartilage (arytenoidectomy), and leukoplakia (potentially precancerous disease of the mouth that is relatively common and involves the formation of white spots on the mucous membranes of the tongue and inside of the mouth).

Other laser applications are in ear disease and cranial base disorders. Lasers are also used to reduce or eradicate vascular lesions and neoplasms that obstruct the airway in infants, children, and adults.

As facial plastic surgeons, otolaryngologists also apply various laser wavelengths to resurface the face, reshape cartilage of the ear and nose, and manage vascular malformations. An example, blepharoplasty, was discussed under cosmetic surgery.

Laser-assisted uvulopalatoplasty (LAUP) is an outpatient treatment to remedy snoring. The procedure originated in France around 1990. The uvula is reduced by laser coagulation, which leaves a larger air passage and reduces the occurrence of snoring. Snoring often results from partially blocked airways.

The following section describes the selective use of lasers for treatment of foot ailments in podiatry.

13.12 Podiatry

In podiatry laser is used as a replacement of the scalpel to remove warts, ingrown nails, callouses, and treat various other foot problems, such as fungus-infected toenails. The main attribute of the laser that is focused on is the sterilization and cauterization of the surrounding areas of the treatment zone.

Various examples of laser therapy on podiatric applications are listed in Table 13.3.

Podiatrists have been looking for ways to treat plantar verruca (warts) and found that the CO_2 laser fits their needs. Other treatments with carbon dioxide laser are onychoplasty and matrixectomy for the treatment of ingrown toenails and laser surgical removal of the bone spurs by a bone-shaving technique. A bone spur is an outcropping of calcium into and around the muscle and attachments of ligaments in the foot. The calcaneal spur grows on the bone, and within the flesh of the foot.

Fungal nail treatment by laser is performed under CW operation at approximately 15 W to cut the nail to the end. Both for fungal infections and for ingrown nails often the complete and permanent removal of the nail plate is required. In this case the matrix cells of the toe are killed by laser heating. Additional removal of hard fibrotic granulation tissue can be performed by laser vaporization. Hematomas can be removed by creating a tiny hole in the nail plate using a pulsed laser, frequently the CO_2 laser, since this laser is now standard operating equipment in the podiatrist's office.

The final section on use of light in urology covers various coagulation options. The photochemical use in urology for PDT was discussed in Chapter 11.

13.13 Urology

Urology has also benefited from the use of laser tissue welding since most urinary tract closures need to be watertight to prevent leakage, with the

TABLE 13.3

Examples of Laser Applications in Podiatry

Excision of Neuromas and cysts	Matrixectomy
Fungus nail treatment	Subungual hematoma
Heel fissures repair	Verrucae plantaris
Heel spur surgery	

associated risks of infections or fistula formation. The groundwork for laser lithotripsy was laid in 1968 by Mulvaney. Clinical experiences with pulsed laser ablation for lithotripsy were first reported by Dretler et al. in 1987. The general concepts of pulsed laser treatment are discussed in Chapter 11.

Benign prostatic hyperplasia (BPH) is a common disorder in men, leading to retention, eventually requiring surgery. The treatment of choice has been transurethral resection; however, laser prostatectomy has shown to be just as effective, with less associated surgical trauma.

Laser lithotripsy is gaining increased acceptance, since various laser systems have proven to fragment biliary calculi; however, the majority of procedures are a last resort only when conventional therapy fails. There is still a long way to go for most medical laser applications, especially when comparing laser lithotripsy with the conventional mechanical lithotripsy (e.g., ultrasound) and basket retrieval that are well established and relatively cheap.

The advantages of laser are the microscopic delivery devices that can operate from a distance just by light delivery to the entire target volume, depending on the choice of laser wavelength in combination with the optical properties of the tissue surrounding those of the target area. The fiberoptic delivery in respect to endoscopic conventional intervention correlates to comparable minimized surgical trauma.

There are numerous practices of laser cancer treatments available in urology; some are more readily available in some countries than in other countries. Examples are the treatment of superficial bladder carcinoma, carcinoma of the penis, urethral condyloma acuminatum, carcinoma of the ureter and renal pelvis, and bladder hemangioma.

Additional applications of lasers in urology fall under the topic of PDT, for example, the treatment of bladder tumors, and under photoablation for the removal of genital tumors.

13.13.1 Lasers in the Treatment of Benign Prostatic Hyperplasia

Abnormal enlargements of the prostate gland, a condition called BPH, occurs in the majority of men above 50 years of age. If they live long enough, almost all men will experience at least minor voiding problems related to prostatic enlargement. Approximately 20–30% of men who live upto age 80 years will undergo a surgical procedure for treatment of BPH.

The prostate is an organ of the male reproductive system. It secretes a thick whitish fluid, part of the semen that transports sperm. Weighing about 20 g (2/3 ounce), the prostate sits just below the front of the rectum and the urethra, a tube through which urine flows from the bladder and out of the body, runs through the prostate.

As the prostate enlarges, it can compress the urethra and obstruct the flow of urine.

The surgical treatment of prostate obstruction dates back to the late nineteenth century. At that time, the only option was an open surgical procedure. Open prostatectomy remains a useful treatment option. In an open procedure, the prostate is approached through an incision in the lower abdomen (suprapubic or retropubic approach) or between the rectum and the scrotum (perineal approach). The thick capsule of the prostate is incised and the obstructing glandular tissue extracted along with the short segment of urethra that runs through the prostate.

Instruments were developed in the early 1900s that allowed the urologist to visualize and remove prostate tissue through a urethral approach. The cystoscope uses a lens system to examine the urethra, prostate, and bladder. In the 1930s the resectoscope was developed. This consisted of a small wire loop that used electric current to remove the obstructing tissue in a piecemeal fashion under direct vision.

The transurethral resection of the prostate (TURP) is the surgical treatment of choice for prostate obstruction and remains the "gold standard" when evaluating new treatments for BPH. The open prostatectomy described earlier is used when the prostate is very large and is not amenable to TURP. If the prostate is very large (over 80–100 g), it takes too long to remove the tissue piecemeal, possibly resulting in excessive blood loss and absorption of the irrigating fluid. This fluid overload can result in "TURP syndrome," producing confusion, nausea, and low blood pressure, as well as seizures, coma, and heart attack.

Table 13.4 lists the types of lasers used in urology. Laser wavelengths that can be delivered through conventional quartz fibers and can therefore be

TABLE 13.4

Types of Lasers used in Endoscopic Urology

Laser	Wavelength (nm)	Depth of Tissue Penetration (mm)	Tissue Effects	Clinical Uses
Nd:YAG	1064	4–6	Coagulation vaporization with contact tip	Prostate ablation Tumor coagulation Urethral strictures
KTP-532	532	0.3–1.0	Vaporization/ coagulation	Urethral strictures
Holmium: YAG	2100	~0.5	Vaporization/ coagulation	Urolithiasis Urothelial tumors Urethral strictures Prostate ablation/ incision/resection/ enucleation
Semiconductor diode	790–980	~10	Coagulation	Interstitial prostate coagulation Tissue welding
Er:YAG	2940	0.001	Vaporization/ ablation	Interstitial prostate coagulation

used in endoscopy include KTP, holmium:YAG, neodymium:YAG, and diode laser. In the mid-infrared, the Er:YAG at the water absorption peak of 3 μm has been used with germanium oxide fibers for ablation of a variety of hard tissues in a flexible endoscope. The types of lasers that have been used to achieve immediate tissue ablation via vaporization are chiefly Nd:YAG used with contact tips or high-power density settings, KTP, holmium:YAG, and Er:YAG.

13.13.1.1 Nd:YAG for Prostate Tissue Ablation

Nd:YAG lasers have been used since 1990 for interstitial and free beam coagulation of the prostate. At the Nd:YAG wavelength the tissue penetration is relatively deep, which makes it suitable for this type of treatment. However, the use of a contact tip or a high-power density technique can turn the laser fiber into a "hot poker."

13.13.1.2 Contact Tip Technology for Prostate Ablation

Contact of the Nd:YAG laser fiber on the tissue causes evaporation of tissue at the surface and coagulation at a deeper plane where less energy has been deposited. Sapphire tips can be fixed to the fiber to withstand the heat generated from the laser interaction at the emission interface. The sapphire material has a very good transmission in the red and near-infrared spectrum. This allows tissue cutting by vaporization for use in bladder neck incision (BNI) and prostatectomy. One complication is that during contact of the sapphire with the tissue, tissue adhesion will occur and this will result in absorption of laser energy in the sapphire tip; as a result the tip becomes very hot.

Despite these advances, contact tip vaporization of larger glands remains a relatively inefficient process.

13.13.1.3 Free Beam KTP and KTP/Nd:YAG for Prostatectomy

The KTP laser has a wavelength of 532 nm and a relatively short penetration depth in tissue ranging from 0.3 to 1.0 mm, which makes it very safe to use. It can be used for free beam vaporization and as a cutting tool for BNI.

The KTP laser has been used as an adjunct to the coagulation properties of Nd:YAG in hybrid laser operating techniques with clinical outcomes similar to TURP in two prospective randomized studies. In both of these studies there was a delay in reaching the maximal clinical improvement presumably because the Nd:YAG coagulation component left tissue slough behind.

13.13.1.4 Holmium:YAG for Prostate Tissue Ablation

The holmium:YAG laser was initially introduced into urology as a tool for stone fragmentation throughout the urinary tract. The holmium wavelength is strongly absorbed by water and has an absorption depth of only 0.4 mm in tissue. It is a pulsed laser with each pulse being in the kW range of power. This creates very precise tissue vaporization and makes it an excellent incisional tool for soft tissue.

The simplest BPH procedure performed using the holmium wavelength is a BNI. This can be performed bilaterally or with a single incision. Holmium-only procedures evolved when it became apparent that this laser wavelength when used alone possessed excellent haemostatic and tissue-cutting properties. Experiments on canine prostates also confirmed the ability of this wavelength to perform acute prostatic tissue vaporization.

Holmium laser ablation of the prostate uses the vaporizing properties of the laser, in a near contact mode with side-firing and end-firing quartz fibers.

Holmium laser resection of prostate (HoLRP) was developed as a way to overcome the inefficiencies of laser ablation while utilizing the precise incisional qualities of the holmium wavelength. The procedure involves the complete resection of the prostatic adenoma, using a piecemeal incisional technique, down to the surgical capsule with the creation of TURP-like cavity.

13.13.1.5 Holmium Laser Enucleation of the Prostate—HoLEP

The most recent step in the evolution of the holmium laser prostatectomy has been the development of a relatively efficient transurethral tissue morcellator. This tool has allowed the enucleation of the median and lateral lobes of the prostate in their entirety by the holmium laser.

The holmium laser fiber acts much like the index finger of the surgeon during an open prostatectomy in shelling out the adenoma. The operative time has been shortened considerably compared with HoLRP. Using this technique, large glands with volumes determined by ultrasound to be in excess of 100 cc can be ablated transurethrally in an efficient and relatively bloodless manner.

13.14 Summary

As the final chapter covering the therapeutic applications of light in medicine, the laser applications that have a major biological impact were discussed. The cardiologic laser coagulation of arrhythmogenic tissue is very important to find a cure for irregular heart rates. Other biological aspects were covered under dermatological applications of laser, as well as various laser treatments in the disciplines of vascular lesions, fetal surgery, gastroenterology, general surgery, gynecology, neurosurgery, ophthalmology, pulmonology, ENT, podiatry, and urology.

Further Reading

Ashinoff, R., Levine, V., Tse, Y., and McClain, S., Removal of pigmented lesions: comparison of the Q-Switched ruby and neodynium: YAG lasers, *Lasers Surg. Med.*, Suppl. 6, 50, 1994.

Aksan, A., McGrath, J.J., and Nielubowicz, D.S., Thermal damage prediction for collagenous tissues, Part I: a clinically relevant numerical simulation incorporating heating rate dependent denaturation, *ASME J. Biomech. Eng.*, 127, 85–97, 2005.

Ashkin, A., Acceleration and trapping of particles by radiation pressure, *Phys. Rev. Lett.*, 24, 156–159, 1970.

Alora, M.B., Stern, R.S., Arndt, K.A., and Dover, J.S., Comparison of the 595 nm long-pulse (1.5 ms) and ultralong-pulse (4 ms) lasers in the treatment of leg veins, *Dermatol. Surg.*, 25, 6, 445–449, 1999.

Barlow, R.J., Walker, N.P.J., and Markey, A.C., Treatment of proliferative haemangiomas with the 585 nm pulsed dye laser, *Br. J. Dermatol.*, 134, 700–704, 1996.

Black, J.F., and Barton, J.K., Time-domain optical and thermal properties of blood undergoing laser photocoagulation, *Proc. SPIE*, 4257, 341–354, 2001.

Brown, B.A., and Brown, P.R., Optical tweezers: theory and current applications, Am. Lab., 33(22), 13–20, 2001.

Carruth, J.A.S., The argon laser in the treatment of vascular naevi, *Br. J. Dermatol.*, 107, 365–368, 1982.

Chan, H.H., Chan, E., Kono, T., Ying, S.Y., and Wai-Sun, H., The use of variable pulse width frequency doubled Nd:YAG 532 nm laser in the treatment of port wine stain in Chinese patients, *Dermatol. Surg.*, 26, 7, 657–661, 2000.

David, L.M., Laser vermilion ablation for actinic cheilitis, *J. Dermatol. Surg. Oncol.*, 11, 605–608, 1985.

Quintero, R.A., Comas, C., Bornick, P.W., Allen, M.H., and Kruger, M., Selective versus non-selective laser photocoagulation of placental vessels in twin-to-twin transfusion syndrome, *Ultrasound Obstet Gynecol.*, 16, 3, 230–236, 2000.

Ross, E.V., and Domankevitz, Y., Laser treatment of leg veins: physical mechanisms and theoretical considerations, *Lasers Surg. Med.*, 36(2), 105–116, 2005.

Senat M.V., Deprest, J., Boulvain, M., Paupe, A., Winer, N., and Ville, Y., Endoscopic laser surgery versus serial amnioreduction for severe twin-to-twin transfusion syndrome, *N. Engl. J. Med.*, 351, 2, 136–144, 2004.

Shafirstein, G., Bäumler, W., Lapidoth, M., Ferguson, S., North, P.E., and Waner, M., A new mathematical approach to the diffusion approximation theory for selective photothermolysis modeling and its implication in laser treatment of port-wine stains, *Lasers Surg. Med.*, 34, 4, 335–347, 2004.

Van Gemert, M.J.C., Welch, A.J., Pickering, J.W., and Tan, O.T., Laser treatment of port wine stains. In: Welch, A.J., Van Gemert, M.J.C., Eds., *Optical-thermal response of laserirradiated tissue*. New York: Plenum Press. 1995.

Ville, Y., Hyett, J.A., Vandenbussche, F.P., and Nicolaides, K.H., Endoscopic laser coagulation of umbilical cord vessels in twin reversed arterial perfusion sequence, *Ultrasound Obstet Gynecol.*, 4, 5, 396–398, 1994.

Problems

1. Give an example of photothermal use of light.
2. List five examples of the use of laser light in cardiovascular applications.
3. Describe the mechanism involved in the laser-assisted removal of port-wine stains.
4. Give three criteria that are crucial in the planning of a photocoagulation procedure.
5. In ophthalmology, one of the alternatives for corrective vision surgery is LASIK surgery. LASIK applies pulsed ultraviolet light from an excimer laser to the cornea. There is no damage to the retina while the retina is directly in the line of propagation of the laser beam. Give reasons.
6. Derive the time-dependent light distribution for a dye laser beam with a 5 mm spot size irradiating a breast with a pulsed delivery protocol of 100 Hz and pulse-width of 5 ms. Ignore edge effects. The laser is operating in continuous mode and is chopped by a slotted disk. The output of the laser is 1.5 W, and the wavelength is set to 560 nm to be used for port-wine stain removal. The optical properties are as follows: $\mu_a = 0.05\ \text{cm}^{-1}$, $\mu_g' = 8.5\ \text{cm}^{-1}$ and $g = 0.973$.
7. You are asked to model the following endoscopic laser therapy in bladder tissue with refractive index n_2=1.7. The radiant energy fluence rate required to excite a particular photosensitizer in the bladder is 0.5 W/cm^2. The laser light is diffusely transmitted out of the tip of a fiber optic (n_1=1.3) in a semi-infinite geometry. The tumor is located 1 mm from the fiber tip. The optical properties of the bladder at the operative wavelength are given as μ_a=3 cm^{-1} and μ_s'=18 cm^{-1}.
   a. Determine r_{21}. Assume the laser light radiance $L(r, s)$ to be given by diffusio-theory and a step function for Fresnel reflection.
   b. Determine the radiant energy fluence rate at the surface of the tumor.
   c. Will the photodynamic therapy be successful?

Section III

Diagnostic Applications of Light

14

Diagnostic Methods Using Light: Photophysical

This chapter deals with certain diagnostic methods using light, which have been developed over the years. A review of technological developments resulting from the increasing demands for resolution and physiological details is given for the selected imaging methodologies. The relevance to the process of gauging patient performance is illustrated as well.

The diagnostic methods described range from fluorescent imaging, fluorescence in microscopy, and various types of optical microscopy to technologically advanced imaging techniques that can look deep inside the tissue. Some examples of optical imaging techniques are optical transillumination tomography and optical coherence tomography (OCT). Various other novel and established diagnostic procedures are presented.

The main optical diagnostic utility using light to probe medical conditions is microscopy, which will be discussed next.

14.1 Optical Microscopy

With the advances in science and technology, microscopes have become indispensable tools for biologists and scientists in related fields. Microscopes have provided not only for the examination of the external morphology of small structures such as flowers and insects, but also for the examination of the smallest unit of life, the cell.

The standard optical microscope was developed by a team of Dutch eyeglass makers, Hans Jansen, Zacharias Jansen, and Hans Lippershey, between 1590 and 1608. This was the culmination of experimentation with lenses since the introduction of first eyeglasses in Florence, Italy between 1282 and 1285 by Alessandro di Spina and was modeled after Roger Bacon's discovery of the magnifying properties of curved glass in 1262. The compound microscope, a microscope using more than one lens, was invented around 1665 reportedly by Robert Hooke (1635–1703) in England

and a similar device was later made by Jan Swammerdam (1637–1680) in the Netherlands around 1675. The Hooke microscope, made by London instrument maker Christopher Cock, reached a magnification of 30 times while the Swammerdam microscope could magnify 150 times. These microscopes were very similar to the microscopes in use today.

The microscope was later perfected by the Dutch lens-maker Antonie van Leeuwenhoek (1632–1723), which is described in a 1716 publication. The microscope was contrived after elementary imaging instruments constructed with pinholes and water droplets, acting as lenses, which were used at the time for the equivalent to modern-day medical pathological examination.

The compound microscope underwent further improvements primarily in England and Italy in the seventeenth and the eighteenth centuries. It was not until 1733 when an amateur optician, Chester Moor Hall, discovered that chromatic aberration could be minimized when lenses made with different refractive materials were combined to form a complex lens. This was primarily due to the discovery of a new lead-containing flint glass, combined with the tried and true crown glass.

The concept of magnification basically started with the simple microscope. The basic microscope is presently known as a magnifying lens. The basic microscope produces an image of an object focused on with the magnifying glass. The magnifying glass is a bi-convex lens, outward curved on either side of the glass medium. Bi-convex lenses are thus thicker at the center than at the periphery. Light that is reflected from the object (specimen) enters the lens and is refracted. The refracted light is now focused by the lens to produce an image, either virtual or real. When the object is closer to the lens than a certain distance from the glass it produces a virtual image that can be observed by the retina of the eye to appear on the same side of the lens as where the object is located. When the object is placed at greater distances it will eventually appear projected on a screen on the opposite side of the lens with respect to the object. The fact that the image cannot be projected on a screen defines it as virtual, whereas when the image can be shown on a screen it is called real. The virtual image seen by the eye is most well-defined when it appears to be in the near point of the eye. Such an image is magnified because the object is perceived as if it were at a greater distance than it really is and the rays of light enter the eye at more oblique angle than seen without the lens. The near point of the eye is the closest distance between an object and the lens of the eye, which can still be seen as a sharp image. The near point is at approximately 250 mm from the cornea.

To understand the function of the lenses of the microscope, it is important to understand some of the basic principles of lens action in image formation. When an object is placed in front of a convex lens at a relatively far away distance, it will be brought to focus at a fixed point behind that lens known as the focal point. The distance from the center of the lens to the focal plane is known as the focal distance f. The image of the object appears inverted at the focal point. The power of the lens is inversely proportional to the focal length of the lens:

$$P = \frac{1}{f} \tag{14.1}$$

The basic focusing action of the lens is described in detail in Section 3.1.4.

Figure 14.1 shows an example of the mode of operation of the compound microscope. The specimen is put onto the stage and light is focused onto it by a light source. The microscope uses a beam of visible light ($\lambda = 400$–700 nm) to illuminate the specimen. An electric lamp is used to produce that light. The white light from the source is projected onto the sample through a condenser so that the light illuminates the entire specimen evenly. The object O in Figure 14.1 is placed just beyond the focal length of the objective lens f_0. Light rays travel from the illumination source to the object then to the objective lens. The rays are diffracted by the lens and an image I_1 is formed by the objective lens just past the focal length of the eyepiece f_e. The first image is real and inverted and is larger than the original object being examined. The image I_1 is magnified by the eyepiece into a very large virtual image I_2 and remains inverted. The second image is then seen with the observer's eye as if it were in the near point of the eye. Since the total length of the microscope is generally less than 25 cm the ultimate image is perceived as if it were beyond the dimensions of the microscope.

In Figure 14.1 d_0 represents the object distance and l the length between the two lenses. The image formed from the objective lens is a factor m_0 greater than the object itself. The object has a height h_0, while the image has a height

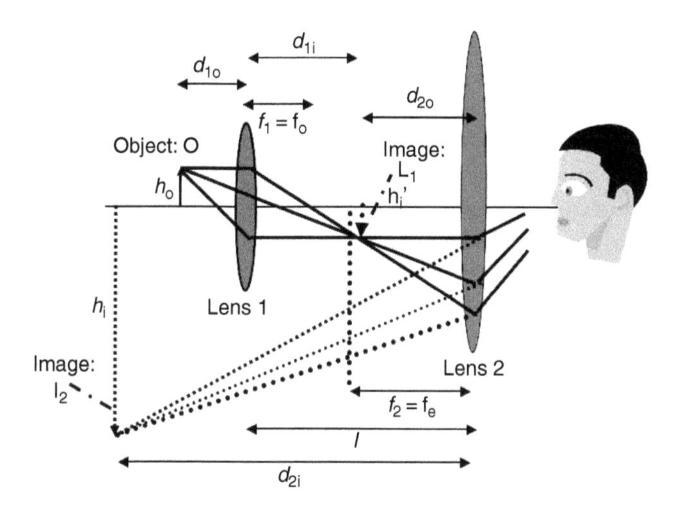

FIGURE 14.1
Diagram of the image formation in a compound microscope.

h_i and is at a distance d_i from the lens. The lateral magnification of a simple lens is calculated by taking the ratio of the image height over the object height, which is equal to the ratio of the respective distances as shown by geometric analysis, and is defined as

$$m_o \equiv \frac{h_i}{h_o} = \frac{d_i}{d_o} \tag{14.2}$$

Assuming that the eye is relaxed the angular magnification M_e can be calculated using the following equation:

$$M_e = \frac{N}{f_e} \tag{14.3}$$

where N is the near point of the normal eye.

The ocular lens, which can be found in the eyepiece of the microscope, is usually selected with a 10 times magnification. The objective lens, which is placed closest to the specimen on the stage, comes in a variety of magnifications depending on the use of the microscope. The two magnification values multiplied together will give the combined magnification for the specimen. The total magnification for the microscope is given by

$$M = m_o M_e = \frac{h_i}{h_o}\frac{N}{f_e} = \frac{d_i}{d_o}\frac{N}{f_e} \tag{14.4}$$

For example, if the ocular has a 10 times magnification and the objective magnifies 40 times then the total magnification will be 400 times.

With the use of the compound microscope, biologists are able to examine specimens that are microns in size. Such specimens include cells and some of their organelles. However, the more questions that are answered in science, the more questions are raised. Thus armed with this new knowledge the need to research finer details of cellular components and of host pathogens such as bacteria grew. The finer cell and even molecular details under examination are beyond the resolution of the compound microscope, as will be discussed next. Different types of microscope used to achieve magnification to resolve finer details will be described following the description of the diffraction limit of the imaging system.

Optical microscopy is predominantly used for in vitro diagnostics of excised pieces of tissue (biopsy or postmortem sectioning). Optical microscopy has a long and time-tested history and has influenced the way other diagnostic techniques were developed. The limited magnification and the lack of physiological information were the main disadvantages of optical microscopy. The limitations in resolution of microscopy will be discussed next.

14.1.1 Diffraction in the Far-Field

In a conventional optical microscope, the incoming and outgoing waves approach is planar in reference to the collection lens of the instrument. Therefore, Fraunhofer diffraction is obtained in the image plane of the system. A schematic of a conventional optical microscope is shown in Figure 14.2.

Light is emitted by the point source, generating a spherical wavefront. This spherical wavefront is sent through a collimating lens, transforming the spherical wavefront into a wavefront that consists of plane waves. The planar wavefront is collected by the collection lens and imaged in the image plane. If the diffraction phenomenon had not existed, then a bright spot with the same dimensions of the collection lens would be imaged in the image plane. However, what is actually seen in the image plane is a bright spot surrounded by secondary maxima and minima in intensity.

The theoretical model will be described on how diffraction limits the resolution in a conventional optical system. The visual aid of Figure 14.3 will show the geometric constraints involved. It shows a plane wave incident on a circular aperture of radius ρ, in an otherwise opaque screen. This aperture can also represent the size of a lens as part of an imaging system.

A wavefront that consists of plane waves is incident upon the circular aperture formed in the opaque screen. According to the Huygens–Fresnel principle (Chapter 3), each portion of the aperture's surface area acts as a secondary coherent point source that emits spherical waves. The Huygens–Fresnel principle states that every unobstructed point of a wavefront, at a given instant, serves as a source of spherical waves. Each

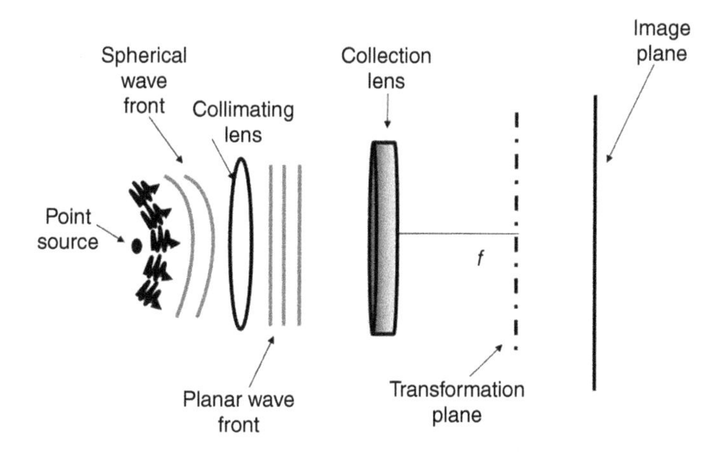

FIGURE 14.2
Schematic of conventional rudimentary far-field optical instrument.

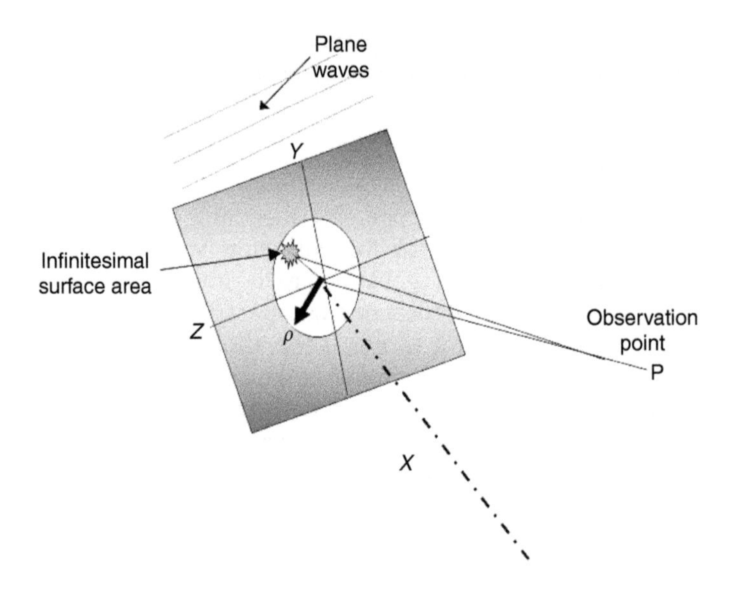

FIGURE 14.3
Schematic of circular aperture illuminated with a planar wavefront.

wave-section has the same frequency as the primary wave. The amplitude of the total field beyond these secondary coherent point sources is the superposition of all the waves emitted by the coherent point sources.

The spherical waves that are emitted by the secondary coherent point sources, which cover the entire surface of the aperture, can be described mathematically as

$$E = \frac{\varepsilon_0}{r} e^{i(kr - \omega t)} \tag{14.5}$$

where ε_0 is the source strength of each coherent point source, r the distance from the point source to a distal observation point, ω the angular frequency, and $k = \lambda / 2\pi$ the wave number.

The incident light interacts with the edges of the aperture, effecting the electric field distribution at some location far away from the aperture. To understand the nature of the field disturbance at a remote observation point, which results from the aperture edge, the contributions of all points within the aperture needs to be investigated. We can start by selecting a small fraction of the secondary coherent point sources on the surface of the aperture. The layout of this process is illustrated in Figure 14.4. The next step will be to combine all secondary sources to derive the total field disturbance at the observation point, which results from each coherent point source. The total effect is derived by integrating across the entire surface area of the aperture.

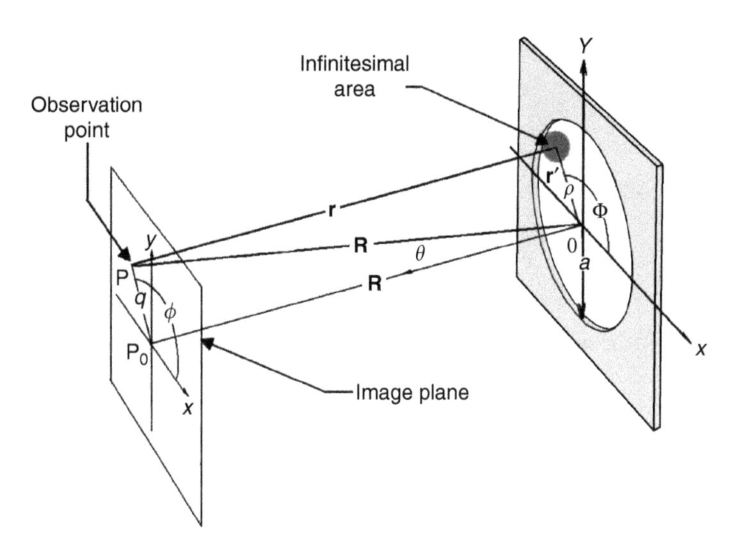

FIGURE 14.4
Vector locations of the infinitesimal area and observation point.

A small fraction of the secondary coherent point sources is enclosed in an infinitesimal surface area ds, with ω_a being the local source strength of the areas ds. This infinitesimal surface area is smaller than the wavelength of the illumination source. Therefore, each wavelength emitted by the enclosed secondary point sources is in phase with each other. This small surface area is shown in Figure 14.3 and Figure 14.4. The field disturbance observed at the observation point is expressed mathematically as

$$\partial E = \frac{\varepsilon_a}{r} e^{i(kr-\omega t)} \tag{14.6}$$

The observation point is located a vector distance **R** from the origin, which is defined at the center of the aperture. The small infinitesimal area ds is located at a vector distance **r**′ from the origin, and a distance **r** from the observation point. The locations of the observation point and the infinitesimal area in reference to each other and the origin are shown in Figure 14.3.

From Figure 14.4, the vector distance **r**′ can be expressed in terms of **r**′ and **R**, using the unit vectors in the x, y, and z directions as e_x, e_y, and e_z, respectively.

First the distance **r**′ is given by

$$\mathbf{r}' = y'e_y + z'e_z \tag{14.7}$$

The location **R** is now given by

$$\mathbf{R} = xe_x + ye_y + ze_z \tag{14.8}$$

The connecting vector **r** can now be defined as

$$\mathbf{r} = \mathbf{R} - \mathbf{r}' \tag{14.9}$$

Equation (14.9) can also be expressed in Cartesian coordinates as

$$\mathbf{r} = xe_x + (y - y')e_y + (z - z')e_z \tag{14.10}$$

The magnitude of **r** is the square root of the dot product of **r** with **R**, as defined by

$$r = \| \mathbf{r} \| = ((\mathbf{R} - \mathbf{r}') \cdot (\mathbf{R} - \mathbf{r}')) \tag{14.11}$$

which, after rewriting, gives

$$r = \| \mathbf{r} \| = (R^2 + r'^2 - 2\mathbf{R} \cdot \mathbf{r}')^{1/2} \tag{14.12}$$

The magnitude $\|\mathbf{r}\|$ is substituted into Equation (14.6), which is rewritten as

$$\partial E = \frac{\varepsilon_a}{(R^2 - r'^2 - 2\mathbf{R} \cdot \mathbf{r}')^{1/2}} e^{i\left(k(R^2 + r'^2 - 2\mathbf{R} \cdot \mathbf{r}')^{1/2} - \omega t\right)} \tag{14.13}$$

The quantity R^2 can be taken out from under the square root radical in both the exponential and the denominator of the field amplitude.

Since the observation point is located very far from the aperture, the quantity **R** is much larger than **r**′. Therefore, the expression for r, the distance between the infinitesimal surface area and the observation point, is expanded into a binomial series.

The denominator term with the distance r is given by

$$r = R\left(1 - \frac{2\mathbf{R} \cdot \mathbf{r}' + \mathbf{r}'^2}{R^2}\right)^{-1/2} \tag{14.14}$$

The binomial expression of Equation (14.14) is given as

$$r = R\left(1 + \frac{1}{2}\left(\frac{2\mathbf{R} \cdot \mathbf{r}' + \mathbf{r}'^2}{R^2}\right) + \frac{3}{8}\left(\frac{2\mathbf{R} \cdot \mathbf{r}' + \mathbf{r}'^2}{R^2}\right)^2 + \cdots\right) \tag{14.15}$$

And for the exponential term with the distance, the expression is given as

$$r = R\left(1+\frac{r'^2 - 2\mathbf{R}\cdot\mathbf{r}'}{R^2}\right)^{1/2} \tag{14.16}$$

Equation (14.16) is developed in a binomial series as well, as shown in the following equation:

$$r = R\left(1+\frac{1}{2}\left(\frac{r'^2 - 2\mathbf{R}\cdot\mathbf{r}'}{R^2}\right)+\frac{3}{8}\left(\frac{r'^2 - 2\mathbf{R}\cdot\mathbf{r}'}{R^2}\right)^2 +\cdots\right) \tag{14.17}$$

The electric field disturbance expression from Equation (14.6) is now transformed into the expression given by

$$\partial E = \frac{\varepsilon_a}{R\left(1+\frac{1}{2}\left(\frac{2\mathbf{R}\cdot\mathbf{r}' + r'^2}{R^2}\right)+\frac{3}{8}\left(\frac{2\mathbf{R}\cdot\mathbf{r}' + r'^2}{R^2}\right)^2 +\cdots\right)} \times \exp\left(\mathrm{i}\left(kR\left(1+\frac{1}{2}\left(\frac{r'^2 - 2\mathbf{R}\cdot\mathbf{r}'}{R^2}\right)+\frac{3}{8}\left(\frac{r'^2 - 2\mathbf{R}\cdot\mathbf{r}'}{R^2}\right)^2 +\cdots\right) - wt\right)\right) \tag{14.18}$$

Since the electric field disturbance generated by the secondary coherent point sources enclosed in the infinitesimal surface area ds occurs very far away from the observation point, in the binomial series only the first two terms in the exponential and only the first term found in the denominator of the field strength are kept.

The electric field disturbance observed in the far field at the observation point is expressed as

$$\partial E = \frac{\varepsilon_a e^{\mathrm{i}(kR - \omega t)} e^{\mathrm{i}(k[(\mathbf{R}\cdot\mathbf{r}'/R) + (r'^2/R)])}}{R} \tag{14.19}$$

When the observation point is moved farther from the aperture: $(r'^2/R)\rightarrow 0$, Equation (14.19) reduces to

$$\partial E = \frac{\varepsilon_a e^{\mathrm{i}(kR - \omega t)} e^{\mathrm{i}(k(\mathbf{R}\cdot\mathbf{r}'/R))}}{R} \tag{14.20}$$

In Cartesian coordinates Equation (14.20) can be rewritten as

$$\partial E = \frac{\varepsilon_a e^{\mathrm{i}(kR - \omega t)} e^{\mathrm{i}(k(y' + z'))}}{R} \tag{14.21}$$

To find the total scalar electric field, due to all the secondary coherent point sources, one would integrate ∂E over the entire surface of the aperture. The integral is shown in the following equation:

$$E_{\mathrm{T}} = \frac{\varepsilon_{\mathrm{a}} \mathrm{e}^{\mathrm{i}(kR - \omega t)}}{R} \iint_{\text{aperture}} \mathrm{e}^{\mathrm{i}(k(y' + z'))} \mathrm{d}z' \, \mathrm{d}y' \tag{14.22}$$

Transformation into spherical or polar coordinates makes the double integral quite solvable. The final solution for the total scalar electric field disturbance at the observation point incorporates a Bessel function J_1, which is given by

$$E = \frac{\varepsilon_{\mathrm{a}} \mathrm{e}^{\mathrm{i}(kR - wt)}}{R} \frac{2\pi}{Kq} R\rho J_1\left(\frac{kq\rho}{R}\right) \tag{14.23}$$

The quantity q is the radial coordinate out to the observation point that is found in the image plane, as shown in Figure 14.4.

The intensity observed at the observation point is equal to the square of the amplitude of the electric field:

$$I = \tfrac{1}{2} E^* E \tag{14.24}$$

where E^* is the complex conjugate of the electric field expression.

The electric field intensity can now be derived and is given by

$$I = \frac{1}{2} \frac{\varepsilon_{\mathrm{a}}^2 A^2}{R^2} \left[\frac{J_1(kq\rho/R)}{kq\rho/R}\right]^2 \tag{14.25}$$

where A is the total area of the aperture.

Equation (14.25) can be rewritten with a suitable substitution for the peak intensity as shown in Equation (14.26). Using $I_0 = \frac{1}{2}(\varepsilon_{\mathrm{a}}^2 A^2 / R^2)$, Equation (14.25) can be written as

$$I = I_0 \left[\frac{J_1(kq\rho/R)}{kq\rho/R}\right]^2 \tag{14.26}$$

A one-dimensional plot of the intensity function, Equation (14.26), is shown in Figure 14.5. It shows the intensity pattern through the central core of the diffraction spot. There is a central maximum flanked on either side by a series of maximal and minimum intensities.

In a two-dimensional geometry the diffraction pattern, or the intensity distribution, is shown in Figure 14.6. This pattern is sometimes referred to as the airy disk.

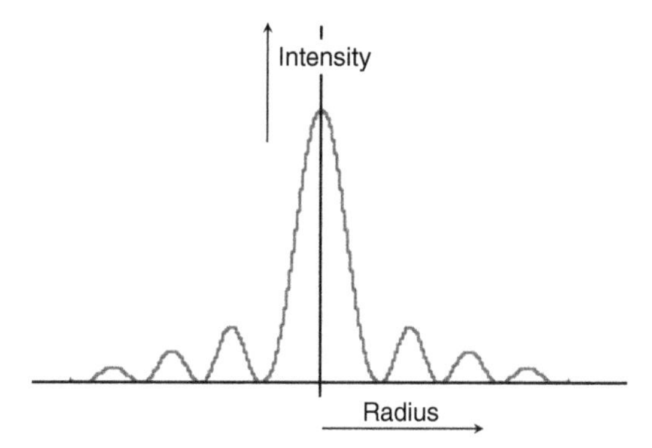

FIGURE 14.5
The plot of the intensity function in radial direction.

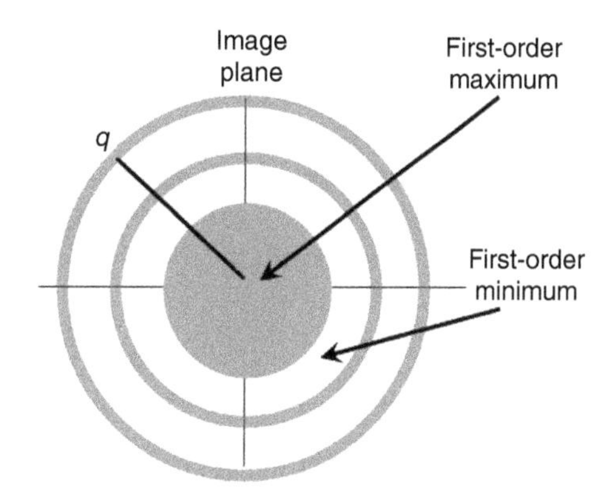

FIGURE 14.6
Image of the aperture formed in the image plane.

A dark ring that corresponds to the first zero in the J_1 Bessel function surrounds the central disk, which corresponds to the central maximum. The first zero in the Bessel function J_1 occurs when the following condition is satisfied:

$$J_1(u) = 0 \tag{14.27}$$

This condition can be checked against tabular lists of Bessel functions and it can be shown that the condition is satisfied for the following value:

$$u = \frac{kq\rho}{R} = 3.83 \tag{14.28}$$

Rewriting Equation (14.27) for q yields the expression correlating the distances as shown in the following equation:

$$q = \frac{1.22\lambda R}{2\rho} \tag{14.29}$$

where 2ρ is the diameter of the aperture.

The radial distance out to this first zero or first-order minimum can now be expressed as

$$q = \frac{1.22\lambda R}{D} \tag{14.30}$$

Reducing this situation more specifically to a conventional optical microscope, the aperture formed in the opaque screen is replaced by a collection lens. Therefore, for a lens focusing on the image plane, the focal length f is approximately equal to the distance **R**, as can be derived from the geometric analysis of Figure 14.4. The radial distance out to the first-order minimum can now be written as

$$q = \frac{1.22\lambda f}{D} \tag{14.31}$$

The ratio of q over the focal length of the lens is equal to the sine of the viewing angle θ, i.e., $\sin(\theta)$, as shown in Figure 14.4. Therefore Equation (14.31) can be rewritten as

$$\sin\theta = \frac{q}{f} = \frac{1.22\lambda}{D} \tag{14.32}$$

If θ is very small, then one can apply the paraxial condition, and Equation (14.32) can be written as

$$\theta \approx \sin\theta = \frac{q}{f} = \frac{1.22\lambda}{D} \tag{14.33}$$

Equation (14.33) is often referred to as the Abbe condition, after the German scientist Ernst Abbe (1840–1905), who worked on several issues for Carl Zeiss. He also made some suggestions for improvements on the state of the art of microscope design.

Having obtained the electric field distribution, taking into account the interaction of light with the edges of the aperture, at some location far from the illuminated aperture, the resolving capabilities of an imaging system can be characterized.

Consider the schematic representation of an optical microscope shown in Figure 14.7. There are two point features: A and B, both located at the same radial displacement from the collection lens. Since this is an optical microscope, the radial displacement from the collection lens is equal to the focal length of the lens. The angular displacement between A and B, in reference to the center of the circular aperture, is θ.

Essentially in Figure 14.7, A and B are two noncoherent point sources. Each point source is imaged in the image plane. What is actually observed in

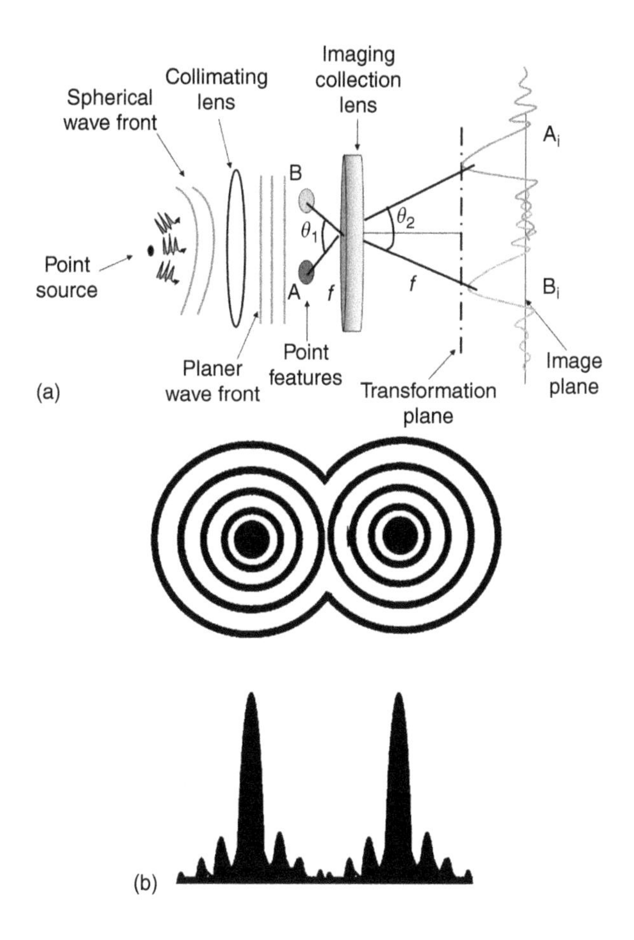

FIGURE 14.7
(a) Schematic of the resolution-limiting composition of the conventional optical microscope. (b) Illustration of the image formed as a result of diffraction of two objects in the focal plane of the objective.

the image plane are two diffraction patterns, which are generated by the point sources. The angular displacement between the two central maxima, in reference to the center of the aperture, is θ.

If the two sources are far enough apart to ensure that their central maxima do not overlap, their images in the image plane can be distinguished and are said to be just resolved. If the sources are too close together, the two central maxima may overlap, and then the images are not resolved.

Lord Rayleigh suggested that "Two image points A and B in the image plane can be just resolved if the central maximum of one image falls on the first diffraction pattern zero of the second." This corresponds to an angular separation between the two point sources, which is equal to the expression in Equation (14.33) for the first minimum of a diffraction pattern of a single source.

Typically, one wants to look at the lateral displacement between two points. Therefore, Equation (14.33) is multiplied by the focal length of the lens, yielding the arc length between the two points A and B. This lateral distance is then put in terms of the angular aperture of the lens. Thus, one has the resolving power of the lens, which is the minimum lateral distance between two points that can be resolved by a lens. The definition of the resolving power, which is also known as the Rayleigh criterion, is shown in the following equation:

$$\text{Resolving power} = \frac{0.61\lambda}{\text{NA}} \tag{14.34}$$

where NA is the numerical aperture of the lens, which is defined as the light-gathering power of an optical fiber. The mathematical definition of numerical aperture is given by the following equation as the index of refraction times the sine of half the acceptance angle of the aperture α:

$$\text{NA} = n \sin \alpha \tag{14.35}$$

Placing a drop of oil between the object and the objective lens increases the resolving power. In this case the wavelength of light is reduced to λ/n where n is the index of refraction of the oil.

The theoretical limit of the resolving power for the microscope is $\lambda/2$. Therefore, since the illumination source produces a beam of visible light with a wavelength as short as 400 nm, no details finer than approximately 200 nm can be clearly distinguished with the compound microscope. This limitation relies purely on the physical nature of light. This diffraction limit places a restriction on the maximum attainable magnification with visible light at 600 times.

However, one can only make the wavelength too small or the numerical aperture too large. This limitation on resolution is known as the diffraction barrier.

The theoretical model that was used to characterize the resolution capabilities of a conventional optical microscope cannot be applied to a near-field device, such as the near-field scanning optical microscope (NSOM). The Fraunhofer diffraction approximation cannot be extended to a near-field device, because many of the assumptions used to generate the characterization of the resolving power are no longer valid.

If one looks back at the binomial series in the equation for the electric field, Equation (14.18), only the first two terms were kept. Since the observation point was located very far away from the aperture, the higher order terms of the series, such as $1/R^2$, $1/R^4$, etc., were considered negligible compared to the $1/R$ term. These higher order terms contain specific information about the features being imaged in the optical system. However, these terms cannot be neglected in the near field.

This is the heart of the NSOM's resolution capability. The higher order spatial terms that are lost in a far-field device are now gained by working within the near field. The NSOM is described in detail in Section 14.3.

The following sections will delve into other available microscopic techniques in greater detail. The first derived microscopic technique discussed will be that of confocal microscopy.

14.2 Various Microscopic Techniques

The optical microscope needed improvement due to its inherent limitations on the resolution; this resulted in the development of the interference microscope that makes use of the wave properties of light and the optical path length while traversing an object with varying indices of refraction. The phase-contrast microscopy also makes use of interference of light waves, and relies on diffraction, with resulting phase retardation. Another type of microscope uses the light dispersion in tissues, which is caused by the naturally occurring birefringence in most tissues, and the loss of the birefringence after the tissue is damaged either by disease or by artificial induction (e.g., coagulation), and is hence called the polarization microscope.

Many other microscopic devices have since then been developed, namely, fluorescent microscopy, confocal microscopy, scanning optical microscopy, two-photon microscopy, and NSOM.

These developments were all prompted by the limitations in resolution of conventional microscopy: the fact that it only allows for two-dimensional examination and that no information is provided on the biological composition of the histology sample.

Other techniques lend their name to the following imaging instrumentation: confocal microscope, and the fluorescent microscope (imaging only fluorescence emission wavelengths), which rejects light that does not originate in the focal plane of the viewing field, thus providing shallow depth resolution and the capability to obtain cross-sectional images below the surface of the sample. Some of the latest high-resolution three-dimensional rendering methods are two-photon microscopy, two-photon laser-scanning microscopy, and multiphoton microscopy, which will also be discussed.

14.2.1 Confocal Microscopy

Confocal microscopy or laser-scanning confocal microscopy (LSCM) has changed some of the limitations placed on scientists by the compound microscope. It has enabled scientists to examine thicker samples and it projects the image as a three-dimensional structure. Confocal microscopy uses lasers as the source of light, which provide a tightly collimated coherent beam; therefore, the probing point is almost perfect.

Laser-scanning confocal microscopy can discriminate between signals originating from in-focus or out-of-focus optical planes to produce digital images. It also uses pinholes to exclude out-of-focus light from an image, therefore reducing background.

A typical LSCM system consists of the conventional microscope and a confocal unit above it. These are both connected to a detector system that collects the signals sent from the unit and ultimately reconstructs those images in three dimensions. A laser beam is produced from above the specimen being examined. Light coming from the laser source will move through an x-y scanner. This scanner will focus the laser light at specific targets on the viewed specimen. The laser light will then move from the x-y scanner onto a dichroic mirror. Dichroic mirrors are designed so that certain wavelengths of light can pass through while others get reflected. Each laser source within the confocal part will have its own dichroic mirror designed to reflect its specific wavelength to the specimen on the stage. The laser beam passes through the scanning mirrors and is turned into a scanning beam. This beam passes through the objectives, which diffract light and focus it onto the specimen. Because this is a laser light source, the light will not spread and will be focused intensely on one single spot on the specimen. The real key is in the pinhole or confocal aperture. An illustration of the confocal mechanism is shown in Figure 14.8.

The aperture in the confocal microscope is a small hole, which only allows light from a single focal plane to be seen by the detector and ultimately taken to the computer for processing and visualization. Both the condenser and objective lenses are focused to a common point. All light from other unfocused planes are eliminated. This process is repeated a number of times along the specimen, reading only small fields within the plane with the laser point in the x-y direction. In this way, the confocal can sweep through a specimen point-by–point, virtually cutting an optical section that is in focus and of a high resolution.

Additionally a specimen can be labeled with a fluorescent probe in which the light is absorbed by the sample and re-emitted at a longer wavelength. The reflected light and emitted fluorescent light from the specimen are captured by the objective lens. The scanning device/scanning mirrors act as a deflection

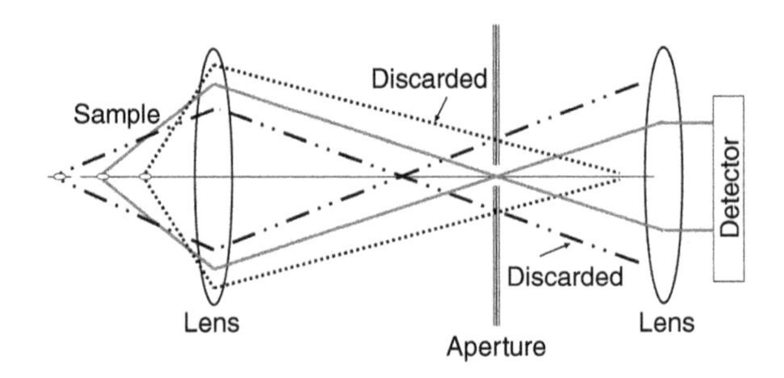

FIGURE 14.8
Representative schematic of a ray diagram of confocal microscope.

mechanism where they convert this light into a static beam. This beam then approaches a dichroic mirror that acts as a beam splitter. This mirror separates the light paths, such that the fluorescent light of longer wavelength is focused while the reflected light of the shorter wavelength is deviated. The light then passes through a detector lens, which bends the light rays accordingly and focuses them onto a pinhole. The role of the dichroic mirror is very important since it is the mechanism by which fluorescent light reaches the pinhole. The pinhole itself collects the fluorescent light, allows it to pass through to the photomultiplier unit, and rejects the light that did not originate from the focal plane. The photomultiplier is placed behind the analyzing hole to produce the video signal, and scanning is accomplished by moving the object. Taking a series of confocal images at successive planes in the specimen and assembling them into computer memory process, three-dimensional image reconstruction can be performed.

The layer slicing algorithm is illustrated in Figure 14.9.

Laser-scanning confocal microscopy offers several advantages: (1) it provides higher resolution of images, (2) it allows the use of higher magnifications, (3) it produces three-dimensional images, and (4) it minimizes stray light because of the small dimension of the illuminating light spot in the focal plane. Although this system provides higher resolution, it is still limited in resolution because this is a limitation in the properties of light itself.

To place the confocal microscopy in perspective the following is an illustration of the biological aspects involved in the field of imaging. Typically to make a cell culture, osteoblasts are isolated from 2-day-old mice. Three million cells are cultured in 75 cm^2 tissue culture flasks at 37°C. Cells reach confluence in approximately 3 days at which point they are split into two new 75 cm^2 tissue culture flasks. Once cells reach approximately 80% confluence, they are ready for the next step in the experimental procedure. To monitor the progress of this mechanism frequent observations are needed. Confocal microscopy is one of the tools that can be used to provide the feedback needed.

Another advantage of this type of microscopy is that you can also cut optical sections in the *z* plane from the top of the three-dimensional

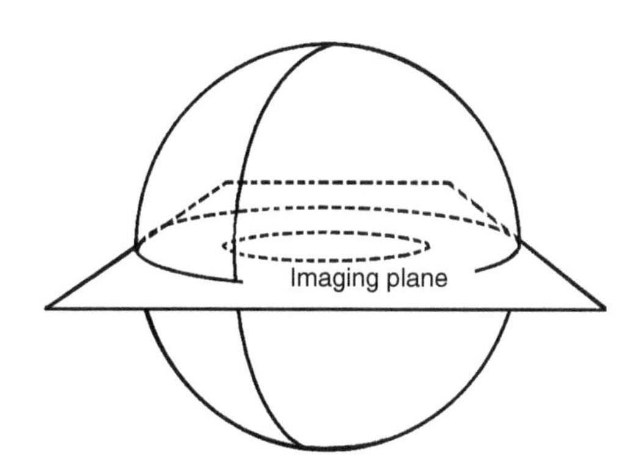

FIGURE 14.9
A diagrammatic representation of optical sectioning through a three-dimensional object performed by confocal microscopy.

specimen to the bottom. In this way, multiple sections can be compiled together to reconstruct the original three-dimensional image. All optical sections in the z plane are high resolution and high focus. In this three-dimensional representation of the specimen, specific localization of the protein can be visualized. All of these attributes make the confocal microscope a very powerful tool to be used in scientific research.

Confocal microscopy has attracted considerable interest because it achieves up to approximately twofold better lateral resolution and better sectioning in longitudinal axis as compared with conventional light microscope. In principle, confocal microscopy should permit the characterization of ion concentrations in situations where wide-field microscopy cannot, such as deep within tissues.

Immunofluorescence can also be applied to confocal microscopy. Confocal microscopy is an ingenious method for detection of fluorescence. This microscope allows a specimen to be optically sectioned into many focal planes that improves the resolution of the resulting micrograph and aides the researcher in determining localization of a given protein within a tissue or cell.

Fluorescent microscope will be discussed in Chapter 15.

Other microscopic techniques use fluorescence as well. In Chapter 16 the use of fluorescence for functional imaging will be discussed as the photobiological application.

The limitations of the short-wavelength penetration in biological media are one of the shortcomings of the various microscopy techniques. In contrast the shorter wavelengths provide higher resolution. A solution to this problem was introduced by the concept of multiphoton imaging, which will be described next.

14.2.2 Multiphoton Imaging

The principles of multiphoton imaging were presented in theory in 1931 by Maria Göppert-Mayer (1906–1972); however, it took 32 years for this theory to be put into practice when P.A. Franken and A.E. Hill started experimental work with second-harmonic-generated laser pulses. Subsequently other research groups applied these principles to study the interaction of two-photon absorption with polycrystalline materials.

14.2.2.1 Two-Photon Microscopy

The principle of two-photon excitation of a fluorochrome was first combined with that of microscopy in 1989 by Watt W. Webb and Winfried Denk, although the principle of two-photon excitation had existed much longer than this development.

The physical principles, advantages, limitations, and future developments of two-photon microscopy will be discussed and will hopefully resolve the question as to which system would be better for most research and clinical applications.

In general the setup of the two-photon microscope is much like that of the confocal microscope. Both systems use laser light emitted from a pump laser to induce the fluorescence of their target molecules. Also these systems are in general comparable to most light microscopes seen in every lab. They possess ocular lenses, objective lenses, stage, etc. In general they collect data in the same way, either through a photon acceptor or through a Charge Coupled Device (CCD) camera. There data is processed through a computer that runs virtually the same programs to produce the images of the specimens being examined. Two-photon and confocal microscopy both are limited to the same basic fluorochrome systems such as tagged antibodies, direct dyes (4′,6-Diamidino-2-phenylindole (DAPI), rhodamine), and quantum dots. These systems will be excited in the same spectrum across each system and yield roughly the same data, with some variation. Structurally and systematically two-photon and confocal microscopy techniques are very similar, but the mechanisms they employ to obtain their respective data are quite dissimilar. It is these differences that impart novelty and great advantage to two-photon microscopy.

The main difference between these two systems is explained in their names, two-photon and single-photon microscopy. For either of these systems to work, a single wavelength of light, or a laser, must be employed to excite fluorochromes to emit light; without this ability these systems would be no more than just normal light microscopes. The emission of light from these fluorochromes involves excitation; this is the key concept in understanding how these two disparate systems work. At a resting, nonexcited state all of the electrons in the fluorochromes are at ground state. Ground state is the lowest energy state that the electrons can achieve. Also it is the most stable, nonreactive state of these molecules. Given these facts, it is important to know that the ground state is the normal state of these molecules, where all the atoms remain when they are not stimulated. Once stimulated, by a photon or other source of energy, one or more of the electrons move to a state of high energy. This excitation can be thought of as an electron moving to a higher orbital away from its counterparts. In this state the electron is said to be ionized or to be at a higher state of energy. Once in this state the molecule is unstable and must return to ground state to regain its stability. To do this the electron must move back from its higher orbital state to its original orbital state. To do this the electron must lose the energy it has picked up from the excitation process. Since energy cannot be created or destroyed, the energy gained by the electron must be converted into another form. There are two possibilities that this energy can be transformed into and thereby removed: heat or photon energy. The latter of the two is necessary to collect data from either scope. Although heat is sometimes a byproduct of the excitation of some fluorochromes, this can have a deleterious effect on many biological samples.

The method of photon elicitation from the fluorochromes is the critical difference between these two systems. Confocal utilizes a single-wavelength laser to excite the fluorochromes into a state of ionization. This laser must be of a specific wavelength to excite a narrow range of fluorochromes, usually a one-to-one ratio. Excitation with a single laser causes the diffuse activation of fluorochromes at points away from the central target. This is because even though the exciting light is a laser, some of the light will diffract at the surface of the object while other light is scattered in its passage through the sample. Since the energy of the light is the entire wavelength necessary to bring out the photon from the fluorochrome, the stray light will cause the excitation of surrounding fluorochromes. The two-photon microscope also employs lasers to release the photon from the fluorochromes, but they are not of the specific wavelength necessary to induce excitation from one specific fluorochrome.

The great advantage of the two-photon microscope is its wide range in targets it can excite. There is no need for a separate laser for each different fluorochrome. The wavelength of the light emitted from the near-infrared laser can be varied in several of the available microscopes. This allows for a range of different excitation energy variants. It is also possible to excite several fluorochromes at the same time. This eliminates the need for several scans of the same image and makes the assessment of colocalization of two or more targets instantaneous, yielding data that would take multiple scans from a confocal.

Typically the two-photon microscope does not use an aperture in the collection of its data. However, the confocal must use this device. The excitation from a confocal single-photon laser causes diffuse excitation of the neighboring fluorochromes; photons from these other out-of-focus excitations can interfere with the data that is collected by the system. Therefore, the aperture is used to focus the photons from the exact target of the laser on the collector. The photons from the exogenous simulated fluorochromes are filtered out by not allowing them to pass through the lens of the aperture. This is typically not needed in the two-photon microscope, but some still do use an aperture to ensure maximum focusing of the excited photons to the detector.

The fluorochromes that are common to either the two-photon or confocal microscopes are finite. Once they have been excited for long they will no longer be able to emit light, and are completely useless for obtaining data. This is a process called photobleaching. The amount of time required to bleach a fluorochrome varies from a few seconds to a few minutes, with most of the fluorochromes bleached in just a few seconds. This is a major problem with the confocal and one-photon excitements, since there is so much excitation of fluorochromes other than just the directly targeted ones. Samples will bleach in just a few scans with confocal, and if the necessary data are not gained at the beginning of the scan, the samples will be useless for further study. However, the two-photon excitation is limited to a very few fluorochromes that are located just in the target area. So the only fluorochromes that will be affected are the direct ones being studied. Therefore, you will have a longer time to examine your samples, since no bleaching of the other fluorochromes will occur.

Generally, the lasers used to excite the fluorochromes in a two-photon microscope are around the near-infrared range (760–940 nm), with the mean wavelength being around 860 nm. Instead of one photon arriving to ionize the fluorochrome, two photons will arrive at the fluorochrome at the same time, within a nanosecond of each other to elicit the photon. The first photon to arrive at the fluorochrome will excite the molecule to some extent. The first photon will impart enough energy to the molecule to take one or more electrons to a partially excited state, which does not contain enough power to emit a detectable photon. However, this minor change in energy will soon be lost if the second photon does not arrive within a nanosecond of the arrival of the first. Therefore, the second photon adds more energy to the partially excited electron or electrons and takes it to a fully ionized or excited state necessary to emit the photon from the fluorochrome upon the electron's return to its ground state.

The excitation area from two-photon excitation is very precise. There is virtually no extra excitation of fluorochromes outside the target area, because of the narrow range of interaction of the photons from the exciting beam. As the photons travel in their specific wave patterns, there are only certain points on the waves where they will interact directly with each other in a way to impart the energy needed to elicit the photon from their target. As with single-photon excitation, there is scattering of the beams when the lasers from the two-photon excitation interact with their target material, but this scattering of light bends the rays away from each other. This does not allow the beams to interact anywhere but at their focal point. The excitation rays that are scattered to other fluorochromes will transfer energy to them, but the fluorochromes will only be partially excited and will lose that energy before the next photon of scattered light can reach them. The nonscattered beam does possess a diameter of its own where excitation could take place, since both photons will be in close proximity. However, it is only at the point where the two photons interact completely that excitation to an ionized state will take place. Taken together, these facts show why there is such a high specificity of excitement from the two-photon microscope. Excitation can be limited to as little as a single fluorochrome.

This tight excitation of the fluorochrome leads to numerous advantages for the two-photon microscope. One of the main advantages of the two-photon microscope is its ability to be semiquantitative. The amount of photons that are emitted from a single area of the specimen can be quantified as

$$n_{\mathrm{a}} \approx \frac{p_0^2 \delta}{\tau_{\mathrm{p}} f_{\mathrm{p}}^2}\left(\frac{\pi \mathrm{NA}^2}{hc\lambda}\right) \tag{14.36}$$

where n_{a} is the number of photons emitted per fluorochrome, p_0 the average laser power, f_{p} the repetition rate, τ_{p} the pulse width, λ the wavelength, NA the numerical aperture, h the Planck's constant, c the speed of light, and δ the two-photon cross-section. The ability to calculate the amount of photons emitted per area provides the opportunity to roughly calculate the highest localization of the specific target that is under examination. This would be of a particular advantage for a scientist looking directly at differential protein localization or something of that nature. This will not be possible using a confocal or single-photon excitation system. The diffuse excitation will not allow tight calculation of the photons emitted from a single fluorochrome.

Two-photon excitation uses a wavelength of light, usually near the infrared spectrum, for excitation that is typically longer than the wavelengths of light used by single-photon excitation. This longer wavelength gives the two-photon microscope several advantages. The longer wavelength carries less energy to the sample. This will keep the amount of heat in the sample down. Too much heat can cause the denaturization of proteins and the destruction of the sample. Therefore, the sample will last much longer than it would if a confocal was used, which causes heat buildup at its target if the laser is left in the same position for more than a few seconds. Also this longer wavelength employed by the two-photon microscope does not scatter as much as a shorter wavelength. Scattering will cause the light to be bent off its original course. This is a problem with the shorter wavelengths used by the confocal microscope, which will scatter significantly more with the molecules in the sample that is being examined. The roughly 800 nm light used for excitation by the two-photon system will not interact with these small molecules so readily. Therefore more penetration by the light is possible. This penetration is another advantage of the two-photon system. The light can penetrate a sample to a mean depth of around 500 μm, which is good enough to see the inner organs of a small animal or a tissue preparation.

Since the light is able to penetrate the sample so well, three-dimensional imaging is inherent. With the resolution and depth of light the two-photon microscope possesses, a better definition of the images of the sample can be gained, and a precise three-dimensional model of the sample is easily obtainable.

14.2.2.2 Advantages of Two-Photon Microscopy

Given the low energy of the exciting laser and the depth profile ability of the two-photon microscope, live samples can be studied, which range from cells in culture to an anesthetized mouse. Also the laser pulse from the system ranges from 11 to around 400 fs. With this pulse width live metabolic action of the specimen can be studied, specifically since most cellular processes are on the order of microseconds. This is not possible with the confocal microscope. To use the confocal microscope the samples must be fixed and their respective fluorochromes attached after the cells are dead. The process of fixing can be used with the two-photon microscope, but in some instances it is not necessary.

An added benefit of the low impact on the fluorochromes of the two-photon excitation allows for a process called lifetime scanning. In this process the fluorochromes are followed till they are bleached. Usually this system is employed in a live specimen where the targets are moving in and out to the excitation range of the initiating photons from the microscope. Some samples can be followed from minutes to hours. With the right computer system having a fast enough processor, actual movies can be made of things ranging from cells interacting with one another, protein production, or cells migrating in a small animal, which is typically a mouse. Lifetime imaging is sure to produce some data that were before unattainable, and it may reveal how several biological processes work.

Another great advantage of the two-photon microscope is the focusing power this system possesses. It is possible to examine a specimen as small as one molecule. Some groups are using two-photon microscopy to measure changes in DNA molecules, protein expression, and diagnosis of certain diseases. One possible example of a cellular process that multiphoton imaging can investigate would be the activation of proteins of several signaling pathways necessary for cell-to-cell communication.

14.2.2.3 Limitations of Two-Photon Microscopy

Given all of the great advantages of the two-photon microscope, it does have several limitations. Some of these limitations are common between both two-photon and confocal microscopy. Although two-photon can be used to examine things inside small animals, it is not good for examining lager animals. The light will not penetrate far enough through the dermal and fat layers of the larger animals. Therefore, no useable data could be obtained from a large animal.

One common problem of the systems is the fact that there are only a few tags available for use. However, several groups and companies specialize in producing the tags necessary to obtain data from these systems. Therefore, this is an ever-shrinking problem. Another shared problem, which has recently been solved, is the slow scan time of each of these machines. The first two-photon and confocal microscopes used only two or one excitation beams, respectively. This made for very slow scanning of the samples, roughly 10 μs per frame. To solve this problem the systems employ a simple beam splitter. This works by taking the initial incidence ray and passing it through a 50% mirror. The resulting beams will then strike a fully reflective mirror, which will send them back through the 50% mirror amplifying their number. This process will happen once more and will result in eight beams of light where there was only one. Several of these devices are used in one microscope across x and y coordinates. The most complex pattern to date forms a grid of 8 by 8 beams, all of which will be pulsed at once. This will cover 64 times more surface area than with the previous setup. Thus, results can be obtained in the microsecond range, allowing the live images to be produced.

Next we will discuss another microscopic technique exceeding the diffraction - limited techniques described so far, the NSOM.

14.3 Near-Field Scanning Optical Microscope

New forms of microscopy have been developed over the past decades to yield resolutions that go far beyond the confines of the diffraction barrier, such as the field emission microscope, the electron microscope, the scanning tunneling microscope, the atomic force microscope, and the NSOM. These huge leaps in resolution capabilities do not come without any cost.

Generally, the two problems in these new forms of microscopy are that there are losses in flexibility in working environments, and there are high demands on sample preparation. For example, the electron microscope requires that a sample be treated with metallic dyes and placed in a vacuum. Suppose that the sample under observation was biological in nature, no aerobic organism would survive either through that treatment or while being placed in that environment. Although high-resolution information about the structure of the microorganism is gained, this is at the expense of observing any functionality of that structure. In contrast, the atomic force microscope provides resolution on the atomic level with little stress induced on the sample under observation. The information that is yielded by an atomic force microscope only describes the surface topography of the specimen, and provides no details on the internal structure.

The problem with the atomic force microscope and some of the other new techniques in microscopy is that there is no optical information that is generated. There is a wealth of information that can be gained with optical microscopy, such as chemical information, molecular bond structure, and other information that can only be generated through light and sample interactions.

The NSOM, also called scanning-near-field optical microscope, utilizes the fact that observations are made from a distance to the object under investigation at less than, or in the order of the interrogating wavelength. The illumination of the sample is also provided by a source with dimensions less than the wavelength. In addition, the NSOM does not provide an instantaneous global view of a section of the sample under investigation; however, the object is probed (scanned) in the near-field of the sample surface, by moving the source or detector across the specimen in nanometer steps, hence the name. A total image is formed with resolution less than the Abbe limit, after a series of line scans is performed, and all the intensity arrays are combined. Even though the NSOM is a relatively new instrument (inception 1982, D.W. Pohl and W. Denk), the concept was described in letter exchanges between Edward Hutchinson Synge and Albert Einstein in 1928.

Near-field scanning optical microscope is an optical microscope that provides high-spatial-resolution capabilities with light sources in the visible spectrum. There is no high demand on sample preparation or confinements on working environments with NSOM.

14.3.1 The Concept of the Near-Field Scanning Optical Microscope

Thus far, it has been shown that working within the near field generates high-spatial-frequency information. However, what is needed in an optical microscope is high-spatial-frequency information on a subwavelength scale.

Most explanations of the diffraction barrier on spatial resolution capabilities take into account just the diffraction of light as it propagates through a collection lens of a far-field optical system. Light will interact with the edges of the circular aperture of the lens causing the deviation of light from its initial line of travel. However, it is not just the interaction of the light with the collection lens that limits the resolution; there exist the effects of the light interacting with detail in the sample structure.

In the initial developmental stages of NSOM the proposition was made for an optical microscope in which light would pass through a subwavelength aperture in an opaque screen, illuminating an object directly below the screen. The screen would be maintained at a constant distance of a few nanometers from the sample surface, placing the sample in the near field. The idea is that the transmitted radiation would remain collimated, as it interacts with the sample. Therefore, a subwavelength beam interacts with a small volume element of the sample, and this light that interacts with the sample is collected by an objective. The sample is basically broken into small pixel areas, as shown in Figure 14.10. The subwavelength light source would be scanned over each pixel area, generating a two-dimensional intensity image.

At this point, one might be a little concerned about how all this ties into resolution. There are two theoretical parameters that give the NSOM the high-spatial-resolution capabilities. The first theoretical boundary condition is that the sample is maintained in the near field by keeping the subwavelength aperture at a constant height above the sample surface. The second condition is that the spot size of the light that interacts with the sample is confined to the dimensions of the subwavelength aperture. The combined effect of these two parameters yields high-spatial-frequency information on the subwavelength scale.

14.3.2 General Design of the Near-Field Scanning Optical Microscope

The NSOM can take on many different forms, depending on the user's specific application. Typically, the NSOM is constructed on top of an inverted microscope. The inverted microscope allows the user to target

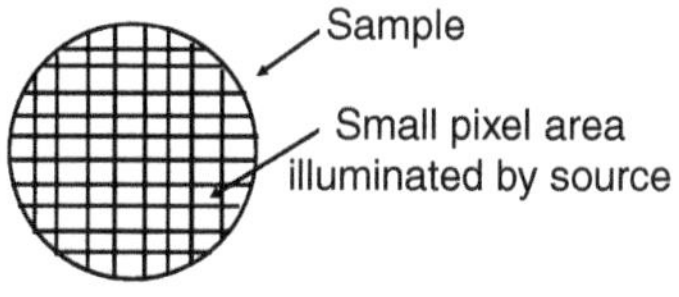

FIGURE 14.10
Illustration of how a sample is broken into small pixel areas, which the sub-wavelength probe scans over.

specific areas on the sample surface before any high-resolution imaging is performed.

First, laser light is passed through a band-pass filter to remove any undesirable wavelengths. The band-pass filter is followed by some wave plates, which are used to control the polarization state of the laser light. The light is then coupled into a single-mode optical fiber via a launcher. The end of this single-mode optical fiber is heated and pulled or chemically etched to form a subwavelength aperture. An illustration of the tapered fiber tip design is given in Figure 14.11. The very end of the single-mode optical fiber, where the subwavelength aperture was fabricated, is mounted onto an atomic force microscope (AFM). The AFM head will be discussed in more detail, but it is part of the mechanism that is used to maintain the subwavelength light source at a constant height above the sample surface. The light is transmitted through the subwavelength aperture and impinges upon the sample that is mounted onto an *xyz* piezoelectric translation stage. This *xyz* translation stage is used to scan the nanoscopic light source over each pixel area described earlier. A graphical representation of the operation of the NSOM is shown in Figure 14.12.

The light interacts with a small volume element of the sample, and then, it is collected by the inverted microscope's objective. The collected light is optically guided out of the inverted microscope where it is sent to a detector. The type of detector that is used at the output of the NSOM depends directly on the desired application in which the system is being used. The optical information from each pixel that is measured by the output detector is sent to a computer. The computer is used to construct a high-resolution image of the sample and to run the electronics involved in scanning.

14.3.3 Near-Field Scanning Optical Microscope Tip

The NSOM tip is the most essential part of the imaging device and requires that great care is taken in its preparation.

Near-field scanning optical microscope tips have many different shapes, sizes, and material make-up. Most of the earlier NSOM tips suffered low transmission of coupled light and poor reproducibility. However, the most widely used NSOM tip design was adopted from Bell Labs. In the early

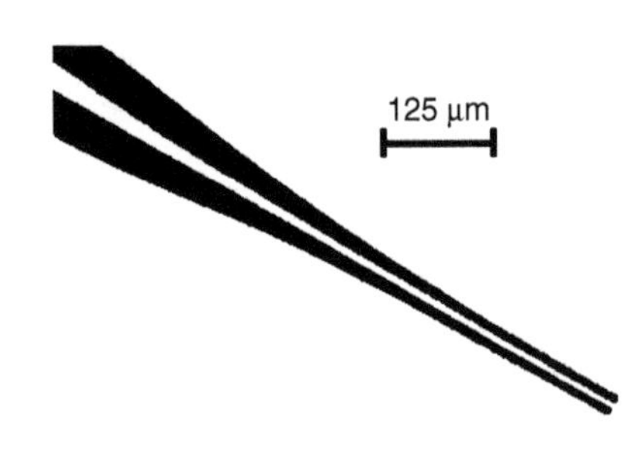

FIGURE 14.11
Illustration of the tapered fiber tip design.

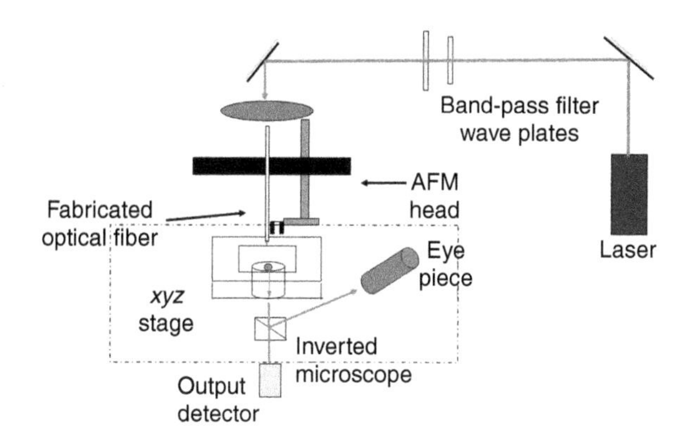

FIGURE 14.12
The general NSOM configuration.

1990s, Bell Labs proposed a tapered single-mode optical fiber with a reflective metal coating as a design for an NSOM tip.

The NSOM tips that are made from optical fiber are fabricated in two different ways: heating and pulling or chemically etching the fiber. In heating and pulling the optical fiber, a commercial micropipette puller is used in the fabrication process. Essentially, a laser is used to heat the optical material, and the fiber is then pulled to a fine point. At the tip of this point, an aperture with a diameter on the order of 50–100 nm is formed. One can carefully control the heating and pulling parameters, making this method of fabrication highly reproducible. In contrast, chemical etching utilizes a solution that consists of hydrofluoric acid (HF) with an organic buffer layer on top to fabricate the optical fiber. The etching process is graphically illustrated in Figure 14.13.

One end of a single-mode optical fiber that has been stripped off its cladding is immersed into the etching solution that consists of a buffer layer on top of HF. A meniscus is formed in two different locations along the fiber, due to electrostatic interactions between the solution's components and the fiber. However, the etching process only occurs at the organic solvent/HF interface. As the diameter of the immersed optical fiber in the HF decreases with time, the height of the meniscus formed at the solvent/HF interface decreases. Over a period of time, the etching process is terminated and a tip is formed. The conical shape of the tip can be adjusted by changing the organic solvent that is used in the etching solution.

Although now there is a subwavelength aperture formed at the end of the single-mode optical fiber, the fabrication process is not complete. There still exists the need to coat the NSOM tip with a reflective metal.

The reason for this metallic coating is related to the mode-field structure that propagates through an optical fiber. The ability of a single-mode optical fiber to guide and confine one wavelength of light is directly related to the dimensions and material composition of the optical fiber. As the diameter of the single-mode optical fiber decreases in the tapered region, the ability of

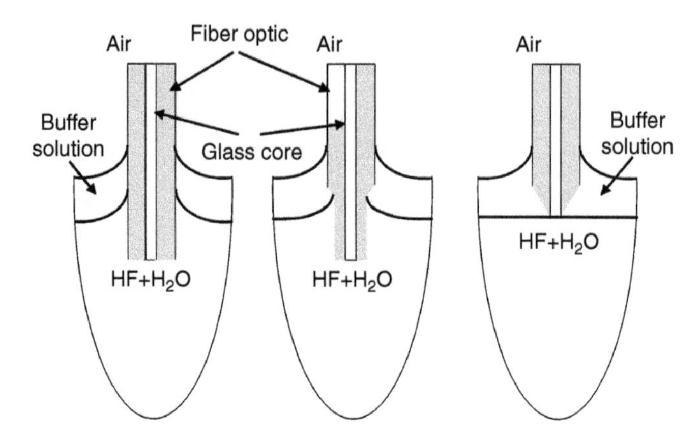

FIGURE 14.13
Schematic geometries of the liquid-layer protection etching procedure.

the fiber to confine and guide the light is lost. Therefore, the light is no longer confined to just propagate out of the subwavelength aperture; it escapes out of the sides of the tapered region. High-spatial-resolution capabilities of the NSOM depend strongly on the fact that the beam spot size is to be confined to the subwavelength aperture.

The metallic coating applied to the tapered region is required to prevent the leakage of light. This metallic coating also attenuates the light before it can escape.

The NSOM tip proposed by Bell Labs, although the most widely accepted, is prone to some inefficiency problems. One such problem has to do with the amount of light that can be transmitted through the subwavelength aperture. A large quantity of light is lost by tip design itself, or the reflection of the light back-up through the fiber from the tapered region. Therefore, only a small fraction of the initial coupled light from the source is transmitted through the aperture. This inefficiency tends to limit the applications of NSOM.

14.3.3.1 Feedback Mechanisms Employed to Maintain a Constant Tip and Sample Separation

To achieve high-resolution imaging capability, an NSOM must employ another specific feature. The NSOM tip needs to be maintained at a constant height of a few nanometers from the sample surface during scanning. There are various techniques that can be employed to maintain a constant distance between the tip and the sample surface. However, only two of these techniques have become widely adopted by NSOM designers.

Shear-force tip feedback and tapping-mode feedback have their foundations in atomic force microscopy. These are the most reliable two methods available to maintain a constant tip and sample separation. Both methods involve the mounting of the NSOM tip to a quartz tuning fork and driving the fork/tip system at mechanical resonance. However, one technique dithers the NSOM tip laterally in reference to the sample surface and the other dithers the tip perpendicularly.

14.3.3.2 Shear-Force-Mode Tip Feedback

A schematic of the tuning fork method used for shear-force tip feedback is shown in Figure 14.14, where the tip is dithered laterally in reference to the sample surface. An NSOM tip is rigidly attached to one arm of a quartz tuning fork that has a well-defined resonance frequency of 32.768 kHz and a quality factor, Q, of around 7500 in air. Typically, the tip is mounted onto the quartz tuning fork using some sort of glue or epoxy. This fork/tip system has a new resonance frequency that is slightly lower than that of the quartz tuning fork, and a quality factor Q that ranges from 150 to 300.

The fork/tip system is driven at mechanical resonance either by directly applying a sinusoidal voltage to the fork or by applying a sinusoidal voltage to a piezoelectric tube that shakes the fork/tip system at resonance. However, we will only discuss the situation in which the fork/tip system is mounted onto a piezoelectric tube that shakes the fork/tip system at mechanical resonance. The quartz tuning fork is piezoelectric in nature. Therefore, the shaking of the fork/tip system at mechanical resonance induces a periodic voltage across the tuning fork, which can be monitored through contacts integrated in each arm of the fork.

As the resonating fork/tip system approaches the sample surface within a few nanometers, the NSOM tip experiences frictional forces. These frictional forces cause the fork/tip system to become damped, and the amount of damping can be measured through the integrated contacts in each arm of the

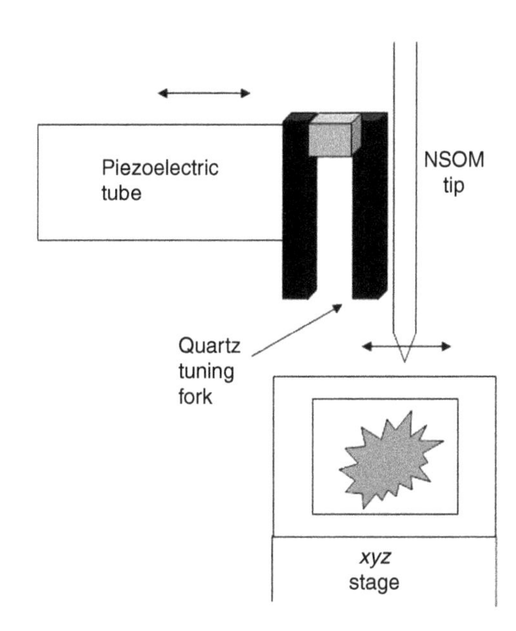

FIGURE 14.14
Schematic of the tuning fork method used for the shear-force tip feedback mechanism.

fork. Essentially, this resonating fork/tip system can be used as a mechanical pickup for the frictional forces that cause damping. This measured voltage across the quartz tuning fork can be used in a closed-loop feedback mechanism, which employs a lock-in amplifier and proportional integral (PI) electronic circuit to maintain the NSOM tip at a constant distance from the sample surface. A configuration of the closed-loop feedback system is shown in Figure 14.15. As the fork/tip system is brought close to the sample surface, frictional forces interacting with the NSOM tip cause the monitored potential across the fork to decrease in amplitude and shift in phase, any shift from 90° out of phase with the reference generates an output voltage from the lock-in. An active feedback voltage is sent from the lock-in to the PI circuit. The PI circuit sends an immediate response voltage to the *xyz* piezoelectric translation stage to adjust the *z* position of the stage, to maintain the monitored potential at resonance.

14.3.3.3 Tapping-Mode Tip Feedback

Tapping-mode feedback has also been employed to maintain the NSOM tip at a constant height above the sample surface. The tapping-mode method utilizes an NSOM tip that has a near 90° bend in its structure, as shown in Figure 14.16.

This method is similar to the shear-force tip feedback method in which the resonating fork/tip system can be used as a mechanical pick-up for frictional forces that interact with the NSOM tip, as it approaches the sample surface. However, the fork/tip system is dithered perpendicularly in reference to the sample surface. The induced potential across the fork is employed in a

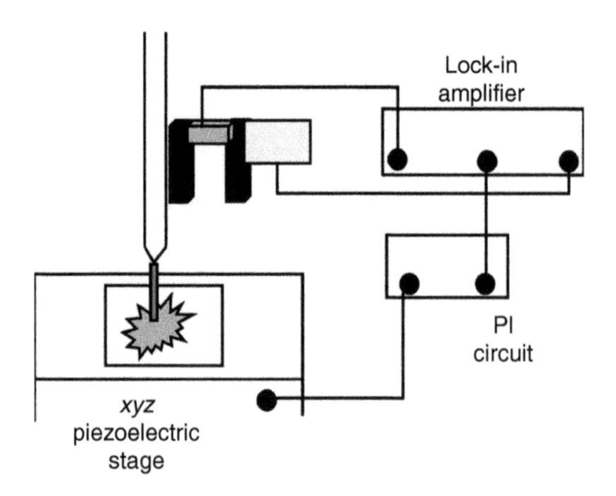

FIGURE 14.15
Shear-force tip feedback mechanism. The fork/tip system is shaken at mechanical resonance by driving the piezoelectric tube with a sinusoidal voltage that has a constant amplitude, frequency, and phase. The sinusoidal voltage is referenced into the lock-in amplifier. The electric potential across the fork is sent to the lock-in input to be monitored. The fork/tip system retracted away from the sample surface has a potential measured across the fork proportional to the reference signal, except there is a phase shift of 90°.

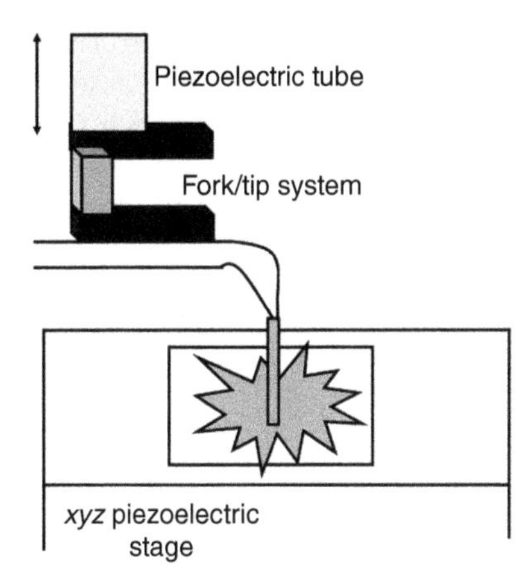

FIGURE 14.16
Schematic of the tuning fork method used for the tapping-mode tip feedback mechanism.

feedback loop, similar to that in Figure 14.15, to maintain the tip at a constant height above the sample surface.

Tapping-mode feedback is a more sensitive mechanical pick-up for the frictional forces that interact with the NSOM tip. However, the bend in the tip reduces the throughput efficiency. Therefore, in certain applications that require a relatively high intensity to interact with the sample, shear-force tip feedback will be chosen over tapping-mode feedback.

The fork/tip system and the piezoelectric tube form the AFM head that was mentioned earlier in the shear-force mechanism. Essentially, an AFM has been integrated into the NSOM. The AFM consists of the lock-in amplifier, PI circuit, and xyz piezoelectric stage, which are employed in a closed-loop feedback mechanism.

As the tip is being scanned across the sample that has some surface variations, the z position of the piezoelectric stage is adjusted to maintain the monitored potential across the fork at resonance. A topography image can be formed serially, pixel by pixel, by monitoring z-voltage that has to be applied to the positioning stage.

The advantage of having the AFM integrated into the NSOM is that intensity and topography images can be built simultaneously.

14.3.3.4 Intensity Imaging

As mentioned earlier, most biological samples that are studied are very thin and transparent. There are not many contrast mechanisms within the sample to exploit. Therefore, samples have to be treated with dyes, even in NSOM imaging.

Typically, samples are stained with dyes that have specific absorption spectrum. The stained samples are deposited on a piece of mica or a microscope cover slip and air-dried. A sample is mounted onto the xyz piezoelectric stage, which scans the sample under the fixed NSOM tip. The light excites fluorescence in the sample, which is collected by the collection lens of the inverted microscope. The collected light is optically guided out of the inverted microscope and passed through a notch filter. The notch filter is used to separate the fluorescence from any laser light that may be propagating through the system. The light is sent to a high-quantum-efficiency detector, such as an avalanche photodiode operating in single-photon count mode. The detector measures the frequency at which photons strike the sensing element in the detector. An image is constructed serially, pixel by pixel.

A topography image of red blood cells is shown in Figure 14.17, while Figure 14.18 shows the intensity image for the same red blood cells.

14.3.3.5 Phase Imaging

Additional work is being done to extend the capabilities of NSOM to include phase imaging. To achieve high-resolution subwavelength phase-contrast images an NSOM is integrated into an interferometer. The ability to construct high-resolution phase images would have a huge impact on the study of biological samples. Samples would no longer need to be treated with dyes to have a contrast mechanism to exploit. This method would be one step closer to providing a system that could be used to image living cells.

Since NSOM makes use of an optical fiber to achieve subwavelength resolution capabilities, a hybrid Mach–Zehnder interferometer utilizing an all-fiber bi-directional coupler represents the system of choice. The Mach–Zehnder interferometer relies on the same phase-sensitive information as described for the other interferometers, for instance in Section 8.8 in

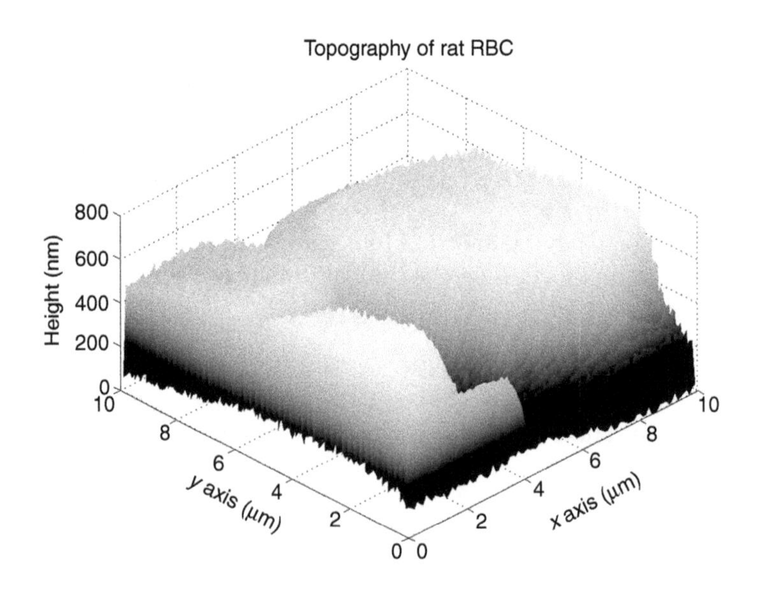

FIGURE 14.17
Shear-force topography image of red blood cells. (Courtesy of Kert Edward, University of North Carnolina at Charlotte, Charlotte, North Carolina.)

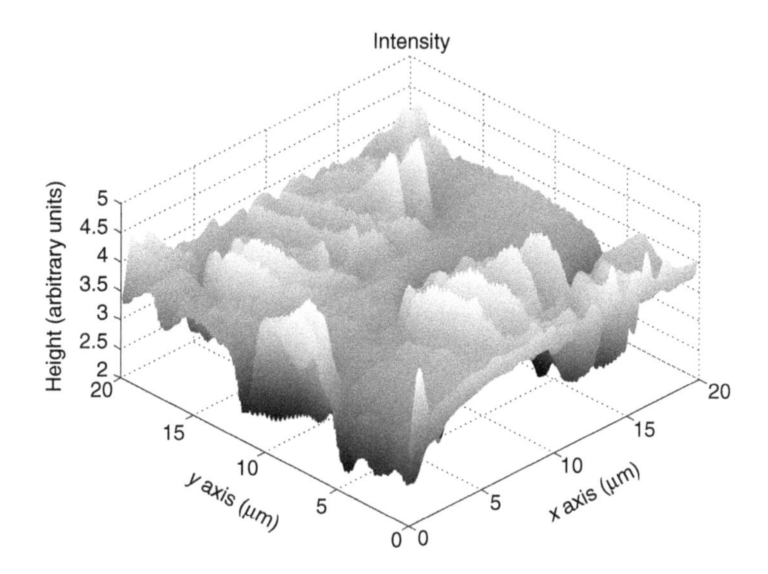

FIGURE 14.18
Intensity image of the same red blood cells as depicted in Figure 14.17. (Courtesy of Kert Edward, University of North Carnolina at Charlotte, Charlotte, North Carolina.)

Chapter 8. Only this time the recombination takes place after transmission instead of reflection. The Mach–Zehnder system is shown in Figure 14.19. The phase image of the red blood cell shown in Figure 14.17 (topography) and Figure 14.18 (intensity) acquired by the Mach–Zehnder interferometer NSOM is shown in Figure 14.20.

The combined application of Raman spectroscopy with NSOM will be discussed in detail in Chapter 15, Section 15.3, under one of the photochemical applications of imaging.

14.4 Spectral Range Diagnostics

A molecule may absorb the energy from an incident photon, exciting the molecule to a higher electronic state, assuming the incident photon carries a sufficient amount of energy. In conserving energy, the re-emitted scattered photon has slightly less energy than the incident photon. The difference in energy between the incident photon and the scattered photon is equal to the amount of energy required to excite the molecule to the higher vibrational state. This is shown in Figure 14.21. The vibrational energy structure of a molecule is a fingerprint of molecular bonding energies within a chemical structure.

In general spectroscopy stands for the spectral analysis of emitted light. When monochromatic light of frequency, v, illuminates a sample, the light scattered from the sample not only contains radiation of the incident frequency, but

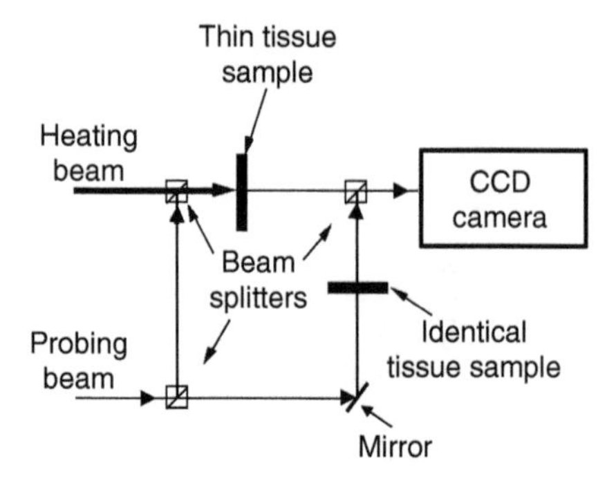

FIGURE 14.19
Schematic of NSOM Mach–Zehnder interferometer system that is used to construct phase images.

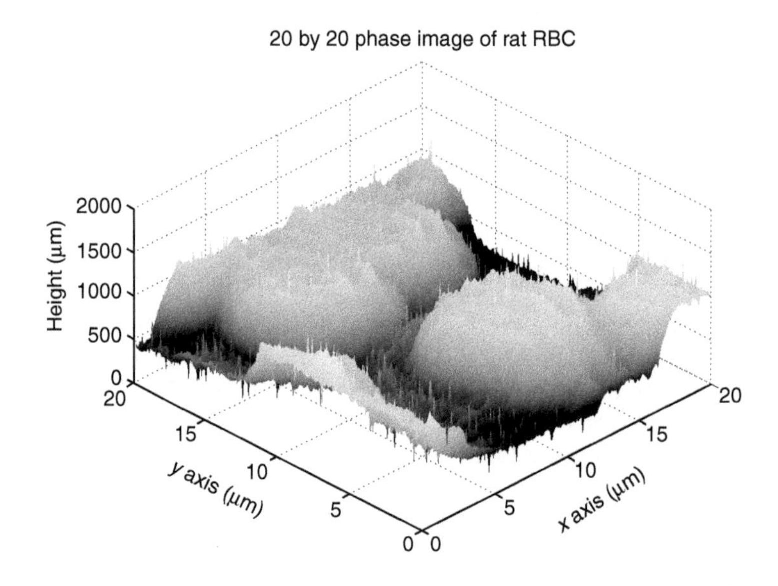

FIGURE 14.20
Phase image of the same red blood cells as depicted in Figure 14.17. (Courtesy of Kert Edward, University of North Carnolina at Charlotte, Charlotte, North Carolina.)

also weaker radiation of frequency $v \pm v'$. The weaker radiation components of the scattered light are generally known as the Raman spectra. A detailed description of Raman spectroscopy is presented in Chapter 8, Section 8.9.

In addition to the spectral frequency shift of scattered photons a broadband source illuminating a medium will provide a different form of spectroscopy, "absorbance spectroscopy". In absorbance spectroscopy the missing frequencies of the light transmitted through a sample will reveal the molecular absorbance, thus providing chemical binding information. Certain biological applications of spectroscopy will be described in Chapter 16.

A special class of absorbance spectroscopy is intracavity spectroscopy and in particular the miniaturized form using a diode laser concept. The principle of miniaturized diode laser spectroscopy is described next under the heading "Lab-on-a-CHIP".

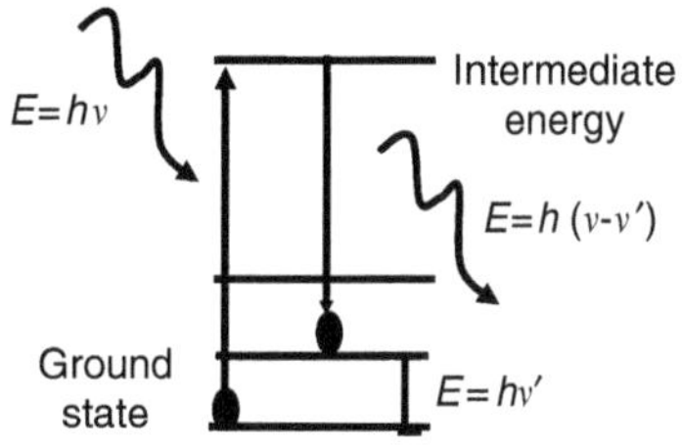

FIGURE 14.21
A molecule being excited by a photon to a higher vibrational state.

14.4.1 Lab-on-a-CHIP

Intracavity laser absorption spectroscopy (ICLAS) is an ultra sensitive technique to perform chemical analysis on gases and liquids. The advantages of intracavity spectroscopy over standard bare beam scattering spectroscopy are the sensitivity to changes in fluence inside the laser cavity as well as the higher energy fluence rate and the spectral multiplexing. An added benefit is a superior signal-to-noise ratio compared to scattered spectroscopy. With the development of vertical external cavity surface emitting semiconductor lasers (VECSEL's) this spectroscopic diagnostic technique can be adapted for *in-situ* blood-gas analysis under *in-vivo* circumstances.

Researchers at the Department of Energy's Los Alamos National Laboratory have developed a method of using certain polymers as luminescent sensors to detect and identify biological and chemical agents, almost instantaneously. The polymers fluoresce in the presence of these agents with the help of molecular intermediaries that bind to the biological and chemical agents' receptor sites. This process is analogous to the ligant antibody carrier technique which will be described in detail in Chapter 16, Section 16.4. Certain polymers transfer their electrons over to electron-accepting molecules when excited by laser light. However, the polymer's luminescence is extinguished (quenched) once the polymer transfers its electrons to the acceptor molecule, which is attached to the polymer.

All pathogens, proteins, viruses and bacteria are equipped with receptor sites, which let specific ligands attach themselves, and thus provide a means for the cells to be infected. By matching the proper ligand with a receptor a fluorescent tag can be formed to identify specific proteins, viruses or bacteria. This process is described as a lock and key mechanism. The ligand part of the molecular package, which is called a quencher-tether-ligand (QTL), forms the "key," that can be attached to the "lock", which is at the receptor site on the biological and chemical species to be detected. When the QTL is removed from the polymer the fluorescence is activated and the detection can be initiated. The spectral absorbance is described by Equation 14.37

$$\alpha(\nu) = Nk_{\nu}\Phi(\nu) \tag{14.37}$$

where N is the measurement of low concentrated species, k_{ν} is the detection of forbidden transitions and $\Phi(\nu)$ is the spectral line-profile. Since the absorption linewidth is limited by the instrumental resolution, sensitivity increases when the spectral resolution increases. This means that the broader the linewidth of the source the more accurate the spectroscopic detection of particular molecular chains will be.

The VECSEL diode laser design uses for instance a GaSb (Gallium-Sb) semiconductor base with Bragg grating mirror design to achieve a several cubic millimeter chip configuration that delivers a tunable laser source. Additional features such as multiple quantum-well and the versatility of both electrical and optical pumping increases the application range of the VECSEL for diagnostic purposes. The VECSEL can be tuned by modulation of the cavity length which is realized by controlling the chip temperature and pump power level. Internal heterodyning can also be implemented by modulation of the Bragg mirror to achieve an ICLAS sensitivity in absorption coefficient determination of better than 10^{-11} cm^{-1} with a spectral resolution in the order of 10^{-6} cm^{-1}. The heterodyning can be applied by injection of a repetitive current ramp which can simultaneously be used to control the detection mechanism. The light pulses passing through the bioagent are sent to the spectrograph where the frequency shifted fringe pattern is averaged over a pulse train. Additional features that can be added to the diode laser design are mode-locking and even dual frequency operation. Due to its size and complex operation the VECSEL "Lab-on-a-CHIP" can be used for *in-situ* trace gas analysis.

Another spectral advantage in imaging is in the short wavelength range of the x-ray. The x-ray wavelengths are in the molecular size regime and offer an entirely different scale of imaging which will be discussed next.

14.4.2 Coherent X-Ray Imaging

In addition to the near and mid infrared tuning capabilities of the Free Electron laser mentioned in Chapter 2, Section 2.4.3.6 there are additional features of the FEL that are providing new biomedical diagnostic opportunities. Under specific tuning conditions the FEL can also be used to produce spectra of shortwave electromagnetic radiation, in particular ultraviolet and Röntgen radiation or x-ray. The ability to produce a coherent x-ray beam with the Free Electron Laser has opened the door to molecular imaging. Extremely intense x-ray radiation with laser-like properties generated by free-electron lasers forms the basis for a new source of diagnostic radiation. The successful operation of an x-ray producing FEL with a wavelength in the order of 0.1 nm or equivalent energy of 12.4 keV depends on the combination of high peak current and a low emittance of the FEL electron beam.

14.4.2.1 Ultrafast X-Ray Pulses Reveal Atoms in Motion

One area of special interest is the potential to image molecular motion. Proteins are invaluable in the cellular process as waste removers, signal generators, chemical catalysts, motors, and pumps. For instance, proteins carry supplies such as nutrients and chemical messengers in addition molecular motion applies to active transport across the cellular membrane in so-called molecular motors. Two molecules, each with a head and a tail form a functional unit, a motor protein. Another field of interest lies in the research of muscle action. X-ray holography is an excellent tool to establishing the molecular processes in the functioning of a muscle fiber during contraction. Here the interaction between the actin filament and myosin molecules can be investigated. Further interest is in the DNA resequencing process and the duplication process of RNA from DNA.

High intensity and short pulse duration achieved by FEL x-ray production allow for momentary data collection, thus immobilizing the protein in time. Thus repeatedly momentarily freezing the motion of the protein in time sequential images can be used to reconstruct an active process in replay. The means to produce the required ultra short x-ray pulses needed has been designed by a group of researchers from the Department of Physics at the University of Michigan. An ultrafast repetition rate laser beam hits on the surface of a birefringent crystal, generating an acoustic pulse that is very short in both time and space. This acoustic pulse modifies the diffraction patterns through the crystal, and it can be used to switch energy from one diffracted beam to another, as such it becomes an ultra fast shutter for x-rays, operating in the femtosecond range. Time-resolved structural studies on diffusive processes in crystalline enzymes can this way be resolved.

The coherent x-ray source provides the opportunity for phase-resolved imaging and lowers the radiation levels below damage threshold allowing in-vitro functional imaging. In the case of x-rays, phase sensitive imaging yields an increased contrast (Figure 14.22). This brings us to phase-contrast imaging using x-ray interferometry. Phase-contrast imaging relies on refraction of x-rays, similar to the way light rays bend when they pass through an interface between two media with different indices of refraction, only now x-rays refract when traveling through objects of varying densities yielding the complex index of refraction of the tissue. This deflection can be measured with the standard interferometry as described in Chapter 8, Section 8.8.1. An expansion on the interferometric methods in x-ray will be discussed next.

14.4.2.2 Free Electron Laser Protein X-Ray Holography

In order to form a rigorous understanding of the protein function in the cell it is not enough to know the elements that make up the protein and the chemical structure of the atomic configuration. The complete geometric structure needs to be known and the covalent binding sequences need to be revealed to fully understand and predict its function. Protein crystallography using incoherent x-ray photons requires special crystal sample preparation and demands several hours of imaging time. In contrast the FEL laser coherent x-ray radiation can be used to excite molecules in a crystalline form producing an x-rays diffraction pattern off the excited molecules in the crystal, providing the data from which the structure of the transient species can

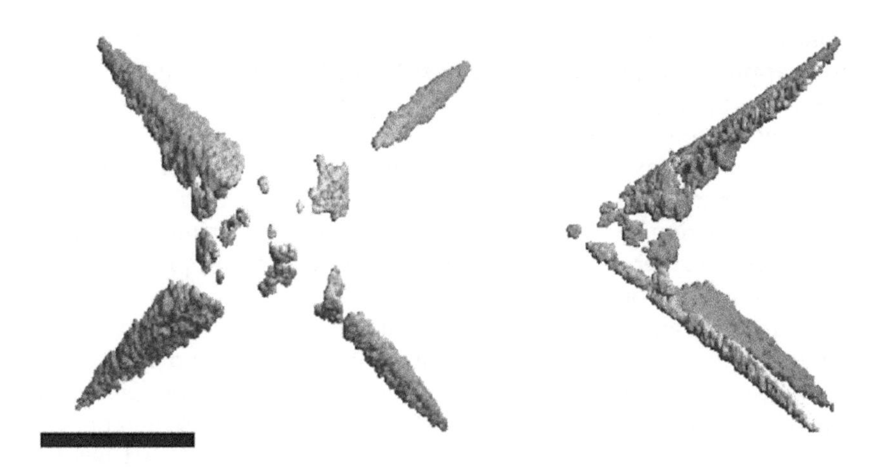

FIGURE 14.22
Views of an iso-surface rendering of a pyramid test object. Image of pyramid object using FEL x-ray coherent diffraction imaging. The black scalebar is 1 micron. (Courtesy: Henry Chapman, Physics and Advanced Technologies, Lawrence Livermore National Lab)

be derived. Applying diffraction imaging using coherent x-ray radiation, the internal structure of complex protein molecules such as enzymes can be decoded in three-dimensional form. The first demonstration of soft x-ray laser holography was carried out by James Trebes et al in 1987.

The process of holographic imaging will be discussed next in greater detail.

14.5 Holographic Imaging

The fact that a great deal of detail on the anatomical structure is contained in the phase information of the wave phenomenon of the photons has been described in several of the previous sections. The concept of holographic imaging relies on the phase of the electromagnetic radiation described in Chapter 2 with regard to the sinusoidal character of the wave phenomenon. Our stereoscopic vision by definition uses phase information to retrieve depth information. The eye processes the incoming electromagnetic radiation with the encoded phase information, and the brain performs a differential analysis of the respective signals from both eyes to obtain the particular distance to the individual eyes of the object in sight.

The initial concept of using this electromagnetic phase information in image reconstruction was suggested by the Hungarian scientist Dennis Gabor in 1948. It was not until the coherent laser light made its entry that phase-sensitive image recordings could be stored on photographic plate.

Holography is interferometry from a three-dimensional object. A laser beam is split into two, forming a reference beam and a probing beam, each by definition in phase with each other. The reference beam is aimed at a photographic grid after diversion through an array of mirrors. The probing beam is also reflected off a similar set of mirrors to create an equal path length after reflection from the target object to the same photographic grid. The phase information is initially recorded by means of photographic oxidation through interference on a three-dimensional open weave grit coated with silver chloride. The small differences in path length caused by the topography of the target object will cause a changing interference pattern to be written on the photographic grid. The redisplayed image also uses the interference from the secondary sources on the photographic screen to retrieve the depth information by appealing to the stereoscopic vision.

The production of holographic dental recordings, for quality control purposes, is a reliable and effective way to store dental records. These holograms can subsequently be used to carve a fitting denture replacement.

14.6 Polarization Imaging

As discussed previously in Chapters 2 and 4, the orientation of the electric field determines the polarization of electromagnetic radiation. Similar to a polaroid sheet, striated tissues will let light with the polarization parallel to the striation pass through the tissue without too much attenuation. The only problem is that the striation is not necessarily linear throughout the tissue. Another disadvantage is that the tissue needs to be relatively thin to at least have only a single cell layer that is relatively parallel in striation. This last criterion virtually limits the use of polarization imaging to histology slides only.

One particular application of polarization imaging is the determination of irreversible damage in microscopic histology of coagulation of muscle tissue. Everywhere the cells are destroyed, the striation is lost, and the input polarized light passes through the normal muscle after lining the polarization orientation up with the cell striation. All the dead cells do not transmit the majority of the polarized light, even at the boundary between normal and visibly destroyed cells, but the irreversible cell damage can be discriminated from the reversible cell damage, since the reversibly damaged cells still maintain their striation. An example of the use of polarized illumination for histological verification of a pathological condition is shown in Figure 14.23, where polarization is the pivotal criterion for diagnosis. The image is a biopsy of pulmonary silicosis stained with hematoxylin and eosin stain.

The next section will delve into some larger scale optical imaging techniques, starting out with various forms of transillumination imaging.

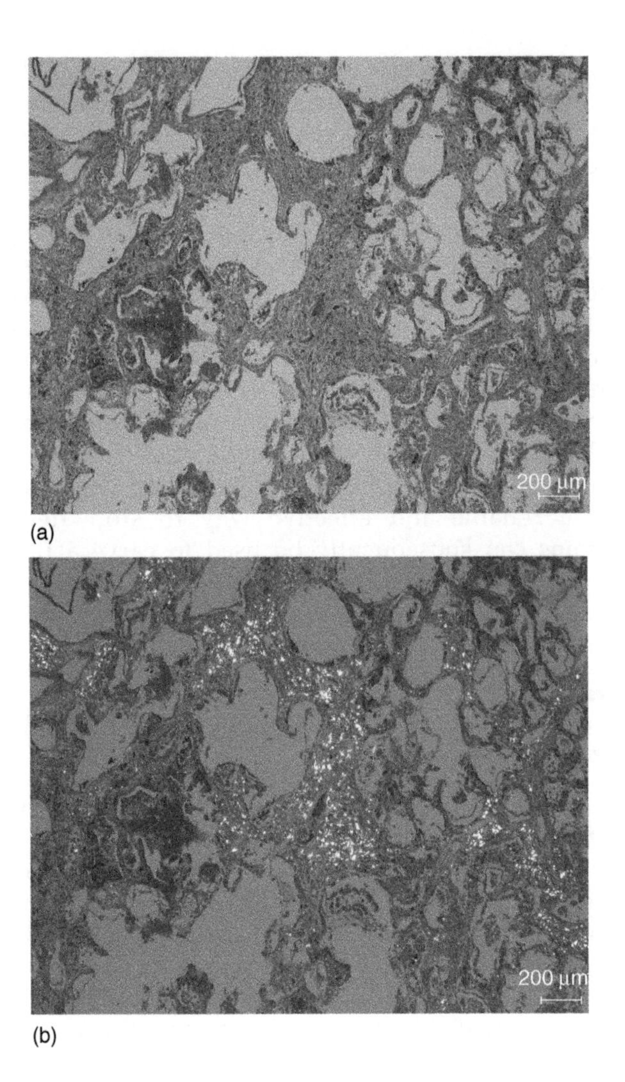

FIGURE 14.23
Microscope histology slide of pulmonary silicosis stained with hematoxylin and eosin and illuminated (a) without aligned polarized light and (b) with polarized light. Pulmonary silicosis has a typical pattern of histiocytic infiltration associated with scarring. Examination with polarized light reveals the strong birefringence of the fibrotic tissue, which cannot be seen under unpolarized light, making the use of polarized light the pivotal criterion for the diagnosis. (Courtesy of Dr. Paulo Sampaio Gutierrez, Heart Institute (Incor) of the University of São Paulo Medical School, São Paulo, Brazil.)

14.7 Transillumination Imaging

A crude but often effective method to get preliminary information on organs that are relatively transparent is transillumination. Especially infants and premature babies are small enough that the total thickness of the organ or in fact the body of the biological specimen is only a few optical penetration depths at virtually all but blue visible wavelengths. Even in infants the bone tissue is not matured enough to induce a significant amount of scattering. Especially for red and near-infrared wavelengths the bone tissue is still relatively transparent.

Transillumination was used extensively before the discovery of x-rays for detection of dental caries. During transillumination a carious lesion appears dark as a result of decreased transmission resulting from the increased scattering and absorption by the caries lesion. Recently there has been renewed interest in this method with the availability of high-intensity fiberoptic-based illumination systems.

Generally the transillumination imaging technique assumes a Beer–Lambert law type of attenuation. Small concessions can be made for scattering, but they are held responsible for image blurring. The main restriction on the usefulness of transillumination imaging is that all tissues need to be relatively transparent. Even for the section on imaging the head of an infant for blood oxygenation distribution, the skull is not fully formed and the scattering contributions from its bone are significantly less than that from a fully developed human skull.

14.7.1 Examination of the Male Genitalia of Infants

Genital examination in newborn boys is necessary in the evaluation of a number of conditions, including hydroceles and undescended testes. To minimize the examination trauma sometimes a simple transillumination examination by means of shadow imaging from a bright white light source can provide a first-order approximation of the severity of the complication, before sending a child off for x-ray examination.

Transillumination may be a useful technique to visualize the contents of an enlarged scrotum but will require additional diagnostic measures, especially in infants. Part of the male genital examination also includes checking for the presence of both testes in the scrotal sac, which can be done under transillumination as well.

14.7.2 Transillumination for Detection of Pneumothorax in Premature Infants

Artificial respiration is now an established therapy for all forms of respiratory failure in infants. Many neonatal infants derive significant benefits from forced respiration assistance.

One significant risk associated with artificial respiration, especially of premature babies, is the potential damage to the lungs. Infants and more so premature infants are vulnerable to the pressure of the ventilator, with the ultimate consequence of rupture of the lung. A pneumothorax is defined as the presence of air or gas in the space between the lung and the chest wall, the pleural space.

A time-tested crude but effective method of determining a pneumothorax of a premature infant on a ventilator is to hold the baby up to a bright light, and look at the shadow pattern in its chest. When the shadow image is suspicious, the infant will subsequently be referred for x-ray radiography. This is a delicate and relatively common problem, which has drawn some attention from researchers.

Although distinct improvements have been made over the years, the incidence of the pneumothorax remains fairly high. Additionally, the onset of diagnosis and therapy often starts relatively late, normally initialized by declining clinical condition often associated with a low arterial oxygen pressure.

14.7.2.1 Continuous Monitoring for Pneumothorax Detection

Several devices have been developed over the years for continuous monitoring of pneumothorax in premature babies. When the chest is illuminated, the pleural and lung will provide backscatter of light. The backscattered intensity will most likely change when air leaks into the pleural space, because the tissue volume reached by the light changes its composition and the partial index matching condition is disturbed. Both the magnitude and direction of re-emitted light will also change. The changing magnitude of backscattered light, remittance, will be available to provide a warning sign.

A complicating issue in the optical detection of backscattered light is the fact that the optical properties of the chest wall and lung change during respiration. Changes in the oxygenation of hemoglobin also lead to changes in absorbance of the chest wall and lung. Additionally, changes in the expansion of the chest cause changes in hemoglobin concentration in the chest wall and lung, resulting in changing absorbance conditions. The extent of the variations in optical properties in the lung will most likely be different for the various phases of ventilation.

An additional consideration resulting in a change in backscattered radiance is the fact that under normal physiological conditions the pleural space is filled with liquid with a thickness of 7–24 μm. When pneumothorax occurs, the redistribution of air in the pleural space during inspiration and expiration also changes the movements of the chest wall with respect to the visceral pleura and the lung. These variations in position can produce artifacts for the detection of backscattered radiance. Especially under continuous monitoring there will be a range of normally occurring configurations.

The alternative for detecting permanent changes is to look for instantaneous changes. When air is introduced between the chest wall and lung, a trauma will occur instantaneously. The resulting detected backscatter signal will display a simultaneous abrupt change in remittance and amplitude. A layer of air that steadily increases in thickness, providing a steadily increasing distance between the lung and the light probe, is generally much harder to detect. The difficult part is in determining the rate of changes and thus establishing the detection protocol. One complication in measurement is that the layer of air should be located directly underneath the probe for maximum sensitivity.

14.7.3 Transillumination of Infant Brain

Research at the University College London, London, Great Britain, involves the development of an optical instrument for imaging the brains of newborn infants using picosecond pulses of light. The device is known as multichannel optoelectronic near-infrared system for time-resolved image reconstruction: MONSTIR. The imaging device that is mounted on the head of an infant is illustrated in Figure 14.24. The device employs pulses of light from a dual-wavelength laser, which are coupled into a series of optical fibers in a predetermined pattern so that the point of illumination on the tissue surface is varied sequentially. Each laser pulse has a pulse width of approximately 1 ps. The head gear employs 32 detectors to record the temporal distribution of near-infrared light as it passes through the brain. The time-of-flight recording between the multiple sources and array of detectors will provide both the temporal and spatial information required for performing signal processing to accomplish an image.

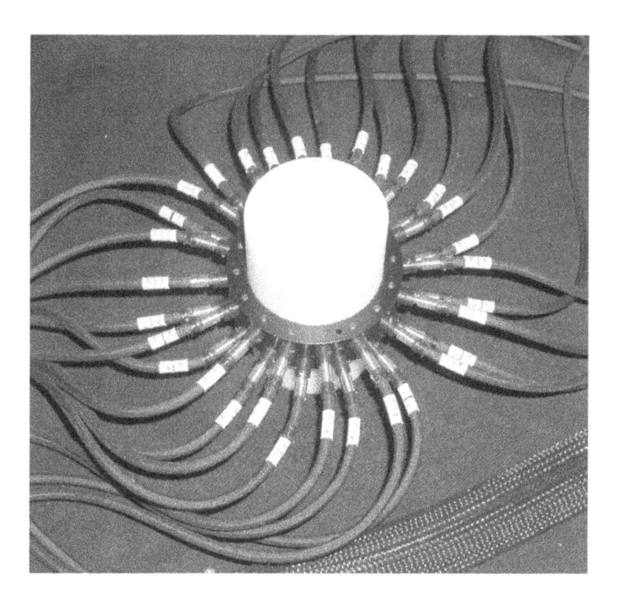

FIGURE 14.24
A schematic of the 32-channel time-resolved brain imaging system. Light delivery is by a pulsed laser coupled to 32-fiberoptic delivery devises, while the scattered light is collected by 32 fiber bundles connected to photomultiplier tubes. The photomultiplier tubes are connected to a data acquisition system and processed by a computer. (Courtesy of Jem Hebden, Ph.D., Department of Medical Physics, University College London, BORL, London, UK.)

The probing laser wavelengths allow for the detection of cerebral oxygenation and hemodynamic abnormalities in infants suspected of suffering from hypoxic–ischemic brain injury during birth. Early diagnosis will offer faster response time and reduce the occurrence of permanent brain damage.

The illumination protocol uses far-field illumination with approximately 6-mm spot size, thus allowing for the use of relatively high radiance without the risk of damage. A representative example of an optical brain oxygenation image acquired by the MONSTIR is shown in Figure 14.25.

The instrument is also being evaluated as a potential means of identifying and specifying breast disease.

The optical interrogation of the breast has specific advantages and resolution that matches or exceeds x-ray imaging, especially for smaller breasts. The principle of diffuse optical tomography is described next.

14.7.4 Diffuse Optical Tomography

Near-infrared diffuse optical imaging may offer significant contrast resolution in breast cancer imaging. The use of near-infrared light has the advantage of being nonionizing when compared to the current standard of

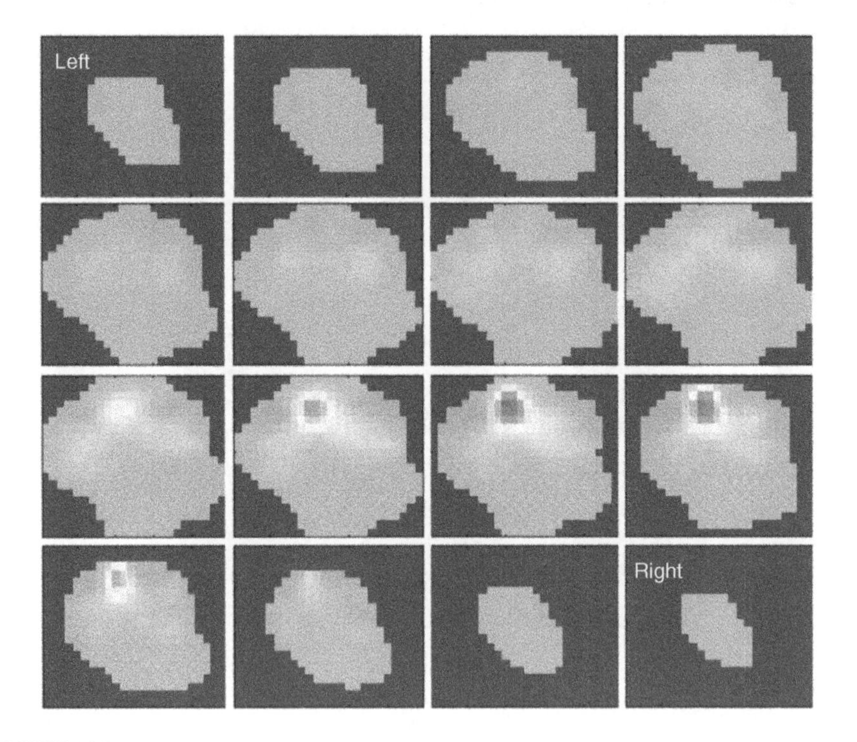

FIGURE 14.25
Recording of spatial distribution of oxygenation in an infant brain made by the MONSTIR device. (Jem Hebden, Ph.D., Dept. of Medical Physics, University College London, BORL, London, UK.)

mammography. Additionally the detection algorithm has been developed to provide resolution exceeding that of the existing imaging technologies. Image reconstruction methods utilizing finite-element solution of photon diffusion theory of the equation of radiative transfer can provide detailed imaging solutions in tissues that experience relatively small scattering in the red and near-infrared spectrum. Other potential applications are in tissues that have a significant contrast between the organs in a section of the body. The group of Huabei Jiang at Clemson University has performed groundbreaking work in the development of a detection algorithm in diffuse optical tomography.

14.7.4.1 Theoretical Background of Diffuse Optical Tomography

In the near-infrared the conditions favor the diffusion approximation in general and in particular for breast tissue when considering the optical properties in the near-infrared spectrum. The theoretical description of diffuse optical tomography involves considerable theoretical analysis and can be found in various specialized books. Starting from the diffusion approximation as given by the following equation, the theoretical derivation of the imaging algorithm will be illustrated,

$$\nabla D(x,y,z)\nabla\Psi(x,y,z)-\mu_{\mathrm{a}}(x,y,z)\Psi(x,y,z)=-S(x,y,z) \qquad (14.38)$$

where $D(x,y,z)$ is the diffusion coefficient with $D=1/3[\mu_{\mathrm{a}}+(1-g)\mu_{\mathrm{s}}]$, $S(x,y,z)$ the source function, $\mu_{\mathrm{a}}(x,y,z)$ the local absorption coefficient, and $\Psi(x,y,z)$ the radiant energy fluence rate.

At the heart of diffuse optical imaging lies the fact that neither the diffusion or absorption coefficient is known, nor is the local radiant energy fluence rate, except for a finite number of measurement sites. The local optical properties are representative of the local tissue condition and will need to be resolved to form an anatomically representative image of the tissue under illumination. The optical parameters are found by curve fitting the discrete measurements of fluence rate to an appropriate light distribution under iteration of the optical properties in a Monte Carlo simulation.

The boundary conditions to Equation (14.38) are given by

$$-D(x,y,z)\nabla\Psi(x,y,z)\cdot\mathbf{n}=\alpha\Psi(x,y,z) \qquad (14.39)$$

where $\mathbf{n}$ is the normal to the surface and α a function of the internal reflection at the surface.

The assumption is made that there is a diffuse point source, as given by

$$S=S_0\delta(x-x_0,y-y_0,z-z_0) \qquad (14.40)$$

As explained in Chapter 6 the diffusion equation can be developed in a discrete number of terms and fitted to a finite-element approximation under the point source from Equation (14.40). The equation-fitting procedure provides the following set of matrix equations in the discrete algorithm, including derived differential equations. The expression in the following equation is the matrix formulation of the radiance energy fluence rate:

$$[A]\{\Psi\} = \{b\} \tag{14.41}$$

where $[A]$ is the discrete component matrix defined as

$$[A] = \begin{bmatrix} a_{11} & a_{12} & a_{13} & \cdot \\ a_{21} & \cdot & \cdot & \cdot \\ a_{31} & \cdot & a_{ij} & \cdot \\ \cdot & \cdot & \cdot & a_{mn} \end{bmatrix} \tag{14.42}$$

where the elements of the matrix A are defined as

$$a_{ij} = \langle -D\nabla\Phi_j \cdot \nabla\Phi_i - \mu_a \Phi_j \Phi_i \rangle \tag{14.43}$$

where Φ_i and Φ_j are locally spatially varying Lagrangian basis functions. Introducing χ as an expression of the local optical parameters, depending on the local boundary conditions, allows the following derivation of Equation (14.41):

$$[\mathbf{A}]\left\{\frac{\partial \Psi}{\partial \chi}\right\} = \left\{\frac{\partial b}{\partial \chi}\right\} - \left[\frac{\partial \mathbf{A}}{\partial \chi}\right]\{\Psi\} \tag{14.44}$$

Finally the expression for the differential measurements between the various detectors and the iterated values from calculations is represented as

$$(\Im^{\mathrm{T}}\Im + \lambda I)\Delta\chi = \Im^{\mathrm{T}}(\Psi^{\mathrm{o}} - \Psi^{\mathrm{c}}) \tag{14.45}$$

where the assumption is made that the absorption is much less than the reduced scattering coefficient: $(1-g)\mu_s \gg \mu_a$. The symbol $\langle\,\rangle$ signifies integration over the three-dimensional vector configuration of the measurement setup; $\Im$ is the Jacobian matrix that is formed from $\partial\Psi/\partial\chi$ based on the measurements at the boundary sites, with $\Im^{\mathrm{T}}$ the transposed matrix of $\Im$; $\Delta\chi = (\Delta D_1, \Delta D_2, \ldots, \Delta D_N, \Delta\mu_{a,1}, \Delta\mu_{a,2}, \ldots, \Delta\mu_{a,N})^{\mathrm{T}}$ is the update vector for the optical properties profile, N is the number of nodes used in the finite-element method; and the following vector notations are introduced: $\Psi^{\mathrm{o}} = (\Psi_1^{\mathrm{o}}, \Psi_2^{\mathrm{o}}, \ldots, \Psi_M^{\mathrm{o}})^{\mathrm{T}}$ and $\Psi^{\mathrm{c}} = (\Psi_1^{\mathrm{c}}, \Psi_2^{\mathrm{c}}, \ldots, \Psi_M^{\mathrm{c}})^{\mathrm{T}}$, where Ψ_i^{o} and Ψ_i^{c} are the observed (o) and calculated (c) data for each of the $i=1,2, \ldots, M$ boundary conditions, respectively.

The actual measured values for the local fluence rate are obtained by means of fiberoptic detectors arranged on a circular pattern at fixed angles with each other. The radius of the circle would need to be adjusted to fit around various size objects and make contact or close contact to satisfy the boundary conditions to the best possible terms.

Additional measurements can be performed by axially moving the ring of (fiberoptic) detectors or by placing identical rings at fixed axial distances. A useful measuring geometry is outlined in Figure 14.26. Combining all simultaneous or successive axial measurements to provide a large three-dimensional data array provides the platform to find the inverse solution.

To form a justified estimation of the diffusion and the absorption coefficients, these will need to be expanded in a similar fashion as a function of the radiant energy fluence rate. The optical parameters will become finite series of Lagrangian basis functions with unknown coefficients.

The image reconstruction is then based on the inverse solution to Equations (14.43) through (14.45), with an emphasis on Equation (14.45), using an iteration process. The iteration process starts with the reasonable initial values for the diffusion coefficient D and absorption coefficient μ_a, obtained from in vitro or in situ measurements. The iteration process uses the measured values of the diffuse light collected at the boundary of the medium at predetermined locations.

The matrix $\Im^T\Im$ is known to be ill-conditioned. Stabilizing or regularizing of the matrix can be accomplished by various standard mathematical techniques.

The greater the number of stages in the diffusion Taylor expansion, the more detail can be retrieved, although standard diffusion approximation only uses the first (polynomial term 1:P_0) and the second terms (polynomial term 2:P_1); this was referred to as the P_1-approximation in Chapter 6.

A least-squares minimization algorithm can be used to determine the previously mentioned optical parameters. The algorithm used to determine the boundary conditions coefficient α is outlined as an example. The minimum of χ^2 by definition corresponds to the correct value of α associated with the tissue. The optical parameters χ^2 is expressed as

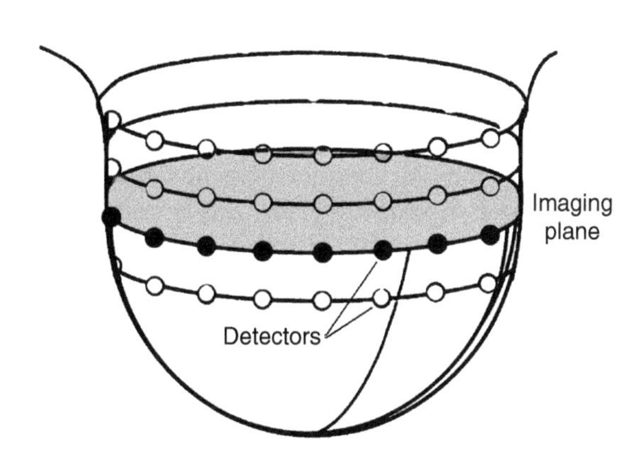

FIGURE 14.26
Cylindrical symmetric detection configuration commonly used in diffuse optical imaging.

$$\chi^2 = \sum_{i=1}^{M} (\Phi_i^{(m)} - \tilde{\Phi}_i^{(c)})^2 \tag{14.46}$$

where M is the number of boundary measurements, $\Phi_i^{(m)}$ the measured photon density, and $\tilde{\Phi}_i^{(c)}$ the computed photon density from a numerical simulation of a homogeneous medium with the same geometry. Applying the same technique can resolve the initial estimates of the optical properties.

Standard heterodyning can be used to collect the average signal, avoiding the vast majority of background noise sources and instabilities of the device.

A near-infrared multichannel frequency-domain optical imaging system powered by intensity-modulated diode laser has already shown the potential of imaging aberration in the female breast structure. Objects of smaller than 10 mm have so far been recognized with this methodology, detecting discrepancies in optical properties between the healthy and the diseased tissues ranging from 20 to 70%.

14.7.4.2 Optical Heterodyning

Heterodyning is the modulation of either amplitude or path length of an optical system, while measuring with the exact same frequency as the modulation frequency, thus providing a common mode rejection of any steady-state signal, or signal with random fluctuations: ambient light and background noise.

The modulation of the input signal, by either mechanical or electronic means, causes a change in frequency pattern of the collected signal. Combining the input and detector signal gives a term that is obviously periodic and is said to occur at the beat frequency between the two signals. The mechanism is analogous to any optical communication system, with the beat frequency serving as the carrier frequency.

Optical heterodyne detection is used in the fiberoptic system shown in Figure 14.27. The translation of the reference mirror causes a change in frequency of the reflected light, proportional to the mirror velocity that creates a Doppler shift (see Section 4.2). If there is a piezoelectric transducer (PZT) in the sample arm it can be modulated at high frequency and stretches the fiber to change the path length of the sample arm. A PZT is a ceramic element that changes its seize under electrical stimulus. A fiber wrapped around a PZT tube can thus be stretched at the rate of the electric current modulation. When the PZT is modulated at a frequency higher than the Doppler frequency generated by the mirror, this has the effect of amplitude modulating the interferometric signal. Combining the reference light with light returned from the sample, which does not undergo a frequency shift, at the detector results in a photocurrent term proportional to the frequency difference between the two waves. This term is obviously periodic and is said to occur at the beat frequency between the two signals.

Combining balanced detection with optical heterodyne detection is a powerful method for reducing the noise that is generated along with the interferometric signal. Consider the OCT system shown in Figure 14.27.

The amplitude of the source is modulated by external means with a frequency f_1, and the detector collects light that is modulated at a frequency f_2 and results from dispersion effects in the medium and optional PZT fiberoptic stretching. The detected signal is recombined with the source signal for reference. The source and detector signals can be written with the frequency ν_0 of the light source as shown in Equations (14.47) and (14.48).

The source signal is represented as a standard sinusoidal expression varying in time as given by

$$A_1(t) = A_{10}\sin[2\pi(\nu_0 + f_1)t] \tag{14.47}$$

Similarly the transmitted signal is given by

$$A_2(t) = A_{20}\sin[2\pi(\nu_0 + f_2)t] \tag{14.48}$$

Applying the superposition principle the mixing of the measured and source signals becomes the sum of the amplitudes as given by

$$A(t) = A_1(t) + A_2(t) = A_{10}\sin[2\pi(\nu_0 + f_1)t] + A_{20}\sin[2\pi(\nu_0 + f_2)t] \tag{14.49}$$

The fluence rate is proportional to the square of the amplitude, thus giving

$$\begin{aligned}\Psi(t) \propto A(t)^2 &= (A_1(t) + A_2(t))^2 \\ &= A_{10}{}^2\sin^2[2\pi(\nu_0 + f_1)t] + A_{20}{}^2\sin^2[2\pi(\nu_0 + f_2)t] \\ &\quad + 2A_{10}A_{20}\sin[2\pi(\nu_0 + f_1)t]\sin[2\pi(\nu_0 + f_2)t]\end{aligned} \tag{14.50}$$

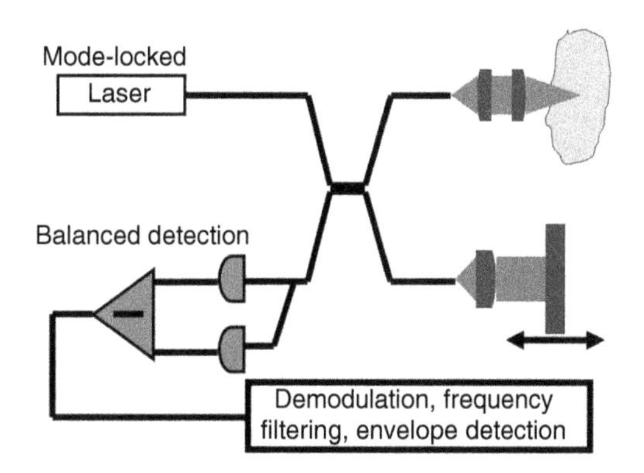

FIGURE 14.27
Illustration of OCT system that combines optical heterodyne detection with balanced detection.

Next a helpful trigonometric identity is applied to simplify the equation for the detected signal, as given by

$$2\sin\theta_1 \sin\theta_2 = \cos(\theta_1 - \theta_2) - \cos(\theta_1 + \theta_2) \tag{14.51}$$

Using Equation (14.51) and substituting in Equation (14.50) the detected signal can now be written as

$$\begin{aligned}\Psi(t) \propto A_{10}^{\ 2} \sin^2[2\pi(\nu_0 + f_1)t] + A_{20}^{\ 2} \sin^2[2\pi(\nu_0 + f_2)t] \\ + A_{10}A_{20}\{\cos[2\pi(f_1 - f_2)t] - \cos[2\pi(2\nu_0 + f_1 - f_2)t]\}\end{aligned} \tag{14.52}$$

The frequency of the light source itself is too high for the photodetector to accurately detect it; in fact, it averages the signal over time, which means that all terms that contain ν_0 will either average or cancel out. The time average of $\sin^2(\theta)$ is 1/2 and the last term in Equation (14.52) averages out to zero; thus, rewriting Equation (14.52), the voltage output of the detector can be expressed as

$$V(t) \propto \frac{A_{10}^{\ 2}}{2} + \frac{A_{20}^{\ 2}}{2} + A_{10}A_{20}\cos[2\pi(f_1 - f_2)t] \tag{14.53}$$

which provides a sinusoidal varying signal with the beat frequency $(f_1 - f_2)$, while eliminating any random functions in the signal resulting from ambient light and electrical disturbances. The signal-to-noise ratio is dramatically enhanced this way. The concept of the signal generated by heterodyning is illustrated in Figure 14.28.

Another approach to the problem of elimination of noise is by modulating the input source and tuning the detector to measure at the same frequency, not just record the signal and process it later. This has a slightly different theoretical approach, but produces the same effect, and increases the signal-to-noise ratio significantly. This second methodology can be achieved by phase-locked-loop detection or by active filtering.

The Doppler-modulated interferometric signal, along with the direct current (dc) components that are the reference signal and the normal reflection signal from the sample, is sent to two detectors D1 and D2. By introducing amplitude noise associated with the superluminescent diode (SLD) or laser source (although such noise is much more significant when a mode-locked laser is used), the total returning signal measured by the detector can be approximated by

$$V_t(t) = V_s(\nu, t) + V_r(\nu, t) + \nu_n(t) \tag{14.54}$$

where P_s and P_r are the optical power reflected from the sample and mirror, respectively, resulting in a voltage amptitude V_s and V_r, and $\nu_n(t)$ is the random noise source associated with the laser or SLD. The Doppler-shifted reference signal occurs at optical frequency ν'. The two photocurrents exhibit the beating term that is necessary for optical heterodyning, which, in the case of the system in Figure 14.27, occurs for instance at 50 kHz with a reference mirror velocity of 20 mm/s. The two photocurrents are sent to a subtracter. The subtracter introduces a phase shift of 180° for the beat note on one of the signals, so that when the two are subtracted the resulting signal is approximated as

$$V_t(t) = 2V_s(t)V_r \cos[2\pi(f - f')t] \tag{14.55}$$

This effectively eliminates ambient light dc terms and the amplitude noise associated with the laser or SLD source. Figure 14.28 illustrates a representative signal acquired under standard heterodyning technique. The remainder of the detection system provides frequency filtering and envelope detection. Such systems have been successful in producing a 93 dB dynamic range in an OCT device that was not optimized for the source spectral width. The decibel scale uses the 10-log of the ratio of the signal divided by the noise, multiplied by 10; thus, when the signal strength is 1000 mV and the noise is 1 mV, the signal-to-noise ratio is 30 dB.

A later system achieved 110 dB of sensitivity by combing optimization of the optics with polarization control and dispersion compensation.

When the signal is demodulated at the sum frequency of the PZT and Doppler modulation, this process isolates the amplitude-modulated interferometric signal from noise outside a small bandwidth, which is

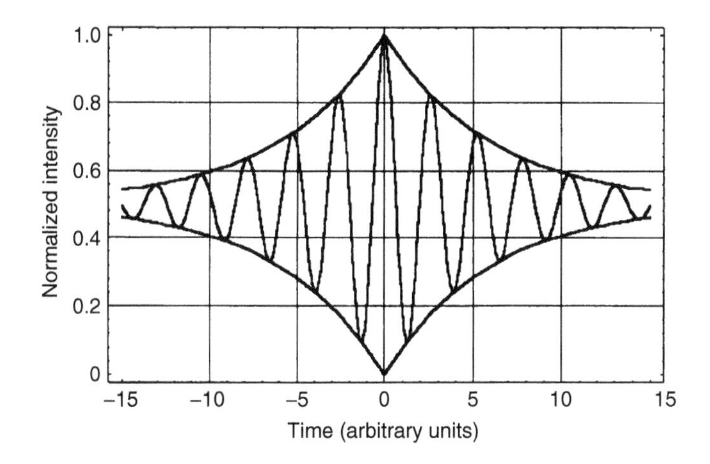

FIGURE 14.28
Signal obtained under heterodyning.

determined by the detection electronics. The demodulator is followed by an envelope detector, which consists of an intermediate frequency (IF) filter, a squarer, a low-pass filter, and a threshold detector. The IF filter isolates the frequency difference term that is generated by the demodulator. Since this term is sinusoidal, a squarer is used to rectify the signal. The low-pass filter further reduces any high-frequency noise that might be on the signal, and the threshold detector measures the IF signal amplitude.

Using this technique, it is possible to detect a returning signal that is a fraction 5×10^{-10} of the incident signal.

Other bulk tissue imaging techniques available are for instance OCT imaging. A detailed description of OCT systems and the corresponding theory is presented next. Applications are discussed, and derivatives of OCT that introduce polarization, and phase and spectral sensitivity are also described.

14.8 Optical Coherence Tomography

Successful treatment of many diseases depends on diagnosis in the early stages of the disease. There are two basic methods for sampling internal tissue to determine the presence and progression of tissue degeneration. A biopsy can be performed, in which tissue is sampled at random and surgically removed for testing. Alternatively, an imaging technique, either invasive or noninvasive, can be used to obtain information on the tissue microstructure. If the images provide sufficient detail and can be obtained for a large enough volume of tissue, the imaging approach is advantageous because it does not require removal of tissue and does not introduce a significant time delay for testing of biological samples.

There are several imaging techniques that can provide near real-time and three-dimensional imaging of biological samples. Among the most prolific are magnetic resonance imaging, x-ray-computed tomography, and high-frequency ultrasound. Each finds applicability for specific imaging problems. However, none of these techniques are capable of achieving resolution on the cellular level, which requires spatial resolution of less than 10 μm. To diagnose certain diseases in their early stages, this level of resolution is required.

A technique called OCT, which is capable of providing high-resolution images has been developed at Massachusetts Institute of Technology. Optical coherence tomography is based on the principle of low-coherence interferometry.

Analogous to computerized x-ray tomography, in OCT an image is reconstructed from multiple adjacent one-dimensional recordings. However the general concept of OCT is more like B-scan ultrasound imaging; that is, an additional feature, the coherence of the wavefront, is utilized in an interferometric approach. The original application was again in ophthalmology, where tomographic images of the retina were acquired with 10 μm resolution.

The major drawback in optical tissue diagnosis has been and continues to be the highly scattering nature of most types of tissues found in humans and animals.

Optical coherence tomography is promising because unlike many of the other techniques, which rely on high-speed optical gates and picosecond laser pulses, it relies on the low-coherence length of a broadband light source to differentiate least scattered light from repeatedly scattered light. It has potential as a low-cost alternative to other types of optical diagnosis since it does not rely on expensive high-power lasers or high-speed gates. In an OCT imaging system, a single-mode optical fiber is used to deliver light to the tissue and collect light that is backscattered from the internal tissue structure. This leads to the necessity of scanning the fiber tip over the surface to build up a three-dimensional image of the tissue under investigation since the numeric aperture of the fiber is very small. Optical coherence tomography systems typically have resolutions of 5–30 μm, depth penetrations of 1–3 mm, and up to real-time imaging rates.

Optical coherence tomography is a noncontact, nondestructive imaging modality that acquires depth-resolved two- and three-dimensional images of biological tissue.

Clinical applications of OCT include early diagnosis of macular edema, a disease that affects the human retina and can lead to blindness. Optical coherence tomography is particularly applicable in this case because the disease causes small holes to form in the retinal layer that cannot be detected by high-frequency ultrasound or other ocular diagnosis techniques. Diagnosis of glaucoma requires precise analysis of ocular structures such as the retinal nerve fiber layer and ganglion cell layer that can be provided by high-resolution OCT.

Other applications of conventional OCT are diagnosis of Barrett's esophagus, imaging of arterial walls as well as cervical and intestinal tissues, and monitoring of arterial stent deployment.

As an optical method, OCT can make use of all the information inherent in performing optical measurements. That is, information on the polarization and phase of the returning light can be used to obtain sample information in addition to that obtained by measuring the amplitude of the returning signal. Because of the broad bandwidth required to generate low-coherence light and the coincidence of this spectral range (typically between 600 and 1500 nm) with important absorption bands in biological tissue, spectroscopic OCT has also become an area of active research.

Initially only conventional OCT systems that examine the amplitude of the reflected signal will be described. Other systems capable of resolving polarization and phase information will subsequently be described.

14.8.1 Conventional Optical Coherence Tomography Systems

Conventional OCT measures the amplitude of the light reflected from a biological sample and determines the longitudinal position of the reflection sites through low-coherence interferometry. Light for performing OCT is typically delivered noninvasively by simple fiber optic and objective lens components, although invasive techniques using hypodermic needles and catheters have been developed. By scanning of the sample or the delivery apparatus, two- and three-dimensional imaging is possible.

Optical coherence tomography is based on the same principles as coherence domain reflectometry (CDR). The operating principle of an OCT system is to measure the reflected signal in a biological sample as a function of depth. This allows for two- or three-dimensional imaging of tissue microstructure, depending on the type of scan that is performed. The original OCT system, developed at MIT by Fujimoto and partners, is depicted graphically in Figure 14.29.

The basic system consists of a fiberoptic Michelson interferometer with a low-coherence (broadband) light source. Optical Coherence Tomography uses optical heterodyning to generate a high-frequency alternating signal with improved signal to noise ratio while the direct two-dimensional imaging method is a measurement that cannot separate the signal of interest from repeatedly scattered light. Optical heterodyning involves mixing the backscattered light with a Doppler-shifted reference signal. The light of interest that is backscattered from a given depth can then be removed from the steady-state continuous illumination domain, which is identified as the repeatedly scattered background signal. Repeatedly scattered light is the light that has undergone many scattering events while in the tissue and which has lost information that it carries about the tissue structure.

Optical power from the light source, shown in Figure 14.29 as a superluminescent diode, is split approximately 50/50 between the two arms using a 2 × 2 fiberoptic demultiplexer. In the sample arm, collimating and focusing optics direct light onto the biological sample to be imaged. Light is reflected from the sample at different depths and is allowed to pass unidirectionally through the fiberoptic splitter, so that it reaches the detector but does not result in optical feedback to the light source, which could result in narrowing of the emission spectrum. Note that light reflected from varying depths within the sample will travel different distances, or equivalently, the light will have a variation in the optical path length. The reference arm contains a mirror, which reflects most of the light transmitted down the

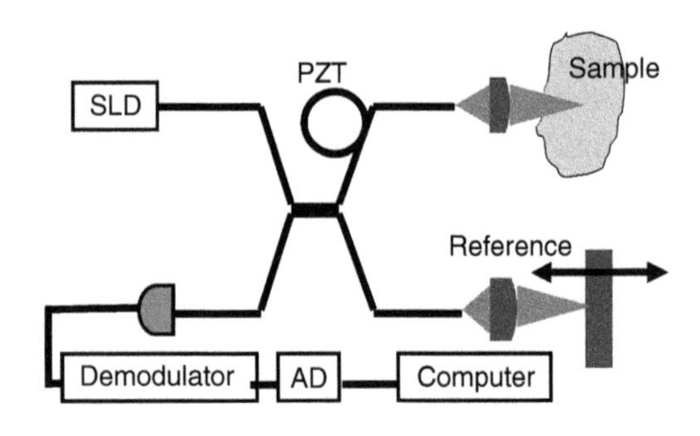

FIGURE 14.29
Diagram of the original Michelson Interferometer OCT configuration. AD, analog-to-digital converter; SLD, superluminescent diode; PZT, piezo-electrictransducer.

reference arm back to the detector. The position of the mirror is modulated to vary the optical path length along the reference arm. This general principle was shown in the section on heterodyning (Section 14.7.4.2). The only time an interference signal is generated at the detector is when the optical path length of the reference arm matches that of the sample arm, to within the coherence length of the source. Scanning the reference mirror thus allows measurement of the amplitude of the signal generated from sample reflection as a function of depth within the sample.

The rudimentary description of OCT operation so far does not capture the physics involved. The basic principles involved in OCT are as follows:

1. Coherence length and the generation of low-coherence light
2. Operation of a low-coherence Michelson interferometer

Since the rigorous physical analysis of partial coherence interferometry is extremely involved, an approximate description that preserves the important features while attempting to provide an intuitive picture will be used. Detection systems will also be discussed in some detail.

14.8.2 Light Sources and Coherence Length

To obtain the high resolution that is achieved using OCT it is necessary to employ a light source that has a relatively long coherence length compared to laser. The axial resolution in OCT is limited to this coherence length. The concept of coherence length will be discussed on the basis of a simplified viewpoint, so that this resolution limit can be understood relatively easily.

The coherence length of a source can be defined as the physical length in space over which one part of the electromagnetic wave train bears a constant phase relationship to another part. For practical measurements, the coherence length of a source is defined as the optical path difference (OPD) (between two electromagnetic waves) over which interference effects can be observed. The coherence length is thus the net delay that can be inserted between two identical waves, which still allows observation of constructive and destructive interference.

There are two important points to make in an elementary discussion of coherence length. The first concerns the relationship between the length of an electromagnetic wave train, which cannot be infinite, and the spectral content of the radiation. The second is the relationship between this spectral content and coherence length of the radiation.

Consider the radiation source shown in Figure 14.30. All electromagnetic radiation must originate from an accelerating electric charge. The source depicted in Figure 14.30 is analogous to an electron that undergoes an atomic transition from an excited state to a lower state, as occurs in a laser or SLDs. While it is not intuitive to say that the electron oscillates during this process, this assumption is made on the theoretical premise explained in quantum mechanics based on the requirement mentioned earlier for the generation of

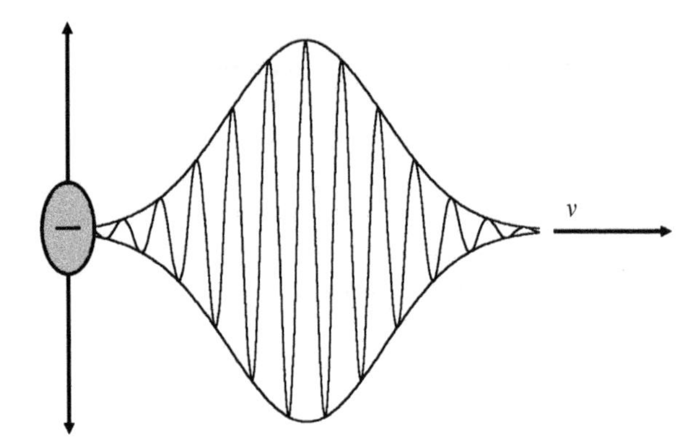

FIGURE 14.30
An electron that oscillates for a finite time period produces a wavepacket that travels away from the source.

electromagnetic radiation. Note that the duration of the wave packet is finite, as required by the finite time period of the electronic transition. This places a requirement on the spectral content of the wave packet that is best understood through an application of Fourier analysis. The Fourier series allows construction of any periodic function through a (infinite) summation of sine and cosine functions with different amplitudes and frequencies. The number of component functions with nonzero amplitude depends on the length and structure of the original waveform. Using the Fourier integral, the explicit relationship between the length of the pulse and its spectral content can be obtained in frequency units as explained in the following equation:

$$\Delta t \, \Delta \nu \geq \frac{1}{4\pi} \tag{14.56}$$

The product of the pulse duration and its spectral width must be equal to a constant, requiring that for a short pulse there will be a large number of component frequencies present. The case as presented in Equation (14.56) describes the minimum constant value that can be obtained as predicted by either Fourier analysis or quantum mechanics, and occurs for a Gaussian distribution of the amplitudes of the component sine and cosine functions. It is important to note the consequences of this relationship for optical sources. The only way to produce a truly single-frequency source is to use an oscillating electric charge that has been in motion for an infinite time period, which is not physically possible. This places fundamental limits on the source coherence length and spectral width.

To obtain the relationship between spectral width and coherence length, consider the effects of introducing a delay between two identical wave packets. Suppose a single wave packet is partially reflected off two mirrors separated by a distance *d*. Each returning pulse will have traveled a different distance upon reaching a detector. Obviously, if the mirror separation d is greater than the length of the wave packet, no interference effects can occur as the two pulses will be detected at different times. This presents the definition of the coherence length of light as the length of the wave packet.

If a series of wave packets are generated, the coherence length depends on whether or not there is a constant phase relationship between the individual wave packets. In the case of a thermal source, atoms emit at random and with random frequency, so there is no relationship between pulses emitted from a single atom, and obviously not between different atoms. In the case of a continuous wave (CW) laser, however, the process of stimulated emission requires subsequent photons (or wave packets) to be emitted with the same frequency, phase, and polarization as the stimulating photon. The coherence length of a CW laser is then limited to by cavity instabilities that allow the laser center frequency or primary mode of oscillation to change.

Each component frequency of the two pulses mentioned earlier produces an interference pattern at the detector, resulting in the superposition of a large number of such patterns. However, if the delay between the two pulses is long enough, one extreme frequency will exhibit an interference maximum at the same point that the other extreme frequency exhibits a minimum, resulting in constant intensity. For values of d greater than this, no interference can occur, either because the pulses will arrive separated by a time delay greater than the duration of the pulse, or because the next, unrelated pulse will have arrived. Analysis of the location of interference maxima and minima as related to source spectral width in units of wavelength gives the following approximate relationship for the coherence length L_c:

$$L_c = \frac{2\log(2)}{\pi} \frac{\lambda^2}{\Delta\lambda} \cong \frac{\lambda^2}{\Delta\lambda} \tag{14.57}$$

The coherence length is proportional to the square of the source center wavelength and inversely proportional to the spectral width. In low-coherence interferometry, described in the next section, the coherence length is the limit of the system's longitudinal resolution.

High longitudinal resolution in an OCT system can be obtained by using a source with a broad spectral width. Superluminescent diodes and mode-locked lasers have become the standard and high-resolution light sources of choice. The two devices achieve broad spectral widths using different principles. An SLD is essentially a diode laser with a broad gain bandwidth that does not have a mirror to produce optical feedback. The output is continuous wave but, as with a thermal source, the individual wave packets are not coherent with one another. A typical SLD can have a spectral width full width half maximum (FWHM) $\Delta\lambda$ as large as 32 nm in the red or near-infrared. This leads to a coherence length in air of approximately 10–15 μm, with the resolution in tissue being the free-space resolution divided by the tissue's refractive index. This resolution is insufficient for imaging individual cells or intracellular structure. Mode-locked lasers, however, are capable of producing extremely short pulses. From Equation (14.56), it can be shown that this generates a broad spectrum. Pulses as short as two optical cycles have been produced, with spectral widths of 350 nm corresponding to a free-space resolution of 1.5 μm, and less than 1 μm in tissue.

14.8.3 Theory of Optical Coherence Tomography

Optical coherence tomography utilizes the low-temporal coherence of a light source to resolve, on the *z*-axis, the position where backscattered light is being measured. The diameter of the target point is related to the numeric aperture of the delivery/detection fiber. A single-mode fiber is generally used because the small numeric aperture reduces the solid angle from which light may be collected. This provides high lateral resolution. Either the tissue or the fiber tip is then scanned in two dimensions to build up a point-by-point two- or three-dimensional image of the specimen. The reflectance of the tissue at a known depth is the physical property that is measured.

The single-mode fiber satisfies the additional requirement of mandating coherence throughout the detection system. A multimode fiber will allow for interference to take place inside the fiber as well, between the multiple modes, before reaching the detector. Internal interference will void valuable measurement information.

Figure 14.31 shows the basic OCT apparatus. Analysis with a spectrum analyzer can then be used to select the modulation frequency and use only this as the source. The interference signal can now be examined separately from the rest of the returning light. Heterodyning produces signal to noise ratio of up to 120 dB. Signal to noise ratios of at least 100 dB are necessary to obtain any useful depth resolution in highly scattering media.

The method selected was to determine the visibility of the fringes after the light has passed through the tissue. Images of the interferometric fringe pattern are correlated with a reference image taken at or close to zero optical

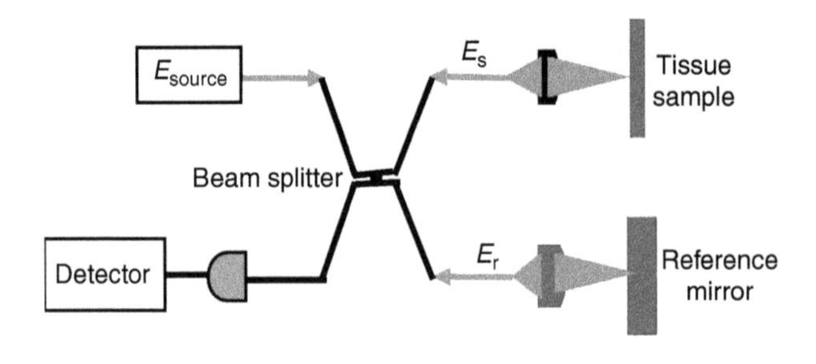

FIGURE 14.31
Fiber optic Michelson Interferometer. Electric fields are defined as follows: E_s: sample signal, E_r: reference signal. The sample has been replaced with single reflecting surface.

path imbalance. The images are processed electronically after they are acquired. The position of the autocorrelation peak is located and the value at that position is saved. Subsequent data are taken from these coordinates, so in fact, it is not the correlation peaks that are being recorded, rather the value at the position where the autocorrelation peak is located. This number allows the determination of the visibility function $V(t)$. It is expected that this function will broaden as the scattering of the medium increases.

14.8.4 Operation of the Fiberoptic Michelson Interferometer

The principles of interferometry are discussed in two parts: the intensity transfer function of the interferometer and the dependence of the detected signal on the OPD between the sample and reference arms. The following approximations are made:

1. Source is monochromatic.
2. Dispersion effects are thus eliminated in analyzing the detected intensity as a function of reference mirror position.

These approximations will be modified qualitatively in later sections. The additional approximation that the source emits plane waves, which is not truly a good approximation to the Gaussian single-mode beam profile, will not be revised for the sake of simplicity.

Initially, consider the fiberoptic interferometer shown in Figure 14.31. The incident wave is written as follows:

$$E_{\text{source}} = E_0 \mathrm{e}^{\mathrm{i}(kz-\omega t)} \tag{14.58}$$

where ω is the angular frequency of the electric field and E_0 the real amplitude of the wave. The quantity k is the (pure real) wave number, which is related to the wavelength of the source as

$$k = \frac{2\pi}{\lambda} n \tag{14.59}$$

Equation (14.58) is specific to an electric wave propagating in the positive z-direction in a nondispersive, dielectric material.

The amplitude of the source electric field is split approximately 50/50 at the fiber-optic coupler. To obtain the intensity transfer function of the interferometer, the expressions for the returning signals are as follows:

The reference signal is defined as

$$E_{\mathrm{r}} = E_{\mathrm{r}0} \mathrm{e}^{\mathrm{i}(kz_{\mathrm{r}}-\omega t)} \tag{14.60}$$

While the sample signal is defined as

$$E_s = E_{s0} e^{i(kz_s - \omega t)} \tag{14.61}$$

where the subscript "r" indicates the returning reference signal and the subscript "s" refers to the returning sample signal. The terms z_r and z_s are the respective distances to and from the reference mirror and a specific reflection site within the sample. The amplitudes of the returning signals are no longer the same, as the total amount of light reflected from the sample will be very small, while nearly all of the light incident upon the reference mirror will be collected. The expression for the total electric field at the detector is given by the superposition of these two returning waves as

$$E_{total} = E_{r0} e^{i(kz_r - \omega t)} + E_{s0} e^{i(kz_s - \omega t)} \tag{14.62}$$

For the electric waves described above, the detected intensity can be calculated from the square of the electric field amplitude as

$$I = \frac{n}{2\mu_0 c} E_{total} E^*_{total} \tag{14.63}$$

where n is the material refractive index (n = 1.5 for glass), μ_0 the magnetic permeability, c the speed of light in vacuum, and the asterisk (*) denotes complex conjugation of the electric field expression. The result after carrying out the complex conjugation and transferring the expression to trigonometric notation for clarity is given by

$$I = \frac{n}{2\mu_0 c} \left(E_{s0}^{\ 2} + E_{r0}^{\ 2} + \sqrt{E_{s0} E_{r0}} \cos(kz_r - kz_s) \right) \tag{14.64}$$

The time dependence vanishes and the resulting expression contains an interference term proportional to the cosine of the wave number k times the path length difference $z_r - z_s$. By substituting Equation (14.59) for k and noting that the sample and reference intensity can be given by Equations (14.65) and (14.66), the detected intensity can be found.

The sample intensity is given as

$$I_{s0} = \frac{n}{2\mu_0 c} E_{s0}^{\ 2} \tag{14.65}$$

The reference intensity is defined as

$$I_{r0} = \frac{n}{2\mu_0 c} E_{r0}^{\;2} \tag{14.66}$$

From these equations the detected intensity can be presented as

$$I = I_{s0} + I_{r0} + 2\sqrt{I_{s0} I_{r0} \cos\left[\frac{2\pi}{\lambda_0}(z_r - z_s)\right]} \tag{14.67}$$

The above equation shows that the detected intensity depends on the relative distance traveled by each of the electric waves because of the interference (cosine) term. For an ideal case in which the returning electric field amplitudes are exactly equal, Equation (14.67) predicts the peak intensity to be four times the intensity of either wave, while the minimum is zero for perfectly destructive interference. In the case of OCT, the interference term is relatively small compared to the background due to the relative amplitudes of the reflected reference and sample signals.

The quantity inside the square brackets in Equation (14.67) is the OPD, which introduces a phase delay between the returned signals. Interference effects are the direct result of adding waves that have a different phase at the detector. For values of the OPD less than the source coherence length, the two returning electric waves will still have a constant phase relationship, and interference effects can be observed. For larger values of OPD, the two waves will have a random phase relationship so that interference effects will be random, resulting in a constant intensity at the detector. In this case, Equation (14.67) does not apply. The period (in distance) over which the intensity varies from maximum to minimum, a variation known as a fringe shift, is of the order of half the source center wavelength.

Equation (14.67) was developed assuming a monochromatic wave. The effect of using a broadband source is to decrease the OPD over which Equation (14.67) applies. Interference effects are still observed within the coherence envelope, but they are blurred as the OPD exceeds the coherence length. In contrast to conventional interferometry, OCT systems do not measure individual fringe shifts within the coherence envelope. Measuring individual fringe shifts allows determination of the change in optical path length of a single arm, typically the arm with the movable mirror. It is required that one arm is held constant as a reference, and the other is used for measurement. However, attempting to measure individual fringe shifts in OCT and obtaining sample information would require using both arms for measurement with no reference. It is assumed that for each transition through the coherence envelope the reflection site within the sample is constant. Fringe shifts then reveal the change in position of the reference mirror and nothing about the reflection site within the sample. However, the mirror position is still used to gauge the sample measurement depth indirectly. The absolute distance to the reference mirror is always known, so that the distance to the reflection site can be calculated by noting the distance to the reference mirror at which interference effects begin to occur. Because interference effects will occur over a finite distance defined by the coherence length, the absolute axial position of the reflection site in the sample would not be known with greater precision than the source coherence length. It can, however, be located repeatedly with much higher precision.

This has implications for determining the structure of the reflecting feature. The minimum interference signal generated will always be the entire interference pattern, even for an infinitely thin reflecting surface, for reasons stated earlier. As the thickness of the sample increases, if light is continuously reflected from different depths throughout the feature the interference pattern will be broadened. This gives information about the thickness of the reflecting feature, but also requires that individual features be spaced further than one coherence length apart to be clearly resolved.

An important feature of OCT is the fact that the axial resolution is limited only by the source coherence length. This is in contrast to confocal microscopy, in which the axial resolution is determined by the numerical aperture of the lens. This allows use of both large numerical aperture optics and fiberoptic micro-lenses, so that highly adaptive imaging devices can be designed.

For the Michelson interferometer the delay time τ is given by

$$\tau = \frac{2d}{c} \tag{14.68}$$

in air and is simply the retardation time of the beam due to the path difference $d=|L_1-L_2|$ divided by the speed of light c. The detected intensity as a function of the lingering time is given by

$$\langle [E(t)+E(t-\tau)]^2 \rangle = \langle E(t)^2 \rangle \{\langle [E(t-\tau)]^2 \rangle + 2\langle E(t-\tau)] \rangle\} \tag{14.69}$$

where the angle brackets denote time averages. The first two terms on the right-hand side give the intensities of the two fields if interference does not occur. If conditions are such that interference occurs, then the last term describes the interference effects. The conditions for interference are as follows:

1. Both beams are either unpolarized or polarized in the same plane
2. The coherence time of the light source is significantly longer than the delay time τ

When this last term is normalized by dividing by the beam intensity, it is known as the normalized autocorrelation function $\gamma(\tau)$. The concept of autocorrelation will be discussed in the following section regarding correlation theory.

So far monochromatic light was used in the derivation. The next section expands on that assumption by introducing the effects resulting from a broadband source.

14.8.5 Correlation Theory

Correlation theory is concerned with quantifying the similarity between two signals. It has wide application from radar signal processing to character recognition.

The autocorrelation function of the optical electric field is defined by

$$\gamma(\tau) = \frac{\langle E(t)E(t-)\rangle}{\langle [E(t)]^2 \rangle} \tag{14.70}$$

and represents the degree to which the field correlates to itself at an earlier time.

For a monochromatic field, the fringe intensity for a Michelson interferometer can be written as

$$S(\tau) = \frac{S_0}{2}[1+\cos(2\pi\nu\tau)] \tag{14.71}$$

where S_0 is the intensity of the input beam and the term $(2\pi\nu\tau)$ represents the interference term as was shown in Equation (14.68).

The intensity of fringes due to a monochromatic source varies sinusoidally with the OPD d, which is contained in the term τ. Therefore, if the interferometer is used to measure distance, once the fringes have passed through one fringe period, the position measurement becomes ambiguous. In this case the interference pattern repeats itself every whole wavelength distance. This is not the case with white light as shall now be shown.

For the broadband light source used in OCT, the electric field consists of a spectrum of frequencies $S(\tau)$ such that the intensity at the output of a Michelson interferometer is expressed as

$$S(\tau) = \int_0^\infty S(\nu)[1+\cos(2\pi\nu\tau)]\mathrm{d}\nu \tag{14.72}$$

The term in the brackets is the Michelson interferometer output intensity for one frequency. The term $S(\nu)$ corresponds to the intensity of the light at a given frequency. Since these terms are integrated over all frequencies, white light is simply the sum of many different frequency components. This integral contains both a constant term and an oscillating term. It is the oscillating term that is of interest because it gives rise to the interference effects.

The solution of the integral in Equation (14.72) yields a function given by the following equation:

$$S(\tau) = \frac{S_0}{2}[1 + \gamma(\tau)] \tag{14.73}$$

where $\gamma(\tau)$ is defined by

$$\gamma(\tau) = \int_0^\infty P(v)\cos(2\pi v\tau)\mathrm{d}v \tag{14.74}$$

where $P(v)$ is the normalized spectral density function defined as

$$P(v) = \frac{S(v)}{S_0} \tag{14.75}$$

When the normalized spectral density function can be represented as a symmetric function centered about a frequency v_0, Equation (14.76) can be written as

$$P(v) = D(v - v_0) \tag{14.76}$$

Since P(v)is normalized, the following equation follows by definition:

$$1 = \int_0^\infty P(v)\mathrm{d}v = \int_0^\infty D(v - v_0)\mathrm{d}v = \int_{-\infty}^\infty D(v - v_0)\mathrm{d}v \tag{14.77}$$

Substituting Equation (14.77) in Equation (14.74) results in the following expression for the parameter $\gamma(\tau)$:

$$\gamma(\tau) = \int_{-\infty}^\infty D(\mu)\cos[2\pi(v_0 + \mu)\tau\,\mathrm{d}\mu \tag{14.78}$$

where $\mu = v - v_0$. When a trigonometric identity is applied to Equation (14.78), the result can be expressed as shown in the following equation:

$$\gamma(\tau) = \cos(2\pi v_0\tau)\int_{-\infty}^\infty D(\mu)\cos(2\pi\mu\tau)\mathrm{d}\mu - \sin(2\pi v_0\tau)\int_{-\infty}^\infty D(\mu)\sin(2\pi\mu\tau)\mathrm{d}\mu \tag{14.79}$$

The functions D_c and D_s can now be defined such that D_c is the integral containing the cosine term and D_s is the integral containing the sine term. When this is done, the autocorrelation function can be written as

$$\gamma(\tau) = D_c \cos(2\pi \nu_0 \tau) + D_s \sin(2\pi \nu_0 \tau) \tag{14.80}$$

This equation can be further reduced by applying the goniometric expression:

$$A \sin(\vartheta) + B \sin(\vartheta) = \frac{A}{\cos(\vartheta_0)} \sin(\vartheta - \vartheta_0) = \frac{A}{\cos(\vartheta_0)} \cos\left(\frac{\pi}{2} - \vartheta + \vartheta_0\right)$$

where $\tan(\vartheta_0) = -B/A$, to yield

$$\gamma(\tau) = U(\tau) \cos[2\pi \nu_0 \tau + \phi(\tau)] \tag{14.81}$$

Thus, the cosinusoidal interference pattern is modulated by an envelope function. For white light, which has a Lorentzian spectral density, the envelope is given by the exponential decay function as

$$U(\tau) = \exp\left[\frac{-d}{L_c}\right] \tag{14.82}$$

where d is the varying optical path imbalance of the interferometer and L_c the coherence length of the light source. For any given source, the coherence length is inversely related to the bandwidth of the source so that a greater bandwidth produces a shorter coherence length.

It is this exponentially decaying envelope that allows unambiguous position measurements and is the reason that white light is frequently used in OCT imaging. The fringes rapidly vanish leaving only the constant intensity ($S_0/2$).

Fringe visibility is a parameter that is directly related to the autocorrelation function. The visibility $V(\tau)$ is defined as the difference in the intensity between the brightest and the darkest fringe divided by the sum of the brightest and darkest fringe intensity as shown in the following equation:

$$V(\tau) = \frac{S(\tau)_{\max} - S(\tau)_{\min}}{S(\tau)_{\max} + S(\tau)_{\min}} \tag{14.83}$$

Substituting Equation (14.81) into Equation (14.73), the visibility intensity of the light can be expressed as

$$V(\tau) = S_0 \{1 + U(\tau) \cos[2\pi \nu_0 \tau + \phi(\tau)]\} \tag{14.84}$$

When this expression is in turn substituted into Equation (14.83), the result is that the visibility function is simply equal to the envelope function as

$$V(\tau) = U(\tau) \tag{14.85}$$

The fringe visibility is thus governed by the envelope function and is consequently a function of the path imbalance, and is therefore also a function of the interferometer delay τ (Equation [14.68]). The visibility function is illustrated in Figure 14.32.

In cases where there are no samples present in the interferometer arms, the visibility function and the autocorrelation function are the same. However, when there are scattering samples in the object arm, the visibility function and the autocorrelation function are no longer the same. The autocorrelation function is a physical property of the light source and so it does not change unless the source changes. Fringe visibility, however, is subject to many parameters; one of which is the scattering of any medium through which the light may pass. In general, visibility is different from the autocorrelation function.

The OCT technique uses correlation theory to determine the visibility function of fringes after light has passed through a scattering medium. A reference image is taken as close to the position where the two optical paths of the interferometer are balanced. Finding the lowest radiance and subtracting it from the brightest radiance determines this position. The paths are balanced when the result of this operation is maximized. As the position of the reference mirror is shifted, images at the different mirror positions are recorded. Each of the succeeding images is correlated individually with the reference image starting with the reference image itself. The values at the position of the original autocorrelation peak are recorded.

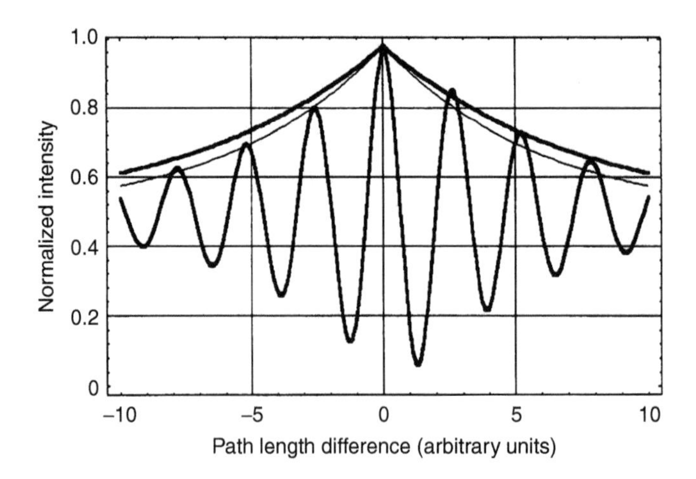

FIGURE 14.32
Graphical representation of the visibility function.

These values are plotted as a function of the position of the reference mirror to obtain the visibility function of the scattered light.

The correlation can then be calculated using a two-dimensional discrete Fourier transform. The actual calculation is most likely performed by a computer algorithm, using the right-hand part of the autocorrelation theorem expressed in the following equation:

$$\mathrm{FT}\left\{\int_{-\infty}^{\infty}\int_{-\infty}^{\infty} g(\xi,\eta)g'(\xi-x,\eta-y)\mathrm{d}\xi\,\mathrm{d}\eta\right\}=\left|G(f_\chi,f_\gamma)\right|^2 \qquad (14.86)$$

In Equation (14.86), the function that is being autocorrelated is $g(\xi, \eta)$, which would be the expression in Equation (14.81) in our case. On the right-hand side, $G(f_x\, f_y)$ is the Fourier transform of $g(\xi, \eta)$. FT stands for the mathematical symbol for the Fourier transform operation. A qualitative understanding of what is happening is that the complex conjugate $g^*(\xi, \eta)$ is swept over $g(\xi, \eta)$. Where the two functions overlap, the autocorrelation represents the area of the product of the two functions. When the functions are displaced by x and y, the area changes and thus the values of the autocorrelation change. Therefore, the autocorrelation as given by Equation (14.86) produces a third two-dimensional function with a maximum at the center of the function. This occurs because when the functions are directly atop each other, i.e., when there is no x or y displacement. The area of overlap will be at its maximum, thus leading to a maximum value at that point.

The mathematical procedure is significantly simplified when using the right-hand term of the autocorrelation theorem in Equation (14.86), which relates the autocorrelation to Fourier analysis to do the actual computation. A derivation of the autocorrelation theorem is given later on in this section. Once the right-hand side of Equation (14.86) has been computed, the inverse Fourier transform can be calculated to retrieve the visibility function for the sample.

To test the idea that correlation of displaced fringes with the image of fringes at zero path imbalance could lead to the recovery of the autocorrelation or visibility function, a computer program was developed that generated a series of simulated interference fringes. The results supported the idea that the autocorrelation function can be obtained by the cross-correlation of images. The intensity of the generated fringes $I(r)$ was described by the formula given in Equation (14.71) or can be rewritten as follows:

$$I(r)=\frac{S}{2}\left[1+\cos\left(\omega(r)+2\varphi\right)\right] \qquad (14.87)$$

where $\omega(r)$ is the frequency of the fringes due to the angle of a simulated mirror and the last term containing the variable φ represented the motion of the fringes. Also, the last term 2φ corresponds to the simulated retardation time τ. As the fringe generating function in the program proceeded, this term shifted the position of the fringes inward as if the reference mirror was changing position.

14.8.6 The Effect of Scattering on the Visibility Function

Scattering by biological media mainly has the effect of scrambling spatial intensity information. Scattered light tends to have a longer path length than light that is not scattered, and thus is less likely to interfere with the light coming from the reference arm of the interferometer. Any fringes formed are therefore due to the least-scattered or unscattered light. This means that the low coherence interferometers may be able to reject repeatedly scattered light from a signal and allow the experimenter to look at only the portion of the light returning from a particular depth that can be determined by varying the path length of the reference arm. For a broadband light source, the depth resolution should be approximately equal to the coherence length L_c of the source given by Equation (14.57) and described as follows:

$$L_c = \frac{\lambda_{avg}^2}{\Delta\lambda} \tag{14.88}$$

where λ_{avg}^2 represents the square of the average wavelength of all available light and $\Delta\lambda$ the bandwidth of the source. Typical values for these variables for white light in general are λ_{avg}=550 nm and $\Delta\lambda$=300 nm leading to a coherence length of L_c=1008 nm. The coherence length of white light therefore implies a micrometer-level depth resolution in a scattering medium.

The transfer function of a dual-beam interferometer illuminated by a partially coherent light source is described by the mean optical frequency of the source combined with the corresponding autocorrelation function. The autocorrelation function $[\gamma(\tau)]$ describes the coherence properties of the source. The source will need to have a certain level of coherence to accurately detect interference. When the interferometer has a path-length difference of ΔL, the interference pattern will be observed if the absolute value of the autocorrelation function has a nonzero value. We now define the time difference τ over the course of the path length as

$$\tau = \frac{\Delta L}{c} \tag{14.89}$$

where c equals the speed of light.

The measured irradiance can now be written using the substitution of Equation (14.89) as follows:

$$\Psi = \Psi_{mean}\left[1 + |\gamma(\Delta L)| \cos\left(\frac{2\pi}{\lambda_0}\Delta L\right)\right] \tag{14.90}$$

where Ψ_{mean} is the average fluence rate of the interferometer output, and λ_0 the center wavelength of the interferometer output, and an autocorrelation function is given by $\gamma(\Delta L)$. The theory of the autocorrelation function was described in detail in Section 14.8.5.

When the source has a Lorentzian spectral distribution, the time average of the autocorrelation function can be written as

$$|\gamma(\Delta L)| = \exp\left[\frac{-|\Delta L|}{L_c}\right] \tag{14.91}$$

where L_c is the coherence length of the source.

The visibility function will drop in proportion to the increasing bandwidth of the light source. The interference fringes will appear from $\Delta L = -\lambda_0/2$to $\lambda_0/2$. The position at $\Delta L = 0$ is called the central fringe of interference. The signal has the maximum amplitude within the entire interference pattern of the system irrespective of the line profile of the source. This key feature determines the location of the central interference fringe and locks the interferometer position to zero path-length imbalance.

The position of zero path imbalance defines the depth in the tissue where the backscattered light was collected, after correction for the tissue index of refraction.

In the formulation of this theoretical description of interferometry it is assumed that the microscopic scattering bodies could be treated in bulk as thin semi-reflective layers. Any light that scattered at extreme angles will have a large path length and will therefore not interfere with the reference light. Only light that passes straight into the medium and is subsequently backscattered almost directly back in the direction that it entered will remain coherent with the reference arm. A simple model will be introduced to illustrate this idea. Consider several layers of thin semi-reflective films stacked on top of each other as shown in Figure 14.33.

Thus, the scattering medium can be modeled as a sandwich of semi-reflecting layers. If the layers are all within the coherence length of the source then interference effects will arise due to the presence of the many layers. The nature of interference can be determined in the usual way by adding the electric fields and then taking the time average intensity.

Consider a simple case with two partially reflective surfaces stacked on top of each other: layer 1, layer 2, etc. The intensity of the fringes is as usual given by the time average of the square of the sum of the fields as expressed in the following equation:

$$\begin{aligned} \langle [E(t) + E(t-\tau_1)]^2 + E(t-\tau_2)]^2 \rangle = {} & \langle E(t) \rangle^2 + \langle [E(t-\tau_1)^2 \rangle \\ & + 2\langle E(t-\tau_1) \rangle \\ & + \langle [E(t-\tau_2)^2 \rangle + 2\langle E(t-\tau_2) \rangle \\ & + 2\langle E(t-\tau_1) E(t-\tau_2) \rangle \end{aligned} \tag{14.92}$$

where τ_1 and τ_2 are the time spend in each respective layer.

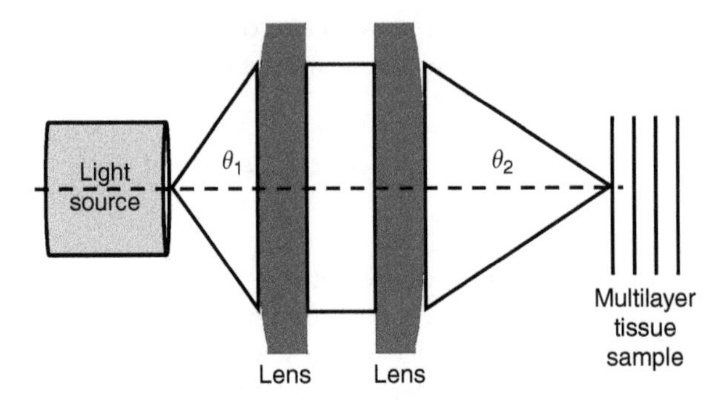

FIGURE 14.33
Close-up of typical OCT sample arm delivery optics. The tissue is illustrated as several layers of thin semi-reflective films stacked on top of each other.

One important point to note is that although there is interference arising between the layers themselves, since the distance between the layers never changes, there is no path-length variation to produce periodic fringes.

Therefore, the final interference term adds a constant intensity just like the first three terms. According to Equation 14.92 the autocorrelation functions add to produce the final effect. Each autocorrelation function simulated is a "white light type" of decaying envelope function with a "coherence length." Adding reflecting layers broadens the visibility function. This phenomenon predicts that for a scattering medium where the "layers" are only microns apart the broadening should be quite noticeable. The degree of scattering should also affect the amount of broadening as well; greater scattering should produce greater broadening.

14.8.7 Image Acquisition Process

Optical coherence tomography measures the amplitude of the light signal reflected from the sample and determines the distance to the reflection point in order to construct two- or three-dimensional images.

In this section a detailed description is provided leading up to the image construction, and some of the issues that arise due to the nature of the measurement process are discussed.

The basic method for constructing an image can be described using Figure 14.33 as an example. Later, systems introduce variations to the signal modulation and detection process that are described in various sections of this book.

The OPD (the bracketed term in Equation [14.67]) is assumed to be greater than the coherence length of the source, so that no interference effects occur. With both arms of the interferometer illuminated, the intensity at the detector is given as

$$I = I_{s0} + I_{r0} \tag{14.93}$$

which is Equation (14.67) without the interference term. The two terms in the equation are the intensity of the light reflected from the sample (that which can be collected, at any rate), and the light reflected from the reference mirror. Light exits the fiberoptic cable, is collimated by a lens and delivered to a microscope objective, and is then focused onto the sample.

Light that returns to the objective lens at an angle greater than the acceptance angle will not be collimated, and will thus fall outside the acceptance angle for the optical fiber.

For example, consider a series of simple plane reflectors. Each layer is partially reflective over the spectral width of the source, and the focal point is set to an arbitrary depth of the first reflecting layer. When the ray optics approximation is introduced, the reflections from the successive layers that result are shown by the dotted lines with arrows in the Figure 14.34.

Because of the law of reflection, all the light that is reflected from the first layer, which is at the focal point, is reflected back to the objective lens and collected by the fiber. When moving further into the sample, the beam begins to diverge. For layers close to the first layer some of the rays that are not purely axial will fall within the acceptance angle of the objective lens, but for deeper layers only the light incident normally will be returned to the objective lens and be collected by the fiber. Even in this simplified case it can be seen that an explicit expression for the amount of light returned to the detector would be difficult to obtain. Variables for this case would include the depth of the reflecting layer, the reflection and transmission coefficients for each layer, the divergence angle of the beam, and so forth. In practical systems, effects such as absorption and the scattering coefficient of the sample material would have to be accounted for, further complicating the analysis. Nevertheless, it is clear that a small amount of light reflected from each plane will be collected by the fiber and detected. It is also clear that the light reflected from each plane travels a different distance, which allows use of the reference signal to determine the longitudinal position of the reflected site.

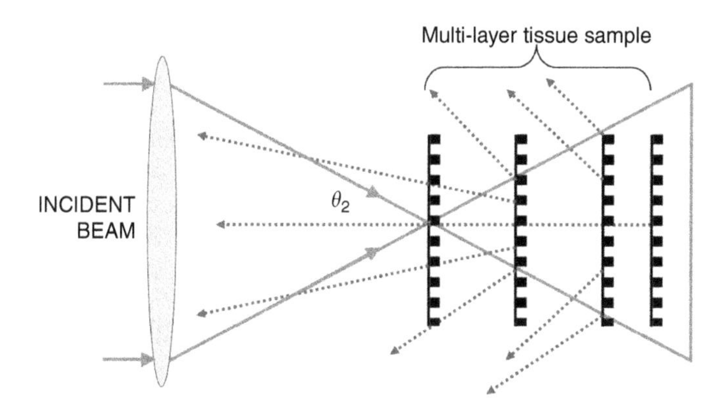

FIGURE 14.34
Examples of reflected rays from a series of planar reflectors. Dotted lines indicate reflected light.

Continuous scanning of the reference mirror modulates the length of the reference arm. Because light is reflected from different depths within the sample, each longitudinal scan will generate an interference signal at several points. Each scan corresponds to a reflection site. The positions of the backscattering sites are mapped by recording the interference envelope and simultaneously the position of the reference mirror. A single longitudinal scan thus generates a "one-dimensional" image of the reflection sites within the sample, with a transverse dimension equivalent to the spot size of the beam at the reflection site. Two- and three-dimensional images are generated by translating the beam horizontally or horizontally and vertically while performing longitudinal scans. For deeper tissue imaging, it is necessary to translate the focal plane inside the sample to maintain high resolution.

The transverse resolution of OCT depends heavily on the choice of optics in the delivery system. In a conventional system as shown in Figure 14.32, the sample is illuminated with light focused by a microscope objective or other large numerical aperture objective lens. The large numerical aperture serves to produce a larger light-gathering area, along with a smaller spot size at the focal plane. A smaller spot size directly implies better transverse resolution and, in fact, the transverse resolution of an OCT system is defined by the spot size at the beam focus. Without considering the data acquisition rate, it can be shown that there is an inherent trade-off between transverse resolution and imaging depth. It is well known that the minimum spot size that can be generated with a focusing lens is proportional to the focal length. If images of deep tissue are to be produced without the use of an invasive probe, a longer focal length lens must be used and the transverse resolution will correspondingly decrease. This problem turns out to be particularly important when imaging the human retina, which is located approximately 2.5 cm behind the outer surface of the eye. Typically, an aspheric lens that can achieve diffraction-limited spot size of ~3 μm requires a focal length of less than 5 mm.

14.8.8 Applications of Optical Coherence Tomography to Physical Problems

The following sections give examples of particular applications of OCT that have the potential of providing a significant contribution to the overall understanding of biological, anatomical, and physiological phenomena.

Two examples of conventional OCT systems will be discussed. First, because OCT and high-frequency ultrasound can be considered as competing technologies for several applications, a comparison will be made between images of deployed arterial stents obtained with each technique to compare the resolution of the two systems. Second, OCT has demonstrated to be extremely important in diagnosis of ocular degeneration. Several standard resolution images of in vivo human esophagus, an in vivo human nailbed, and an in vivo human finger pad are shown in Figure 14.35A, Figure

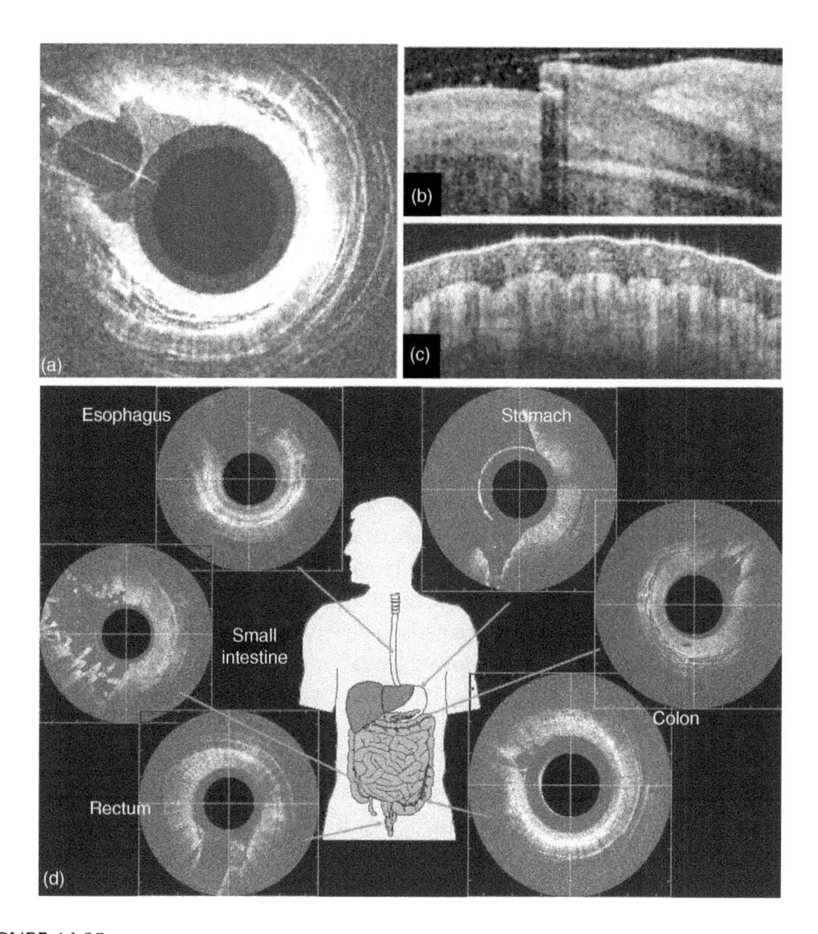

FIGURE 14.35
Standard resolution images obtained by Michael Choma during his Ph.D. work at Duke university. (a) In vivo human esophagus, (b) in vivo human nail bed, (c) in vivo human finger pad, (Courtesy of Dr. Michael A. Choma, Ph.D. thesis, Duke University, 2004), and (d) in vivo endoscopic OCT. (Courtesy of Bioptigen, Inc., Durham, NC.)

14.35B, and Figure 14.35C, respectively. The images in Figure 14.35 were collected with a system similar to that shown in Figure 14.32. The optical coherence images shown in Figure 14.35 were collected by Dr. Michael A. Choma during his Ph.D. research at Duke University under the guidance of Dr. Joseph Izatt. Additionally, an endoscopic OCT image from Dr. Izatt at Bioptigen is shown in Figure 14.35D.

14.8.8.1 Optical Coherence Tomography in Dentistry

In dentistry, it has great potential for the detection and diagnosis of periodontal diseases, oral cancer, and dental caries. The intensity of backscattered light is measured as a function of its axial position in the tissue using low-coherence light-gating methods. Enamel only weakly scatters near-infrared light. Therefore, OCT has an imaging-depth capability of greater than 2–3 mm. This is sufficient penetration for the direct imaging of early occlusal and approximal carious lesions. Demineralized enamel highly scatters light in the visible and near-infrared, and incident polarized light is rapidly depolarized. Polarization-sensitive OCT can be used to discriminate between carious and noncarious enamel and dentin, and it is anticipated that in the future it will enable measurement of the extent of demineralization in depth, thus providing the clinician with diagnostic information about lesion severity.

14.8.8.2 Polarization-Sensitive Optical Coherence Tomography

Several classifications of biological tissues, such as muscles, tendons, and cartilage, have organized fibrous structures that lead to birefringence, or an anisotropy in refractive index that can be resolved using light of different polarizations to measure tissue properties. Partial loss of birefringence in biological tissues is known to be an indicator of tissue thermal damage. In cases where it is necessary to perform laser irradiation therapy to eliminate diseased or degenerating tissue sites, polarization-sensitive imaging could provide the necessary dosage measurement to allow successful treatment without excess damage to surrounding, healthy tissue. Other applications of such an imaging system would include measurement of burn depths to aid in treatment or removal of burned tissue. A variation of conventional OCT has been developed that shows promise in this capacity.

Circularly polarized light is incident upon the sample and is returned to the detector in an elliptical state determined by sample birefringence. When combined with linearly polarized light in the reference arm, the resultant projected polarization state can be resolved into horizontal and vertical components through phase-sensitive detection, and OCT imaging can be performed on each state (Figure 14.36). The resulting interference patterns for each polarization depend on the product of imaging depth (which is known from the mirror position to within the 10 μm axial resolution of the system) and refractive index. This allows measurement of the sample birefringence.

14.8.8.3 Phase-Resolved Optical Coherence Tomography

A technique called optical Doppler tomography or Doppler OCT has been developed that allows simultaneous tomographic imaging and measurement of blood flow velocity. Measurements of blood flow velocity are important for treating burns and determining burn depth, evaluating effectiveness of laser therapy and photodynamic therapy, and evaluating brain injuries. Flow velocity can be calculated by measuring the Doppler shift in the fringe frequency of the interference pattern. The Fourier transform of the frequency shift then gives information on the velocity. However, the minimum velocity sensitivity is inversely proportional to the Fourier transform window size, so

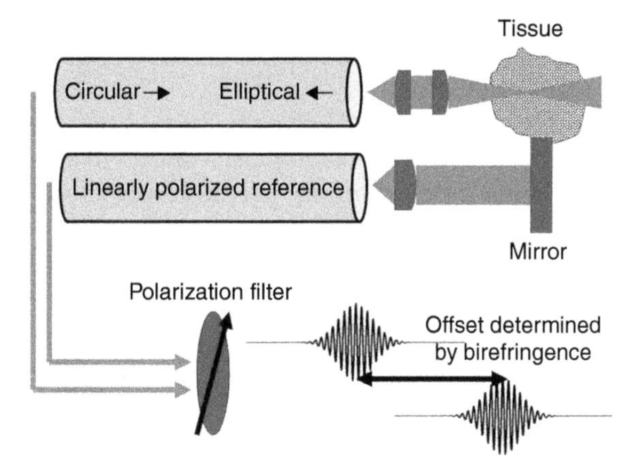

FIGURE 14.36
Diagram of polarization-sensitive OCT setup. The offset in patterns for a given depth (as determined by the reference mirror) is determined by birefringence.

that for high-velocity sensitivity the scan rate must be slow or, equivalently, the spatial resolution must be poor. A new technique has been developed that decouples the velocity and spatial resolution. Rapid successive scans of the same transverse position are used to determine the phase change in the interference pattern between scans, which allows calculation of the Doppler frequency shift and consequently the flow velocity. It is possible to detect blood flow velocities as low as 10 μm/s with a spatial resolution of 10 μm in this manner.

Another catheter-based system has been developed for measuring blood flow in deeper tissues. The noninvasive probe in the system described above is limited to conventional OCT imaging depths of 2–3 mm; while the difficulties of measuring blood flow inside highly scattered arterial and venous structures limit this application to approximately 1 mm. A catheter-based system would thus be beneficial for measuring flow velocity in deeper tissues and internal organs.

14.8.8.4 Spectroscopic Optical Coherence Tomography

Perhaps the most valuable extension of conventional OCT is the development of spectroscopic OCT. Optical spectroscopy can provide detailed information on the biological and chemical structure of tissues and cellular microstructure by examining frequency shifts in the reflected light as well as the intensity of preferentially absorbed and reflected wavelengths. Spectroscopic measurements can be used to obtain information about water-absorption bands within a sample and, for example, hemoglobin oxygen saturation. Broadband emission generated by mode locking of solid-state lasers overlaps several important biological absorption bands, including water at 1450 nm, oxy- and deoxyhemoglobin between 650 and 1000 nm, and others at 1.3 and 1.5 μm. The availability of broadband sources centered at 800 nm permits imaging in highly vitreous samples, because of the low absorption of water in this wavelength range.

Combining spectroscopy with OCT imaging allows two- or three-dimensional spectroscopic imaging of live biological samples in a natural environment at extremely high resolution. To perform spectroscopic OCT, it is necessary to measure the entire interference pattern, however, rather than the envelope as in conventional OCT. The apparatus is the same except for the detection system, which must be designed to measure the full interference signal and perform image processing to extract the spectroscopic information. Typically, information is extracted for a specific frequency, although it is also possible to obtain spectroscopic information within a given bandwidth or at multiple discrete frequencies.

14.8.8.5 Time-Domain Optical Coherence Tomography

The key feature of time-domain OCT interferometer design is the scanning delay line in the reference arm.

The broadband light source in time-domain OCT is spatially coherent and need not be pulsed. That a pulsed source is not a prerequisite for OCT systems is noteworthy. As a rule, nonpulsed broadband sources are easier to manufacture than their pulsed counterparts.

The delay line in time-domain OCT systems scans along the optical axis (i.e., the axial direction) in the tissue sample. As the position of the reference arm reflector is scanned, fringes are produced if and only if the position of the reference-arm reflector matches that of the sample arm to within a coherence length. The frequency of these fringes is related to the Doppler shift that light acquires reflecting off the moving reference reflector (in the case of grating-based rapid scanning delay lines, the fringe frequency, or more specifically, the phase velocity, is actually a free parameter that is uncoupled from the group velocity generated by the delay line). The amplitude of the interferometric signal is proportional to the sample reflectivity. When the sample is a perfect reflector at zero displacement, the detector response represents the time-domain impulse response or point-spread function of the imaging system. If, however, a narrowband source is used such as a CW laser, the interferometer has no ability to perform depth ranging because the coherence length has a width several orders of magnitude greater than all microstructural features in a biological sample.

Time-domain OCT can acquire functional as well as structural information about a sample. One of the first functional extensions of OCT was color-Doppler OCT (CD-OCT).

14.8.8.6 Color-Doppler Time-Domain Optical Coherence Tomography

The principle of CD-OCT can be understood as follows. If scatterers (e.g., red blood cells) in the sample arm have a component of motion along the optical axis, the interferometric fringe frequency will be up- or downshifted in proportion to the magnitude of this component. Therefore, by calculating the interferometric fringe frequency as function of axial depth, flow profiles can be developed. Because amplitude scans (A-scan) are nonstationary (i.e., their frequency content is a function of axial position), time-frequency analysis is needed to extract flow information as a function of depth. Time-frequency algorithms that previously have been used in CD-OCT include the short-time Fourier transform and the Kasai algorithm. This approach has been dubbed "axial scan processing" because frequency content along the optical axis is extracted. An alternate approach for extracting Doppler-flow information is to monitor the phase at a particular depth location over several A-scans, so-called "sequential scan processing." Since phase is the integral of Doppler shift, phase accumulates over the course of several A-scans at depths corresponding to the location of moving scatterers. The rate of change of this accumulation is, therefore, the Doppler shift.

Two techniques that have seen recent application in CD-OCT are wavelet analysis and model-based spectral estimators. The signal and image processing are described in various textbooks. Wavelets are a type of time-frequency analysis that are particularly adept at identifying discrete jumps in frequency content in a nonstationary signal. This is promising for CD-OCT because the presence of a blood vessel in an A-scan leads to a discrete jump in A-scan-frequency content at the tissue–blood vessel interface. Unlike Fourier and wavelet transforms, which decompose a signal onto a set of basis functions and which obey Parseval's identity, a model-based spectral estimator extracts parameter values from a signal that characterize certain features of the signal. These parameters are then used to fit an assumed model to the measured signal. A particular strength of model-based spectral estimators is that they can exploit a priori knowledge of the measured signal. For example, eigenfrequency spectral estimators assume that the signal is composed of N sinusoids, where N is a user-defined parameter. The analysis determines the frequency and amplitude of these N sinusoids. In the case where $N = 1$, eigenfrequency spectral estimators become a particularly interesting method for CD-OCT because it is of interest to measure the frequency of the single sinusoid that modulates the interferometric envelope. To apply eigenfrequency analysis to CD-OCT, Michael Choma in his Ph.D. thesis work at Duke University (2004) adapted this model-based approach to a short-time framework, much as the Fourier transform is in the short-time Fourier transform.

14.8.8.7 Spectral-Domain Optical Coherence Tomography or Fourier-Domain Optical Coherence Tomography

A conceptually different class of OCT interferometer design was introduced by Fercher et al. in the year 1995. The design initially was referred to as "backscattering spectral interferometry" and the name has changed to Fourier-domain OCT since then.

Just as in time-domain systems, Fourier-domain OCT uses a Michelson interferometer topology with a broadband source. The design of the Fourier-domain OCT has introduced modifications in both the reference and detector arms. In Fourier-domain OCT the reference arm reflector is moving with a fixed phase velocity and no dispersion. In the detector arm the single photodiode has been replaced by a spectrometer. The spectrometer records the interferometric signal as a function of optical wavelength.

In the case of a single-sample reflector, the interferometric signal is typically cosinusoidal and numeric interpolation is used to catalog the wave number-indexed signal, instead of the usual wave of wavelength indexing. The sample position is found from the frequency of the cosine function of the heterodyning technique as usual. The amplitude of the cosine is proportional to the sample reflectivity. The wave number-indexed signal is Fourier transformed to a spatially indexed signal to access the frequency, which represents the position and the amplitude that identifies the reflectivity.

In the Fourier transform of the wave number-indexed signal the high-frequency fringes in the spectral domain are far away from , while low-frequency fringes remain close to $x = 0$. Otherwise the cosine amplitude transforms to peak height. The relationship between source bandwidth and amplitude scan axial resolution is identical for Fourier-domain OCT and time-domain OCT as described in Section 14.8.

Fourier-domain OCT system that presently give moderate quality images require considerably less imaging time than time-domain OCT systems. This suggests that Fourier-domain OCT is indeed more sensitive than time-domain OCT. There also is evidence that swept-source OCT is more sensitive than time-domain OCT.

In general time-domain OCT systems require 100–1000 nW of source power to achieve the aforementioned sensitivity. Both Fourier-domain and swept-source OCT systems require approximately 100- to 1000-fold less power to achieve similar sensitivity.

The Fourier-domain and swept-source OCT system sensitivity is independent of the scan depth and source bandwidth in contrast to time-domain OCT. An illustration of the work on Fourier OCT imaging by Dr. Joseph Izatt is presented in Figure 14.37.

In the past there has been some difficulty constructing swept-laser sources with the sweep rates and output powers needed for OCT imaging. Examples of the currently available sources include current-tuned laser diodes and various external cavity-dispersive (i.e., grating or prism) tuning mechanisms.

The spectral-domain OCT A-scan however has a source of signal artifact that results in a certain amount of ambiguity. While there is only one reflector in the sample arm there are still three amplitude peaks in the recorded A-scan. The peak at $x = 0$ is the Fourier transform of the source signal. This central peak however obscures the reflector's position at $x = 0$. This phenomenon is called complex conjugate ambiguity and refers to the symmetry in the A-scan algorithm. This symmetry will not discriminate between the reflector position at Δx and mathe, $-\Delta x$. This ambiguity is the result of the fact that the spectral-domain signal is real-valued.

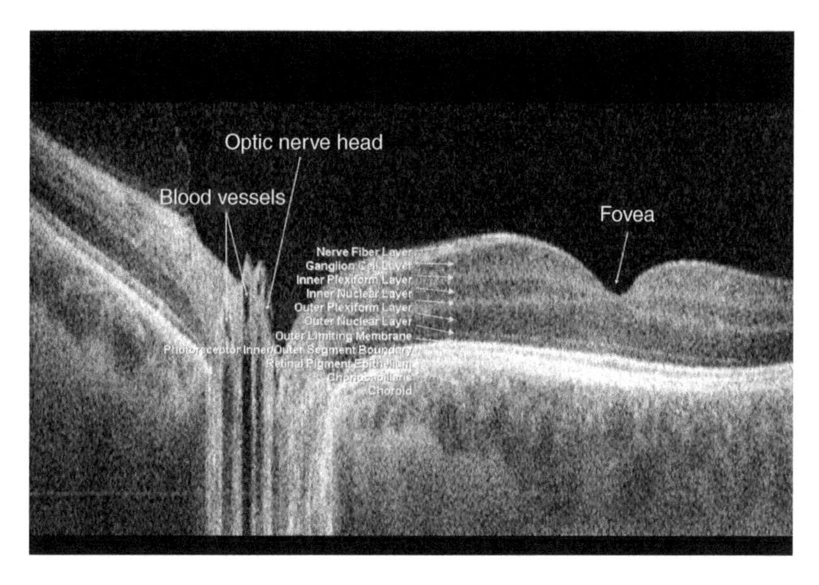

FIGURE 14.37
Fourier-enhanced human retinal image from new commercial Fourier-domain OCT system. (Courtesy of Bioptigen, Inc., Durham, NC.)

Another type of imaging using the unscattered photons this time will be discussed in the next section under ballistic photon imaging.

14.9 Ballistic Photon Imaging

One promising noninvasive, nonionizing imaging modality being developed by several research groups is the so-called shadow imaging. Since Victorian times physicians have tried imaging breast tumors with the available light, candle light at first, and better and brighter sources as they became available. The early form of transillumination imaging was called diaphanography.

The introduction of high-speed detection techniques is providing a high-resolution adversary. Since light is scattered while passing through a scattering tissue, the scattering probability increases as the distance increases. Light passing through a turbid medium can be captured on the path of the incident beam with an aperture placed on the optical axis, allowing only photons to reach the detector that emerge from the tissue within a small finite solid angle. The emerging photons will be a combination of unscattered, coherently scattered, and multiple-scattered photons. The unscattered photons or ballistic photons will travel the shortest distance and will thus require the least amount of time to traverse the tissue. The photons will retain all of their original characteristics, and thus carry the most significant information about the media they traveled through. Another group of photons, the forward-scattered or 'snake' photons provide a compliment in information. The third group of photons, the diffusively forward-scattered photons, will take the longest time to traverse the medium, and will have lost most their characteristic properties. In contrast, these scattered photons will have gained tissue-scattering characteristics and will also be useful in the diagnostic imaging application. An illustration of the different types of paths followed by photons passing through a scattering medium is shown in Figure 14.38.

The ballistic photons can be collected by means of time gating. Time gating in the pico- and femtosecond range will require the use of an autocorrelator to reconstruct the average number of photons that will pass through the medium as a function of time. An illustration of the time distribution of passing photons is given in Figure 14.39. The time delay between the ballistic, snake, and diffuse photons will gradually increase with increasing thickness of the scattering medium. Another complication is the fact that less photons will appear on the transmission side when the thickness increases, since the probability that a photon will not encounter a scattering nucleus will become increasingly negligible.

The ratio of the coherent radiance (ballistic) [I_c] versus the diffuse collected light (snake and diffusively scattered) [I_d] passing through a slab of thickness d is given by assuming scattering dominating over absorption in a Beer–Lambert law attenuation mechanism yielding an expression given by the following equation:

$$\frac{I_c}{I_d} \propto \frac{1}{\Omega\left\{\frac{\left[\exp(\mu_s d)-1\right]}{1+\mu_{tr}}\right\}} \tag{14.94}$$

where Ω represents the solid angle for collection of the transmitted photons.

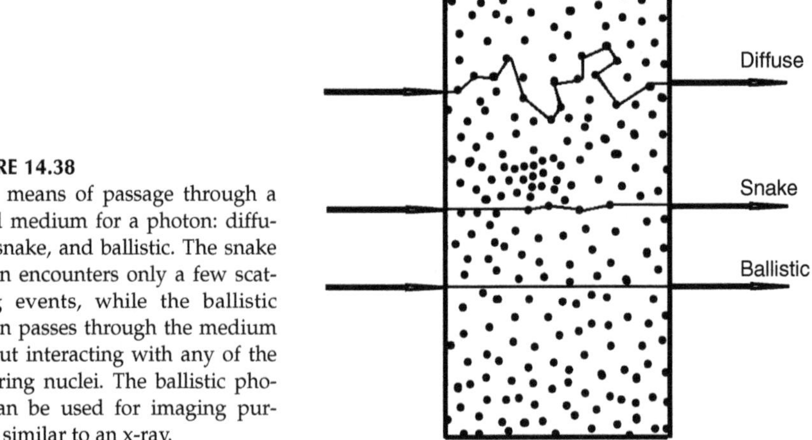

FIGURE 14.38
Three means of passage through a turbid medium for a photon: diffusive, snake, and ballistic. The snake photon encounters only a few scattering events, while the ballistic photon passes through the medium without interacting with any of the scattering nuclei. The ballistic photon can be used for imaging purposes similar to an x-ray.

The temporal profile of the collected incoherent light as measured by a photodetector can be approximated by the diffusion approximation as outlined in the following equation to give an expression for the collected fluence rate:

$$I(t) \propto \frac{D}{\pi d} \sum_{m=1}^{\infty} m \left\{ \left(\frac{\pi d}{d+1.42\delta} \right)^2 \sin\left(\frac{m\pi d}{d+1.42\delta} \right) \right\} \exp\left(-Dt \left(\frac{m\pi}{d+1.42\delta} \right)^2 \right) \tag{14.95}$$

with $\delta = \sqrt{1/3\,\mu_a\,(\mu_a + (1-g)\,\mu_s)}$ as the mean free path length and D thediffusion coefficient as defined in Chapter 4.

Theoretically, it is now possible to derive the reduced optical properties from Equation (14.95) by separating the ballistic from the diffuse photons.

Figure 14.39a illustrates the time spread of the ballistic, snake, and diffusively scattered photons for a medium thickness of a few optical mean free path lengths. Figure 14.39b is the same as Figure 14.39a, only for a relatively thick turbid medium. This can mean either a large physical thickness or thin but with a very high concentration of scatterers. The ballistic photons will reach the detector first, followed by the snake photons and finally the diffusively scattered photons take the longest time to reach the detector.

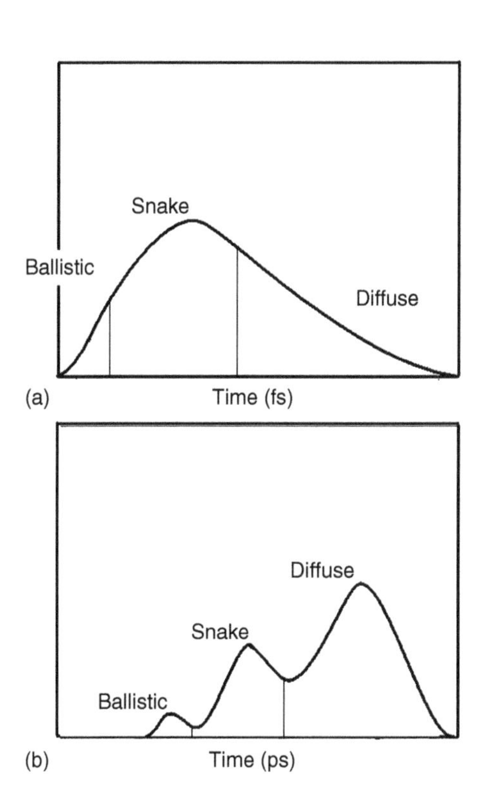

FIGURE 14.39
(a) Time spread of the ballistic, snake, and diffusively scattered photons for a medium thickness of a few optical mean free path lengths. (b) Same as (a) only for a relatively thick turbid medium.

In the realm of optical diagnostic modalities there are various other techniques available. The next section will describe the determination of blood oxygenation from spectral absorption curves.

Other optical methods in diagnostic can be purely technical, such as the noninvasive determination of optical properties. Reflectometry is one of those diagnostic techniques and will be referred briefly next.

14.10 Reflectometry

In the interest of real time monitoring of the progress of a laser treatment, and the initial assessment of the irradiation protocol, the in situ optical properties will need to be known and tracked real time in vivo. Backscatter or reflection spectral measurements of light offer a simple but effective method to accomplish this.

The lion share of reflectometry was discussed in Chapter 7, Section 7.2.2 under the topic "Noninvasive Techniques," and also in other chapters various derived techniques are described under erythema determination (Chapter 16) and even microscopy (Section 14.1).

Another derived technique uses the Maxwell interaction at an optical interface to collect detailed but shallow tissue information. This technique uses the wave-guide action of light entering an interface at critical angle and subsequent re-emittance (shallow reflection) of the spectral content. The technique referred to as evanescent wave interaction used for diagnostics is discussed next.

14.11 Evanescent Wave Imaging Applications

Evanescent wave diagnostic and therapeutic applications rely on the interaction of infrared light with tissue at the interface between transparent, high-refractive index optical materials (prisms and fibers), and tissue. When using laser light, the use of evanescent optical waves offers a precise approach to examining tissues in the infrared (2–10 μm). The near-infrared spectrum is of interest for several reasons, one being the substantial difference in the absorption spectra of water-rich and lipid-rich tissues.

For low light levels, such as light from a spectrophotometer, a novel approach was necessary to interrogate the tissue optical properties in vivo, in particular the resolve for the refractive index and extinction coefficient in the near-infrared wavelength range. Knowledge of these optical properties is critical for understanding the laser–tissue interaction at the optic–tissue interface.

New fibers (hollow wave guides and chalcogenide glass) are promising for efficient transport of infrared wavelengths. Catheter-based surgical devices that have diagnostic and therapeutic capabilities using the evanescent wave technology are currently being developed for various applications.

14.11.1 Evanescent Optical Wave Device Designs

The evanescent wave-device design depends largely on whether one wants a handheld probe or a small catheter (1–3 mm in diameter) suitable for minimally invasive procedures. Several device designs have been proposed and constructed. All of the device designs incorporate the general evanescent optical-wave approach described in Section 8.7. An evanescent optical wave is launched into the tissue when the refractive index of the optic n_1 is greater than the refractive index of the tissue n_2, and the angle of incidence of the laser light θ_i at the optic–tissue interface is equal to or greater than the critical angle θ_c, defined as the angle for total internal reflection.

The critical angle as a function of wavelength ranging from 2 to 10 μm for evanescent wave generation at a sapphire–water or a ZnS–water interface is shown in Figure 14.40. Also shown in Figure 14.40 is the 45° fiber interface angle for reference.

One particularly useful device design incorporates a 45° angle of incidence for the laser beam at the optic–tissue interface. This geometry is useful because the light reflected from one side of the tip is still at a 45° angle of incidence on the other side. Figure 14.40 plots the critical angle for a sapphire–water and a ZnS–water interface as a function of wavelength from 2 to 10 μm, and compares the critical angle to the 45° fiber interface angle. ZnS and diamond are good replacements for the sapphire tip for wavelengths longer than about 5 μm, where transmission drops to 40% over 8 mm. For sapphire, only wavelengths from 2.70 to 2.86 μm will be evanescent, whereas, all wavelengths are evanescent at a ZnS–water and a diamond–water interface with a 45° interface. This is due to the larger refractive index for ZnS (2.24) and diamond (2.38) as compared with that of sapphire (1.70). One

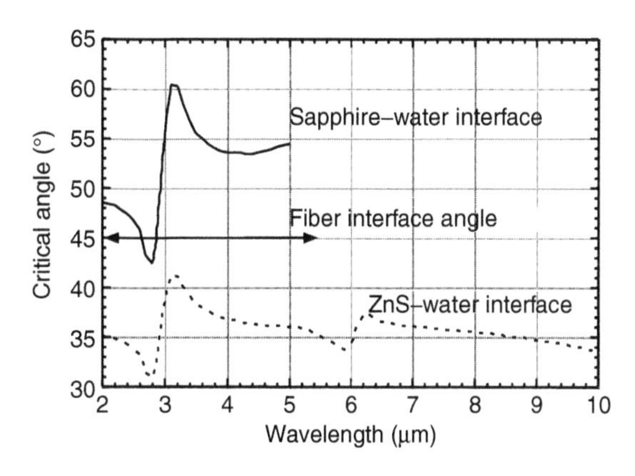

FIGURE 14.40
Critical angle θ_c for evanescent wave generation at a sapphire-water or a ZnS-water interface as a function of wavelength from 2 to 10 μm. Also shown is the 45° fiber interface angle for reference. (Reprinted with permission Elsevier: Hooper, B.A., LaVerde, G.C., and Von Ramm, O.T., Design and construction of an evanescent optical wave device for the recanalization of vessels, *Nucl. Instr.* and *Meth.*, A 475, 645–649, 2001)

design suitable for either application simply consists of an optical fiber with the end polished at 45° angles in the shape of a screwdriver tip. This screwdriver-tip fiber (STF) "retro-reflector" device design is similar to a right-angle prism. A probe laser (HeNe) can also be launched into the device to monitor the dynamics of the ablation at the interface, since after interrogation at the optic–tissue interface the probe retro-reflects out of the fiber.

A catheter device will need to be 1–2 m in length, have high transmission of the laser light through this length of fiber, and be flexible. Another design consists of a 1–2 m long, 1–2 mm outside diameter hollow glass wave guide (HGW) or photonic band gap (PBG) fiber with a high-refractive index optic (sapphire, ZnS, or diamond) attached to the end. This device design is shown in Figure 14.41 with two HWGs and a ZnS conic optic tip. The optic tip needs to be only a few millimeters long and a few millimeters in diameter with a conical polish at 45°. For wavelengths longer than about 5 μm, ZnS and diamond can be used as a replacement for a sapphire optic tip.

The important parameters of the evanescent wave are the depth of penetration and the required incident energy for ablation $E_{required}$, and are plotted in Figure 14.42 as a function of wavelength from 2 to 10 μm. The short, solid lines are for sapphire and the dashed lines are for ZnS. The evanescent wave penetration depth is between 0.1 and 1 μm over this wavelength range, and the required incident energy for ablation is about 1 mJ for sapphire and varies from 0.2 to 100 mJ for ZnS.

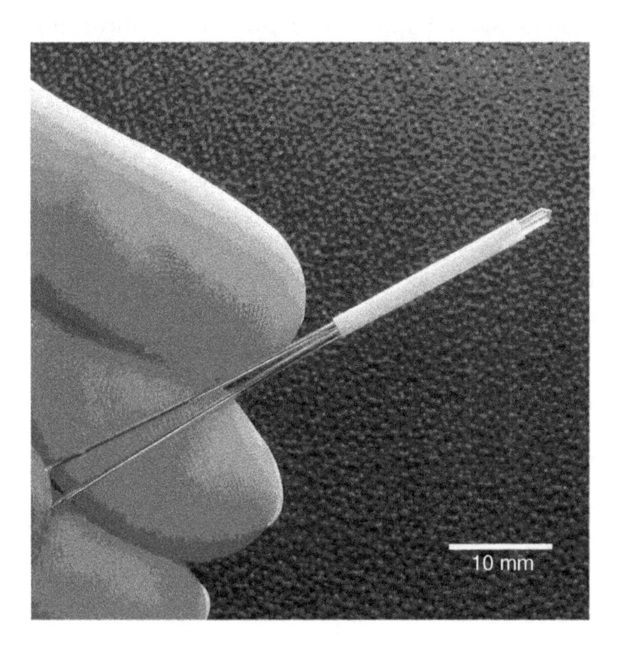

FIGURE 14.41
A 1-m long catheter with two HWGs for light delivery and collection and a zinc sulfide (ZnS) 45° conical optic tip for launching the evanescent waves. Also shown is a Delrin coupling sheath for holding and aligning the HWGs to the optic.

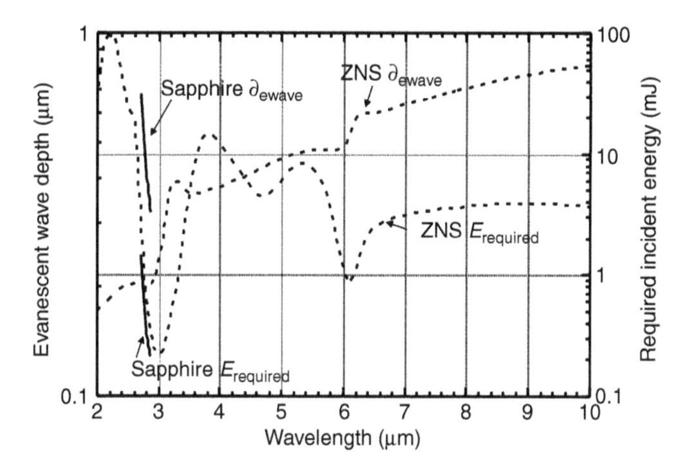

FIGURE 14.42
Evanescent wave depth of penetration ∂_{ewave} and required incident energy for ablation $E_{required}$ at the sapphire-water and ZnS-water interfaces as a function of wavelength from 2 to 10 μm. The value of $E_{required}$ is calculated from the latent heat of vaporization of water; $2.5\times10^{9}\ Jm^{-3}\times\partial_{ewave}\left[\dfrac{\pi\omega_0^{\,2}/2}{\cos(45^0)\times(1-R)}\right]$, where $(1-R)$ is the fraction of incident laser energy absorbed in the tissue. (Reprinted with permission Elsevier: Hooper, B.A., LaVerde, G.C., and Von Ramm, O.T., Design and construction of an evanescent optical wave device for the recanalization of vessels, *Nucl. Instr. and Meth.*, A 475, 645–649, 2001).

Another field of interest in the infrared spectrum is thermographic imaging, which will be discussed next.

14.12 Medical Thermography

The heat balance in an organism is determined by the amount of heat production, mostly due to metabolic reactions, minus the effective work performed which equals the heat loss as a result of evaporation. Other factors in the heat equation are convective, conductive, and radiative heating (or cooling), and the rate of heat stored in the organism. The radiative heat loss from a body is expressed by the Stefan–Boltzmann equation, which provides the emission of power density at a particular wavelength in W/m^2 as

$$W_b = \varepsilon\sigma T^4 \tag{14.96}$$

where ε describes the emittance of the body (skin emittance is between 0.96 and 0.98, depending on whether the skin is dry or moist), σ the Stefan–Boltzman constant (5.7×10^{-8} W/m^{2}K^4), and T the absolute temperature in K. The average skin surface of the human body is 1.8 m^2, while about 300 W is continuously emitted by the average human being at a body temperature of 37°C or approximately 310 K.

The conductive and convective processes that are relevant to the particular situation of light–tissue interaction have been described in various sections of this book for cases where they have a significant influence on the total effect.

Every object that has a temperature above the absolute zero will radiate energy. Temperature is the average kinetic energy of all molecules and atoms in motion that make up the composition of an object or an organism in the context of this book. Since these atoms and molecules have charges associated with them, it can be made plausible on the basis of the fact that accelerating charges generate electromagnetic radiation and temperature results in the emission of light. The higher the temperature, the more energy is stored in the organism, and thus higher is the expected energy emitted as electromagnetic radiation. This heat emission is evident from electromagnetic radiation that can be detected.

In general a blackbody is assumed for convenience of the derivation of the phenomenon. A blackbody has 100% emissivity; all energy will leave the surface with maximum efficiency. Water has all the characteristics of a blackbody, thus leaving the surface of the object under examination moist will make the observations approach the theoretical optimal conditions.

The emission spectrum is explained by the quantum theory of photons. The basis of quantum theory is that each photon carries an energy package E_v, which is described by the fact that the photon has a frequency of electromagnetic radiation (or for that matter the wavelength) that defines the energy as given by the following equation:

$$E_v = hv = h\frac{c}{\lambda} \tag{14.97}$$

where h is the Planck constant ($=6.6 \times 10^{-34}$ J s), c the speed of light, ν the frequency, and λ the wavelength. However, the emission of ultraviolet light would mean that an infinite amount of energy would be needed. The solution was found by combining the concept of quantization with the Stefan–Boltzmann equation. This combination formulated the Planck's law of radiation, where the spectral emission as a function of energy is described as

$$W = \frac{2\pi hc^2}{\lambda^5(e^{hc/\lambda kT} - 1)}10^{-6} \tag{14.98}$$

where the integral of Planck's law gives the Stefan–Boltzman equation (Equation [14.96]). Note that the Planck's law of radiation uses the wavelength in nanometers. A blackbody emission spectrum produced by a body at 400 K is shown in Figure 14.43. Another interesting feature of the spectral emission of a blackbody as a function of temperature is that the peak emission shifts to higher energy levels in addition to the fact that the total area under the curve

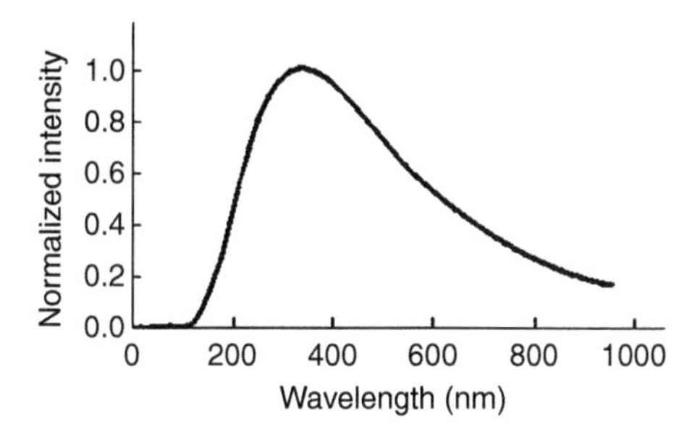

FIGURE 14.43
Optical spectrum produced by a blackbody (human tissue) at 400 K.

increases as well. The shift in peak emission toward shorter wavelengths is given by Wien's displacement law as

$$\lambda = \frac{2.898 \times 10^{-3}}{T} \tag{14.99}$$

All objects at room temperature will emit in the near-infrared only. Only at temperatures of several hundred degree Kelvin the emissions will become visible, e.g., electric heater coil glows red and melting iron glows white. The white light emission is the result of the wide range of emission frequencies that can be detected by human eye ranging from blue to red, combining to produce white light. Based on both the experimentally derived Equations (14.99) and (14.98) the thermographic camera can be used to perform detailed temperature measurements, and with a thermographic Chagre Coupled Device (CCD) camera a temperature distribution can be displayed of an entire area imaged by the CCD elements.

Corrections will be needed in medical imaging for areas under observation that do not fulfill the criteria of a blackbody, such as dry skin.

Thermographic imaging can thus provide information of local metabolic nuances, while keeping in mind the physical limitations imposed by blood circulation and ambient air-flow. Figure 14.44 shows a section of the left ventricle of 12 mm thickness while irradiated from the endocardial side by a neodymium:YAG laser of 30 W for 30 s at 1064 nm. Note the deformed shape of the elevated temperatures on the epicardium resulting from coronary vessels, which act as heat sinks.

In general a body will be in equilibrium, meaning the temperature throughout the organism is constant and similar. However, the inner medium will be shielded from external influences, providing a very homogeneous temperature distribution. On the outer shell of the organism the local vascularity and the depth of the blood vessels to the surface will affect the cooling effects from the external conditions such as outside temperature and direct air-flow.

Increased vascularity will result in higher temperatures since the blood is arriving from the core of the body, which is not subject to cooling effects. Other increases in temperature may result from increased cellular

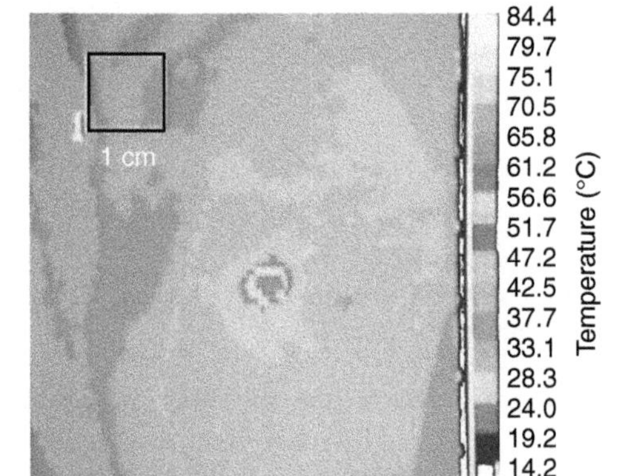

FIGURE 14.44
Temperature of a 12 mm thick section of heart muscle recorded by thermographic imaging at the end of a 30 s exposure to 30 W Nd:YAG laser irradiation operating at 1064 nm from the inside of the left chamber. Note the temperature profile following the blood vessel acting as a heat sink.

metabolism. Both factors are often a sign of inflammation. Additional conditions that provide localized increased blood perfusion are pregnancy and cancer.

Thermography offers a significant contribution in tracking inflammatory reactions of postoperative incisions. Another application is in the recognition of warning signs for the potential of breast cancer sites.

The final topic in this chapter is the use of laser light to generate acoustic waves. The principle of photoacoustic imaging will be described and the use of its diagnostic capabilities will be discussed.

14.13 Photoacoustic Imaging

The principle of light absorption will generate elevated temperatures. Higher temperatures will lead to increased pressure and from the Boyle–Gay-Lussac law it can be seen that a net increase in volume can also be expected. Provided several conditions are met, this temperature rise can produce mechanical shock waves under pulsed laser irradiation. Combining the light penetration with the generated acoustic shock wave, an imaging system can be developed that may surpass both optical and acoustical imaging.

The photoacoustic effect is a function of three equations with a range of unknowns. The first relevant equation is the light distribution inside the target medium, second the temperature generation and associated thermal effects (damage), and third the acoustic wave propagation. Solving this problem relies on cleaver choices of approximations when possible.

Selection of nano- or femtosecond irradiation creates several theoretically advantageous conditions.

The laser beam produces a light distribution inside the tissue according to a solution to the transport theory as discussed in Chapter 5.

In most cases of laser irradiation the source function $S(\mathbf{r},\mathbf{s})$ contains the beam profile of radiation entering the interface between the clear medium (air, water, or fiber optic) and the turbid medium, and it has the following standard Gaussian characteristics:

$$S(\mathbf{r},\mathbf{s}) = I_0 \exp\left[\frac{-2r^2}{w^2}\right] \tag{14.100}$$

where w is the beam waist, r the radius from center of the beam, and I_0 the radiance at the center of the beam.

Generally, the time-dependent equation of radiative transfer is solved by separation of the anisotropic radiance in a diffuse term and a direction-dependent term as shown below:

$$L(\mathbf{r},\hat{\mathbf{s}},t) = \frac{1}{4\pi}\Psi(\mathbf{r},t) + \frac{3}{4\pi}\varpi F(\mathbf{r},\hat{\mathbf{s}}') \tag{14.101}$$

Here, the symbol F denotes the direction-dependent radiance. The fraction of energy scattered to that removed from the radiance stream is indicated by ϖ, for pure absorption $\varpi = 0$, for conservative scattering $\varpi = 1$, and ϖ is a function of optical depth called the albedo. The assumption of separation in homogeneously diffuse and directional term (first two terms of polynomial series development) allows one to solve the equation of radiative transfer independently for the diffuse and directional portion. The diffuse fluence rate follows from the diffusion approximation as

$$\begin{aligned}&\frac{1}{c}\frac{\partial\Psi(\mathbf{r},t)}{\partial t} - \mathbf{s}\cdot\nabla\Psi(\mathbf{r},t) + (\mu_\mathrm{a}(\mathbf{r}) + \mu_\mathrm{s}(\mathbf{r}))\Psi(\mathbf{r},t)\\ &-\mu_\mathrm{s}(\mathbf{r})\int_{4\pi} p(\mathbf{s},\mathbf{s}')\Psi(\mathbf{r},t)\mathrm{d}\Omega' = S(\mathbf{r},\mathbf{s}')\end{aligned} \tag{14.102}$$

For a short pulse the source $S(\mathbf{r},\mathbf{s}',t)$ can be written as

$$S(\mathbf{r},\mathbf{s}',t) = \delta(\mathbf{r})\delta(\mathbf{s})\delta(t) \tag{14.103}$$

This would be the case for nano and femtosecond pulsed laser irradiation, since the thermal relaxation time is much longer than the pulsed duration (See also Section 8.2.2 in Chapter 8).

Even with speed of light assumed to be at c/n= 299792458/1.37=218,826,611 m/s, within the laser pulse duration of 100 fs the photons can travel only 22 μm. Assuming isotropic scattering this may still result in a significant dispersion; however, the scatter will be mostly in the forward direction with a scattering anisotropy factor of $g > 0.95$, for most tissues in the red and near-infrared.

With a output energy of only 9 nJ, as is commonly found for Q-switched lasers, the peak power will come to 9000 W, or at 0.0076 J/cm^2 energy density this comes to 760 kW/m^2. It can be shown that scattering saturation will occur and the majority of light will pass through in a slightly diverging beam. However, there may still be a risk of plasma formation associated with these high levels of irradiance.

Generally, under these conditions only a slight temperature rise will be seen, and vaporization and cracking can be avoided.

A laser beam will heat up the focal volume. The heat will tend to flow in the direction of any temperature gradient inside the sample. The sharper the temperature gradient, the more efficient will be the conversion of energy from optical to acoustical energy.

The instantaneous local temperature rise of tissue is given by

$$\Delta T = \frac{\mu_a \psi}{\rho C_p} \tag{14.104}$$

The temperature rise is directly proportional to the absorbed local light fluence rate $\mu_a\Psi$ divided by the local specific heat C_p times the density ρ, where Ψ is the local fluence rate and μ_a the local optical absorption coefficient of the medium.

The bioheat equation describes the temperature evolution T, resulting from the heat source, ignoring the influence of blood flow in this case. It is shown as follows:

$$\nabla(k \nabla T(r,t)) = \rho C_p \left(\frac{\partial T(r,t)}{\partial t} + \tau \frac{\partial^2 T(r,t)}{\partial t^2} \right) - \mu_a \Psi(r,t) \tag{14.105}$$

where Ψ(r,t) is the solution to the transfer equation (Equation [14.102]), and the absorption coefficient times the fluence rate describes the amount of absorbed light power, κ is the thermal conductivity and α_T the thermal diffusivity, which is defined as follows:

$$\alpha_T = \frac{\kappa}{C_p \rho} \tag{14.106}$$

Introducing next is the thermal relaxation time τ, which places the laser time phenomena in context with the response times of the medium. The thermal diffusion or relaxation time for axial heat conduction is given by

$$\tau = \frac{1}{4k\mu_a^{\ 2}} \tag{14.107}$$

Femtosecond laser pulses are much shorter than the acoustic transit time δ/ν, with δ the optical penetration depth and ν the acoustic velocity. Therefore, it is reasonable to assume that the time component of the source function can be approximated with a delta distribution.

It can be shown that it is justified to use a Dirac delta distribution for modeling of the temporal excitation (due to the short laser pulse width compared to characteristic times for reaching thermal equilibrium) and a Gaussian spatial distribution for laser excitation, which are the initial assumptions. As a result the second-order terms with respect to time become negligible.

The temperature distribution is modeled by the heat transfer equation as shown below

$$\nabla^2 T(\mathbf{r},t) - a^2 \frac{\partial T(\mathbf{r},t)}{\partial t} = -\frac{S(\mathbf{r},t)}{\kappa} \quad \text{and} \quad T(\mathbf{r},0) = T_0(\mathbf{r}) \tag{14.108}$$

The source term $f(\mathbf{r},t) = -S/\kappa$ describes the absorption of the laser energy, where $S \sim \mu_a J$, where J represents the local fluence. The solution to Equation (14.108) is given in terms of the Green's function as

$$T(\mathbf{r},t) = \int_0^t \int_V G(x,t;\xi,\tau) f(\xi,\tau) \mathrm{d}\tau \, \mathrm{d}V(\xi) - a^2 \int_V G(x,t;\xi,0) T_0(\xi) \mathrm{d}V(\xi) \tag{14.109}$$

Here, the heat source $f(\xi,\tau)$ is modeled as a spatial Gaussian distribution. Again, using the Dirac delta distribution for time, Equation (14.109) can be rewritten as

$$f(\xi,\tau) = -\frac{(1-R)}{\kappa \tau_\mathrm{P}} \mu_a J \exp[-\mu_a z] \exp\left[-\frac{2\rho^2}{\omega^2}\right] \delta(t) \tag{14.110}$$

with

$$\omega(z) = \omega_0 \left(1 + \left(\frac{\lambda(z - z_0)}{\pi \omega_0^2}\right)^2\right)^{1/2} \tag{14.111}$$

where R is the surface reflectivity, j the laser flux (J/cm^2), w the beam radius (cm), z_0 the location of the minimum waist, λ the wavelength, and τ_p the laser pulse width.

The Green function solution for a point source is given by the following equation:

$$G(x,t;\xi,\tau) = -\left[\frac{a^2}{4\pi(t-\tau)}\right]^{3/2} \frac{1}{a^2} \exp\left[-\frac{a^2\,|\mathbf{r}-\boldsymbol{\rho}|}{t-\tau}\right] \tag{14.112}$$

where $\boldsymbol{\rho}$ is the radius in the "Greens medium."

Finally, after substitution of Equations (14.110) and (14.112) into Equation (14.109), the solution for the temperature following femtosecond laser pulse excitation is given by

$$T = T_0 + \frac{c_2\sqrt{\pi}d_1}{4b_1^{3/2}} \exp\left[\frac{d_1^2}{4b_1}\right] \tag{14.113}$$

where $b_1 = (a^2/4t) + (2/w^2), d_1 = a^2 r/2t$, and $c_2 = (1\text{-}R)\mu_a j e^{-\alpha z} \quad \exp\{-a^2r^2/4t\}/2\sqrt{\pi} tar\ \kappa\tau_p$.

Regarding the short laser pulse in the equation of radiative transfer, one can develop the fluence rate Ψ in a Fourier series. This will only be necessary when the repetition rate of the laser pulses is in the same order of magnitude as the pulse duration (MHz), which is not the case for Q-switched laser irradiation.

The following two separate cases can be distinguished:

1. Absorption > scattering
2. Scattering > absorption

In the first situation all the light emitted by the laser will be absorbed on the surface of the medium being imaged, and there will be no need for solving the radiative transfer equation.

In the second situation it can be shown that under irradiation in excess of 600 kW/cm^2 there is saturation of absorption and scattering nuclei, giving a small divergence within the angle of the anisotropy factor, which will be on the order of 0.98. Ordinarily, 99% of the light will propagate through the medium within an 11° planar angle, or in solid angle $4\pi/33 = 0.38$ sr, or less.

Under these conditions the light will channel through the medium until a preferable situation is encountered where the first condition applies, at this point the shock wave is initiated.

Under near-infrared femtosecond irradiation, mentioned earlier under micrometer spot size, the thermal relaxation time ($\tau_T \approx 1\mu s$) is much greater than the pulse duration for the size tissue and the optical and mechanical tissue characteristics, thus making the thermal diffusion a linear event, rendering the second-order derivatives to be zero.

Introducing the thermal relaxation length ℓ_T, as

$$\ell_T = \tau_T v \tag{14.114}$$

where v is the thermal speed of propagation. The diffusion length l_D is defined as the square root of the thermal diffusivity α_T divided by the angular frequency of repetitive heating (the laser repetition rate) as shown in the following equation:

$$\ell_D = \sqrt{\left[\frac{2\alpha_T}{v}\right]} \tag{14.115}$$

This provides a platform to relate the passive and active processes in both time and spacial confinement.

It can be shown now that the thermal expansion length is generally much smaller than the diffusion length and additionally the thermal diffusion length is also much smaller than the thermal relaxation length; the second-order derivatives will in most equations become zero.

This can also be seen when making Taylor expansions of all the parameters, yielding infinitesimally small second- and higher-order terms.

Neglecting the thermal diffusion/propagation in the heat equation (after Fourier expansion of the quadratic initial heat equation), which can be shown to be appropriate from thermal relaxation time and thermal diffusion time for ceramic/porcelain under femtosecond excitation, Equation (14.108) reduces to

$$\rho C_p \frac{\partial T}{\partial t} = \mu_a \psi(\mathbf{r}, t)\delta(t) \tag{14.116}$$

The third declaration describes how the temperature influences the displacement and as such creates a pressure wave, which is shown as

$$\rho \frac{\partial^2 u}{\partial t^2} - \frac{E}{2(1+\sigma)} \nabla^2 u - \frac{E}{2(1+\sigma)(1-2\sigma)} \nabla(\nabla u) = -\frac{E\beta}{3(1-2\sigma)} \nabla \theta \tag{14.117}$$

where E is the Young's modulus, u the displacement, and β the cubic expansion coefficient (which is three times the linear expansion coefficient 3α).

The parameter σ is the Poison ratio, which divides the transverse contraction strain on the longitudinal extension strain in the direction of the stretching force and is a material constant as shown in the following equation:

$$\sigma = \frac{-\varepsilon_{\text{trans}}}{\varepsilon_{\text{longitudinal}}} \tag{14.118}$$

As an acoustic wave is transmitted through a medium with a reactive nonlinearity, it is distorted along the propagation path. The near-field laser spot incident on the medium will rapidly disperse to produce a far-field wave. In the theoretical analysis of the ultrasound photoacoustic image formation the following assumptions will be made, and these will be refined later on and expanded to include the observed deviations from these approximations:

- All analyses are made in the Fraunhofer region, i.e., in the far field.
- The waves will be approximated as plane waves in the far field.
- The square of the received amplitude is a linear measure of the scattered energy.
- The scattered pressure is weak relative to the incident pressure.

Equation (14.117) may also be rewritten in a better manner in terms of pressure instead of displacement. The pressure is directly proportional to the square of the displacement.

The resonance frequency ν of the sound wave, created by local heating and subsequent expansion, depends on the value of Young's modulus of the tissue and the local dimensions. The resonance frequency ν of the material excited by the laser pulse can be written for an arbitrary-shape tissue block as

$$\nu = C\sqrt{\frac{E}{\rho}}\frac{d}{w \times h} \tag{14.119}$$

where d represents the depth of the medium, w the width, and h the height of the structure of the medium, respectively. C denotes a constant that depends on the geometric nature of a particular medium (which is related to the moment of inertia of the structure and the absorption depth of the laser beam).

Equation (14.119) can be rewritten in a more generalized form as

$$\nu = A\sqrt{\frac{EI}{m_1 L^4}} \tag{14.120}$$

where I equals the moment of inertia of the volume where the acoustic stimulus takes place, L the length of the sides of a cube of the medium, m_1 the mass per unit length, and A a constant that depends on the mode of excitation. The moment of inertia is appropriate in this case since during the tissue expansion a vortex behavior will be displayed.

Equation (14.120) can be reduced to a more convenient form as shown below

$$\nu = A'\sqrt{\frac{E}{\rho L^2}} \tag{14.121}$$

The strain (ε) generated as a consequence of the volume expansion is related to the change in volume (ΔV) with respect to the initial volume (V) as

$$\varepsilon = \frac{\Delta V}{V} \tag{14.122}$$

It is also related to the temperature jump as described by the volume expansion coefficient β ($\Delta V = \beta V \Delta T$), giving

$$\varepsilon = \beta \Delta T \tag{14.123}$$

The stress equals the force per unit area, which also describes pressure. The pressure generated by this strain is expressed as

$$P = -K\varepsilon \tag{14.124}$$

Here, K is the bulk modulus that specifies the pressure per unit strain.

Using these specifications a representation of the acoustic wave resulting from the pulsed laser irradiation can be derived.

14.13.1 Acoustic Wave

The acoustic wave equation in the medium that is generated by the light absorption (Figure 14.45) can be derived from the momentum equation, the equation of continuity, and the equation of state to give the following three conditions, respectively:

$$\frac{\partial \rho(\mathbf{r},t)}{\partial \tau} + \nabla \cdot (\rho \mathrm{V}(\mathbf{r},t)) = -\int_{t_0}^{t_1} \frac{S(\mathbf{r},t)}{\mathbf{r}^2} \delta\tau \tag{14.125}$$

$$\frac{\beta(\mathbf{r})}{\rho_0 C_0^{\;3}} \rho \frac{\partial V(\mathbf{r},t)}{\partial \tau} - (V(\mathbf{r},t) \cdot \nabla) V(\mathbf{r},t) + \nabla \cdot P(\mathbf{r},t) = \nabla S(\mathbf{r},t) \tag{14.126}$$

$$P(\mathbf{r},t) = K\rho^{\gamma} \tag{14.127}$$

Here, $V(\mathbf{r},t)$ is the displacement velocity and $P(\mathbf{r},t)$ the sound pressure amplitude of the primary wave.

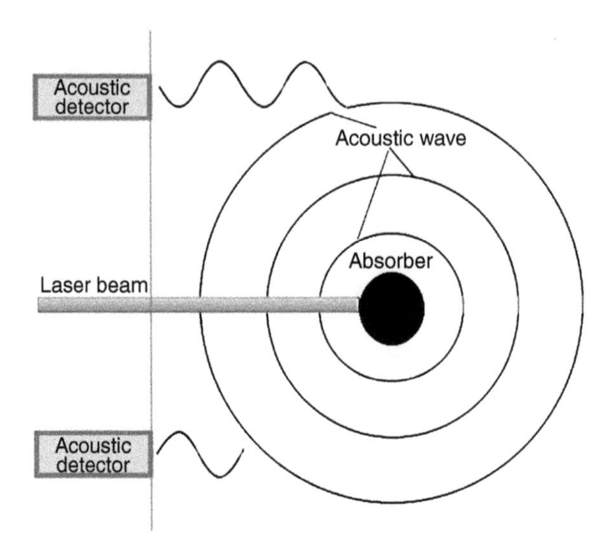

FIGURE 14.45
Illustration of the photoacoustic principle.

Combining Equations (14.104), (14.124), and (14.127) results in an expression for the initial pressure $P_0(\mathbf{r},t)$, generated under laser absorption as:

$$P_0(\mathbf{r},t) = G_r K \beta \frac{\mu_a \psi}{\rho C_p} \tag{14.128}$$

The pressure equation shows the correction factor G_r, which expresses the fraction of thermal energy from the absorbed light that is converted into mechanical expansion. The ramifications of this correction factor were described in Section 8.3.2.2.

14.13.2 Medical Imaging Applications

Photoacoustic imaging is a relatively new member in the family of technologies used for imaging biological media for diagnostic purposes. However, this trend will soon be changing to an imaging technology that may rival MRI, x-ray, ultrasound, and many other technologies currently in use according to current literature. Never before have two technologies been married in quite this way to produce an effect that is so useful for imaging biological media. Using photoluminescent techniques with that of current acoustic (ultrasound) image processing techniques is a promising solution to many imaging problems today. With photoacoustics, biological media such as blood vessels can be mapped, which before often required the use of a radioactive chemical being inserted into the patient (e.g., x-ray radiography).

For photoacoustic imaging, there are currently two known methods for inducing a measurable signal that is useful in imaging. These two techniques are known as reverse-mode detection and forward-mode detection.

Reverse-mode detection is very similar to how current ultrasound is used. In reverse mode, the light source is on the same side of the medium to be imaged as the detector. Therefore, the detector actually picks up the inverse image, or the waves that are bounced back from the acoustical interaction. This method of imaging is relatively computationally intensive and requires the use of image and signal processing techniques to recover a useful image. In forward mode the acoustic detector is on the opposite side of the laser source. Compared to the reverse mode only the geometry has changed, making it nonetheless, not in the least, a less computational problem.

14.13.2.1 Acoustooptical Imaging of Teeth

As an example of the order of magnitude of the values obtained under photoacoustic imaging of a hard medium, the tooth is taken as a representative.

Figure 14.46 illustrates the schematic for photoacoustic imaging of the tooth. Optical radiation is delivered through a fiber optic, detected by the acoustic sensors, and processed by a personal computer.

The physical properties of tooth tissue are listed in Table 14.1. The tooth is irradiated by a titanium:sapphire laser tuned to $\lambda = 750$ nm, operating with 60-fs laser pulses with an energy per pulse of 8 nJ, and a beam waist of 35 μm. The Rayleigh range is 5.13 mm to focus the laser beam over the entire depth of the tooth. Figure 14.47 illustrates the focusing schematic for calculations of temperatures in the tooth at the various anatomical locations. The following locations are identified: A is the enamel, B the dentin, and C the pulp of the tooth.

The maximum temperature rise is directly calculated from Equation (14.113) for r $\geq \omega$, and it appears as

$$t_{\max} = \frac{a^2\omega^2}{8}\left(\frac{4}{3}\frac{r^2}{\omega^2} - 1\right)$$

At the spatial location $r = \omega_0$, the maximum temperature rise occurs at $t_{scaled} = 0.33$ (see Figure 14.47 for Pulp [C]). When inserting the value for a^2, for the focal waist $\omega_0 = 35$ μm the time at which the maximum temperature is reached occurs at $t_{max} \cong 300$ μs.

Under tighter focusing to a waist of $\omega_0 = 1$ μm, the temperature maximum occurs much earlier at $t_{max} \cong 0.3$ μs. Figure 14.48 shows the results of the computed temperature rise for irradiation of the tooth surface by a titanium:sapphire laser tuned to $\lambda = 750$ nm with a 35 μm waist, and a pulse width of 60 fs.

A minimal temperature rise induced by the laser radiation is desirable for application of the proposed instrument in dental practices. Tighter focusing will allow for an increase in the spatial resolution of the imaging, yet higher peak irradiances and temperature maxima will result. For laser pulses of 8 nJ pulse energy, 60 fs pulse width, focused to a beam waist of

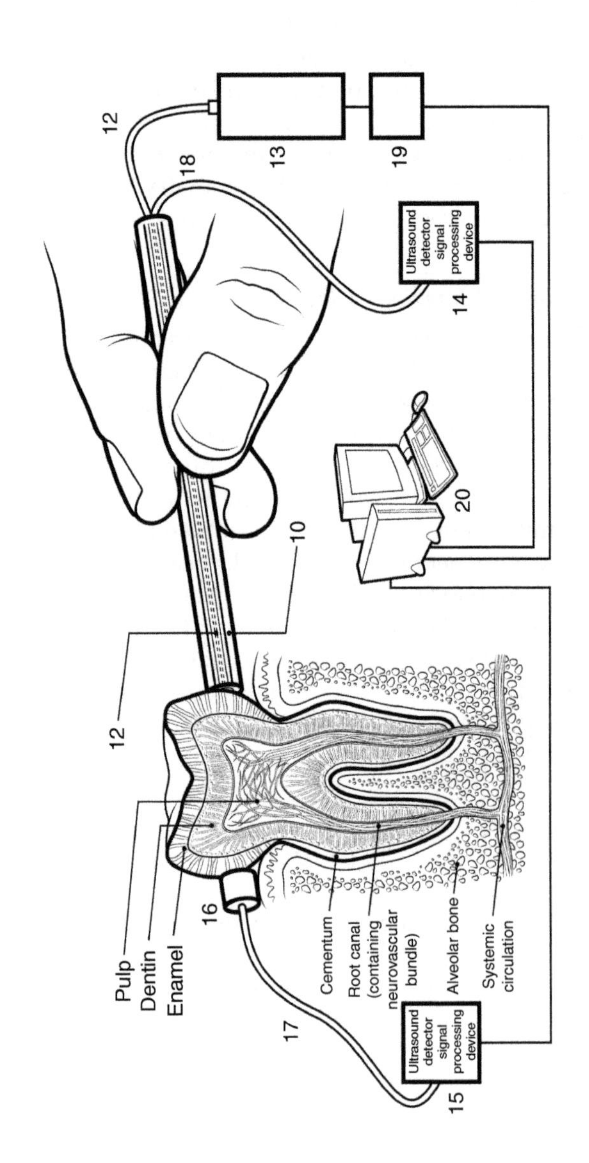

FIGURE 14.46
Schematic for photoacoustic imaging of the tooth. Optical radiation is delivered through a fiber optic, detected by the acoustic sensors, and processed by a personal computer.

TABLE 14.1

Thermophysical Properties of Dental Tissue

	Enamel	Dentin	Pulp
Thickness (mm)	1.5	1.0	5.0
Density (g/cm^3)	3.0	2.0	1.027
Thermal conductivity (J/s cm K)	9.2×10^{-3}	6.3×10^{-3}	0.572×10^{-2}
Specific heat (J/g K)	1.1	1.17	3.93
Optical absorption at 750 nm (cm^{-1})	1.0	6.0	4.5

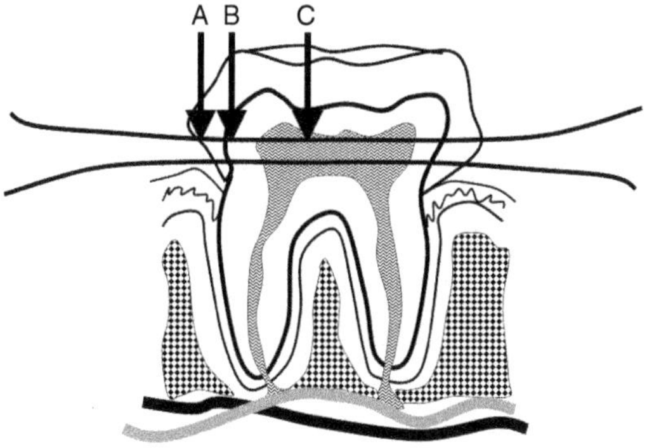

FIGURE 14.47
Focusing schematic for calculation of temperatures in the tooth at the various anatomical locations. The following locations are identified: A is the enamel, B the dentin, and C the pulp of the tooth.

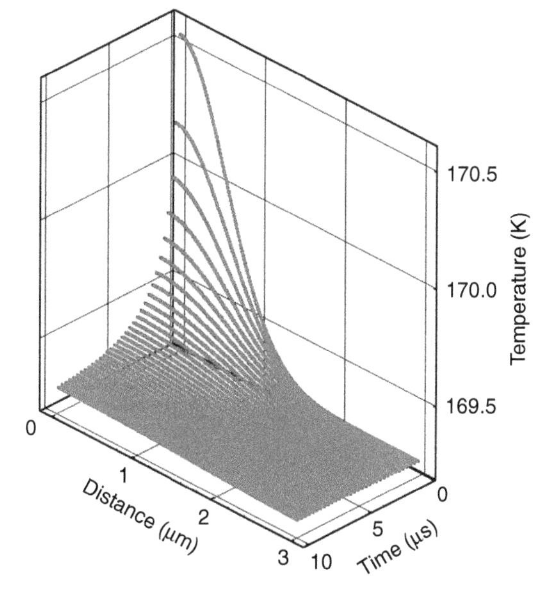

FIGURE 14.48
Computed bulk-temperature rise for the surface of the tooth from a Gaussian beam profile. Temperature calculation for irradiation of the tooth surface by a titanium:sapphire laser tuned to λ = 750 nm with a 35 μm waist and a pulse width of 60 fs. The energy per pulse is taken at the experimental value of 8 nJ. The scaling used for time is $a^2w^2/8$. (Research data from Dr. Pavlina Pike and Dr. Christian Parigger at the University of Tennessee Space Institute, Tullahoma, TN, USA.)

35 μm, the peak irradiance amounts to ~10^9 W/cm^2, with the maximum temperature rise on the order of 0.1K–1K. Initial theoretical/numerical studies show bulk temperature rise on the order of 1° for a single femtosecond laser pulse focused to a waist of 10 μm. This result is obtained by applying the heat transfer equation, including the use of Green's function solution.

In contrast, focusing to a beam waist of 1 μm shows peak irradiances of $\sim 10^{13}$ W/cm^2 and a maximum temperature rise on the order of 100–1000K (inferred from the above-presented low-irradiance model) for the respective locations indicated in Figure 14.47. This kind of temperature rise is definitely undesirable for tooth imaging, or for any hard tisue. The temperature rise in excess of 100° and the rate at which this temperature rise occurs will result in vaporization as described in Chapter 5 in addition to the formation of cracks.

14.13.2.2 Photoacoustic Excitation and Wave Propagation

The range of acoustic imaging frequencies that will be obtained in photoacoustic imaging will depend in particular on the material constants of the tissues involved as described earlier. For an enamel cube with 7 mm edges the resonance frequency can be calculated from Equation (14.119) using the material properties listed in Table 14.2, yielding

$$f = C\sqrt{\frac{10\times10^8\ \text{kg/s}^2\text{m}}{3\ \text{kg/m}^3}\frac{7\times10^{-3}}{7\times10^{-3}\times7\times10^{-3}}} = C\sqrt{3}\frac{10^4}{7\times10^{-3}}$$
$$= C\frac{1.7}{7}10^7 = C\times2.43\times10^6\ \text{Hz} \quad (14.129)$$

For a cube the constant is C = 0.926. This gives a resonance frequency of approximately 2.25 MHz.

In an experimental collaboration effort between the Carolinas HealthCare System, Erythema Diagnostics, LLC, University of Tennessee Space Institute and Analytica Sciences, Inc., a block of enamel has been irradiated with a Q-switched titanium:sapphire laser tuned to 750 nm. The acoustic signal obtained from the tooth was collected by a polyvinylidene fluoride acoustic transducer and analyzed by Fourier transformation. The Fourier spectrum of the detected frequencies resulting from laser excitation is presented in Figure 14.49.

TABLE 14.2

Material Properties of Tooth Tissues

Parameter	Material	Value range	Units
Young's modulus E	Enamel	85–130	GPa (1 kPa =
	Dentin	14.7	101.97 kg/s^2 m)
Poison ratio σ	Enamel	0.31	
	Dentin	0.31	
Volume expansion	Enamel	120–1500	1 K^{-1}
coefficient β	Dentin	60–120	
Density ρ	Enamel	2.97	10^3 kg/m^3
	Dentin	2.14	
Specific heat C_p	Enamel	1172–1591	J/kg °C
	Dentin	0.18	
Thermal conductivity	Enamel	0.974	J/s °C m
κ	Dentin	0.569	

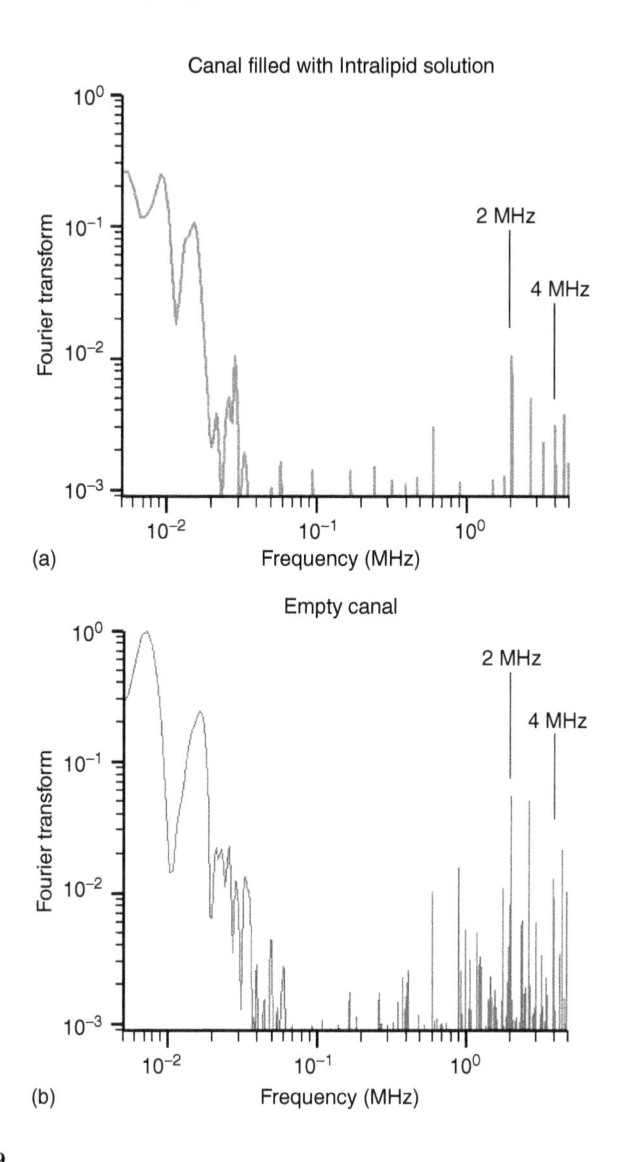

FIGURE 14.49
Fourier spectrum for a 7 mm cube of ceramic crown material under pulsed laser excitation. (Research data from Dr. Pavlina Pike and Dr. Christian Parigger at the University of Tennessee Space Institute, Tullahoma, TN, USA.) (a) A 2 mm diameter canal at a distance 2 mm from the edge of cube filled with scattering solution (Intralipid) and (b) empty canal in ceramic block under pulsed laser irradiation.

Figure 14.49a shows the Fourier spectrum of a cubic ceramic block with a 2 mm diameter hole filled with soymilk, while in Figure 14.49b the hole is left empty, providing a local acoustic impedance discontinuity. The frequency spectrum from the ceramic block filled is notably different from the ceramic block under pulsed laser irradiation with air in the canal.

From Figure 14.49 it can be seen that the resonance frequency is clearly present. Other frequencies that can be recognized in the Fourier spectrum can be attributed to dimensions involved with a well-defined duct running the length of the cube.

Another recent development in biomedical imaging is terahertz imaging, which will be described next.

14.14 Terahertz Imaging

Terahertz imaging research has been going on for over 15 years and is only recently starting to make its impression on medical imaging.

To overcome some of the complications of visible and near-infrared electromagnetic interaction of light with tissue, a new approach in imaging produces a sampling frequency with a wavelength longer than infrared and shorter than microwave, in the frequency range of 300 GHz to 30 THz. Even though a visible or near-infrared laser source is used initially, after pulsing this laser beam in the terahertz regime, the pulsing frequency becomes the probing frequency, with a significant reduction in scattering, since scattering decreases with the forth power of the wavelength. Generally terahertz systems use ultrafast red or near-infrared lasers, e.g., Ti:sapphire operating at pulse widths of less than 70 fs, and a bandwidth in excess of 10 THz.

It is also well known that microwave radiation, operating in the gigahertz frequency range, has a high penetration in tissue, and is in that manner used for thorough heating.

Due to the short pulse duration and high pulse rate, the laser line width is broadened to become a broadband source.

The index of refraction n for this virtual frequency is determined by the relationship between the time delay TD for transmission and the sample thickness x as

$$\mathrm{TD} = \frac{(n-1)x}{c} \tag{14.130}$$

where c is the speed of light in vacuum.

Applying a curve fit to the time delay as a function of thickness provides a slope m, which corresponds to

$$m = \frac{(n-1)}{c} \tag{14.131}$$

After rewriting Equation (14.131) for the index of refraction we get the following expression:

$$n = 1 + mc \tag{14.132}$$

The other parameter of concern is the absorption coefficient at the virtual wavelength, which will be dependent on the real wavelength and the virtual refractive index.

The absorption coefficient can also be derived from a curve fit using the Beer–Lambert law applied to the transmitted fluence I as a function of thickness of the medium, while taking into account the reflections at the front and back interface of the slab of medium as shown in the following equation:

$$\ln\left(\frac{I}{I_0'}\right) = -\alpha x \quad (14.133)$$

where I'_0 equals the incident fluence minus the backscatter shown as

$$I_0' = (1 - R)I_0 \quad (14.134)$$

Even though there is still significant absorption in the red and near-infrared spectrum, the apparent absence of scattering under terahertz imaging provides an opportunity for deep-tissue imaging.

In terahertz detection the signal is Fourier transformed to produce a spectrum: the signal intensity across the range of frequencies (power spectrum).

The imaging protocol includes information in both the time and frequency domains.

The time domain mainly uses the following parameters:

- Amplitude: property of the transmitted/reflected signal
- Pulse time delay: used to derive the refractive index in combination with the sample thickness
- Pulse width
- The electric field strength: yields detailed dielectric tissue information

The frequency domain uses the information retrieved on the following parameters:

- Transmission: to apply the Beer–Lambert identity as a function of frequency
- Spectral information: arising from the pulse broadening under this kind of modulation. Especially the relationship between the spectrum of the incident and the detected pulse provides specific tissue information
- The pulse phase: derived from the Fourier transform
- Multiple frequency probing: to resolve the power spectrum for each wavelength simultaneously allowing to formulate multiple equations with several unknowns

Medical terahertz imaging has become a more likely candidate for clinical application with the introduction of new developments in the technologies of coherent generation and detection of terahertz radiation. Various commercial imaging systems are now available. Some of the medical applications are in the assessment of wound healing and monitoring cancer progression.

The main disadvantage of terahertz imaging is the fact that the spatial resolution is generally lower than commercial infrared imaging techniques. In contrast, the main advantage is the differences in attenuation with frequency between healthy and diseased tissues.

14.15 Summary

In this first chapter on the diagnostic use of optical techniques, the tried and true optical microscope and its limitations were described. Subsequently various derived techniques were presented such as confocal microscopy, multiphoton imaging, and NSOM. Other techniques using various physical concepts for obtaining information described were spectroscopy and the derived technique of erythema detection. Other physical imaging techniques described were holography, predominantly used for dentistry; polarization imaging, again combined with optical microscopy; and multiple applications of transillumination imaging. Several techniques relying on the equation of radiative transfer for image formation were also introduced, such as diffuse light tomography, OCT, and ballistic photon imaging. To provide a complete review the correlation theory and visibility of these imaging techniques was also described. The topics discussed toward the end of the chapter were evanescent wave diagnostics, thermography, photoacoustic imaging, and terahertz imaging.

Further Reading

Chen, Q. and Zhang, X.-C., Electro-optic THz imaging, in *Ultrafast Lasers: Technology and Applications*, Fermann, Galvanauskas, Sucha, Eds., Marcel Dekker, Inc., New York, 2002, chap. 11.

Choma, M.A., *Optical Coherence Tomography: Applications and Next-Generation System Design*, Ph.D. Dissertation, Duke University, Durham, NC, 2004.

Dunn, R.C., Near-field scanning optical microscopy, *Chem. Rev.*, 99, 2391–2927, 1999.

Fercher, A.F., et al., Measurement of intraocular distances by backscattering spectral interferometry, *Opt. Commun.*, 117, 43–48, 1995.

Gibson, A.P., Austin, T., Everdell, N.L., Schweiger, M., Arridge, S.R., Meek, J.H., Wyatt, J.S., Delpy, D.T., and Hebden, J.C., Three-dimensional whole-head optical tomography of passive motor evoked responses in the neonate, *Neuroimage* 30, 521–528, 2006.

Huang, D., Swanson, E.A., Lin, C.P., Schuman, J.S., Stinson, W.G., Chang, W., Hee, M.R., Flotte, T., Gregory, K., Puliafito, C.A., and Fujimoto, J.G., Optical coherence tomography, *Science*, 254, 1178–1181, 1991.

Jiang, H., Frequency-domain fluorescent diffusion tomography: a finite element algorithm and simulations, *Appl. Opt.*, 37, 5337–5343, 1998.

Jiang, H., Paulsen, K.D., Osterberg, U.L., and Patterson M.S., Frequency-domain optical image reconstruction for breast imaging: initial evaluation in multi-target tissue-like phantoms, *Medical Physics* 25, 183–193, 1998.

Jiang, H., Paulsen, K.D., Osterberg, U.L, and Patterson, M.S., Enhanced optical image reconstruction using DC data: initial study of detectability and evaluation in multi-target tissue-like phantoms, *Phys. Med. Biol.*, 43, 675–693, 1998.

Luo, M.S.C., Chuang, S.L., Planken, P.C.M., Brener, I., Roskos, H.G., and Nuss, M.C., Generation of terahertz electromagnetic pulses from quantum-well structures, *IEEE J. Quantum Electr.* 30, 1478, 1994.

Mirabella, F., Internal reflection spectroscopy, in *Practical Spectroscopy Series*, Vol. 15, Marcel Dekker, New York, 1992.

Najarian, K., and Splinter, R., *Biomedical Signal and Image Processing*, CRC-Press, Boca Raton, 2005.

Rost, F.W.D., *Fluorescence Microscopy*, Cambridge University Press, Cambridge, 1992.

Simhi, R, Gotshal, Y., Bunimovich, D., Sela, A., and Katzir, A., Fiber-optic evanescent-wave spectroscopy for fast multicomponent analysis of human blood, *Appl. Opt.*, 39, 3421–3425, 1996.

Problems

1. Consider the average wavelength of visible light used in conventional optical microscopy to be 550 nm.
   (a) Calculate the maximum attainable resolution for an objective with a 3 mm aperture.
   (b) Calculate the maximum useful magnification of an optical microscope based on this limitation.
2. Name five different optical microscopic techniques and briefly explain their mode of operation.
3. Describe the fundamental advantage of confocal microscopy over conventional microscopy.
4. Explain the limitations of two-photon imaging.
5. Why does water present a good medium for Raman spectroscopy?
6. Explain how near-field scanning optical microscopy is the only microscopic technique that is not limited by the Rayleigh criterion.
7. A pneumothorax is a collapsed lung.
   (a) Give a brief description using Snell's law on the operating mechanism of a continuous optical pneumothorax detector.
   (b) Describe some of the complications that need to be overcome in this detection scheme.

8. What are the limiting factors for resolution in optical coherence imaging and how exactly do they influence the resolution?
9. Explain how a shorter laser pulse will increase the bandwidth of the emitted light.
10. Calculate the bandwidth of a 100 ps titanium:sapphire laser tuned to a center wavelength of 760 nm at a repetition rate of 76 MHz.
11. Illustrate how the bandwidth of a light source affects the coherence length.
12. Calculate the fringe intensity ratio between a 20 nm bandwidth source (e.g., SLD) and a 0.5 nm bandwidth source (e.g., continuous laser).
13. Calculate the coherence length for a stable laser source operating at 632.8 nm with a 1.7 GHz line width, which corresponds to a nanometer line width and a superluminescent diode operating at 830 nm with a bandwidth of 25 nm.
14. Elaborate on the commonalities and differences of ballistic photon imaging and x-ray imaging.
15. How can polarization provide additional selection criteria in optical coherence tomography?
16. Give two examples of diagnostic evanescent wave applications.
17. Explain how medical thermography works and how it can be used in medical diagnostics.
18. What is the shift in blackbody emission peak when the local temperature changes from 36.5 to 37.5°C?
19. How can photoacoustic imaging be used to detect anatomical anomalies in a biological medium?
20. Calculate the axial temperature profile of a thullium/holmium/chromium:YAG Gaussian beam operating at 2.06 μm hitting plaque stenosis in arterial vessel with a spot size created by 26 fibers with 100 μm core diameter arranged in a tightly packed circular array. The laser has a pulse width of 250 μs and pulse energy of 350 mJ operating at 5 kHz (assume that the heat diffusion is negligible during each pulse) The optical properties for plaque at the Th/Ho/Chr:YAG wavelength are μ_a = 1260 cm^{-1}, μ_s= 18.5 cm^{-1}, g =0.982 , with the density of plaque ρ = 1453 kg/m^3 and the specific heat equals C_p = 837 J/kg K.
21. Placing a drop of oil between the object and the objective lens increases the resolving power. How does it accomplish this?
22. According to Lord Rayleigh, what is the criterion for resolving two image points A and B that lie in the image plane? What is the angular separation between the two point sources that this criterion corresponds to?
23. How does confocal microscopy perform its optical sectioning in tissues that improves resolution and aides the researcher in determining locations of important cellular components?
24. What are the advantages and limitations of two-photon microscopy?
25. An optical coherence tomography imaging system requires a low coherence light source for high spatial resolution. How does low coherence relate to the spectral bandwidth of the light source? What bandwidth is needed for micron resolution at the short wavelength limit in the visible around 400 nm?
26. Discuss ballistic, snake, and diffuse light as it exits a thick sample of tissue. Draw the scattering profile of each of these paths and plot the temporal dependence, as well.
27. What is evanescent wave spectroscopy uniquely suited to measure? In what wavelength region does this apply?

15

Diagnostic Methods Using Light: Photochemical

Certain diagnostic methods that use light-activated chemical reactions for diagnostic purposes and that have been developed over the years are discussed in the next section. A review of technological developments resulting from the increasing demands for resolution and physiological details is presented for the selected diagnostic methodologies. The relevance to the process of gauging patient performance is illustrated as well.

The diagnostic methods described range from fluorescent imaging, fluorescence in microscopy, and various types of optical imaging techniques that can reveal metabolic or anatomic details. Some examples of optical imaging techniques that rely on the interaction of light with chemical compounds to produce diagnostic information are Raman spectroscopy and a technique referred to as an optical tongue. Various other novel and established diagnostic procedures are presented.

The first method described is the most simple as well fluorescence imaging.

15.1 Fluorescence Imaging

Fluorescent imaging is a good first-order indication of bacterial infections, fungal infections, and alterations in pigmentation to name but a few diagnostic procedures done under elementary ultraviolet illumination. The diagnostic fluorescence can generally be achieved by illumination from an ultraviolet lamp: a Wood's lamp. Normal healthy skin will not fluoresce under ultraviolet illumination.

The Wood's lamp is frequently used to identify superficial fungal or bacterial infections. The Wood's lamp can be used to detect the presence of a fungal scalp infection or various skin and nail infections. The clinical examination process is somewhat similar to examining the progress of bacterial growth in a Petri dish, which has been done by fluorescent emission inspection for decades.

Under ultraviolet light normal skin will not fluoresce. However, various bacterial and fungal infections will be revealed by a certain color spectrum, for instance, leprosy will fluoresce blue–white, tuberous sclerosis will show the appearance of ash-leaf shaped spots, a yeast infection fungus of the skin called tinea versicolor will show a golden yellow glow, and *Trichophyton schoenleinii* is a fungus causing favus in humans. Favus is a chronic, scarring form of an infection of the scalp characterized by saucer-shaped crusted lesions and resulting in permanent hair loss. This disease pattern is characterized by a pale greenish yellow fluorescence under Wood's lamp illumination.

15.1.1 Fluorescence Molecular Explanation

Electrons of molecules or atoms can occur in different energy states: ground state (minimal energy) and excitation state (higher energy) in addition to rotational and vibrational states. Prior to excitation, the electron configuration of a molecule is in the ground stage. On excitation by particular wavelengths the electrons may be raised to a higher energy and vibrational excitation state. The excitation and fluorescence mechanism is outlined in Figure 15.1.

When electrons absorb energy, they get excited to a higher electronic energy and vibrational state. In fluorescence, the electrons lose some vibrational energy and return to the lowest excited singlet state. Furthermore, the electrons drop back to the ground state with simultaneous emission of fluorescent light. The final return to the ground state occurs with the loss of energy in several forms:

1. Emitted light (fluorescence)
2. Generation of heat
3. Molecule may undergo a chemical reaction such as polymerization, decomposition, etc.

The emitted light is always of longer wavelengths than the excitation light (Stokes law, Section 8.9.6 in Chapter 8). However, if the electron reaches the forbidden level called metastable state, it leads to phosphorescence, as illustrated in Figure 15.2. In fluorescence microscopy the separation of excitation and emission wavelength is achieved by the proper use of filters to block or pass specific wavelengths of the spectrum.

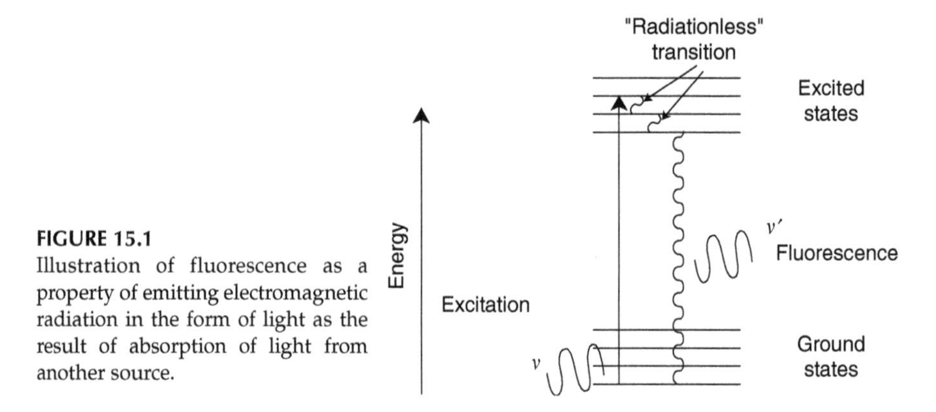

FIGURE 15.1
Illustration of fluorescence as a property of emitting electromagnetic radiation in the form of light as the result of absorption of light from another source.

The ability of the fluorochrome (excitable molecular chain) to absorb the excitation light is known as the extinction coefficient. The greater the extinction coefficient, the more likely is the absorption of light, which is a prerequisite for fluorescent emission. The yield is referred to as the quantum yield, the ratio of the number of quanta emitted to the number of quanta absorbed. Quantum yields below 1 are the result of the loss of energy through nonradiative pathways (heat or photochemical reaction) rather than the re-radiative pathways of fluorescence.

Extinction coefficient, quantum yield, mean luminous intensity of the light source, and fluorescence lifetime are all important factors contributing to the intensity and utility of fluorescence emission.

The lifetime of an electron in an excited state is very short. Absorption is 10^{-13} s, and emission is approximately 10^{-9} s. Generally, molecules absorbing energy obey the quantum theory as defined in the following equation, where the energetic change, ΔE, is proportional to the incident photon energy,

$$\Delta E = h\nu = \frac{hc}{\lambda} \tag{15.1}$$

where ν is the frequency of the photon, h the Planck's constant, c the speed of light, and λ the wavelength of light.

The fluorescence microscope differs from the compound microscope in having two types of filters and a special light source of short wavelength (Figure 15.3). A high-pressure mercury arc lamp is commonly used to illuminate the sample at a short wavelength such as ultraviolet or blue. The first filter, the excitation filter, is located between the lamp and the specimen and provides light over a narrow band of wavelength corresponding to maximum absorption of

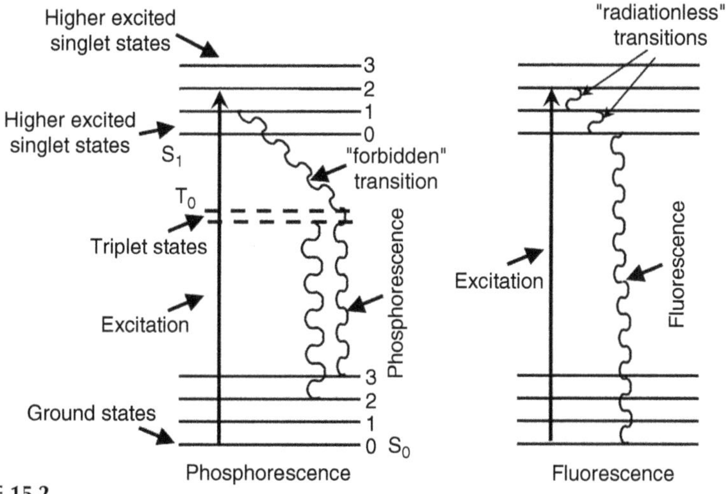

FIGURE 15.2
Outline of the excitation and fluorescence mechanism. Schematic of the fluorescence and phosphorescence processes.

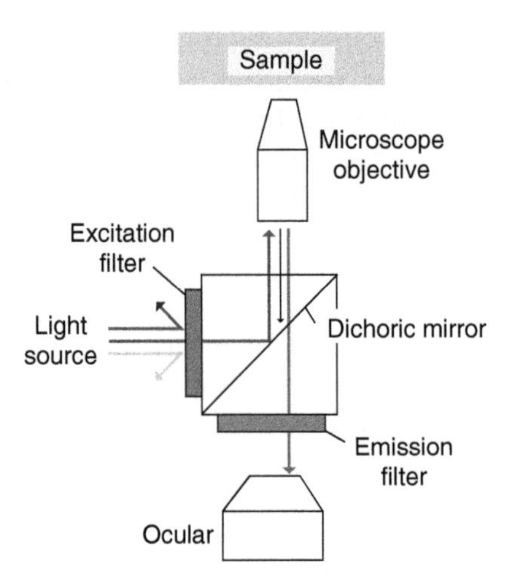

FIGURE 15.3
Diagram of fluorescent microscope.

the fluorescent substance used. In addition, lasers can be used that operate at specific excitation wavelength, thus reducing the amount of ambient light and creating a more efficient excitation.

Fluorescence is weak; therefore, to enable it to be seen despite the strong illumination from the light source, a second filter, the barrier filter, is placed between the specimen and the eyepiece. In principle, the barrier filter is a high-pass filter, as illustrated in Figure 15.4. It should prevent the shorter wavelength from reaching the eye or other detector, while allowing longer wavelengths (fluorescence) to pass through. The fluorescent object is seen as a bright image against a dark background.

An example of the use of fluorescence microscopy is given for a bacterial culture growth. Suppose a known number of cells are grown on chamber slides. Upon treatment of the cells with the bacterium, the cells are fixed with 4% paraformaldehyde solution. An optional step is background reduction followed by probing with the proper primary antibody. Background reduction is achieved by adding sodium borohydride or sodium cyanoborohydride. The secondary antibody is added after washing the cells. This wash step is necessary to clear any primary antibody that did not bind to cellular proteins, and to prevent competitive binding of the secondary antibody to unbound primary antibody. The sample is now ready for imaging. A difference between preparing samples for fluorescence microscopy and confocal microscopy is in the mounting. When using the confocal microscope, the samples are mounted with antifading agents such as Prolong or Slowfade. This step minimizes fading by more than 50% because of the use of the laser. For example, flourescein isothiocyanate (FITC) fades relatively fast.

The imaging process continues as follows. After 1-min exposure to 50% laser power, the intensity is reduced approximately 5%. Therefore after 15-min exposure to 50% laser power, the intensity is reduced by 75%. When using antifading agents, after 15 min of exposure to 50% laser power, only 20–40% reduction in intensity is observed. During the entire process the fluorescence can be recorded and vital information about the biological processes can be uncovered.

15.1.2 Fluorescent Molecules

Fluorescence microscopy is most often used to detect specific proteins or other molecules in cells and tissues. A specific and widely used tool is to couple fluorescent molecules to specific fluorescent dyes, which in turn are attached to antibody molecules. The antibodies serve as highly specific and versatile staining reagents that bind selectively to the particular macromolecules that they recognize in cells or in the extracellular matrix.

Two commonly used fluorescent dyes in biological research are FITC, which emits an intense green fluorescence when excited with blue light, and rhodamine, which emits a deep red fluorescence when excited with green–yellow light. Fluorescein, commonly called FITC, is generally used to tag antibodies, which in turn can be used to find specific localization of proteins in tissues and cells.

By coupling one antibody to FITC and another to rhodamine, the distributions of different molecules can be measured in the same cell; the two molecules are visualized separately in the microscope by switching back and forth between two sets of filters, each specific for one dye.

Figure 15.4 shows the excitation and emission spectra of a hypothetical fluorophore. The curve on the left (toward the shorter wavelength) is the excitation curve with a peak in the blue wavelength range. The curve on the right is the emission curve with a peak in the green wavelength and a long tail extending to red. An excitation filter with a narrow band pass in the blue provides optimal excitation. Such a filter would allow only light in the peak excitation spectrum of the fluorophore to pass through the specimen. An ideal barrier filter (represented by the dotted curve to the right above the emission curve) should be fully opaque at the excitation wavelength and

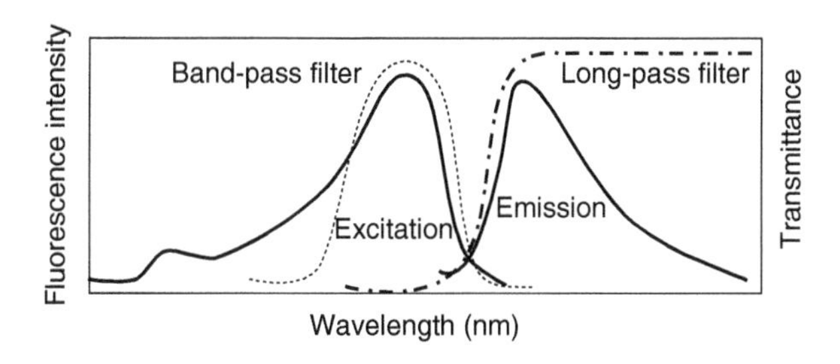

FIGURE 15.4
Absorption and emission spectrum and the use of specific filters to block the excitation light from washing out the fluorescence image.

TABLE 15.1

Examples of Commonly Used Fluorochromes in Research and the Electromagnetic Spectrum of Wavelengths

Fluorochrome	Absorbed Wavelength (nm)	Emitted Wavelength (nm)
Fluorescein (FITC)	490	525
Propidium iodide	536	617
Calcein	370	435

fully transparent to the longer wavelengths of the emission spectrum. Table 15.1 shows a list of several commonly used fluorochromes.

While the fluorescent microscope has allowed biologists to examine more detail than the compound microscope, there still exists the limitation of the resolution since it is a limitation in the physical nature of light. In addition, images produced are two-dimensional, and at that level of detail, scientists are eager to see more with better reconstruction of the images.

This process will be explained in more detail in the following section. As shown in the chart, FITC atoms will become excited when light at the argon laser wavelength of 488 nm, or blue light, illuminates the sample. It will then emit at a wavelength of 525 nm and will register as green. Propidium iodide, or PI, is a DNA stain that is most commonly used to examine the nucleus in normal cells, or the fragmented nucleus in apoptotic cells. Calcien is a water-soluble dye, which is taken into cells and cleaved by a certain enzyme. This enzyme ensures that the molecule cannot leave the cytoplasm and thus can be used as an indicator of live versus dead cells in a population. This dye can also be used for cell swelling assays in live cells where the calcien, and the signal, will be more dilute if more water is introduced into the cell.

15.1.3 Fading

There are conditions that may affect the re-radiation of light and thus reduce the intensity of fluorescence. This reduction of emission intensity is called fading. This can be further divided into quenching and bleaching. Bleaching is the irreversible decomposition of the fluorescent molecules because of light intensity in the presence of oxygen molecule. Quenching also results in reduced fluorescence intensity and frequently comes about as a result of oxidizing agents or the presence of salts of heavy metals or halogen compounds.

Sometimes the quenching results from transfer of energy to other so-called acceptor molecules physically close to the excited fluorophores. This phenomenon is known as resonance energy transfer (RET). The occurrence of bleaching has led to a technique known as FRAP: Fluorescence recovery after photobleaching. This technique is based upon bleaching by short laser bursts and subsequent recovery of fluorescence caused by diffusion of fluorophores into the bleached area.

15.2 Ratio Fluorescence Microscopy

Quantitative fluorescence microscopy is becoming an increasingly important tool in the study of cell biology. Fluorescence microscopy has long been used for qualitative characterizations of subcellular distributions of proteins, lipids, nucleic acids, and ions, but quantifying these distributions are complicated varieties of optical, biological, and physical factors. Many factors that complicate quantification of fluorescence in cells can be circumvented by analyzing fluorescence ratios derived from pairs of fluorescent images.

Fluorescence ratio images may be collected by sequentially exciting the sample with two different wavelengths of light and sequentially collecting the different images. The images are created by exciting the sample with a single wavelength of light and collecting images formed from light of two different emission wavelengths, or by exciting the sample with two different wavelengths and collecting emission of two different wavelengths. Ion indicators have been developed for excitation ratio microscopy (Fura-2 for calcium and FITC for pH) and emission ratio microscopy (Indo-1 for calcium and SNARF-1 [carboxy-seminaphthorhodafluor] for pH).

Ratio fluorescence microscopy has been most commonly applied to the characterization of intracellular ion concentrations. The lasers used in most laser scanning microscopes emit at only a few wavelengths, and these wavelengths are inappropriate for most excitation ratio ion indicators. A newly developed pH indicator, NERF (Cl-NERF-Dextran), is exceptional in that its excitation spectra at the two principal emission lines of the argon laser, 480 and 514 nm, are differentially sensitive to pH. The dissociation constant (pK_a) of NERF is relatively low, making it useful in characterizing acidic organelles. The pK_a is a measure of the length of an acid or base and can be used to determine the molecular charge under respective acidity (pH).

15.2.1 Fluorescence Resonance Energy Transfer (FRET) Microscopy

Within the living cell, interacting proteins are assembled in the molecular machines that function to control cellular homeostasis. These protein assemblies are traditionally studied using biophysical or biochemical methods such as affinity chromatography or co-immunoprecipitation. These in vitro screening methods have the advantage of providing direct access to the genetic information that encodes unknown protein partners. These techniques do not allow direct access to interactions of these protein partners in their natural environment inside the living cell, but using the approach of fluorescence resonance energy transfer (FRET) microscopy, this information can be obtained from single living cells with nanometer resolution.

FRET microscopy relies on the ability to capture weak and transient fluorescent signals efficiently and rapidly from the interactions of labeled molecules in single living or fixed cells.

One particular mode of operation of FRET is RET. In RET an excited fluorescent donor molecule, rather than emitting light, transfers this energy via dipole–dipole interaction to an acceptor molecule in close proximity. If the acceptor is fluorescent, then the decrease in donor fluorescence due to RET is accompanied by an increase in acceptor fluorescence (sensitized emission). The amount of RET depends strongly on the distance between the donor dipole and the acceptor molecule. Typically the efficacy of RET decreases with the sixth power of the distance. In this way the fluorophores can directly report on the phenomena occurring on the scale of few nanometers, well below the resolution of optical microscopes.

Among other things, RET has been used to map molecular distances and to study aggregation states, membrane dynamics, and DNA hybridization. Implementation of RET in the fluorescence microscope has been limited at present, though some of the examples include lipid probes to investigate sorting of lipid and lectin in cells, an assay c-AMP-dependent kinase activity in which catalytic and regulatory subunits of the kinase were labeled with donor and acceptor, respectively, and fluorogenic substrate for alkaline phosphatase that allows microscope imaging of enzyme activity.

15.2.2 Mechanism of Fluorescent Resonant Energy Transfer Imaging

Fluorescence resonance energy transfer is a process involving the radiationless transfer of energy from a donor fluorophore to an appropriately positioned acceptor fluorophore. FRET can occur when the emission spectrum of a donor fluorophore significantly overlaps (>30%) the absorption spectrum of an acceptor, provided that the donor and acceptor fluorophores dipoles are in favorable mutual orientation. Because the efficiency of energy transfer varies inversely with the sixth power of the distance separating the donor and acceptor fluorophores, the distance over which FRET can occur is limited to between 1 and 10 nm. When the spectral, dipole orientation, and distance criteria are satisfied, illumination of the donor fluorophore results in sensitized fluorescence emission from the acceptor, indicating that the tagged proteins are separated by less than 10 nm.

The energy transfer efficiency (E_{tr}), the rate of energy transfer (K_t), and the distance between the donor and acceptor molecules (r) are calculated using Equations (15.2), (15.3), and (15.4), respectively:

The energy transfer efficiency is given as

$$E_{tr} = \frac{R_0^6}{R_0^6 + r^6} \tag{15.2}$$

while the rate of energy transfer is defined as

$$K_t = 1 - \left(\frac{I_{DA}}{I_D} \right) \tag{15.3}$$

where the distance r between the donor and acceptor molecules is a function of the energy itself as shown in the following equation:

$$r = R_0\left[\left(\frac{1}{E_{\mathrm{tr}}}\right) - 1\right]^{1/6} \tag{15.4}$$

The molecular distance constant R_0 is defined by the following equation:

$$R_0 = 0.21[\kappa^2 n^{-4} Q_{\mathrm{p}} J(\lambda)]^{1/6} \tag{15.5}$$

In Equation (15.3), I_D and I_{DA} are, respectively, the radiance of the donor in the absence and the presence of the acceptor. The distance between the donor and the acceptor, R_0, is often referred to as the Förster distance. At the Förster distance half the excitation energy of the donor is transferred to the acceptor, and the other half of the energy is dissipated by all other processes, including light emission. The quantum yield of the donor is given by the parameter Q_p, and n is the index of refraction. The parameter κ describes the relative dipole orientation.

The overlap energy integral $J(\lambda)$ expresses the degree of spectral overlap between the acceptor absorption and the donor emission. The overlap integral is defined by the following equation:

$$J(\lambda) = \frac{\int_0^\infty f_{\mathrm{d}}(\lambda)\varepsilon_{\mathrm{A}}(\lambda)\lambda^4 \mathrm{d}\lambda}{\int_0^\infty f_{\mathrm{d}}(\lambda)\varepsilon_{\mathrm{A}}(\lambda)\mathrm{d}\lambda} \tag{15.6}$$

where $f_d(\lambda)$ is the corrected fluorescence radiance of the donor wavelength over the range λ to $\lambda + \Delta\lambda$ with the total radiance normalized to unity and ε_A is the extinction coefficient of the acceptor at λ, units (M^{-1} cm^{-1}).

The occurrence of FRET signal (sensitized signal) can be verified by acquiring the two-emission signal bands of the double-labeled cells excited with donor wavelength. If FRET occurs, the donor channel signal will be quenched and the acceptor channel signal will be sensitized or increased. In principle, the measurement of FRET in a microscope can provide the same information that is available from the more common macroscopic solution measurements of FRET; however, FRET microscopy has the additional advantage that the spatial distribution of FRET efficiency can be visualized throughout the image, rather than registering only an average over the entire cell or population. Because energy transfer occurs over distances of 1–10 nm, a FRET signal corresponding to a particular location within a microscope image provides an additional magnification surpassing the optical resolution (~0.25 mm) of the light microscope. Thus, within a voxel of microscopic resolution, FRET resolves average donor–acceptor distances beyond the microscopic limit down to the molecular scale. This is one of the principal and unique benefits of FRET for microscopic imaging: not only colocalization of the donor- and acceptor-labeled probes can be seen, but also intimate interactions of molecules labeled with donor and acceptor can be demonstrated. Several FRET techniques exist based on wide-field, confocal, and two-photon microscopy as well as FRET/FLIM (flourescent-lifetime imaging microscopy), each with its own advantage and disadvantage. All FRET microscopy systems require neutral density filters to control the excitation light intensity, a stable excitation light source (Hg or Xe or combination arc lamp; UV, visible, or infrared lasers), a heated stage or a chamber to maintain the cell viability, and appropriate filter sets (excitation, emission, and dichroic) for the selected fluorophore pair. It is important to carefully select filter combinations that reduce the spectral bleed through (SBT) to improve the signal-to-noise ratio for the FRET signals.

15.2.3 Fluorescence Resonance Energy Transfer Pair

The widely used donor and acceptor fluorophores for FRET studies come from a class of autofluorescent proteins called green fluorescent proteins (GFPs). The spectroscopic properties that are carefully considered in selecting GFPs as workable FRET pairs include sufficient separation in excitation spectra for selective stimulation of the donor GFP, an overlap (>30%) between the emission spectrum of the donor and the excitation spectrum of the acceptor to obtain efficient energy transfer, and reasonable separation in emission spectra between donor and acceptor GFPs to allow independent measurement of the fluorescence of each fluorophore. GFP-based FRET imaging methods have been instrumental in determining the compartmentalization and functional organization of living cells and for tracing the movement of proteins inside cells.

A number of combinations of FRET pairs can be used depending on the biological applications. Selected popular FRET pair fluorophores are CFP–YFP, CFP–dsRED, BFP–GFP, GFP– or YFP–dsRED, Cy3–Cy5, Alexa488–Alexa555, Alexa488–Cy3, FITC–rhodamine (TRITC), YFP–TRITC or Cy3, etc.

Concentration-dependent properties of molecules that alter the spectroscopic properties of fluorophores have a wide range of application in fluorescence ratio microscopy. An example includes metachromatic behavior of organic dyes such as acridine orange and pyranine; acridine has been used as a sensitive pH probe. The monoamine dye accumulates in acidic organelles, and multimerization creates a complex whose emission maximum is red shifted. This allows a relatively straightforward fluorescence emission ratio application to qualitatively estimate pH differences in organelles.

Another concentration-dependent spectroscopic alteration results when some fluorophores interact to form excited state dimers, which results in a red shift from steady-state monomer emission fluorescence and a reduction in monomer excited-state lifetimes. The decay of red-shifted fluorescence reflects the characteristic reaction kinetics of the excited state. A methodology has been designed to develop cells with pyrene excimer-forming probes, and by ratio microscopy excimer distribution has been obtained in fibroblasts. Another work includes generation of a range of fluorescent analogs to facilitate the study of specific accumulation of ceramide in Golgi apparatus.

Quantitative fluorescence microscopy has been used for characterizing the kinetics of internalization of endocytic ligands, for sorting of recycling from lysosomally directed ligands, for maturation of sorting endosomes into late endosomes, and for recycling of endocytic receptors, i.e., sorting of transfer in receptor and recycle to cell surface.

Cytogenetic application of ratio fluorescence microscopy includes identification and localization of specific DNA sequences in both interphase nuclei and metaphase chromosomes through fluorescence in situ suppression hybridization. This technique has been used to physically map the location of particular sequences and to identify chromosomal abnormalities.

The factors affecting fluorescence quantification include spatial and spectral distribution of illumination and sensitivity of detector, optics of microscope system, and number and quality of optical elements of microscope. Each of the improvements in imaging hardware, optics, and especially probe chemistry will enhance the speed, sensitivity, and efficiency of fluorescence detection for the increased exciting applications.

15.2.4 Problems with Fluorescence Resonance Energy Transfer Microscopy Imaging

One of the important conditions for FRET to occur is the overlap of the emission spectrum of the donor with the absorption spectrum of the acceptor. As a result of spectral overlap, the FRET signal is always contaminated by donor emission into the acceptor channel and by the excitation of acceptor molecules by the donor excitation wavelength. Both of these signals are termed SBT signal into the acceptor channel. In principle, the SBT signal is same for 1p- or 2p-FRET microscopy. In addition to SBT, the FRET signals in the acceptor channel also require correction for spectral sensitivity variations in donor and acceptor channels, autofluorescence, and detector and optical noise, which contaminate the FRET signal. How to correct the contaminated signals is explained in data process part.

Other microscopic techniques use fluorescence as well. In the next section the use of fluorescence for functional imaging is discussed.

15.3 Raman Spectroscopy with Near-Field Scanning Optical Microscopy (NSOM) Employed

Spectroscopy is the measurement and interpretation of electromagnetic radiation that is absorbed or emitted when that radiation produces changes in the energy of the molecules, atoms, or ions of a sample.

Since NSOM has single molecule detection capabilities, Raman spectroscopy is done on the molecular level. For example, the structure of proteins can be studied at the molecular level. Most spectroscopy measurements take into account a large population of proteins. Therefore, the behavior that is observed is an average of a large population.

For spectroscopic NSOM the sample preparation is slightly more involved. Microscope cover slips are etched to make the surface of the glass rough. A thin coat of aluminum is evaporated onto the cover slips. The sample is deposited on one of the cover slips and air-dried. Since the Raman effect is very weak in nature, surface-enhanced Raman spectroscopy is employed in conjunction with NSOM. Another way of approaching the origin of Raman is by using electromagnetic theory. The Raman effect is very closely related to the theory of dispersion. The time-changing electric field from the incoming photon interacts with the molecule, causing a time-changing polarization effect in the molecule to occur. The molecule becomes an oscillating dipole, hence an excited vibrational state. The aluminum coating enhances the amplitude of the time-changing electric field that interacts with the sample. Therefore, the Raman effect is enhanced.

The sample/coated cover slip is mounted onto the *xyz* piezoelectric translation stage, and the sample is scanned under the NSOM tip. The light may excite the Raman spectra to be emitted, but conditions depend on quite a few parameters. The transmitted light is collected by the collection lens of the inverted microscope and optically guided through a notch filter. The notch filter removes any residual laser light from the source. Any fluorescence that occurs is sent with the Raman spectra to a spectrometer.

15.3.1 Fluorescence Resonance Emission Transfer

This type of single molecule detection research done with NSOM will be discussed briefly, but it is considered to have the maximum potential among other NSOM research projects. Consider a protein that a polypeptide chain formed from a collection of linked amino acids. A fluorescent probe that binds to a specific amino acid in the polypeptide chain is introduced into the protein sample. The emission spectrum of the probe overlaps the absorption spectrum of the individual amino acid. When the sample is illuminated with an excitation source that overlaps the probes absorption spectrum, the illumination will excite fluorescence in the probe. The fluorescence generated by the excited probe will excite the individual amino acid to fluoresce. This gives the ability to map out a polypeptide chain amino acid by amino acid.

In most of the imaging applications discussed so far the spectral information has been used to indicate certain specific attributes of either the cell or a particular molecule. The concept of spectral analysis is a specialty of its own, called spectroscopy. The basic principles of spectroscopy have been discussed in Chapter 8. The next section will describe a few selected examples of spectral diagnostic imaging.

15.3.2 Applications in Biology

Microscopy carried out by biologists usually involves samples that are sufficiently thin and transparent. Conventional optical microscopes could never be used to resolve thick samples. Therefore, biologists turn to other forms of microscopy to image their samples. These alternate forms of microscopy may require the sample to be treated with dyes, sliced into small cross-sections that are supported in some medium, and placed in environments in which no aerobic organism could survive. Essentially the living sample must be destroyed to gain detailed images.

To date, NSOM is not a technique that provides high-resolution imaging of biological samples in vitro. Biological samples are deposited on microscope cover slips and air-dried before any high-resolution imaging is performed using NSOM. The removal of biological membranes from their natural aqueous environments tends to kill the sample. The necessity of having the samples dried and fixed to the microscope cover slips is required for the feedback mechanism, which maintains a constant tip and sample separation, to function properly.

NSOM provides the biologist with an imaging method that generates subwavelength resolution optical information. One might ask the question, "Why use NSOM over the confocal microscope that provides resolution capabilities slightly less than the wavelength?" The confocal microscope does provide a means to generate high-resolution optical images without all the complications that are involved in an NSOM system. No confocal technique will yield the lateral resolution that is achieved through NSOM. However, it is the ability of NSOM to detect single molecules that makes it a valuable tool.

The fact that several wavelengths can be involved in the diagnostic imaging has already been discussed, and the spectral content has been analyzed to obtain additional information about the molecular structures and processes in the cell.

15.4 Optical "Tongue"

Several research groups are experimenting with what could be called an "optical tongue," using a combination of new technologies. A silicon chip with an array of microbeads chemically formulated to respond to different analytes such as the tongue responds to salty, sweet, sour, and bitter taste stimuli. The technology available has the potential to make "taste buds" for any analyte. Even though the development is still in its early stages, the food and drug industries believe that a commercially developed "tongue" will replace human taste testers. Such an optical "tongue" will be able to create an archive of successful taste patterns based on RGB (red, green, blue pixel) data files recorded with a charge-coupled device (CCD) color camera.

15.4.1 Mechanism of Operation

There have been a number of different methods that have been described by McDevitt and co-workers for the signal transduction and analyte binding events. In one of the initial and most simple strategies, indicator molecules are used to serve both the binding and transduction themes. For complex analytes such as proteins and toxins, antibody reagents and dye conjugates are used. For many other analytes, dye displacement methods are employed. Here the microspheres contain light-sensitive indicators and ligands that change color in the presence of specific food substances. Since foods contain multiple taste stimuli, it was necessary to form microspheres from a combinatorial library of polymeric monomers so that RGB data files can accurately represent a pattern of the taste stimuli composition.

The biological taste process will be described followed by an outline of the photochemical process that mimics this course of action.

15.4.2 Taste Stimuli Transduction

The compounds that induce the stimulus at the taste receptor cells do not necessarily have functional groups representative of only one of the elementary tastes. This phenomenon generates two important properties of the gustatory system:

1. Individual classes of taste stimuli or even individual compounds do not rely on one single transduction mechanism.
2. Overlap may exist among various classes of stimuli on a single transduction pathway.

The influence on transduction by these gustatory properties is twofold. First, they may contribute to the many subtle tastes present in food-containing mixtures of individual chemical compounds. Second, an overlap of transduction pathways may account for the ability of individual taste stimuli to be described as having more than a single taste, such as NaCl being described as both salty and sour, frequently depending on the concentration.

15.4.3 Taste Transduction Mechanisms

Many of the taste transduction mechanisms are well described. The chemical transduction of salts involves the permeation of sodium ions through amiloride-sensitive sodium channels that are present on the apical membrane of taste cells. Sodium ions also diffuse through the tight junctions of the paracellular pathway.

The sour (acids) taste stimulus pathway is very similar to salt. Protons directly affect a variety of cellular targets and alter intracellular pH. It is therefore not surprising that a number of transduction mechanisms have been proposed for acids.

Quinine, which produces a bitter taste, inhibits outward potassium currents similarly to protons from acids. The bitter stimulus is coupled to gustducin by a transmembrane receptor, which activates a phospho-inositide-specific phospholipase C enzyme called phospholipase C-β. In a split second, the enzyme cleaves PI bisphosphate to generate two products: inositol trisphosphate and diacylglycerol. Inositol trisphosphate is a water-soluble compound that diffuses rapidly through the cytosol. It generates the release of intracellular Ca^{2+} initiating neurotransmitter release for afferent nerve excitation.

Sweets also generate a G-protein-mediated taste cell response. Although the pathway is only slightly different, the stimulus binds to the G-protein receptor. The G_s α chain (α_s) binds to GTP and activates adenylyl cyclase to form cAMP. Elevated levels of cAMP alter the membrane conductance of taste cells by closing the basolateral resting potassium conductance channels.

Taste cells have receptors on the membrane surface that are specific for individual amino acids. The amino acid binds to its receptor and generates cell depolarization. Intracellular Ca^{2+} is released, which initiates an action potential that opens Ca^{2+} influx channels and allows the increased Ca^{2+} concentration to release neurotransmitter for afferent nerve excitation. Free fatty acids utilize a somewhat different nerve excitation pathway.

One can begin to realize that the various pathways of the basic elementary taste stimuli present a highly sophisticated and complex information translation/transduction system. Yet, the body deciphers, transports, and responds to minute concentrations of sensory input rather effortlessly.

15.4.4 Taste Processes

In comparison with the human taste process, there are two basic, albeit very important, concepts that have been learned from the biochemical and biophysical study of the gustatory system:

1. The sense of taste is not based on highly specific receptors, but is based on pattern recognition.
2. Biological action potentials are produced from the following chemical properties: pH (acidic or basic), metal ion content (+2 or +3 oxidation state), chemical compound functional groups, and sugar content.

To accomplish a similar but quantifiable effect by artificial means, selected chemical processes were designed that can be measured optically.

15.4.5 Combinatorial Libraries

Artificial chemoreceptors representing human taste buds, for instance, are composed of polymeric microspheres. The microspheres change color in the presence of specific food substances. The initial criterion was that the polymeric monomers chosen for the microspheres had to contain ligands that would associate the desired optically active reporter molecules responsible for analyte detection. Second, the combination of monomers had to be random to generate reproducible results. A "split lot" technique was chosen to produce the wide diversity necessary for the microspheres. Split lot is a simple but highly effective method of combining a number of different polymeric monomers, dividing them into lots, and recombining them in all possible combinations to form a pool of a large number of combinations.

By way of an example, the nine dimers (shown in Figure 15.5) can be divided and coupled to form 3^3 trimers and so on. This process is repeated ten times or 3^{10} forming a library of over 59,000 different oligomeric trimers. The libraries combine target binding with an optically active structure that can change color or fluorescence, creating a definite RGB detection pattern. The trimers bind optically active chemicals such as FITC. When analytes of interest are present, they displace the dye and evoke the fluorescence signal. In the other cases where indicator systems are used, octacalcium phosphate (OCT) can be used to detect the presence of Ca^{2+}. This compleximetric dye deprotonates at a pK_a of 10.1, and changes to a royal purple color (Figure 15.5). If the pH is raised to 10 without the presence of Ca^{2+}, OCP will deprotonate without the color change. Therefore, the purple color change denotes the presence of Ca^{2+} as well as

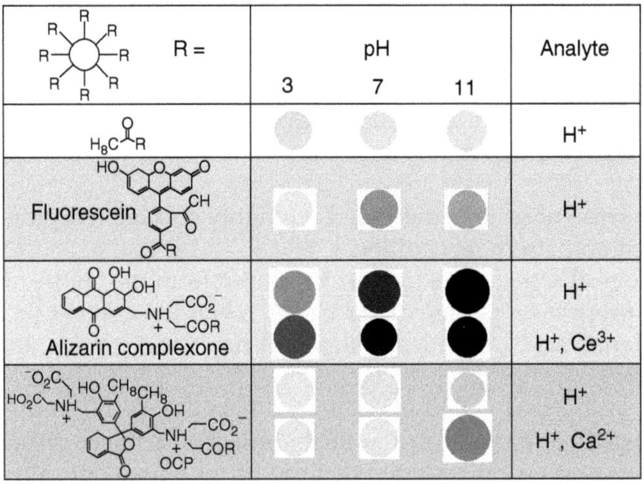

R =	pH 3	pH 7	pH 11	Analyte
H_8C–C(=O)–R				H^+
Fluorescein				H^+
Alizarin complexone				H^+
				H^+, Ce^{3+}
OCP				H^+
				H^+, Ca^{2+}

FIGURE 15.5
Green effective absorbance versus Ca^{2+} concentration calibration plot for a single OCP-coated microsphere over the concentration range of 10–500 μM. The data here are acquired with a stopped-flow method whereby two drops of the analysis fluid are introduced into the chip in each case. OCPC-derivatized microspheres detect Ca^{2+} *o*-cresolphthalein complexone colorimetric response to Ca^{2+}. (Courtesy of Dean Neikirk and John McDevitt, McDevitt Research Laboratory, Austin, TX, USA.)

other divalent and many trivalent cations. As seen in Figure 15.6, changes in the color of other optically active compounds are shown at various pH (H^+) levels, with and without metal ions present.

Certain specific attributes, such as carbon chains (sweet), acidity, or metal ions (salt) food substances, can make colorimetric or fluorometric changes in the microbeads. The chemoreceptor beads are placed in cavities etched into a silicon chip. The micromachined four-sided cavities are localized on silicon wafers that have the rear face terminated with a Si_3N_4–SiO_2–Si_3N_4 transparent membrane structure. The membrane acts as a window that enables studies of the optical properties of the sensing elements upon their exposure to analyte-carrying aqueous solutions. The buckets (~100 μm across) are micromachined in an array 500 μm apart. The beads are positioned in the buckets and then a wire mesh is attached over them. The bottom membrane allows light to pass through each cell and can be detected as reflected white light (top illumination), transmitted white light (bottom illumination), or fluorescence (epi-detection).

The unique combination of carefully chosen reporter molecules with water-permeable microspheres enables simultaneous detection and quantification of colorimetric or fluorometric changes in the receptor molecules that are covalently bound to amine termination sites on the polymeric microspheres. Each bead represents the chemical taste in their individual types of responses given by the FITC reporter molecule.

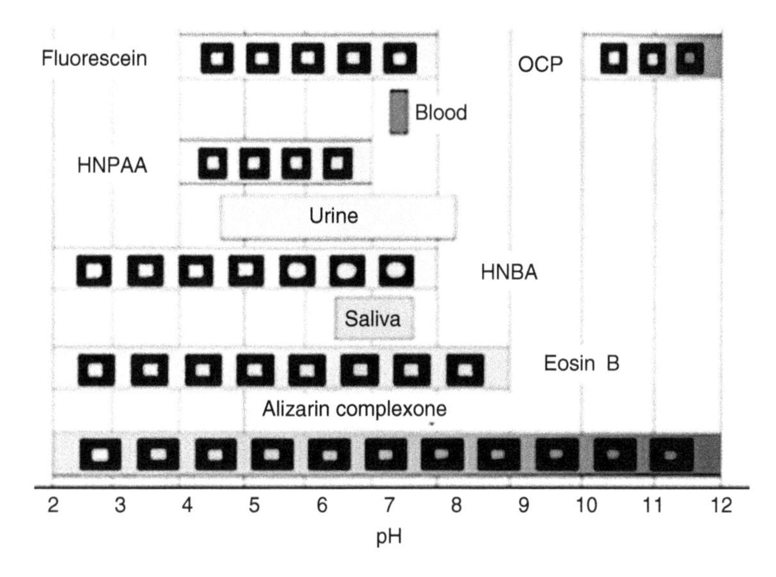

FIGURE 15.6
Colorimetry of various indicator molecules that have been used to signal the presence of sour (i.e., pH) and salt (i.e., electrolyte) stimuli. (Courtesy of Dean Neikirk and John McDevitt, McDevitt Research Laboratory, Austin, TX, USA.)

15.4.6 Charge-Coupled Device Detection

With the "taste bud" beads mounted onto silicon wafers using microelectromechanical systems technology, the bead's color can be recorded using a CCD color video camera. These cameras use a prism block incorporating dichroic filters. These filters are coatings of only a fraction of a light wavelength thickness. The thickness determines which colors are passed or reflected. Little light is lost, resulting in truer colors and greater sensitivity. Each camera has three identical chips attached to the prism block and aligned. The three respective chips will register the three basic colors—red, green, and blue—to represent a range of color representations. Since each CCD matrix array coordinate (pixel) produces only one color, all components of the signal are at full system resolution.

The resulting fluorescent luminescence gives an analog output. The camera RGB data files obtained from bead illumination are then analyzed by a pattern-recognition software to reveal the elements of the substance being tested. Reproducibility studies of the bead optical characteristics indicated that the colorimetric illumination RGB files from several different beads produced a digital output resolution having a standard deviation of 1.6% for blue and 2.3% for green.

Another critical property is the bead's ability to revert to its original color when the stimulus is abated. For instance, when pH is returned to the original

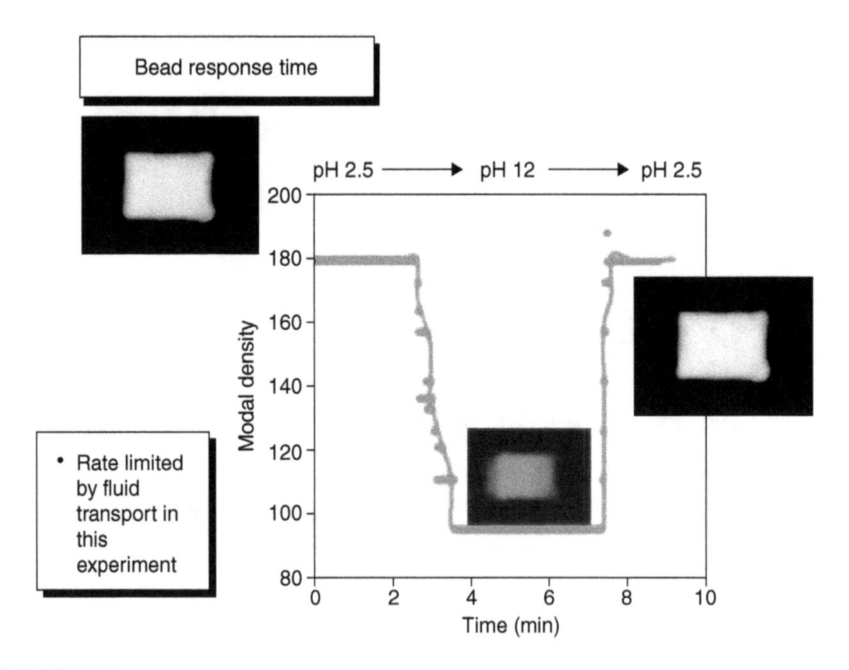

FIGURE 15.7
Reversibility and response time of reporter beads. (Courtesy of John McDevitt and Dean Neikirk Research Group, McDevitt Research Laboratory, Austin, TX, USA.)

value the luminescence follows suit. Figure 15.7 illustrates that as follows: a pH assay was run for 10 min; during this time the pH was systematically raised from 2.5 to 12 and then lowered back to 2.5. As seen in Figure 15.7, the choice of reporting molecules and the combinatorial libraries generated beads that had excellent sensitivity, stability, reversibility, and response time to aqueous solutions representing elementary food taste stimuli.

15.5 Summary

This chapter presented the optical diagnostic methods that can be arranged under photochemistry primarily on the basis of the interaction mechanism. It started with fluorescent detection, the fluorescence mechanism, and the derived spectroscopic aspects. One application of particular interest in fluorescence diagnostics was the description of the "optical tongue," designed to replace human taste testers.

Further Reading

Ali, M.F., Kirby, R., Goodey, A.P., Rodriguez, M.D., Ellington, A.D., Neikirk, D.P., and McDevitt, J.T., DNA Hybridization and discrimination of single-nucleotide mismatches using chip-based microbead arrays, *Anal. Chem.*, 75(18), 4732–4739, 2003.

Carey, P.R., *Biochemical Applications of Raman and Resonance Raman Spectroscopy*, Academic Press, New York, 1982, pp. 2–6, 8–9, 11–24, 48–59, 63–67.

Christodoulides, N., Tran, M., Floriano, P., Rodriguez, M., Goodey, A., Ali, M., Neikirk, D., and McDevitt, J.T., A novel microchip-based multi-analyte assay System for the assessment of cardiac risk, *Anal. Chem.*, 74, 3030–3036, 2002.

Curey, T.E., Goodey, A., Tsao, A., Lavigne, J., Sohn, Y., McDevitt, J.T., Anslyn, E.V., Neikirk, D., and Shear, J.B., Characterization of multicomponent monosaccharide solutions using an enzyme-based sensor array, *Anal. Biochem.*, 293, 2, 178–184, 2001.

Goodey, A., Lavigne, J.J., Savoy, S.M., Rodriguez, M., Curey, T., Tsao, A., Simmons, G., Wright, J., Yoo, S.J., Sohn, Y., Anslyn, E.V., Shear, J.B., Neikirk, D.P., and McDevitt, J.T., Development of multi analyte sensor arrays composed of chemically derivatized polymeric Microspheres Localized in Micromachined Cavities, *J. Am. Chem. Soc.*, 123, 11, 2559–2570, 2001.

Goodey, A.P., and McDevitt, J.T., Multishell microspheres with integrated chromatographic and detection layers for use in array sensors, *J. Am. Chem. Soc.*, 125(10), 2870–2871, 2003.

Kirby, R., Cho, E.J., Gehrke, B., Bayer, T., Park, Y.S., Neikirk, D.P., McDevitt, J.T., and Ellington, A.D., Aptamer-based sensor arrays for the detection and quantitation of proteins, *Anal. Chem.*, 76, 14, 4066–4075, 2004.

McCleskey, S.C., Floriano, P.N., Wiskur, S.L., Anslyn, E.V., and McDevitt, J.T., Citrate and calcium determination in flavored vodkas using artificial neural networks, *Tetrahedron,* 59(50), 10,089 -10,092, 2003.

McCleskey, S.C., Griffin, M.J., Schneider, S.E., McDevitt, J.T., and Anslyn, E.V., Differential receptors create patterns diagnostic for ATP and GTP, *J. Am. Chem. Soc.*, (Communication), 125, 5, 1114-1115, 2003.

Ploem, J.S., and Tanke, H.J., *Introduction to Fluorescence Microscopy*, Oxford University Press, Royal Microscopical Society, Oxford, 1987.

Richards-Kortum, R, Rava, RP, Fitzmaurice, M, Tong, LL, Ratliff, NB, Kramer, JR, Feld, MS, A One-Layer model of laser-induced fluorescence for diagnosis of disease in human tissue: applications to atherosclerosis, *IEEE Trans. Biomed. Eng.*, 36, 12, 1222-1232, 1989.

Rost, F.W.D., *Fluorescence Microscopy*, Cambridge University Press, Cambridge, 1992.

Sheryl, L. et al., A multicomponent sensing ensemble in solution: differentiation between structurally similar analytes, *Angew. Chem. Int. Ed.*, 42(18), 2070–2072, 2003.

Wilson, G., Hecht, L., and Barron, L.D., Residual structure in unfolded proteins revealed by Raman optical activity, Biochemistry, 35(38), 12518–12525, 1996.

Wiskur, S.L., Floriano, P.N., Anslyn, E.V., and McDevitt, J.T., A Multicomponent Sensing Ensemble in Solution: Differentiation between Structurally Similar Analytes, *Angewandte Chemie International Edition,* 42, 18, 2070-2072, 2003.

Wiskur, S., Metzger, A., Lavigne, J., Schneider, S., Anslyn, E., McDevitt, J.T., Neikirk, D., and Shear, J., Mimicking the mammalian sense of taste through single component and multicomponent analyte sensors, in Chemistry of Taste: Mechanisms, Behaviors, and Mimics, *Am. Chem. Soc. Symp.* Series No. 825, Given, P., and Paredes, D., Eds., 2002, Chap. 21.

Problems

1. Describe why 536-nm light cannot produce blue fluorescence.
2. Fluorescence has four important characteristics that make it a useful tool for examining the finer details of biological specimens. Describe them.
3. Electrons of molecules or atoms can occur in different energy states: ground state (minimal energy) and excitation state (higher energy) in addition to rotational and vibrational states. When electrons absorb energy, they get excited to a higher electronic and vibrational energy state. The return to ground-state energy level occurs with the loss of energy in which forms?
4. Fluorescence resonance energy transfer is a process involving the radiationless transfer of energy from a donor fluorophore to an appropriately positioned acceptor fluorophore. What are some common fluorophores used in FRET? What are some of the advantages and disadvantages of FRET?

16

Diagnostic Methods Using Light: Photobiological

This chapter focuses on the biological information that can be obtained by optical means. The list of items is only a selection; new applications are added daily. Some of the examples of photobiological diagnostics are spectroscopic imaging and immunofluorescence, various biosensors, and possibly the oldest of all diagnostic methods, i.e., blood oxygenation determination by attenuation spectroscopy. Another relatively new method using quantum dots for labeling is also described. Although fluorescence was described in Chapter 15, the particular diagnostic application discussed in this chapter is definitely of biological interest.

The first method describes the uses of fluorescence for functional imaging, thus combining the photophysical and photochemical effects to provide biological information.

16.1 Immunostaining ("Functional Imaging")

General immunostaining with antibodies is possible and the cells can then be viewed using a compound microscope. A primary antibody is incubated within the cells upon staining. Once the antibodies bind to their specific proteins on or in the cells, a secondary antibody is incubated with the sample. The secondary antibody is specific for the primary antibody and is conjugated to an enzyme such as horseradish peroxidase (HRP). A substrate is then added that is cleaved by the HRP, which results in a color change, and the specimen is observed with a compound microscope. The stained portions of the cells appear darker than unstained portions.

Immunofluorescence is another technique used to examine cell-surface molecules as well as other molecules.

16.2 Immunofluorescence

Immunofluoresence is the use of antibodies to label a specific protein within a sectioned tissue or a population of cells. Prior to describing the process of antibody labeling a brief explanation is presented about the immunology that led scientists to develop this type of assay.

The body comprises the immune system. It is a series of cell types, duct systems (lymphatic duct system), and primary and secondary organs, which lead the defense of the body against foreign invaders and diseases. This system can be divided into two specific systems, which have different jobs in defense. The first system is cell-mediated immunity. This type of immunity is generally used for intracellular pathogens, or bacteria and viruses that are found within the cells. The process is led by the T cell, and usually destroys the invaded cells to stop the spread of infection. The second arm of immunity is called humoral immunity. A cell type called the B cell leads to this type of immunity. The B cell is known for producing soluble proteins that are called antibodies. Antibodies are used to tag foreign particles in the body so that the rest of the immune system can recognize them and get rid of them in an organized manner.

The immune system is formed through a complex process known as positive and negative selection. During this process, the immune cells are "taught" to recognize all of the proteins in the body as self. This means that self-proteins are safe. The antibodies will tag anything not recognized as self. For this, the antibodies must be very specific for the protein they are made to recognize. Antibodies have a specific protein motif, which is illustrated in Figure 16.1. It is made of two light chains at the top and two heavy chains at the bottom, named on the basis of the different centrifugation technique, which founded them.

In general, a donor organism is required to generate an antibody for an unknown assault. The donor organism frequently used is the rabbit. The F_c or F constant region in the antibody is highly conserved among species. This means that all rabbits have an immunologically similar constant region in all of their antibodies synthesized. The second region is the F_{ab} region or Fragment antigen binding. The F_{ab} region is the real active end of the antibody, which binds to the protein that the formed antibody is against.

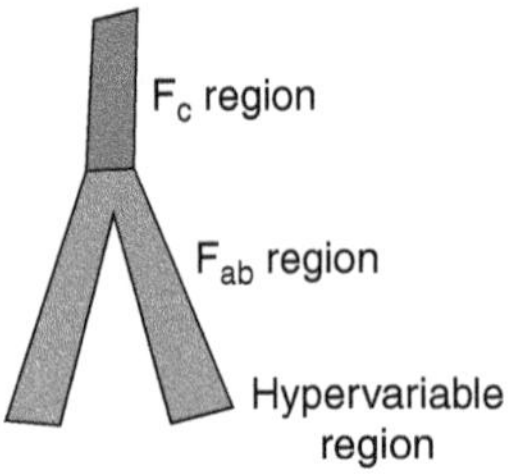

FIGURE 16.1
Schematic of an antibody with regions labeled. The light chain is the "F_c" region and the dark chains are the "F_{ab}" region.

In general, the fluorescent microscope is used to examine individual cells. Once the cells are ready, a known number of cells are plated on cover slips in 24-well tissue culture plates. The cells are allowed to adhere and then treated with bacteria. At the appropriate time points, the media is washed off the cells and the cells are fixed with a 4% formalin solution. Following fixation of the cells, they are stained with the appropriate primary antibody, which will bind to the proper surface protein, in question. An illustration of the antibody labeling is shown in Figure 16.2. A secondary antibody is then incubated with the cells following primary antibody. The secondary antibody is conjugated to a fluorochrome. Some of the commonly used fluorescent molecules are fluorescein isothiocyanate (FITC), tetramethylrhodamine isothiocyanate (TRITC), and phycoerythrin (PE). Both FITC and TRITC appear green, and PE appears red. The choice of which one to use is a matter of preference and commercial availability.

Scientists have learned to exploit the specificity of these antibodies to find proteins of interest. An overview of the process can be seen in Figure 16.2. The specimen is fixed to begin. The fixative is used to preserve the architecture of the tissue so that native localization can be determined. The section is then placed in paraffin to preserve it and sectioned to be put onto slides. Once the specimen is affixed to the slide a solution of diluted antibody is added on top and allowed to incubate. The first antibody added is called the primary antibody because it is specific for the protein. These antibodies are made in a simple but ingenious fashion. If one is looking for a mouse protein, such as AQP-1, one can simply purify this protein and inject it into an animal such as a rabbit. Because the rabbit will not recognize the mouse protein as self, it will make antibodies against this particle. The researcher will then bleed the rabbit and purify the antibodies specific for AQP-1. These antibodies will now be known as rabbit anti-mouse AQP-1 antibodies. Because antibodies are proteins, they may not necessarily be recognized under visual examination, so the next step is to add a secondary antibody, which is conjugated with a specific fluorochrome (which was discussed earlier). The primary antibody will be washed off the specimen to eliminate nonspecific binding. The secondary antibody is added and again allowed to incubate. It is made in much the same way as the primary antibody. When the rabbit primary antibody is injected

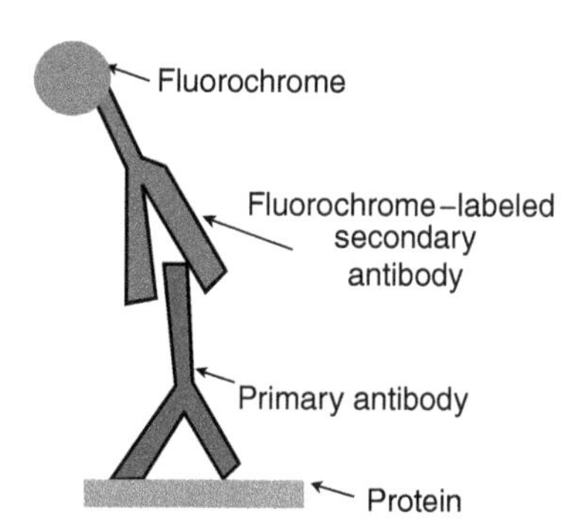

FIGURE 16.2
Schematic illustration of antibody and fluorescent marker configuration.

into another animal such as a goat, it recognizes the rabbit antibody as foreign and makes antibodies against that antibody. These antibodies will become goat anti-rabbit F_C antibodies. The process of localizing proteins in this way is generally referred to as indirect immunodetection because the fluorochrome is indirectly attached to the protein you are looking for. The biggest advantage of indirect immunodetection is the amplification that occurs by using a secondary antibody. When these antibodies are made, we get multiple antibodies, which are specific for different parts of the protein. So when groups of antibodies are applied we get multiple antibodies binding to the protein, making the signal stronger and easier to visualize.

Several techniques have been developed over the years that make use of the spectral analysis of scattered light.

16.3 Diagnostic Applications of Spectroscopy

Applications of spectroscopic imaging include, for instance, the in vivo determination of chemical blood-gasses, blood glucose sensing, and also the detection of toxic substances in the blood stream. The latter technique can be called Raman spectroscopy-assisted histology.

A few examples of spectroscopic diagnostics are caries detection and ischemic muscle quantification. The caries detection will be discussed first.

16.3.1 Detection of Dental Cavities and Caries

The most difficult to detect and the most common early enamel lesions are occlusal (biting surfaces), pit and fissure, and approximal (contact surfaces between teeth) lesions in posterior teeth. If such lesions are detected early enough, it is likely that they can be reversed, or at least brought to a standstill, by topical fluoride or by low-intensity laser irradiation without surgical intervention.

Radiographic methods do not have the sensitivity for early lesions, particularly pit and fissure lesions, because they are obscured by the convoluted topography of the crowns. Furthermore, by the time the lesion is radiolucent, it has extensively progressed from the enamel to the dentin at which point surgical intervention is often necessary.

Clinicians usually base their diagnosis and treatment planning on the lesion color and texture. This can be misleading because lesion color does not provide sufficient information about the state of the lesion, whether it is progressing or arrested, and pigmentation can be due to staining from diet and environment and not by infection from microorganisms.

Teeth naturally fluoresce upon irradiation with ultraviolet (UV) and visible light, and laser-induced fluorescence (LIF) of endogenous fluorophores in human teeth can be used as a basis for discrimination between carious and noncarious tissue. Carious/demineralized areas appear dark upon illumination with near-UV and visible light followed by imaging of the emitted fluorescence in the range 600–700 nm. The origin of the endogenous fluorescence in normal teeth has not been determined. In fact, the molecular identity of neither the chromophores nor the fluorophores has yet been established. It is known that bacterial plaque fluoresces due to the presence of native porphyrins.

Recently, a commercial caries detection system based on LIF was introduced in Europe for the detection of pits and fissures and approximal caries. It uses a diode laser and a fiberoptic probe to detect the fluorescence from bacterial fluorophores in carious lesions.

A laser scanning technique for the early detection of tooth decay uses the tooth's naturally occurring fluorophores, while diseased teeth fluoresce differently from healthy tooth bone, directly underneath the tooth's enamel, currently down to 100 μm, and the inventors are aiming to increase the imaging depth to 0.5 mm. The detection mechanism uses a mode-locked Cr:LiSAF diode laser at 850 nm wavelength for two-photon excitation fluorescence.

It has been shown that spectroscopy can effectively quantify ischemic muscle. Additional research has been aimed at the potential for in vivo determination of chemical blood gases, blood glucose sensing, and the detection of toxic substances in the blood stream, or Raman spectroscopy-assisted histology.

A different type of spectroscopic analysis can be used to perform quantitative colorimetry. In this case the spectral shape of transmitted or reflected light can be used to obtain information about the prevalent chromophores in a section of tissue and derive the clinical background indirectly. One example of this technique is in the optical detection of dental erythema, which is discussed next.

16.3.2 Optical Detection of Erythema

Periodontal disease is a term used to describe an inflammatory disease affecting the tissues surrounding and supporting the teeth. Periodontal diseases are some of the most common chronic disorders, which affect humans in all parts of the world. In the United States alone three out of four people have been diagnosed as having some form of periodontal disease. In gingival diseases, only the soft gum tissue is affected and it is commonly associated with erythema, or a redness and swelling of the gums due to an underlying bacterial infection or trauma. At this early stage, if treated, the prognosis is good and the periodontal disease is reversible. However, if not detected early, the disease can progress to periodontitis and result in active destruction of the supporting tissues around the teeth resulting in tooth mobility and loss of teeth.

Other forms of erythema exist; for example, sunburn is a condition that may be all too familiar.

Increasing amount of redness yield a higher erythema index. An oral erythema meter will produce a higher erythema index in patients with more severe mucositis compared to patients with only mild, moderate forms for the disease, by means of the noninvasive technique of reflectance spectrophotometry. This technique employs reflective scattering, which occurs when electromagnetic radiation is projected into a tissue and the particles involved in the interaction are either reflected or absorbed. Various wavelengths of electromagnetic radiation have different measurements of absorbency in biological tissue. In the epidermis the main chromophores for visible light are melanin and those in the dermis are hemoglobin. Oxyhemoglobin's maximal absorption in the green light spectral range is between 520 and 580 nm. As the wavelength gets longer the absorbency quickly falls so that absorption is minimal in the red light spectrum. Erythema results when blood volume in the subpapillary plexus increases causing a greater sum of green light to be absorbed and a reduced amount to be reflected. In addition, the red light that is absorbed or reflected only reveals a minor change during vasodilation. This concept allows the erythema meter to measure the difference in the reflective intensity of the two different wavelengths of the electromagnetic radiation. The reflectance measurements of red and green light are then compared in a reflectance index, which indicates whether or not a patient has erythema.

One instrument, a cutaneous erythema meter, quantifies erythema at several parallel cutaneous test points. The instrument is a fiberoptic dual-wavelength reflectance meter that consists of a fiberoptic measurement head and two PIN photodiode detectors. Light-emitting diodes (LED) light sources using two peak wavelengths are utilized to emit light onto the skin to measure their reflectance. Reflectance measurements at the 555-nm blood-hemoglobin absorption wavelength are compared to the 660-nm reference wave in a reflectance index to determine the presence of erythema. In clinical trials the cutaneous erythema meter proved to be able to measure at least one order of magnitude higher than that of the human eye. The erythema value is defined as the ratio of the backscattered light from a blue light source over the backscattered red light from another small bandwidth source as

$$\text{Erythema index} = \log 10 \frac{[\text{intensity of blue component of reflected light}]}{[\text{intensity of red component of reflected light}]} \tag{16.1}$$

An example of this detection method is described in the work by Davis et al. described in US 6862542, the phantom absorption spectrum, and the selected center wavelengths of the probing sources. Table 16.1 illustrates the recordings from the skin phantom gel mixture with food coloring added in various concentrations. In water a concentration of 3%

TABLE 16.1

Collected Erythema Value Data from Phantom Dye Solutions[a]

Gel	Blue	Green	Red	White	Black
Blue:Red	0.036 ± 0.008	0.075 ± 0.008	0.208 ± 0.004	0.32	1.9

[a]Recordings were derived from skin phantom gel mixtures composed of soymilk and food coloring added in various concentrations. In water a concentration of 3% Intralipid was created to resemble the scattering of dermis; in addition, the following food color dyes were added to match the absorption spectrum of the skin: yellow food coloring no. 5, green food coloring no. 40, and blue food coloring no.1.

Intralipid was created to resemble the scattering of dermis; in addition, the following food color dyes were added to match the absorption spectrum of the skin: yellow food coloring no. 5, green food coloring no. 40, and blue food coloring no. 1. The food coloring additives were administered in the following concentrations:

- 0.24% yellow food coloring no. 5
- 0.35% red food coloring no. 40
- 0.26% blue food coloring no. 1

The red food color concentration was altered by 5 and 0.5%, representing the conditions for added redness resulting from a hypothetical infection, and the erythema value of each concentration was determined. The optical was able to detect a 0.5% concentration change, whereas the eye could only distinguish minimally 5% color changes. Figure 16.3 illustrates the recorded erythema values by the optical spectral analysis, and it is obvious that a 0.5% change in "redness" can reproducibly be detected by this method. Literature review reveals that the human eye is reportedly capable of detecting a 5% change in visible redness, let alone correlating the follow-up visits compared to the treatment efficacy.

So far we have only focused on the wavelength parameter of electromagnetic radiation; however, the phase information has several beneficial attributes for diagnostic imaging as well. The first example of phase information imaging is described under holography.

The next section describes an entirely different group of optical diagnostic devices: fiberoptic sensors. These sensors can be used to obtain localized biological information by optical means.

16.4 Fiberoptic Sensors

One area of interest in biomedical optics is the design of minimally invasive probes to monitor biological functions and offer high-resolution diagnostic

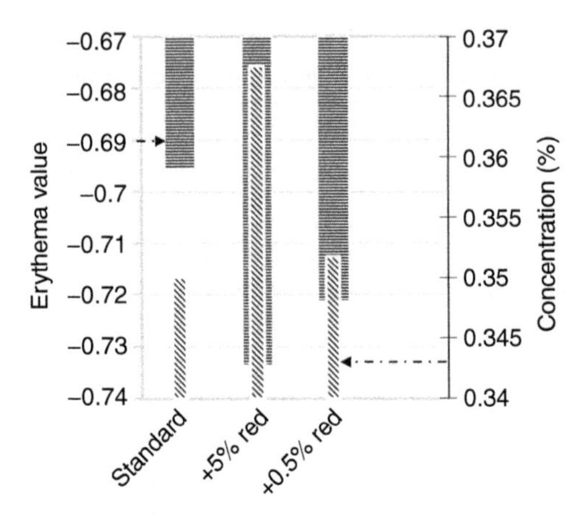

FIGURE 16.3
Erythema results from changes in dye concentration measured by the optical backscatter spectral ratio detection. Standard Dermis Phantom Solution is made with: 3% Intralipid; 0.24% Yellow foodcoloring No. 5; 0.35% Red foodcoloring No. 40; 0.26% Blue foodcoloring No. 1. The red concentration is illustrated, with the erythema value in the far left hanging bar. Subsequent measurements were made after changing the red foodcoloring concentration by either 5% (2nd bar) or 0.5% (3rd bar) respectively.

capabilities. Fiberoptic light guides have been adapted to offer detection solutions for biohazard testing, as well as blood-gas testing and additional analytical assignments.

16.4.1 Biosensors

Biosensors are analytical devices consisting of a biological sensing element in contact with a physical transducer. The biological sensing element can be enzymes, antibodies, or microbes, and the physical transducer can be optical, mass, or electrochemical. The reactive analytical component (analyte) is a specific substance, component, or organism in material derived from a vertebrate species. The analyte will attach to antibodies or oligonucleotides (short DNA sequences) to produce a detectable reaction. In general the fiber optic can be coated with a sensor layer that changes its optical characteristics under the influence of the desired chemical. The sensor coating on the outside surface of the fiber optic either changes index of refraction or produces fluorescence or phosphorescence under the chemical presence of the reagent. The coating is referred to as an optochemical-transducing layer. Biosensor can detect the presence of analytes by means of antibodies or short DNA sequences. The potential for diagnostic applications ranges from the detection of cancer-specific antibodies to groundwater contamination analysis.

Interactions other than purely chemical are also possible, such as pressure, temperature, and strain. All of these mechanisms can alter certain material characteristics of the fiberoptic design and hence change the optical behavior of the probe. The change in optical performance can be detected, and thus provides a sensing mechanism.

Direct observation of viruses is more challenging and to date still requires detection of short DNA sequences.

16.4.2 FiberOptic Biosensor Design

Most optical biosensors now use tapered optical fibers, similar to the design discussed in the section on near-field scanning optical microscopy in Chapter 14. An illustration of a tapered fiber optic was shown earlier in Figure 14.11. In these biosensor applications, an optical fiber is tapered from its normal diameter to a smaller diameter. This increases the electromagnetic field intensity at the surface of the fiber where the optochemical-transducing layer is applied.

This group of fiberoptic sensors relies on the evanescent field properties of the modes propagated within the fiberoptic wave guide. The modal field—this is where the evanescent waves are confined—must be exposed for the intrinsic optical fiber sensor to interact with the surrounding environment. The use of fiber for sensing applications results in a compact design with no requirements for optical alignment.

The use of multimode fibers is often preferred because of their high power-transmitting capacity. The disadvantage is that they require rather bulky optical components in the light-coupling end at the laser, which may easily become misaligned. Single-mode fiber biosensors may soon replace them, since they are more compact and robust.

The principle of fluorescent imaging is described in detail in Section 6.4.

16.4.2.1 Fiberoptic Fluorescence Sensors

An example of a fluorescence-sensing fiberoptic is a detector for monitoring changes in the body's concentrations of calcium ions. This detector may be useful in diagnosing disease or exposure to chemical warfare agents. This biosensor consists of an optical fiber to which a synthesized hybrid molecule is attached. One half of the hybrid molecule binds calcium ions and the other half fluoresces when calcium ions are bound to the molecule.

16.4.2.2 Fiberoptic Sensors in Gastrointestinal Applications

The optical diagnostic applications of gastrointestinal (GI) disease diagnosis are in the diffuse reflectance spectroscopy and fluorescent spectroscopy. A more traditional fiberoptic measurement in the GI tract is the Doppler flow velocimetry. The spectroscopic utilization is predominantly geared to distinguishing between malignant and benign tissues, and measuring local drug concentrations. The optical characterization of polyp types is important for both clinical relevance as well as in situ study of neoplasia and dysplasia. So far in vitro fluorescent spectroscopic classification of polyps is rivaling the accuracy obtained with clinical pathology. Further refinements are in progress to go to in vivo clinical applications.

16.4.2.3 Fiberoptic Immunosensors

Immunosensors act on the principle that the immune response of certain biological species (usually bacteria) to contaminants will produce antibodies as described in Section 6.3 and shown schematically in Figure 16.4.

For instance, an antibody that reacts specifically with the carcinogen benzo(*a*)pyrene (B*a*P) can be attached to the end of an optical fiber. This end of the fiber is then immersed in a sample of groundwater. The antibody is allowed to bind the B*a*P in the groundwater sample. The antibody–B*a*P reaction product in turn produces fluorescence and is detected.

Antibodies can be produced against bacteria, complex carbohydrates, and even smaller organic molecules, which may be cancer-risk factors.

16.4.2.4 Medical Immunosensor

One of the most critical diagnostic stages in medicine is to quantify the amount of thrombin in blood during surgery and after heart attacks in particular. The thrombin plays a critical role in thrombosis and hemostasis by converting fibrinogen into clottable fibrin. Early quantification of thrombin formation can significantly reduce the morbidity and mortality as a result of appropriate treatment planning.

Proper selection of a reagent and design features of a biosensor has made it possible to build an instrument for the direct determination of thrombin in vitro. The signal transduction mechanism for molecular recognition is based on fluorescence resonance energy transfer (FRET) and the chemical alterations in specific molecular chains. These specific molecules act like a switch that is normally closed. In this case the fluorophore is chemically

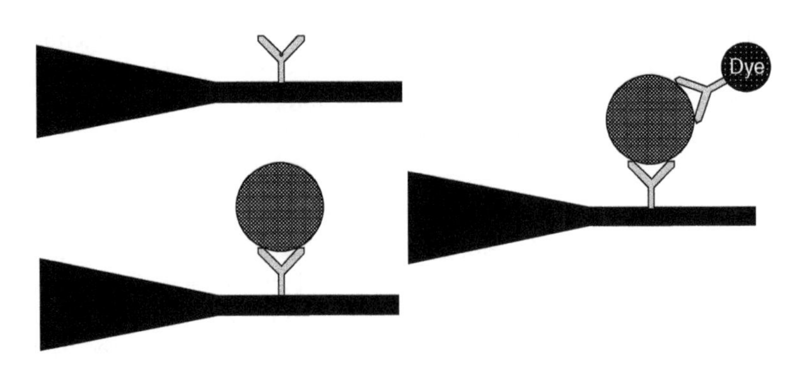

FIGURE 16.4
Antibody sensor based on fluorescence through fluorescent tagging of the analyte and illumination within the absorption band of the fluorescent dye (not to scale).

bound with no fluorescence. When binding to the target molecule in the biological system, it undergoes a change in the chemical conformation that separates the fluorophore from the main structure. At this moment the fluorescence will occur. Certain analytes are, for instance, DNA or RNA oligonucleotides isolated to bind to various biomolecules with high specificity. One specific example of analyte is the 15-mer fluorophore-labeled DNA aptamer, which may be covalently bound to the surface of a fiberoptic sensing probe. This type of ligand–DNA interactions includes highly selective analytes to perform thrombin determinations. The sensor design uses evanescent wave excitation of the labeled DNA aptamer to reduce photobleaching.

16.4.2.5 Environmental Biosensor

A fiberoptic biosensor has been used to detect the presence of an estrogen-mimicking substance in pond water having over 65% of frogs with genetic deviations. Water from the pond was analyzed using a fiberoptic sensor employing estrogen receptor as the sensing element. Binding was inhibited to the estrogen-coated fiber optic in water from ponds that exhibited malformations in the frog population. Water from ponds that did not demonstrate frog genetic disfigurations did not inhibit binding between receptor and sensor. This offers an indirect method to establish conditions that provide the environment for malformations to occur in the frog population in a particular pond.

16.4.2.6 Public Health

Evanescent wave biosensors have been developed for the detection of various DNA strands. One particular application involves the detection of DNA fragments found in the pathogen responsible for the infectious disease cholera (*Vibrio cholerae*). Upon binding to the molecule, a fluorescently labeled analyte molecule is captured within the evanescent field and the fluorescence is excited. This technology might be adapted to the detection of a many other pathogens including microorganisms in food and water. Applications of this technology include on-site use of portable biosensor by hospital personnel or health inspectors to evaluate suspicious medical or environmental samples.

16.4.3 Distributed Fiberoptic Sensors

Distributed fiberoptic sensors use their entire length as chemical transducer. This means that the sensor design is continuous as compared to a multiplexed design of a discrete number of fiberoptic sensor tips. The entire length can be probed using optical time-of-flight (OTOF) chemical detection. Time-of-flight-distributed fiberoptic sensors are intrinsically designed to spatially resolve analyte concentration along the entire length of the fiberoptic probe.

When multiple fiberoptic strands are focused onto a charged-coupled device (CCD), it allows for spatial resolution of signals emanating from different analytes, thus allowing multichannel detection of different analytes in the same specimen or calibration of remote sensors in field environments.

16.4.3.1 Time-of-Flight Measurement Using Laser Light

Optical time-of-flight (OTOF) sensing uses short laser pulses launched into a fiber optic. The sensing fiber has a coating containing specific fluorophore sensitive to the target analyte of interest. As the light pulse propagates down the length of the fiber, it stimulates fluorescence from the fluorophore. The intensity of this fluorescence is dependent on both the concentration of the analyte as well as the stimulus light. The fluorescence is conducted back up the fiberoptic and is detected at the proximal end. The analyte concentration can be determined from the fluorescence intensity. By monitoring the time delay between the stimulus pulse and the returning fluorescence, the analyte concentration can be spatially mapped along the length of the sensor. Spatial resolution is limited by the fluorescence lifetime of the fluorophore and the laser pulse width.

16.4.3.2 Time-of-Flight Measurement Using LED Light

The second type of instrumentation being developed for OTOF chemical sensors uses LEDs as an alternative light sources. Obtaining picosecond laser pulses requires expensive instrumentation and cumbersome optical devices. Using LEDs as light sources may provide an alternative solution.

However, it is very difficult to produce short enough pulses from an LED to be useful for OTOF sensing. To remedy this technical limitation the detection protocol is redesigned. The LEDs are driven by a current pulse protocol that produces close to square-wave optical pulses. The detection algorithm tracks the trailing edge of the LED pulses. The resulting effect is that a decrease in fluorescence is observed as the trailing edge of the excitation light passes through the active region. This temporal resolved detection mechanism will provide spatial information as well. The disadvantage of the broadband LED as a light source is the reduced contrast. The excitation source may significantly overlap with the fluorescence emission spectrum.

16.4.4 Plastic-Clad Fiber

By chemical modification of plastic-clad silica fibers it has been made possible to incorporate the well-known principle of oxygen detection by quenching of fluorescence. These plastic-clad fiberoptic oxygen sensors are providing new opportunities for medical applications. However, the sensor needs to be immune to the influence of both water vapor and most common anesthetics. The fluorescent reagent used for oxygen detection is a porphyrin complex. This fluorophore can be excited with both green and yellow LEDs with emission in the red.

16.4.5 Limitations of Fiberoptic Sensors

One of the limitations of fiberoptic sensors is the fact that the antibody–antigen binding is not readily reversible. The methods used to release the antigen result in the denaturation of the antibodies with an inherent loss of signal. Another limitation is the requirement for direct contact.

The next section will delve into some other biological applications of optical imaging techniques discussed in the previous chapters. The first application involves the diagnostic use of optical coherence tomography (OCT).

16.5 Optical Coherence Tomography in Dentistry

In dentistry, it has great potential for the detection and diagnosis of periodontal diseases, oral cancer, and dental caries. The intensity of backscattered light is measured as a function of its axial position in the tissue using low-coherence light-gating methods. Enamel only weakly scatters near-infrared light. Therefore, it has an imaging depth capability of more than 2–3 mm. This is sufficient penetration for the direct imaging of early occlusal and approximal carious lesions. Demineralized enamel highly scatters light in the visible and near-infrared, and incident polarized light is rapidly depolarized. Polarization-sensitive OCT can be used to discriminate between carious and noncarious enamel and dentin, and it is anticipated that in the future it will enable measurement of the extent of demineralization in depth, thus providing the clinician with diagnostic information about lesion severity.

Time-domain OCT can acquire functional as well as structural information about a sample. Figure 16.5a shows color Doppler OCT images of in vivo bidirectional color Doppler flow imaging of picoliter blood volumes using OCT, and Figure 16.5b shows the biological importance of Doppler OCT imaging by revealing the heart beat of a tadpole.

In the next section a spin-off from OCT is described that can be used to determine tissue optical properties in situ, and as such can identify biological media, as a definite biological application.

16.6 Optical Biopsy

Building on the same principle as OCT the optical properties of the tissue as a function of depth can be derived noninvasively, thus providing indirect anatomic information of the tissues underneath the fiberoptic or other interferometric configuration targeting the organ.

Optical diffusion theory assumes the objective material to be uniform and homogeneous, which is not the case with the biological tissues. It is therefore

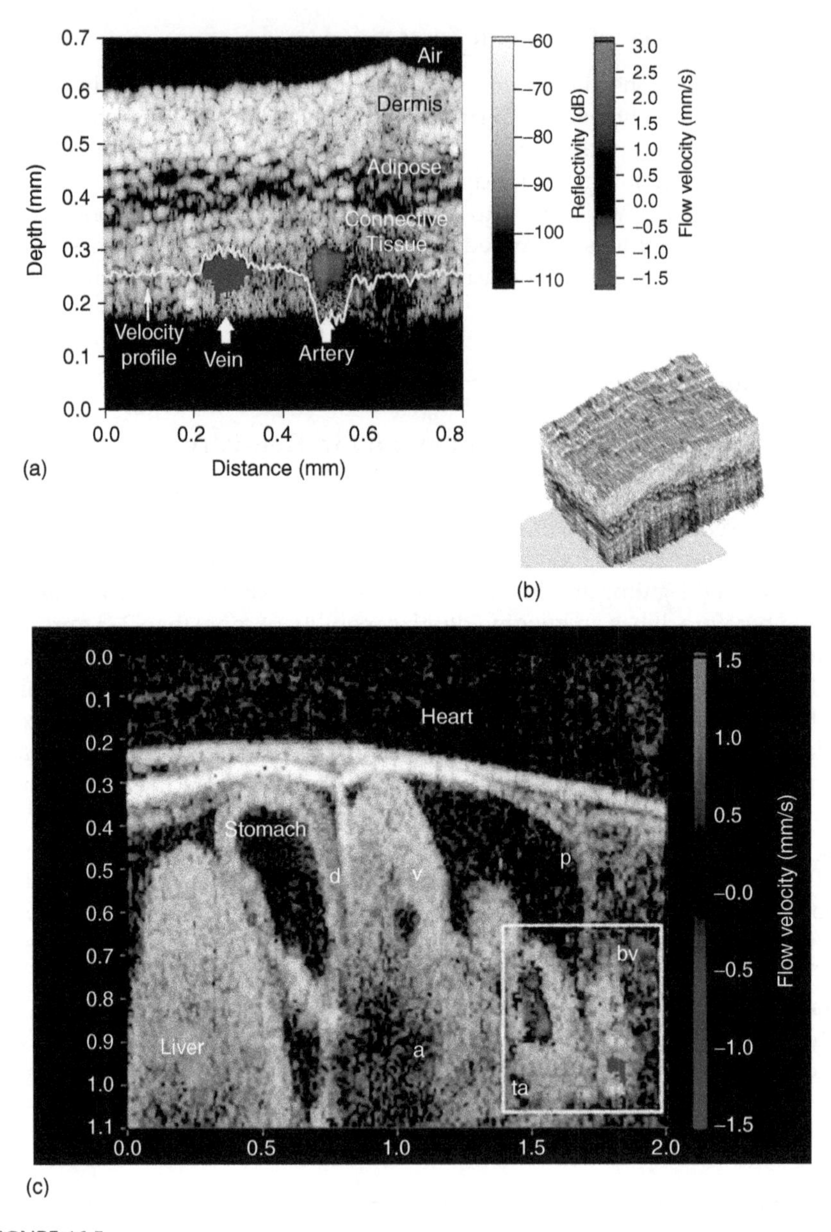

FIGURE 16.5
Doppler OCT. (Courtesy of Dr. John Izatt, Duke University.) (a) Flow velocity profile measured by OCT imaging. (From Izatt, J.A., Kulkarni, M.D., Barton, J.K. and Welch, A.J., In vivo bidirectional color Doppler flow imaging of picoliter blood volumes using optical coherence tomography, *Opt. Lett.*, 22, 1439–1441, 1997. Reprinted with permission.) (b) Three-dimensional representation of the volumetric velocity profile. (Courtesy of Bioptigen, Inc., Durham, NC.) (c) High-resolution imaging of in vivo cardiac dynamics using color Doppler OCT from a tadpole heart.

important to identify the physiology and morphology of intact, diseased, and coagulated tissues so that careful examination and interpretation of the experimental data can be performed.

The determination of the local optical parameters is of essential clinical significance in the optimization of laser delivery in the clinical setting of dosimetry. In addition, the treatment parameters need to be optimized for in vivo applications, while most of the current knowledge is still based on preemptive in vitro measurements. To verify optimal light-treatment conditions the optical and thermal tissue parameters will need to be evaluated in situ. In addition, they will need to be monitored during the entire treatment process to adjust for any variance in the optical design of the treatment volume. A significantly detailed section on noninvasive detection of optical properties was presented in Chapter 7, Section 7.2.2.

The primary purpose of optical biopsy in the optical anatomic identification of biological media is twofold: (1) perform noninvasive in vivo tissue characterization and potentially recognize diseased tissues without the actual need to perform a physical biopsy and (2) assess dosimetry during therapeutic laser irradiation or tissue diagnosis, and monitor the progress of the tissue damage.

Monitoring the volume and degree of laser photocoagulation or tumor death will be the deciding factor in controlling the success of several of the treatment options described in Chapter 11, 12 and 13. Specifying appropriate dosimetry pertinent to the target area on the basis of the local optical configuration is paramount for effective treatment.

Similar to the OCT imaging methodology, broadband interferometry can be used to measure the optical properties as a function of depth based on the measured axial light distribution inside the tissue. The measured light distribution as a function of depth underneath one arm of a Michelson interferometer can yield the decay function of either a broad or a narrow beam of light. The broad beam will produce a light distribution that will satisfy an exponential decay corresponding to the effective attenuation coefficient. Under narrow beam (fiber optic) light delivery the conditions are considerably more complicated, and a curve-fitting procedure will be required to match a Monte Carlo simulation of various combinations of optical properties in an iteration sequence to discover the optical composition of the tissue(s) underneath the aperture of the fiber optic.

The interferometer technique is based on the Michelson design as described in Section 8.8. Figure 16.6 illustrates the free-space Michelson interferometer that can theoretically be used to obtain the optical information of a large tissue volume at once.

Generally the amplitude and phase of a broadband source will vary in time in a random fashion, which requires the detection of the statistical average of the radiance rather than the instantaneous radiance.

In Figure 16.7 the interference signal collected from in vitro optical measurements on cardiac muscle are shown in comparison with the light distribution calculated for the various tissues that can be found in an unhealthy and denatured heart under similar operating conditions as the actual interferometer.

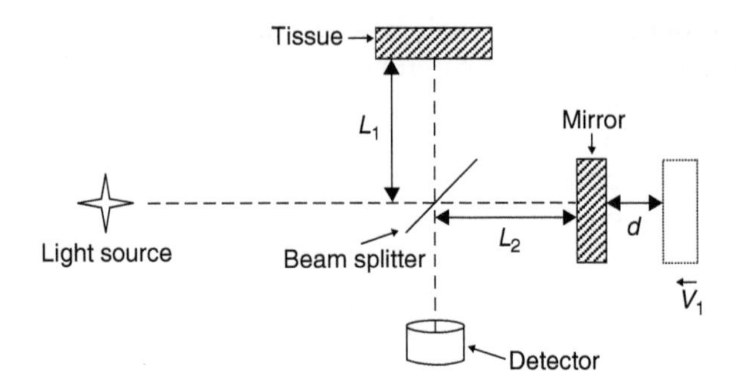

FIGURE 16.6
Free-space Michelson interferometer. A beam splitter splits a single broadband light source in two identical sources that are coherent with each other. The mirror is translated over a distance d to collect light from inside the tissue.

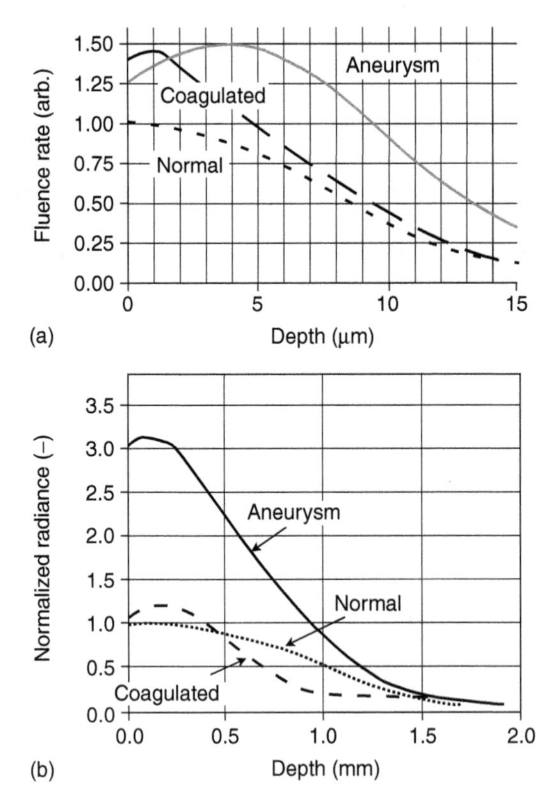

FIGURE 16.7
(a) Radiant intensity as a function of depth measured by fiberoptic interferometer. Measurements of healthy, scar, and denatured heart muscle by fiberoptic biopsy, normalized to healthy heart muscle. (b) Monte Carlo computer simulations of healthy myocardium, aneurysm, and coagulated cardiac muscle from optical properties determined in vitro with integrating sphere measurements.

Another type of imaging using the unscattered photons will be discussed in the next section under ballistic photon imaging.

In the realm of optical diagnostic modalities there are various other techniques available. The next section will describe biological implications of the determination of blood oxygenation from spectral absorption curves.

16.7 Determination of Blood Oxygenation

Oxygen sensors work on the principle that light absorption in blood changes with blood oxygen content as a function of wavelength. The spectral absorption curves for oxyhemoglobin and deoxyhemoglobin have significantly different profiles: at approximately 630 nm the absorption of deoxyhemoglobin is substantially greater than for oxyhemoglobin, while at 960 nm they have the greatest separation with oxyhemoglobin displaying the highest absorption. The absorption spectra of oxyhemoglobin and deoxyhemoglobin are illustrated in Figure 16.8. One complication is the increasing water absorption starting at 960 nm and longer wavelengths; for this reason the sample wavelength is usually chosen around 930 nm. Oxygen sensors contain an emitter, an LED, which emits light at two different wavelengths (630 and 930 nm) depending on the polarity of the applied voltage, and a detector, a photocell, which measures incident light intensity. In practice, the emitter and detector are placed on either side of a bodily appendage, i.e., finger or an ear lobe, and the attenuation (absorption) of light emitted at each wavelength by the intervening tissue is measured. These results are then used to calculate the oxygen saturation level using an experimentally determined algorithm.

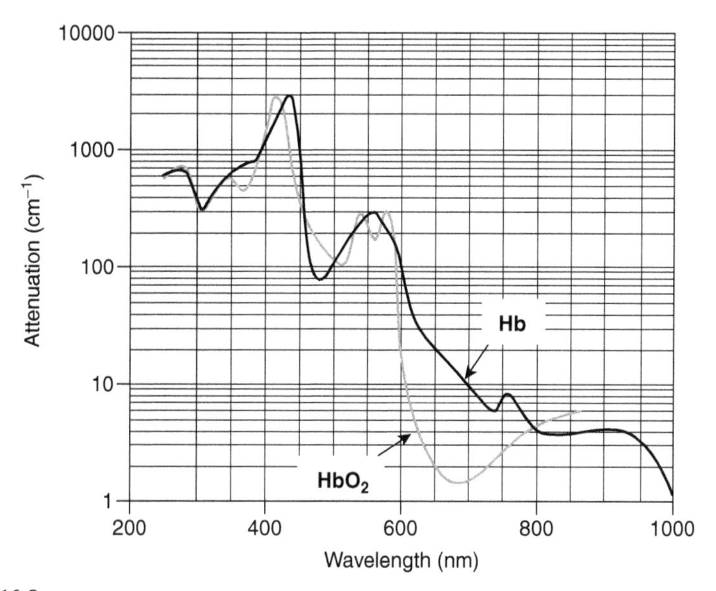

FIGURE 16.8
Oxyhemoglobin and deoxyhemoglobin spectral performance in the visible and near-infrared spectra. (Courtesy of Gratzer, W.B., Medical Research Council Labs, Holly Hill, London and Kollias, N., Wellman Laboratories, Harvard Medical School, Boston.)

16.7.1 Pulse Oximetry

The pulse oximeter is an optical device that estimates the functional oxygen saturation of blood. The blood oxygenation value is useful under anesthesia, critical care, and for monitoring a person's vital signs while performing critical tasks, e.g., fighter-jet piloting.

The pulse oximeter applies a measurement of the attenuation ratio of two wavelengths determined at alternating intervals through a section of tissue containing blood. The principle behind the measurement is the fact that oxygenated and deoxygenated hemoglobin have distinct attenuation spectra that are not identical over the visible and near-infrared, except for one single point around 805 nm where they intersect. The light source used for the transillumination measurement is usually driven by an alternating source that turns one wavelength on while the other wavelength is turned off.

The measured attenuation at any time is the combined attenuation from hemoglobin (Hb), oxyhemoglobin (HbO_2), hemiglobin (Hi), carboxyhemoglobin (HbCO), and the tissue layers surrounding the blood vessels (skin, muscle, etc.). This is expressed by the following definition:

$$\mu_{a,\text{total}}(\lambda) = \mu_{a,\text{Hb}}[\text{Hb}] + \mu_{a,\text{HbO}}[\text{HbO}_2] + \mu_{a,\text{Hi}}[\text{Hi}] + \mu_{a,\text{HbCO}}[\text{HbCO}] + \mu_{a,\text{skin}} + \mu_{a,\text{muscle}} \quad (16.2)$$

where the concentrations of the respective components are expressed in square brackets.

The measured radiance satisfies the Beer–Lambert law of attenuation as

$$I = I_0 e^{-\mu_{a,\text{total}}(\lambda) z} \quad (16.3)$$

with Equation (16.2) substituted for the attenuation coefficient.

Note that the attenuation coefficient is a function of the wavelength and depends on the hemoglobin and oxyhemoglobin concentrations.

The oxygen saturation is defined as the percentage of blood that can release oxygen. This is seen with respect to the total blood content: oxygenated and deoxygenated. In mathematical expression the oxygen saturation SaO_2 is represented as

$$\text{SaO}_2 = \frac{[\text{HbO}_2]}{[\text{HbO}_2] + [\text{Hb}]} \times 100\% \quad (16.4)$$

Since the saturation is directly related to both the concentrations and the attenuation coefficients of the hemoglobin and oxyhemoglobin, an attenuation measurement can provide a good first-order approximation of the oxygen content of the blood. In this approximation, scattering is not considered

to reintroduce light into the direction of propagation, and adds to the total attenuation only. In addition, the presence of carboxyhemoglobin is not considered to form a significant contribution to the measurement, which may actually be more significant than initially assumed. The absorption peak of carboxyhemoglobin and oxyhemoglobin coincide, but carboxyhemoglobin does not contribute to the oxygenation of the tissue. In terms of measured radiance I for the red and infrared wavelengths this translates into an expression for the pulsed measurement of the oxygen saturation level SpO_2.
The pulse oxymeter is looking for the maximum and minimum radiance at each of the selected wavelength to discriminate steady-state blood flow (venous) from pulsating blood flow (arterial). Thus, the maximum (I_{max}) and minimum (I_{min}) radiance values correspond to the diastolic and systolic arterial pressure, respectively, arithmetically filtering out the steady-state venous attenuation, since this will act as a background for both the diastolic and systolic measurements. This derivation of pulsed oxygen saturation is expressed as

$$SpO_2 = f\left[\frac{\ln\left(I_{min}/I_{max}\right)_{630}}{\ln\left(I_{min}/I_{max}\right)_{960}}\right]\times 100\% \tag{16.5}$$

The factor f in Equation (16.5) is a calibration factor taking into account bone, pigmentation, thickness of the various layers of tissue and potentially nail polish for a finger pulse oximeter.
In case the attenuation coefficients for oxyhemoglobin at, for instance, 960 nm and the attenuation for reduced hemoglobin at, for instance, 630 nm are μ_{a,HbO_960} and μ_{a,Hb_630}, respectively, the oxygen saturation from Equation (16.4) can be written as follows:

$$SaO_2 = \frac{\mu_{a,Hb\_960}SpO_2 - \mu_{a,Hb\_630}}{(\mu_{a,Hb\_960} - \mu_{a,HbO\_960})SpO_2 - (\mu_{a,Hb\_630} - \mu_{a,HbO\_630})}\times 100\% \tag{16.6}$$

The oxygen concentration is found by multiplying the oxyhemoglobin saturation with the total hemoglobin concentration, which can be derived from the hematocrit. The hematocrit is the volume percentage of red blood cells in whole blood. The hematocrit (hct) is expressed as a percentage. For example, a hematocrit of 40% means that there is 40 mL of erythrocytes in 100 mL of blood.

The oxygen concentration can thus be found from Equation (16.6) and the hematocrit [Hb] from the following equation:

$$[O_2] = 1.39[Hb]\frac{SaO_2}{100} \tag{16.7}$$

The coefficient 1.39 in Equation (16.7) is called Hüfner's number, and this represents the volume of oxygen that can be collected per gram hemoglobin.

An entirely different biological application of the use of optics will be discussed next. In this case emitted light from an electroluminescent dye is used to monitor cellular electrical depolarization on a bulk-tissue continuous format instead of the regular discrete electrode measurement. The technique is called electroluminescent electrophysiologic mapping.

16.8 Electroluminescent Electrophysiologic Mapping

Recording spatial distribution of electrical activity in the heart is essential to understand the mechanisms of normal and pathological cardiac rhythms. The method of collecting electrical information on the functioning of the heart is called electrophysiologic (EP) mapping, or performing an electrophysiology study. Clinical electrophysiologic mapping can be performed by electrode placement on the chest or by catheters inserted in the blood vessels attached to the heart, as well as inside the heart itself.

Electrical mapping generally uses extracellular electrodes to record extracellular potentials from multiple sites of the heart surface. In research settings, glass electrodes can be used to record transmembrane potential by impaling single heart cells. Maintenance of a sufficient number of simultaneous impalements to map the spatial distribution of transmembrane potential is impractical, primarily because the glass electrodes are extremely fragile.

As an alternative to invasive electrode placement optical mapping with voltage-sensitive dyes has been introduced to study spatial distributions of the electrical activity in the heart.

Electroluminescent EP mapping uses fluorescent dye molecules that can be excited by light while electrical current is flowing nearby. In the process of exciting the fluorescent dye molecules, the fluorescence molecules absorb photons and raise electrons to higher electronic states. These electrons can stay in the higher states only for a very short period. After this period they return to the ground state, and photons are emitted. Not all the energy absorbed by the fluorescent dye molecules is released by the emitting photons. For example, some energy is lost through collisions between the molecules. Because of the lost energy during the de-excitation process, the emitted photons normally have lower energy (i.e., longer wavelength) than the absorbed photons. The quantum yield of a fluorescent dye is defined as the ratio of emitted photons to the absorbed photons. Optical mapping uses a photodetector to record the fluorescence signals from multisites on the heart surface. Fluorescent signals are collected by a photodiode array, a charge-coupled device image array (CCD camera) or a photomultiplier tube. An illustration of the electroluminescent measurement technique is outlined in Figure 16.9.

Optical mapping techniques used to study spatial distributions of cardiac activity can be divided into two categories:

1. Broad-field excitation
2. Laser scanning

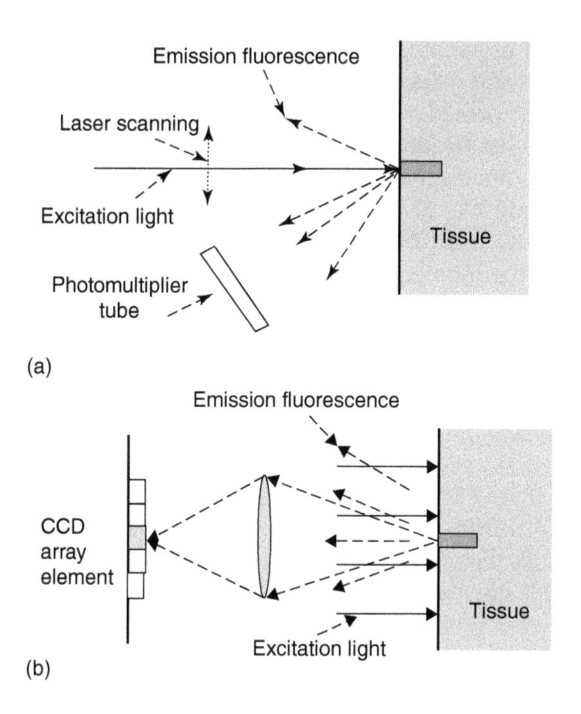

FIGURE 16.9
Diagram of two types of noninvasive electroluminescent electrocardiogram recording. (a) This method uses a scanning laser beam to excite the electroluminescent dye; as the depolarization wavefront proceeds to migrate the shell of the heart muscle, all fluorescence is captured by a photomultiplier tube that collects light emitted from the excitation location. (b) Broadbeam excitation of the entire epicardial wall facing the source is collected by a CCD camera, over viewing the entire field of view spread out over the CCD elements. Each successive image captures a different view of the propagation of the depolarization wavefront (Courtesy of Lei Ding, Cirrus Logic, Austin, Texas.)

During broad-field excitation the heart stained with voltage- or calcium-sensitive dyes is illuminated with broad-field excitation light and fluorescence is collected by image or photodiode arrays. During laser scanning excitation a scanning laser beam crosses the heart surface in a predetermined pattern and fluorescence is collected with a photomultiplier tube.

The spatial localization of the fluorescence signal for these two methods depends upon light absorption and scattering at both excitation and emission wavelengths.

The emission fluorescence spectrum of the photovoltaic dyes shifts with different transmembrane potential. Because of the linear relationship between the fluorescence signal and the transmembrane potential, optical mapping has the unique ability to record the transmembrane potential distribution in the heart. And in the process of imaging with multiple detectors or a CCD camera an instantaneous map of the cellular depolarization wavefront can be recorded.

However, a fluorescence signal may contain fluorescence resulting from areas outside the target section in addition to the cells that are directly targeted.

Fluorescence from tissue outside of this location will be included due to scattering of excitation and emission light.

It is important to consider the spatial localization to correctly interpret the optical recordings, as needed to understand the underlying mechanism of electrical activities in the heart. Selected dyes have various excitation and emission wavelengths. Examples of voltage-sensitive dye are Rh237 (Molecular Probes, Inc., Eugene, OR, U.S.A.) and pyridinium, 4-(2-(6-(dibutylamino)-2-naphthalenyl)-1(3-sulfopropyl)-hydroxide, and inner salt (di-4-ANEPPS). An example of calcium-sensitive dye is Oregon Green 488 BAPTA 1. Each dye has its own excitation and sensitization spectra; di-4-ANEPPS has an excitation wavelength of 488 nm and the emission fluorescence wavelength of 632 nm. The 488 nm excitation wavelength has the ability to excite both voltage- and calcium-sensitive dyes and allow emission ratiometry. RH237 is excited at around 505 nm and emits at approximately 715 nm. Oregon Green 488 BAPTA 1 absorbs at 488 nm and emits at 425 nm. Emission maxima of di-4-ANEPPS or Rh237 dissolved in MeOH occur at 705 or 782 nm, respectively, while they occur at approximately 630 or 696 nm when di-4-ANEPPS or Rh237 are attached to the cell membrane in hearts.

The main concern when using electroluminescent mapping techniques is to quantify the spatial localization of optical mapping.
The process of quantifying the electroluminescence involves answering the following two questions:

1. Given a slab of tissue and light source, what is the amount of light reaching a certain location in the tissue?
2. How much of fluorescence emitted at each location is recorded by the photo detector?

Both questions require an in-depth appreciation of light interaction with biological tissues at all the pertinent wavelengths involved. The actual microscopic optical properties describe single absorption or scattering events. They can be measured directly from samples thin enough that multiple scattering events are negligible and will affect the overall optical mapping accuracy. All the known optical properties can be entered in the equation of radiative transfer for a detailed analysis of the excitation pattern as well as the emission pattern at the respective wavelengths and the appropriate wavelength-dependent optical parameters.

Quantum dots were introduced in Chapter 4; in the next section the use of quantum dots as diagnostic tools is briefly described.

16.9 Quantum Dots as Biological Fluorescent Markers

Quantum dots (QDs) are a tremendous tool to assist developmental biologists understand how tissues and organs are formed from stem cells, and may allow these studies to extend beyond the current models of development. Current methodologies are limited to the physical monitoring of dividing cells by the trained eye.

Quantum dots are used as an effective method to deliver drugs in the pharmaceutical industry.

Another important consideration when using QDs is the targeting of specific proteins in cells and tissues. This is commonly accomplished through immunodetection. Traditionally, an organic dye is coupled to an antibody, which recognizes a specific antigen on the target protein of interest. Thus, it is intuitive that coupling QDs to the same molecules would be an attractive application. In addition, the use of QDs to locate a protein found in breast cancer cells called Her2 has shown the potential to assist pathologists to identify breast cancer tumor cells, which are likely to respond to anticancer drug therapy.

16.9.1 Future Considerations of Quantum Dot Imaging

While QDs have demonstrated a practical application on the bench top in the laboratory of biologists, there still remain concerns that need to be addressed. One is the propensity of QDs to self-aggregate in aqueous environments. Surface modifications must continue to be optimized to prevent QDs from aggregating together, a risk for possibly disrupting cellular processes and inhibiting proper segregation to the target protein within a cell or tissue. Another concern is the size of QDs. Although only nanometers in diameter the QD is about 10-fold larger than the organic dyes. This may limit the ability of QDs to intimately associate with all biological structures and substructures. A study of the half-life (how long it takes for an organism to remove half of the amount administered) of QDs in higher organisms would be of great interest to investigators. In addition, studies on the long-term toxicity of QDs must be undertaken.

We will conclude this chapter with a brief overview of the optical requirements for optimal therapeutic and diagnostic efficacy.

16.10 Compilation of the Optical Requirements for Wavelength Selection Based on the Desired Effects

Table 16.2 illustrates the minimum requirements to obtain the best results when using light for various clinical and diagnostic applications.

From all the discussions in this book it is clear that the majority of applications of light in medical treatment and diagnostics require deep-tissue penetration, with low scattering and low absorption. These requirements will remain the main problem in biophotonics and will place restrictions on the overall performance of optical methods for medical use.

TABLE 16.2

General Requirements for Selecting the Wavelength to Tailor the Efficiency and Precision of Optical Methods in Medical Treatment Processes and Detection Modalities

Desired Effect	Requirements
Coagulation	Low absorption Low scattering/high forward scattering Deep penetration
Imaging	Low absorption Low scattering for ballistic photon imaging and OCT Moderate backscatter in optical biopsy Deep penetration
Photodisruption	High/moderate absorption Low scattering Moderate penetration
Photochemical (photodynamic)	Low absorption Scattering requirements based on desired volume Deep penetration
Vaporization	High absorption Low scattering Shallow penetration
Welding	Moderate absorption Moderate scattering Moderate penetration

Generally the scattering decreases with the fourth power of the wavelength, according to the Rayleigh scattering definition, while beyond 1200 nm the absorption dramatically increases resulting from the many water vibration configurations. However, even in the near-infrared and visible spectrum water has several distinct resonance peaks that need to be taken into consideration, in addition to the oxy and deoxyhemoglobin absorption peaks.

16.11 Summary

This final chapter of the diagnostic applications of optical methods described the functional imaging of immunofluorescence, the detection of caries, and the optical detection of erythema. Other topics included the design of various fiberoptic sensors, in particular diagnostic applications of biological interest. Optical coherence tomography was briefly revisited and a derived technique of optical biopsy was described. Other biological diagnostic methods discussed covered the noninvasive blood oxygenation determination and the use of electroluminescent dye to map electrical depolarization of the heart muscle. The chapter concluded with a concise return to QDs and a summary of the biophotonics considerations for the various specialized applications discussed in this book.

Further Reading

Cohen, L.B. and Salzberg, B.M., Optical measurement of membrane potential, *Rev. Physiol. Biochem. Pharmacol.*, 83, 35–88, 1978.

Diffey, B.L., Oliver, R.J. and Farr, P.M., A portable instrument for quantifying erythema induced by ultraviolet radiation, *Br. J. Dermatol.*, III, 663–672, 1984.

Ding, L., Splinter, R., and Knisley, S.B., Quantifying spatial localization of optical mapping using Monte Carlo simulations, *IEEE Trans. Biomed. Eng.*, 48(10), 1098–1107, 2001.

Ding, L., Splinter, R., and Knisley, S.B., Determination of the cardiac tissue sample volume for optical mapping using Monte Carlo simulation, *Ann. Biomed. Eng.*, 29 (Suppl. 1), S–56, 2001.

Farr, P.M. and Diffey, B.L., Quantitative studies on cutaneous erythema induced by ultraviolet radiation, *Br. J. Dermatol.*, III, 673–682, 1984.

Knisley, S.B., Blitchington, T.F., Hill, B.C., Grant, A.O., Smith, W.M., Pilkington, T.C., and Ideker, R.E., Optical measurements of transmembrane potential changes during electrical field stimulation of ventricular cells, *Circ. Res.*, 72, 255–270, 1993.

Kopola, H., Lahti, A., Myllyla, R.A., and Hannuksela, M., Two channel fiber optic skin erythema meter, *Opt. Eng.*, 32(2), 222–226, 1993.

Problems

1. Show how time-resolved spectroscopy can be used to determine the number of molecules or atoms of an analyte (or the number of molecules of the tagged molecule). (Hint: use the same technique as used in radioactive dating.)
2. Describe the fundamental operating mode of calcium-sensing fiberoptic sensors.
3. Explain how time-of-flight measurement can determine the location of a particular sensory input on a multisensor fiberoptic sensor.
4. Describe the analogy between immunofluorescence microscopy and the "optical tongue."
5. What wavelengths are best suited for transillumination of the human body?
6. Prove that the pulse oximeter saturation can be written with background DC venous blood as

$$\mathrm{SpO_2} = f\left[\frac{\ln\left(\left(I_{\mathrm{min\,arterial}} + I_{\mathrm{min\,venous}}\right)/I_{\mathrm{max}}\right)_{\lambda_1}}{\ln\left(\left(I_{\mathrm{min\,arterial}} + I_{\mathrm{min\,venous}}\right)/I_{\mathrm{max}}\right)_{\lambda_2}}\right] \times 100\%$$

Further Reading

Index

A
Abbe constant, 436
Abbe limit, 448
Abbe, Ernst, 436
Ablation 224, 229, 231, 266, 350, 393
  Depth, 223–226, 231, 358, 379
  Efficiency, 231, 410
  Energy, 226, 260, 341, 350
  Energy density, 223
  Laer, 228, 232, 350, 404
  Photo, 234
  Power, 224
  Threshold, 152, 223, 229–232, 266, 356
  Velocity, 223–226
Absorbance, 259, 278, 338, 466, 546
Absorption, 11, 28, 46, 95, 141, 193, 301, 393
  Coefficient, 49, 99, 148, 168, 197, 210, 514
  Cross-section, 159, 168, 172, 230
  Dominated, 228
  Lorentz model, 98
  Spectroscopy, 280, 458
  Spectrum, 220, 538
  Wavelength, 298, 310, 336, 380, 556
Abu Ali al-Hasan ibn Haytham, see Alhazen,
Abu Zayd Hunayn ibn Ishaq al-Ibadi, 5
Acceptance angle, 282, 438, 495
Acoustic, 233, 461, 514, 524
  Effect, 233, 234, 512
  Wave, 233–235, 241, 354, 512
Acridine orange, 15, 540
Actin, 461
Actinic cheilitis, 403, 404, 408, 421
Actinic cheilosis, 403, 404
Acoustic impedance, 240, 241
Adenosine diphosphate, 290
Adenosine triphosphate , 290
Adiabatic, 350
Adipose tissue, 134, 564
Aerobic, 318, 448, 543
AF, see Atrial fibrillation
Age related macular degeneration, 305
ALA, see Aminolaevulinic acid
Albedo, 146, 159, 205, 358, 513
  Multiple scattering events, 161
  Single scattering, 159
Albumin, 135, 270, 334
Alexandrite, 58, 60–62, 381
Algorithm, Kasai, 501
Alhazen, 5
al-Kindi , Yaqub ibn Ishaq, 5
al-Razi, Mohammad inb Zakarija, see Rhazes
Aluminum-Gallium Arsenide (ALGaAs), 56
Alzheimer's disease, 270
AMD, see Age related macular degeneration
Amino acid, 255, 286, 321, 542
Aminolaevulinic acid, 295, 300, 380, 386
Aminovulinic acid, 300, 379
Amniotic, 409
Ampere, Andre-Marie, 11, 31, 33, 45
Amplification, 9, 46, 52, 320, 348
Amplitude, 24–29, 37, 40, 90, 106, 147
Analysis, 4, 39
  Blood-gas, 459
  Fourier, 90, 480, 491
  Spectral, 269, 457, 542, 554
  Spectroscopic, 555
Anastomosis, 334, 337
Anemometer, flow, 116
Anesthesia, 414, 568
Anesthetize, 446
Aneurysm, 390, 391, 566
Angiogenesis, 313, 356
Angioplasty, 306, 354, 375, 389
Angioplasty, laser, 19, 62, 375, 389
Angle,
  Acceptance, 282, 438, 495
  Critical, 80, 92, 257–261, 507

Angular, 138, 142, 156, 167, 244
Displacement, 436, 437
Distribution, 156, 166–169, 189
Frequency, 26, 38, 64, 430
Magnification, 428
Velocity, 22, 25
Anisotropy factor, scattering, 101, 133, 147, 197
Anterior, 363, 413
Antibody, 339, 459, 534, 551
Anti-Stokes, 114, 272–275
Aorta, 265–267, 328, 355, 395
Aperture, 429–437, 440, 444, 449–452
Aperture, numerical, 54, 122, 147, 282
Apoptosis, 295, 300
Approximal, 498, 554, 563
Approximation, diffusion, 158, 166, 197, 469, 505
Aqueous, 114, 280, 363, 543
Aqueous humor, 5
Arago, Dominique, 10
Arc lamp, 18, 45, 58, 540
Mercury, 101, 533
Source, 15
Xenon, 310
Ar, see Argon
Area, 31, 71, 184, 223, 320, 366
Argon, 57, 363, 381, 383, 385, 408
Aristocle, 4
Aristotle, 4, 5, 22
Arndt-Schultz effect, 297, 320, 388, 421
Arrhythmia, 389–390, 396, 401
Arthroscopic, 220
Arthroscopy, 282, 283
Arthrotomy, 283
Assyrians, 6
Astronomy, 4, 10, 13, 44, 138
Astrophysics, 7, 20, 137, 146
Atherosclerotic, 266, 354, 390
Atherosclerosis, 306, 550
Atom, 15, 28, 278, 532
Atomic force microscopy, 448, 450, 452
Atrial fibrillation, 390, 391
Attenuation coefficient, 175, 568, 569
Effective, 142, 158, 205, 301, 346
Reduced, 145
Total, 139, 145, 200, 205, 211
Autocorrelation, 483, 486, 491
Autocorrelation function, 486–494
Avalanche photodiode, 456
Axis, 83, 115
Extraordinary, 243, 244
Optical, 87, 115, 125, 250, 369, 503
Ordinary, 243
Principle, 83, 87, 89, 126, 245

B

Baby, premature, 18
Backing, black, 170, 171, 201
Bacon, Roger, 6, 425
Bacterial, 115, 336, 531
Bacteriochlorin, 299
Bacteriorhodopsin, 336, 337
Bacterium, 115, 380, 534
Ballistic photon, 150, 503–505
Ballistic photon imaging, 18, 503, 528–530, 566, 574
Band, Soret, 380
Basal cell carcinoma, 307, 386
Beam, 18, 52, 71, 126, 150, 171
Broad, 300, 315, 346–349, 405
Electron, 61, 460
Flattened Gaussian, 127
Free, 263, 267, 419, 459
Gaussian, 125–128, 149–152, 215, 260, 330
Laser, 173, 199–205, 366–373, 518–521
Photon, 167, 190, 195
Splitter, 207, 280, 441, 447, 482
Super-Gaussian, 127
Top-hat, 127, 217, 221
X-ray, 460
Beat frequency, 472, 474
Beer-Lambert-Bouguer law, 99, 142, 166, 200, 228
Beer-Lambert law, 99
Beer, August, 99
Beer's law, 99, 166
Bell, Alexander Graham, 233
Bell lab, 450–452
Benign, 286, 340, 382, 404, 559
Benign pigmented lesion, 285, 388, 403
Benign Protatic Hyperplasia, 417
Benzoporphyrin derivative, 299
Bernoulli, 351, 353
Bessel function, 155, 434, 435
Bilayer, 296
Bilirubin, 291
Binocular vision, 5
Binomial distribution, 126

Binomial series, 432, 433, 438
Bioheat equation, 211–213, 236, 331, 398, 405
Biomedical optics, 3, 4, 9, 12, 13, 18
Biopsy, 197, 282, 340
Biopsy, optical, 206, 563–566, 574
Biosensor, 320, 551, 558–561
Biostimulation, 253, 288, 312–314, 320
Biot, Jean, 10
Birefringence, 17, 243, 244, 252, 464
Birefringent, 252, 461
Black backing, 171, 201
Black body, 12
Bladder, 283, 306–308, 407, 417–419
  Gall, 19
  Tumor, 283, 306–308, 417
  Urinary, 307, 384
Bleaching, 444, 536
  Photo, 304, 338, 444, 536, 561
  Rate, 302
Blepharoplasty, 378, 381, 402, 415
Blood, 336, 341, 375, 404–406, 511, 560
  Absorption, 403, 567
  Alcohol level, 18
  Brain barrier, 386
  Cell, 116, 132, 313, 406, 458, 569
  denatured, 407
  Flow, 61, 136, 212, 313, 354, 409
  Gas analysis, 18, 137, 459, 554, 558
  Glucose level, 18, 375, 554, 555
  Oxygenation, 18, 465, 506, 566–568, 574
  Perfusion, 211, 212, 296, 356
  Pressure, 418
  Vessel, 15, 196, 211, 296, 334, 511
bloodless, 402, 420
Bohr atom model, 44, 51, 337
Boltzmann constant, 29, 275, 322, 324
Boltzmann distribution, 275
Boltzmann equation, 28, 138, 145, 275
Boltzmann equation, Stefan-, 509, 510
Boltzmann, Ludwig, 28, 322
Bond, 109, 242, 281
  Breaking, 228–232, 326, 327
  Chemical, 43, 102, 228–230, 269, 377
  Molecular, 103, 228–230, 281, 322, 488
Bone, 241, 253, 350, 404, 522
Bouguer, Pierre, 99
Boulnois, 284, 333, 340
Boundary, 29, 70, 190, 234, 348, 463
Boundary condition, 50, 76, 80, 205, 300, 330
Boundary, hitting, 183, 191–194
Boyle-Gay-Lussac law, 233, 512
Bozzini, Phillip, 282
BPD, see benzoporphyrin derivative
BHP, see Benign Protatic Hyperplasia
Bradley, James, 7
Bragg grating, 18, 460
Bragg-mirror, 460
Brain, 5, 151, 253, 319, 462
  Damage, 468
  Tissue, 151, 410
  Tumor, 270, 306, 386
Breast, 18, 270, 422, 468
Breast cancer, 306, 503, 512, 573
Broad beam excitation, 300, 330
Bronchoscopy, 283, 415
Bulk reflection coefficient, 176, 347
Bunsen, Robert, 11
Burn, 15, 312, 313, 327, 349
Burn injury, 213, 327

C

C, see Carbon
Ca, see Calcium
Café-au-lait, 382, 385, 403
Calcein, 536
Calcium, 19, 317, 355, 416, 537
Calcium sensitive dye, 571, 572, 575
Cancer, 15, 18, 151, 233, 270, 292
Cancerous, 14, 44, 270, 300, 384
Capillary, 15, 280, 314
Carbogen, 310, 560
Carbohydrate, 270, 290, 321, 560
Carbon, 255, 256, 290, 298
Carbon monoxide, 53, 57
Carbonate, 362
Carbondioxide, 53, 57, 62, 73, 285
Carbonize, 134, 284, 327, 359
Carcinogen, 310, 560
Carcinoma, 307
  Basal cell, 307, 386
  Nasopharynx, 306, 308, 309
  Squamous cell, 307
Cardiac death, sudden, 390
Cardiology, 19, 283, 319, 349, 373, 389
Caries, 310, 359, 498, 563, 574
Carious lesion, 360, 465, 554, 563
Carious tissue, 360, 498, 555

CARS, see coherent anti-Stokes Raman spectroscopy
Cartesian, 30, 32, 66, 86, 122, 432
Cataracts, 5, 363, 378, 412, 413
Cathode, 17, 58
Cauchy theorem, 39, 41, 63, 156
Cauchy, Augustin Louis, 39, 63
Cauterize, 411
Cavitation, 264–267, 296, 378
Cavity, 240, 281, 308, 362, 420
  Abdominal, 411
  Bladder, 307
  Laser, 20, 51–53, 55, 121, 125
  Nasal, 308, 309
  Nasopharyngeal, 308
  Oral, 281, 359
  Resonant, 47
  Ventricular, 356
CCD, see Charge Coupled Device
CCD camera, 203–205, 458, 511, 571
Charge Coupled Device, 443, 543, 548, 570
Cell, 15, 212, 214, 234, 267, 293
  Adhesion, 313, 318
  Death, 289, 295, 296, 299, 302
  Epithelial, 266, 269, 270
  Membrane, 212, 322, 461, 544, 572
  Muscle, 266, 306, 390
  Proliferation, 313, 320
Cervical tissue, 270, 477
Chandrasekhar, Subrahmanyah, 14, 149, 178
Characteristics, 26, 72, 109, 115, 326, 503
  Flow, 116
  Laser, 18, 53, 253
  Optical, 135, 177, 197, 203, 548, 558
  Particle, 269
  Thermal, 332
  Tissue, 121, 517
  Wave, 69, 86
Chemical bond, 102, 103, 228, 230, 273, 321
Chemical laser, 20, 53, 54, 63, 228
Chemokines, 313
Chemoreceptor, 545, 547
Chlorine, 53, 298
Chlorophyll, 289–291, 298, 300, 337
Chloroplast, 290, 291
Cholesterol, 265, 291
Chopping, 55, 234
Chromatic, 540
Chromatic aberration, 426
Chromatic dispersion, 146
Chromium, 55, 56, 58, 60
Chromophore, 220, 269, 279, 334–337, 378, 556
Cinchonamine, 14
Cl, see Chlorine
Clemson University, 469
CO, see Carbonmonooxide
CO_2, see Carbondioxide
Coagulation, 135, 212, 226, 328, 394, 406, 463
Coagulation depth, 226, 341
Cock, Christopher, 426
Coefficient, 127, 135, 237
  Absorption, 49, 99, 141, 148, 211, 230,
  Effective attenuation, 142, 148, 158, 205, 301
  Einstein, 49, 51
  Gruneisen, 239
  Reduced scattering, 139, 145, 151, 178, 202
  Reduced total attenuation, 139, 145, 205
  Scattering, 133, 148, 167, 172, 197
  Total attenuation, 139, 191, 200, 211, 311
  Transport, 162, 205, 209
Coherence, 36, 47
  Anti-Stokes Raman spectroscopy, 279, 280
  Length, 93, 477, 479–481, 492–494, 500
  Spatial, 37
  Temporal, 36, 37, 482
Collagen, 252, 306, 313, 321, 326, 388
Collimated, 130, 132, 159, 160, 449
  Attenuation, 166, 199
  Beam, 123, 130, 173, 201, 215, 308
  Irradiation, 132, 178, 347, 348
  Transmission, 145, 197, 199, 200
Colocalization, 444, 540
Colon, 270, 283, 284, 497
Color, 7, 12, 43, 44, 85, 133
Color Doppler, 564
Color-Doppler OCT, 500, 563, 564
Colorimetry, 547, 555
Combinatorial library, 544, 545, 549
Common mode rejection, 472

Compound microscope, 7, 425–428
Computed tomography, x-ray, 476
Conduction, heat, 227, 330, 331, 515
Conductivity, 235
Conductivity, thermal, 136, 211, 235, 350, 404
Confocal, 440, 441–444, 446, 543
  Microscope, 439–447, 486, 528, 534
  Microscope, laser scanning, 18, 439–441
Conservation law, 350, 352
Conservation of energy, 78, 96, 137, 171, 181, 245, 302
Conservation of mass, 350, 351
Conservation of momentum, 350–352, 370
Constant, 33, 109, 127, 129
  Boltzmann, 29, 275, 328, 509
  Damage, 327
  Dissociation, 537
  Elasto-optic, 248
  Excitation, 518
  Material, 517, 524
  Molecular distance, 539
  Planck, 28, 86, 122, 324, 510
  Rate, 213, 230, 303, 326
  Reaction, 325, 392
  Spring, 102, 237
  Time, 227
  Universal gas, 213, 323, 350
Continuity, equation, 242, 519
Continuous, 47, 126, 281, 359, 466
Contrast, phase, 456, 461
Convolution, 183, 184, 190, 195
Copper-vapor, 57, 381, 383, 385, 408
Cornea, 5, 62, 252, 259, 363, 412, 426
Corpuscular theory of light, 9, 10
Cosmetic, 17, 306, 312, 378, 380–382, 388
Cosmetic surgery, 62, 378, 379, 415
Cosmic dust, 13
Cosmic ray, 24, 43, 44
Cr, see Chromium
Crater, 224, 357
Crater-depth, 230–232
Criterion, Rayleigh, 77, 78, 438
Critical angle, 80, 128, 257–259, 506
Cross product, 36, 277
Cross-section, 27, 49, 52, 129, 138
  Absorption, 159, 168, 230
  Attenuation, 159
  Beam, 121, 125, 129
  Scattering, 100, 101, 168, 172, 367
  Transport, 162
Crystal, 6, 11, 51, 58, 62
Crystal, nano, 338, 339
Crystalline, 57, 336, 379
  Lens, 5
  Poly-, 442
Crystallography, 461
Curve-fit, 397, 565
Cut, 19, 257, 373, 410, 441
Cutaneous, 285, 291, 339, 378
Cut-off frequency, 118
Cylindrical, 124, 223
  Lesion, 395, 398, 400
  Polarized, 114, 124
  Symmetry, 156, 158, 162, 186, 193
Cystoscope, 282, 308, 418
Cytochrome-c oxidase, 318
Cytokines, 313
Cytoskeleton, 296
Cytosol, 298, 545
Cytotoxicity, 15
Cytotoxin, 293

D

da Modena, Tommaso, 7
da Vinci, Leonardo, 6,7
Damage, 4, 18, 227
  Collateral, 358, 400
  Constant, 327
  Endothelial, 305
  Irreversible, 322, 327, 339, 397, 412
  Optical, 267
  Peripheral, 219, 263, 398, 411
  Reversible, 324, 337, 463
  State, 214
  Sun, 403
  Thermal, 130, 214, 253, 264, 339
  Threshold, 461
Damped, 453
DAPI, see 4′,6-Diamidino-2-phenylindole
Davisson, Clinton Joseph, 13
dB, see Decibel
DC, see direct current
de Broglie, 13
de Broglie orbital restriction, 44, 51, 86
de Broglie, Louis Victor Pierre Raymond duc, 13

de Maricourt, Pierre, 11
de Mondeville, Henri, 14
Death, cell, 15, 289, 302, 392, 411
Death, sudden cardiac, 390
Debridement, valve, 354, 355, 361, 409
Decay, 207, 210, 220, 257, 293
  Coefficient, 206
  Curve, 207, 494, 565
  Exponential, 205, 263, 489
  Level, 49, 50
  Molecular, 57, 97, 293
  Radial, 204–206
  Rate, 49, 50, 338
Decibel, 146, 475, 482, 564
Delay, 208, 329, 391, 419, 479
  Line, 500
  Phase, 485
  Time, 96, 200, 337, 476, 526
Delta function, 144, 209, 515
Delta-Dirac, 163, 223, 229
Delta-Eddington, 144
Demineralize, 359, 498, 563
Denaturation, 322–327, 388, 400, 421, 563
Denature, 15, 214, 265, 323, 379, 407
Denatured state, 114, 213, 392, 401
Denaturization, 20, 565, 566
Dendrite, 336
Denk, Winfried, 442, 448
Density, 100, 134, 148, 212, 514, 530
  Ablation energy, 223, 351, 356, 357
  Bond, 230
  Energy, 35, 50, 129, 224, 320, 372, 514
  Flow, 36
  Flux, 147
  Lipid, 291
  Mass, 134
  Material, 239
  Neutral, 540
  Optical, 118
  Photon, 182–184, 231, 472
  Photon current, 184
  Photon hitting, 183–186
  Photon power, 141
  Power, 122, 125, 141, 151, 223, 347
  Radiant energy, 27, 148
  Radiative flux, 64, 148
  Resonant energy, 50, 51
  Spectral, 488, 489
  Spectral energy, 51
  Tissue, 211, 223, 350, 393
Dental, 359–361, 463, 521
  Caries, 465, 497, 554, 563
  Drill, 359–361
  Enamel, 310, 347
  Erythema, 555
  Hard tissue, 359–362
  Plaque, 362
  Prosthesis, 361
  Pulp, 360
Dentin, 241, 359, 498, 521, 563
Dentistry, 61, 349, 373, 528
De-oxygeated, 136, 568
De-oxyhemoglobin, 137, 216, 499, 574
Depopulation, 302, 303
Depth, 148, 176, 203, 224, 299, 397
  Ablation, 223226, 231, 262, 358
  Absorption, 350, 420, 518
  Axial, 501
  Bond-breaking, 231
  Coagulation, 129, 226
  Crater, 230–232
  Dependent, 302
  Evanescent, 261, 264, 509
  Focal point, 372
  Imaging, 496, 499, 563
  Information, 462, 463
  Lesion, 129
  Optical, 513
  Penetration, 140, 148, 166, 258–267, 305, 418
  Resolution, 439, 477, 482, 492
  Resolved, 253
Dermatology, 19, 61, 292, 327, 389
Dermis, 15, 265, 307, 556
Descartes law, 6
Descartes, Rene, 6, 7, 9, 21
Detector, 116, 183, 206, 440, 467
  Arm, 116, 502
  Acoustic, 241, 520
  Fiberoptic, 206, 471, 559
  Photo, 140, 198, 258, 474
  Piezo-electric, 241
Deterministic method, 181
Deuterium Fluoride, 53, 54
DF, see Deuterium Fluoride
Di Popozo, 6
Di Spina, Alessandro, 425
Diagnostic, 3, 14, 17, 423, 531

Diaphanography, 503
Differential, 140, 334, 446, 537
  Equation, 86, 105, 155, 182, 329, 470
  Form, 32, 35
Differentiate, 24, 35, 237, 477
Differentiation, 25, 210, 318
Diffraction, 9, 74, 148, 247, 269, 367, 434
  Barrier, 438, 448, 449
  Far field, 429
  Fraunhofer, 76, 91, 429, 438
  Fresnel, 75, 76
  Law of, 70, 73
  Limit, 152, 249, 368, 371, 428
Diffuse, 132, 183, 211, 307, 347, 469
  Activation, 444
  Backard radiance, 347
  Backward flux, 347
  Energy fluence, 142
  Excitation, 444, 446
  Forward flux, 347
  Forward remittance, 207
  Forward scattering, 207, 504
  Illumination, 307, 347, 349
  Irradiance, 348
  Optical imaging, 469
  Optical tomography, 468, 469, 528
  Photon, 505
  Probe, 201
  Radiance, 349
  Reflectance, 206, 215
  Reflectance spectroscopy, 559
  Reflection, 71, 131, 145, 166, 348
  Spatial light distribution, 208
  Transmission, 203, 207, 211
  Transport attenuation, 203
Diffuser, 296, 308, 309, 354, 402
Diffusion, 405, 471, 536
  Approximation, 158, 160, 186, 197, 201, 301
  Coefficient, 158, 166, 301, 469, 471
  Coefficient, heat, 394
  Coefficient, thermal, 235
  Drug, 302
  Equation, 158, 182, 208, 320, 347
  Heat, 226, 227
  Length, 517
  Length, thermal, 234, 235, 406
  Oxygen, 295
  Rate, 304
  Theory, 145, 155, 158, 177, 181, 563
  Thermal, 517
  Time, thermal, 374, 404, 517
Diffusivity, thermal, 136, 235, 264, 358, 514
Dimer, 57, 540, 546
Diode laser, 53, 61, 363, 378, 401
Diode, superluminescent, 474, 478, 481
Dipole, 110, 114, 271, 367, 538
  Amplitude, 40
  Approximation, 98, 372
  Contribution, 40
  -Dipole interaction, 293, 538
  Energy, 372
  Image source, 184
  Moment, 109–114
  Orientation, 110, 538, 539
  Oscillating, 542
  Polarization, 373
Dirac-delta distribution, 223, 229
Direct-current, 474, 475
Discrete ordinate, 166
Disease, 310, 439
  Cardiovascular, 390, 391
  Heart, 390, 391
  Periodontal, 497, 555, 563
Dispersion, 38, 85, 124, 146, 242, 439
Dispersive, 38, 122, 137, 502
Displacement, 34, 102, 189, 236, 500
  Angular, 436, 437
  Dye, 544
  Function, 104, 108, 112
  Lateral, 438
  Law, Wien's, 511
  Radial, 436
  Velocity, 242, 519
Dissipation, 187, 213, 373
Dissociation, 228, 232, 244, 253
Distal, 122, 128, 206, 332, 397, 430
Distance, Förster, 539
Distribution, 182
  Angular, 101, 142, 156, 167, 189
  Binomial, 126
  Boltzmann, 275
  Concentration, 182, 465
  Dirac-delta
  Electric field, 31, 430, 436
  Electron, 114
  Energy, 28, 29, 138
  Gaussian, 126, 223, 480, 515
  Homogeneous, 157, 169, 189

Light, 19, 141, 177, 187
Lorentzian, 128, 493
Membrane potential, 571
Particle size, 12
Phase, 208
Photon, 95, 142, 183, 393
Probability, 126, 143, 158
Rate, 138, 177
Size, 12, 270
Spatial, 186, 223, 539, 570
Temperature, 213, 227, 245, 331, 345
Uniform 157, 169
DNA, 114, 270, 367, 447, 536, 559
Photosynthesis, 115, 271, 290, 336
Doppler effect, 95, 115, 118
Doppler OCT, see Doppler Optical Coherence Tomography
Doppler Optical Coherence Tomography, 498, 500, 563
Doppler shift, 116, 472, 498, 501
Doppler, Christian, 115, 118
Dot product, 31, 140, 432
Dot, quantum, 317, 335, 346, 443, 572
Double integrating sphere, 197, 198, 434
Draper, John William, 27
Dretler, 417
Drill, dental, 359–361
Drug, 114, 297, 300, 317, 349
Concentration, 559
Delivery, 62, 299, 302, 573
Diffusion, 302
Light-sensitive, 292, 296
Localization, 386
Photochemical, 384
Photoreagent, 298
Photosensitive, 292, 296
Photosensitizing, 384
Release, 354
Resistant, 391, 394
Resistant, 391, 394
Retention, 302
Therapy, 573
Dye, 15, 51, 62, 138, 334, 443
Acidity sensitive, 546
Aniline, 291
Calcium sensitive, 571, 572
Electroluminescent, 570, 571, 574
Fluorescent, 133, 535, 546, 560
Hemoglobin derivative, 292
Industry, 4, 13
Laser, 20, 53, 62, 216, 381
Laser, pulsed, 329, 382
Medium, 55
Organic, 337, 540, 572
Photochemical, 384
Photosensitive, 296, 298, 312, 384
Photovoltaic, 571
Tattoo, 379
Voltage sensitive, 570–572
Dysplasia, 270, 410, 559

E

Edema, 220, 296, 345, 382, 409
Edema, macular, 477
Effect, Arndt-Schultz, 297
Effect, Doppler, 95, 115, 118
Effect, photoacoustic, 233, 234, 236, 512
Eigenvalue, 50, 51
Eikonal equation, 80
Einstein coefficient, 49, 51
Einstein formula, 61
Einstein population inversion, 47
Einstein, Albert, 9, 12, 23, 63, 448
Elastic, 39, 102
Layer, 266, 267
Medium, 39
Modulus, 237, 239, 519
Scatter, 97, 99, 269, 271
Elastin, 306
Electric, 11, 24
Charge, 24, 42, 56, 479
Current, 46, 418, 472, 518
Dipole, 111, 271
Energy, 45, 336
Field, 11, 29–33, 47, 78, 109, 277, 436
Flux, 31, 32
Permittivity, 34
Polarizability, 109–112, 570, 574
Electricity, 11, 282, 345, 427, 511
Electroluminescence, 572
Electroluminescent, 570–572
Electroluminescent dye, 571, 574
Electromagnetic, 11, 13, 23
Energy, 13, 27, 158
Exposure, 333
Field, 48, 51, 58, 98, 242
Law, 63
Phase information, 462
Radiation, 460, 479, 510, 556

Spectrum, 13, 18, 23, 42, 53, 536
Wave, 11, 23, 26, 35, 98, 366
Wave theory, 30, 69, 114, 542
Electromechanical, 219, 242, 346, 548
Electron, 13, 17, 43, 58, 95, 274
Elner, Victor, 282
Embryonic tumor, 407, 409
Emission, 5, 47, 271, 419, 540, 562
Bandwidth, 46, 499, 538
Coherent anti-Stokes, 279
Curve, 28, 338, 535
Field, 58
Fluorescent, 531, 538, 546, 571
Gaussian, 337
Heat, 510
Lifetime, 49, 533
Line, 28, 338
Photon, 45, 57, 97, 274, 510
Radiation, 28
Rate, 11, 271
Spectrum, 86, 410, 510, 535
Spontaneous, 46, 49, 51, 52
Stimulated, 9, 48, 96, 481
Thermionic, 58
Wavelength, 46, 53, 57, 60, 439, 532
Emittance, 460, 506, 509
Enamel, 241, 359–362, 498, 521–524
Endocardial, 391, 394, 397, 401
Endogenous, 291, 295, 299, 334, 384
Endoscope, 46, 85,
Endoscopic, 151, 152, 217, 263, 335, 410
Endoscopy, 281–284, 410, 419
endoscopy, gastrointestinal, 283
Endothelial, 266, 296, 305, 313
Energy, 35, 43, 56, 85, 130, 141, 187, 224
Ablation, 223, 225, 226, 232
Absorbed, 393, 532, 533
Activation, 213
Change, anti-Stokes line, 272–274
Change, Stokes line, 272–274
Chemical, 290, 291, 346
Configuration, 48, 50, 97, 103
Conservation of, 70, 78, 96, 139, 171, 181, 211, 350
Density, 35, 65, 129, 224, 261, 320, 356
Diagram
Dipole, 372, 538
Distribution, 28, 29, 48, 50, 125, 135, 138
Electron, 61, 97, 119, 336, 532
Excitation, 539
Fluence rate
Fluence rate, radiant
Fluorescence, 532, 570
Flux
Gibb's free, 322–324
Internal, 321, 322
Kinetic, 15, 45, 56, 86, 97, 104
Laser, 353, 358, 409, 515
Level, 42, 44–48, 86, 96, 228
Level, excited state, 272, 275, 298, 333, 532
Level, ground state, 274, 277, 293, 322, 335, 443
Mechanical, 346, 514
Molecular bonding, 228, 272, 321, 327, 336, 536
Photon, 15, 22, 28, 42, 95, 140
Potential, 86, 102–106
Production, 318
Pulsed
Radiant, 27, 78, 114, 141, 181, 201
Resonance, 50, 536
Rest, 61, 97, 98, 119
Rotation, 272
Singlet, 298
Source, 338, 391, 443
Spectral, 51
Thermal, 15, 46, 53, 132, 211
Threshold, 223, 225, 230, 261
Transfer, 272, 274
Triplet-triplet, 293
Vibrational, 272, 275, 457, 532
Enthalpy, 321–324, 327
Entropy, 28, 321–323, 326, 327
Envelope, 38, 70, 173, 473, 489
Coherence, 485
Function, 489, 490, 494
Wave, 38, 70
Environment, 6, 14, 38, 338, 554, 559
Environmental variable, 256
Environmental, natural, 256, 273, 393, 500, 537
Environmental, working, 134, 256, 448, 543
Eosin, 15, 254, 267, 292, 463, 547
Epicardial, 394, 401, 571
Epicardium, 511
Epidermis, 307, 328, 349, 380, 404
Epithelial cell, 269, 270, 386
Epithelium, 270

Equation, 14, 24, 25
  Bioheat, 211, 236, 331, 514
  Continuity,
  Maxwell, 33
  Momentum, 242, 519
  Reaction, 213
Erbium-YtriumAlminumGarnet, 58, 60, 62, 263, 352, 360
Er:YAG, see Erbium-YtriumAlminumGarnet
Erythema, 382, 411, 506, 555, 574
  Dental, 555
  Index, 556
  Meter, 556
  Value, 557, 558
Escape function, 183–185
Esophageal cancer, 410, 414
Esophagus, 270, 283, 477, 496
Etching, 361
  Acid, 361
  Chemical, 451, 452
  Corneal, 363, 412
  Laser, 361, 362
Euler, 222
Evanescent wave, 220, 257–262, 265, 507, 559
  Ablation, 259, 261–263, 267
  Depth, 261, 509
  Diagnostics, 18, 257, 506, 528
  Imaging, 506
  Interaction, 219, 257, 264, 267
  Optical, 257–262
  Region, 257
Evaporation, 135, 212, 333, 379, 419, 509
Excimer laser, 57, 62, 128, 259, 263, 352
Excitation, 46–48, 56, 86, 114, 190
  Acoustic wave, 234, 239, 524
  Broad beam, 570, 571
  Chromophore, 335, 444, 445, 533, 541
  Electron, 96, 98, 337, 532
  Energy, 298, 444, 570
  Evanescent wave, 561
  Filter, 533, 534, 535
  Laser scanning, 570, 571
  Nerve, 545
  Pulse, 271
  Shifted, 114
  Single photon, 445, 446
  Temporal, 515–517
  Two-photon, 442, 444–446, 555
  Wavelength, 289, 294, 300, 338, 380, 532
Excited dimmer, 57
Excited state, 46–49, 53, 56, 228, 271, 293
Exogenous, 295, 299, 444
Expansion, 350, 353, 461, 466, 518
  Coefficient, 232, 239
  Coefficient, linear, 239, 517
  Coefficient, volume, 239, 240, 519, 524
  Fourier, 224, 517
  Gas, 146, 234
  Linear, 239, 517
  Mechanical, 520
  Polynomial, 158
  Taylor, 41, 155, 158, 471
  Thermal, 220, 233, 346, 517
  Volume, 232, 238–240, 350, 518
Extramission, 5,6
Eyelid surgery, 378
Eyring rate process, 324

F

F, see Fluoride
Fabry-Perot, 176
Fabry-Perot interferometer, 176
Fabry-Perot interferometer, low finesse, 241
Fading, 534–536
Far field, 76, 238, 429, 433, 449
Faraday, Michael, 11, 32
Faraday equation, 34, 35
Faraday's law30, 32, 33
Far-field, 76, 238, 518
Far-field irradiation, 128, 395, 468
Fat, 134, 188, 196, 266, 378, 447
Fe, see Iron
FEL, see Free Electron Laser
Fercher, 501
Fermat principle, 78
Ferrochelatase, 300
Fetoscope, 409
Fetoscopy, 283
Fiber, 54, 129, 166, 257, 356, 452, 506
  Multimode, 482
  Optical, 54, 135, 186, 206, 386, 461
  Single-mode, 207, 296, 450, 482

Tapered, 450, 559
Tissue, 252, 362, 461, 477
Fiberoptic, 18, 128, 262, 356, 409, 521
Cable, 121, 129, 296, 495
Catheter, 355, 393, 394
Illumination, 16, 220, 345, 465, 565
Imaging, 85
Interferometer, 207, 478, 483, 566
Probe, 203, 206, 412, 482
Sensor, 557, 559, 560–563, 574
Fibrillation, atrial, 39, 390
Fibroblast, 313, 317–319, 541
Fibroblast proliferation, 313, 319
Fibrotic, 395, 416, 464
Fick's law, 160, 185, 209
Finsen, Niels Ryberg, 14
FITC, see Fluorescein isothiocyanate
Fizeau, Armand Hippolyte Louis, 7, 8, 12, 20
Flattened Gaussian, 127
FLIM, see fluorescent lifetime imaging microscopy
Fluoresce, 532, 554, 559
Fluorescein, 56, 546, 547, 553
Fluorescein isothiocyanate, 536
Fluorescence, 17, 95, 138, 203, 269, 293
Auto, 541
Laser induced, 554
Microscopy, 439, 533–538, 541
Ratio, 537, 540
Recovery, 536
Resonance energy transfer, 537, 540–542, 560
Sensor, 559
Fluorescent, 46, 440, 531, 534
Chromophore, 279, 535, 540, 553, 570
Dye, 133, 535, 560, 570
Imaging, 18, 425, 439, 531, 559
Immuno, 442, 551, 560, 574
Lifetime, 271, 338, 533, 562
Spectroscopy, 559, 560, 562, 571
Transfer infrared spectroscopy, 270
Fluoride, 53, 57, 73, 524, 554
Fluorochrome, 442–447, 533, 536, 553
Flux, 27, 31, 167, 176, 234
One, 160, 201
Three, 13, 165, 173, 176, 197, 347
Two, 166, 169, 199, 201
Focal length, 84, 88, 363, 412
Focus, 282, 426, 440, 521
Food-coloring, 462, 556–558
Force, 102–106, 109, 236, 364–367, 517
Electromotive, 32
Lorentz, 98
Mechanical, 335
Shear, 452–456
Förster distance, 539
Foscan, 300
Foucault, Jean Bernard Leon, 8
4′,6-Diamidino-2-phenylindole, 443
Fourier OCT, 501–503
Fourier series, 222, 480, 516
Fourier transform, 91, 270, 491, 498
Fourier, Jean Baptiste, 10
Franken, P.A., 4, 42
FRAP, see Fluorescence recovery after photobleaching
Fraunhofer diffraction, 76, 429, 438
Fraunhofer lines, 11
Fraunhofer, far field, 238, 518
Fraunhofer, Josef von, 11
Freckles, 382, 385, 403
Free radical, 294, 318, 379, 380
Free-beam, 263, 267
Free-Electron-Laser, 53, 58, 258, 264–267, 460–462
Frequency, 25, 29, 38, 48, 85
Angular, 26, 38, 483, 517
Bandwidth, 37, 526
Beat, 472
Chopping, 234, 526
Domain, 44, 472
Doubled, 58, 62, 228, 363, 403
Fringe, 491, 498, 500–502
Modulation, 472, 474
Natural, 40, 472, 488, 533
Range, 38, 42–44, 526
Resonance, 238, 240, 453, 518, 524
Sampling, 526
Shift, 101, 106, 115, 273–275, 460
Spatial, 90, 91, 449
Spectrum, 458, 472, 501
Tripled, 228
Vibration, 106, 114, 275, 279
Fresnel diffraction, 75, 76
Fresnel lens, 363, 413
Fresnel theory, 10, 429
Fresnel, Augustin, 10, 75
Friction, 102, 453–455

Fringe, 67, 208, 460, 482
  Frequency, 491, 498, 500–502
  Visibility, 482, 489, 490
FTIR, see Fourier transform infrared spectroscopy
Function, 24, 39
  Autocorrelation, 486–494
  Bessel, 155, 434, 435
  Envelope, 489, 490, 494
  Escape, 183–185
  Impulse response, 184
  Source, 158, 160, 163, 222, 301, 513
  Spectral density, 488
Fungal, 416, 531
Fused silica, 261
Fused quartz, 73

G
g, see Scattering anisotropy factor
GaAs, see Gallium Arsenide
Gabor, Dennis, 462
Gain, 52, 97, 159, 169, 443
GaInAsP, see Gallium-Indium-Arsenide-Phosphorus
Galactic, 4, 137
Galaxy, 4
Galen, see Galenius, Claudius
Galenius, Claudius, 5
Galilei, Galileo, 7
Gallium Arsenide, 54
Gallium-Nitride, 56
Gallium-Indium-Arsenide-Phosphorus, 54, 56
Gamma, 23, 42–44, 69, 98
Gas, 9, 12, 28, 73, 99, 232
  Constant, universal, 213, 323, 350
  Laser, 9, 20, 51, 53, 56
  Phase, 280
Gastrointestinal endoscopy, 283, 389, 409, 411
Gastroscope, 282
Gauss, Karl, Friedrich, 31, 87
Gauss's law, 30–32
Gaussian, 32
  Beam profile, 125–129, 142, 260, 330, 368
  Distribution, 126, 127, 223, 480, 515
  Function, 127, 128
  Optics, 87
  Profile, 125–127
  Profile, flattened, 127
  Profile, super, 127
Genitalia, 465
Geometric, 62, 128, 237, 307, 428, 518
Geometric optics, 5, 69–75, 82, 101, 367
Germer, Lester Halbert, 13
Gersho, 234
GFP, see green fluorescent protein
Gibb's free energy, 323, 324
Gingival, 319, 555
Glass, 6, 17, 73, 426
  Laser, 51, 57, 58, 60, 413
  Magnifying, 7
Glaucoma, 62, 363, 412, 477
Globulin, 114, 270
Glucose, 18, 256, 257, 290, 554
Gold-vapor, 57
Göppert-Mayer, Maria, 442
Gordon, 318
Gorich, 377
Gould, Gordon, 318
Gray smooth muscle, 266, 267
Green fluorescent protein, 540
Grimaldi, Francesco, 10, 73
Ground state, 45, 47–49, 97, 271, 293–295
Group velocity, 38, 41, 500
Gruneisen coefficient, 239
Gustatory system, 544, 545

H
H, see Hydrogen
Hb, see Hemoglobin
HbO, see Oxyhemoglobin
HBr, see Hydrogen-Bromide
HF, see HydrogenFluoride
Hair removal, 15, 17, 61, 378, 402
Hall, Chester Moor, 426
Halogen, 57, 536
Harmonics, spherical, 155, 222
Haytham, Abu Ali al-Hasan ibn, see Alhazen
Head, 307, 414, 465, 467
Heart, 61, 283, 319, 390, 563
Heat, 17, 28, 46, 135, 227
  Coagulation, 212, 393
  Conduction, 227, 330, 515
  Latent 212, 393
  Sink, 137, 212, 333, 511, 512
  Source, 211, 244, 309, 405, 414
  Specific, 136, 212, 235, 352, 404

Transfer, 14, 136, 211, 333, 414, 515
Vaporization, 212, 222, 259, 509
Heisenberg uncertainty principle, 122, 228
Helmholtz, Hermann Ludwig Ferdinand von, 8
Hematoporphyrin (IX), 202, 298
Hematoporphyrin derivative, 292, 298
Hematoxylin, 254, 267, 463, 464
Heme, 136, 295, 300
Hemodynamic, 468
Henry, Joseph 11
Henyey-Greenstein, 144, 189
Heterodyne, 460, 472–475, 478, 502
Hill, A.E., 442
Hippocrates, 281
Histology, 253, 265, 266, 322, 390, 439
Histology, Raman assisted, 554, 555
Histopathology, 270
Hitting boundary, 194
Hollow-wave-guide, 265, 267, 282, 508
Holmium-YtriumAlminumGarnet, 58, 60, 62, 418–420
Holography, 82, 461–463, 557
Holomorphic, 39
Homopolar, 281
Hooke, Robert, 425
Hooke's law of motion, 102, 103, 106, 237
Horseradish peroxidase, 551
Ho:YAG, see Holmium-YtriumAlminumGarnet
HpD, see Hematoporphyrin derivative
HRP, see Horseradish peroxidase
Hüfner number, 569
Huijgens, Christiaan, 9, 69
Huijgens' principle, 70, 82
Humoral immunity, 552
Hund's law, 338
HVD, see hydroxyethylvinyldeuteroporphyrin
HWG, see Hollow-Wave-Guide
Hydrocarbon, 270
Hydroceles, 465
Hydrocephaly, 412
Hydrofeuric acid, 451
Hydrogen, 255, 276, 290, 294, 392
-Bromide, 53
-Fluoride, 53
Peroxide, 53
Hydrophilic, 296, 299
Hydrophobic, 295, 299, 339
Hydroxyethylvinyldeuteroporphyrin, 298
Hyperpigmentation, 382
Hyperthermia, 135, 392, 412
Hypoxic, 294
Hypoxic-ischemic brain, 468

I

Iceland spar, 10, 11
I, see Iodine
I, see Irradiance
ICLAS, see IntraCavity Laser Absorption Spectroscopy, 459, 460
Idiopathic vulvodynia, 411
Ill-conditioned, 471
Image, 74, 77, 82, 241, 249, 427
Acquisition, 494
Molecular motion, 460
Source dipole, 184
Imaging, 18, 203
Ballistic photon, 18, 503–505, 528, 574
Evanescent optical wave, 506
Schlieren, 244–246, 248, 249
Shadow, 465, 466, 503
Terahertz, 18
Immonostain, 551
Immunofluorescence, 442, 551, 552, 574, 575
Impulse response function, 184, 190, 195
Incandescent, 12, 18, 27, 42, 45, 46, 362, 412
Incoherent, 18, 19, 45, 53, 461, 505
Index of refraction, 29, 39, 41, 73, 79, 130, 134, 245
Indo cyanine green, 334, 380
Infant, 404, 465–468
Infarction, 319, 390, 391, 394, 395
Infinite slab,176, 183, 393
Infinite wide beam, 125, 177, 393
Inflammation, 313, 335, 395, 512
Inflammatory, 12, 270, 296, 380, 512, 555
Infrared, 9, 42, 58, 100, 257,270, 511
Injury, burn, 213
Inorganic, 17, 138
Integral, 31, 366, 434

Integral, damage, 213, 214, 325–327, 339, 400, 406
Integrating sphere, 197, 198, 566
Integrating sphere, double, 197, 198
Integrodifferential equation, 155
Interaction time, 41, 228
Interference, 75, 268, 481, 484–489
Interferometer, 207, 208, 268, 456, 483
  Mach-Zehnder, 207, 208, 456–458
  Michelson, 207, 253, 478, 483, 487, 502, 565, 566
Interferometry, 268, 461, 476, 483
Interstitial,296, 308, 310, 412, 418
Intracavitary, 296
Intracavity, 458, 459
Intracavity laser absorption spectroscopy, 459, 460
Iodine, 53, 54
Irradiance, 27, 99, 126, 147, 159, 223, 347
Iridotomy, 363, 413
Iris, 5, 363, 412, 413
Iron, 55, 56, 300, 511
Irreversible, 14, 213, 312, 322, 324, 327, 392
Ischemic brain, 468
Ischemic muscle, 356, 554, 555
Ishaq, Abu Zayd Hunayn ibn Ishaq, 5
Isotropic scattering, 13, 143, 144, 158, 514

J
Jablonski diagram, 293, 294
Jacobian, 470
Jain, 377
Jako, Geza, 413
Jansen, Hans, 7, 425
Jansen, Zacharias, 7, 425
Jesionek, 292
Jiang, Huabei, 469

K
Kajiya, James, 14
Kaposi carcinoma, 403
Kasai algorithm, 501
Kepler, Johannes, 6
Keratotomy, laser radiative, 243
Keratotomy, radial, 363, 412
Kirchhoff, Gustav, 11, 27
Kottler, Friedrich, 13
Krishnan, K.S., 102
Krypton-ion laser, 57, 60, 363, 403, 412
Kubelka, Paul, 27, 145, 164, 197, 347
Kusch, Polykarp, 9

L
Lab-on-a-chip, 458–460
Lagrange's equation, 105
Lagrangian basis function, 470, 471
Laguerre polynomials, 155
Lambert, Johann Heinrich, 99
Lamp, Wood's, 380, 531, 532
Laparoscope, 411
Laparoscopy, 62, 284, 411
Laplace operator, 86
Laplace, Pierre, 10
Laser, 46–53, 55
Laser assisted in-situ keratectomy, 363, 412
Laser assisted in-situ keratoplasty, 363, 412
Laser assisted in-situ keratotomy, 363, 412
Laser assisted in-situ keratomileusis, 363, 412
Laser, vertical external cavity surface emitting semiconductor, 459, 460
LASIK, see Laser ASsisted In-situ Keratectomy
Latent heat, 334, 357
Law, 5, 28, 70
  Ampere, 31, 33
  Arndt-Schiltz, 320
  Beer-Lambert-Bouguer, 99, 142, 166, 224, 330
  Boyle-Gay-Lussac, 233, 322, 512
  Conservation of energy, 70, 78, 272, 350
  Conservation of mass, 350
  Conservation of momentum, 350
  Diffraction, 70, 73
  Faraday, 30, 32
  Fick, 160, 185, 209
  Gauss, 30, 32
  Geometrical optics, 70
  Hooke's, 102, 103, 237
  Huijgens, 11
  Hund's, 338
  Kirchhoff, 12, 27, 28
  Lenz, 34
  Malus, 11
  Maxwell, 30, 33

Newton's second, 98, 106
Planck, 28, 510
Radiation, first, 27, 28
Radiation, second, 27
Radiation, third, 11, 28
Rectilinear propagation, 70, 71, 79
Reflection, 70, 71
Refraction, 70, 72, 80
Stokes, 253, 272, 279, 532
Thermodynamics, first, 136, 321, 350
Thermodynamics, second, 28, 322
Wien's displacement, 28, 511
LED, see Light emitting diode
Legendre polynomials, 155, 156, 163
Length, coherence, 479–481, 485, 489, 492, 493
Lens makers' equation, 82–84
Lenz's law, 34
Leprosy, 532
Lesion, carious, 465, 498, 555, 563
Library, combinatorial, 544, 545, 549
Lichtleiter, 282
Lifetime, 45, 47–49, 271, 533
Lifetime scanning, 446
Ligand, 459, 541, 544, 546, 561
Light, 3, 42, 53
Emitting diode, 46, 297, 310–313, 349, 556, 567
Scattering spectroscopy, 268, 269
-gating, 498, 563
Line-pair, 250
Linewidth, 51, 460
Lipid, 220, 257, 291, 506, 537
High density, 291
Low density, 291
Lippershey, Hans, 7, 425
Lithotripsy, 19, 243, 283, 417
LLLT, see Low-Level-Light-Therapy
Lopes, Almeida, 319
Lorentz force, 98
Lorentz model, 98
Lorentzian distribution, 128, 489, 493
Low finesse Fabry-Perot interferometer, 241
Low-level-light-therapy, 317, 318
Lumen, 310, 354, 355
Luminescence, 459
Luminescence, fluorescent, 548
Luminescent sensor, 459
Lysosome, 293, 296, 541

M

Mach trail, 247
Mach-Zehnder interferometer, 207, 208, 268, 456–458
Macrophages, 296, 313, 317
Macular edema, 477
Magnetic field, 30–32, 34–36, 48
Magnetic permeability, 34, 138, 243, 364, 484
Magnetic resonance imaging, 476, 520
Magnetism, 11
Magnification, 250, 426, 428, 438
Magnify, 6, 425, 426
Maiman, Theodore H., 9
Malus, Etienne Louis, 10, 11
Mammography, 469
Maraldi, Giacomo Filippo, 10
Marcacci, 14
MARK III, 261, 265, 266
Marshall318
Maser, 9
Mass, conservation of, 350, 351
Maxwell equations, 13, 30, 33–35, 81, 101
Maxwell, James Clerk, 11, 46
Maze, 401
McKenzie, William, 14
Mean free optical path, 145, 205, 405, 406
Medium, 9, 12, 242
Active, 47, 52, 55–57, 255
Biological, 135, 219, 321
Elastic, 39
Laser, 46–48, 50–52
Optical, 187
Refractive, 147
Melanin, 329, 336, 378, 556
Melasma, 382, 385, 403
Melatonin, 313, 318
Mercury arc lamp, 101, 533
Metabolic process, 213, 289, 295, 304, 509
Metal ion, 545, 547
Metal-vapor, 57
Metastable state, 47, 50, 532
Methoxy, 299, 338
Meyer-Schwickerath, Gerd, 15

Michelson interferometer, 253, 269, 486, 502, 566
Michelson interferometer, fiberoptic, 207, 478, 482, 483
Michelson, Albert Abraham, 8,12
Microfracture, 233
Microscope, 7, 436–438
  Atomic force, 448, 450
  Compound, 7, 425–428
  Confocal, 439, 440, 442, 543
  Electron, 13, 287, 448
  Fluorescent, 439, 533, 534, 536, 553
  Multi-photon, 443–447
  Near-field scanning optical, 448–451
  Optical, 425, 429, 436–439
  Raman spectroscopy near-field scanning optical
  Ratio fluorescence, 537–542
Microsurgery, 61, 335
Microwave, 9, 24, 42–44, 526
Mie scattering, 12, 45, 101, 133, 367, 368, 371
Mie, Gustav, 12
Millikan, Robert A., 13
Mismatch, index, 176, 191, 201, 402
Mitochondria, 290, 293, 295, 318
Mitosis, 296
Mode, tapping, 452, 454, 455
Mode-locked, 55, 133, 278, 279, 481, 482
Modulate, 116, 472, 473, 489
Modulation, amplitude, 207
Modulus, bulk, 239, 519
Modulus, compression, 237
Modulus, Young's, 236–239, 517, 518, 524
Mohammad inb Zakarija al-Razi, see Rhazes
Moire, 268
Molecular, 28, 53, 102
  Bond, 103, 322, 228–230, 448
  Bondbraking, 231
  Bonding energy, 457
  Polarizability, 112
Molecule
Momentum, 26, 35, 87, 122, 364, 366, 368
Momentum equation, 242, 519
Momentum, conservation of, 350–352
Monochromatic, 46, 121, 197, 234, 296, 313, 483
Monochromator, 279, 280
Monomer, 540, 544, 546
MONSTIR, see multichannel optoelectronic near-infrared system for time-resolved image reconstruction
Monte Carlo, 177, 181, 182
  Simulation, 182–195, 399, 469, 565
  Technique, 155, 181, 182
Motion, 98, 102, 337, 460, 491
Motor, 460, 461
Mouse, 326, 446, 447, 553
MRI, see Magnetic Resonance Imaging
M-tetradydroxyphenylchlorine, 300
mTHPC, see m-tetradydroxyphenylchlorine
Mucosa, 270
Mucosal tissue, 270
Mucositis, 556
Multichannel optoelectronic near-infrared system for time-resolved image reconstruction, 467, 468
Multimode fiber, 482, 559
Multi-photon imaging, 18, 442, 447
Multi-photon microscopy, 439
Multiple scattering, 145, 161, 167, 572
Multiple sclerosis, 270
Mulvaney, 417
Munk, Franz, 13
Muscle, 61, 188, 250, 357, 400, 401
  Birefringence, 252, 463
  Ischemic, 61, 391, 554–556, 568, 571
  Smooth, 266, 267, 306, 354
Myocardium, 204, 390, 397, 398, 566
Myosin, 252, 461

N

NA, see Numerical aperture
Nail-bed, 496, 497
Napthalocyanine, 299
Nasal polyp, 281, 307–309
Nasopharynx carcinoma, 306, 308, 309
Naevus flanmaeus, see Port-wine stain
Nd:YVO_4, see Neodymium-Vandate
Nd:YAG, see Neodymium-YtriumAluminumGarnet
Near-field, 76
  Irradiation, 128, 221, 518
  Scanning optical microscopy, 438, 448–457
Near-infrared, 42, 56, 259, 318, 445, 467
Necrosis, 295, 302, 304, 326, 384, 392

Neodymium-Vandate,
Neodymium-YAG, see Neodymium-YtriumAluminumGarnet
Neodymium-YtriumAluminumGarnet,
Neonatal
NERF, see [(ethylamino)-7-methyl-3-oxo-3H-xanthen-9-yl]-1,3-benzene-di carboxylic acid, 537
Neuroendoscopy, 284
Neutron physics, 137, 138
Nevi, 382, 385, 403
Newton, Isaac, 9, 10, 46, 69
Newton's second law, 98, 106
N, see Nitrogen
Nitrogen, 56, 57, 279
Noise, 275, 360, 473–476, 541
Non dimensional 73
Noninvasive, 197, 203
  Detection, 270, 565
  Imaging, 503
  Measurement, 197, 499, 506, 563, 571
  Technique, 137, 195, 203, 556
Nonlinear, 238, 242, 295, 397–399
Nonthermal, 220, 317, 320, 333, 335, 346
NSOM, see microscopy, near-field scanning optical
Nuclei, 109, 504, 516
Nucleus, 44, 86, 98, 293, 322
Number
  Hüfner's, 569
  Péclet, 352, 353
  Wave, 25, 38, 122, 148, 368, 502
Numerial, 127, 146, 155, 181
  Aperture, 56, 122, 147, 200, 282, 438, 445
  Method, 127, 177, 181
  Technique, 128, 155, 182, 206, 472, 523

O

O, see Oxygen
Occlusal, 498, 554, 563
Occlusion, 305
OCT, see Optical coherence tomography
  Color-Doppler, 500, 563, 564
  Doppler, 498, 563, 564
  Fourier, 501–503
  Phase-resolved, 498
  Polarization sensitive, 498, 499, 563
  Spectral-domain, 501, 502
  Spectroscopic, 477, 499, 500
  Time-domain, 500, 502, 563
Oersted, Hans Christian, 11
Offset printing, 4, 14
Oligohydramnios, 409
Oligonucleotide, 558, 561
One dimensional, 91, 164, 221, 244, 305, 476
  Attenuation, 142
  Boundary condition, 221
  Image, 496
  Slab, 166
Oncology, 346, 384
Opaque, 10, 27, 37, 249, 429
OPD, see Optical path-difference
Operator,
  Hamiltonian, 86
  Laplace, 86
  Momentum, 86, 87
  Scattering, 100
Ophthalmology, 3, 8, 19, 61, 62, 243, 305, 349, 362, 393, 412
Ophthalmoscope, 85
Opsin, 336, 337
Optic nerve, 5, 336, 503
Optical, 3, 4
  Biopsy, 206, 563, 565
  Instrument, 6 74, 84
  Path, mean free, 145, 147
  Path-difference, 79, 80, 485, 489
  Properties, 133, 137, 145, 329, 337, 469
  Properties derivation, 195, 197, 201, 203, 205–207, 209, 505
  Time-of-flight, 467, 561, 562
  Tweezer, 364, 373
Optical coherence tomography, 61, 117, 207, 476, 477
  Color-Doppler, 500
  Fourier, 501
  Phase resolved, 498
  Polarization, 498
  Spectral, 499
  Time-domain, 500
Optics, Gaussian, 87, 88
Optoacoustic, 18
Optochemical, 558, 559
Opus Majus, 6

Oral, 270, 281, 310, 354, 497
  Cavity, 281, 359
  Erythema meter, 555
Ordinate, discrete, 166
Oregon green, 572
Organ, 20, 135, 220, 297, 464, 503
Organic, 44, 255, 338
Organism, 3, 252, 255, 291, 367, 509
Orthopedic, 62, 282, 389
Oscillation, 24–26, 28, 29, 147, 480
Oscillator, 52, 98, 103
Oscillatory, 34, 42
Otolaryngology, 284, 389, 414, 415
Oxygen (O), 53, 290, 294
  Diffusion, 304
  Reactive, 293, 313, 319
  Saturation, 137, 568
  Singlet, 292, 293, 294, 298, 304, 384
  Triplet, 293, 294, 303, 304
Oxygenation, 18, 466, 468, 567
Oxyhemoglobin, 403, 556–558
Oxymeter, pulse, 568

P

P_0 approximation, 157, 471
P_1 approximation, 158, 471
Papule, 382
Paraformaldehyde, 534
Paramecia, 15, 291
Paraxial, 436
Parseval's identity, 501
Path-difference, optical, 74, 479, 486
Pathogen, 428, 459, 552, 561
Pauli exclusion principle, 44, 51
Pauli, Wolfgang, 44
PDT, see Photodynamic therapy
PE, see Phycoerythrin
Péclet number, 352, 353
Penetration depth, 142, 148, 166, 259, 400, 515
Pennes, 333
Perfluorochemical, 295
Perfusion, 211, 296, 354, 512
Perfusion rate, 137
Pergrinius, Petri, see de Maricourt, Pierre
Periodical, 8, 234
Periodicity, 24, 58, 226
Periodontal, 497, 563
Periodontal disease, 310, 497, 555
Periodontitis, 555
Peripheral iridotomy, 363, 413
pH, 537, 540, 544–549
Pharmacokinetics, 297, 354
Pharynx, 281, 308, 309
Phase, 19, 25, 26, 37, 91, 565
  Angle, 25, 148, 279
  Contrast, 439, 456, 461

  Function, 143, 144, 157, 193
  Function, scattering, 101, 147
  Imaging, 456, 458, 462
  Interference, 268
  Locked-loop, 474
  Mineral, 361, 362
  Resolved, 461, 477, 490
  Shift, 106, 236, 244, 439, 454, 475, 485
  Transition, 135, 188, 212, 213, 400
  Velocity, 24, 38, 272, 500, 502
Pheophorbide –a, 298–300
Phosphate, 290, 546
Phospholipaeses, 296, 545
Phospholipid, 296
Phosphorylation, 318, 319
Photodynamic therapy, 61, 291, 312, 333, 384
Photoablation, 19, 220, 232, 375, 411, 417
Photoacoustic, 233–236, 241, 512, 518–524
Photobiological, 317,
  Diagnostic, 320, 333, 335, 346
  Mechanical, 389
  Therapeutic, 551
Photobiomodulation, 313, 318
Photobleeching, 304, 338, 444, 536, 561
Photochemical, 14
  Diagnostic, 533, 544–548
  Mechanical, 228, 229, 289–313
  Therapeutic, 346, 377–387
Photochemotherapy, 292
Photocoagulation, 15, 187, 213, 320–324, 391
Photocuring, 472, 475
Photodetector, 140, 258, 474, 505, 570
Photodiode, 265, 502, 556, 570, 571
Photodiode, avalanche, 456
Photodisruption, 19, 219, 220, 363, 412

Photodithazine, 298
Photodynamic action, 15, 292
PhotoDynamic Therapy, 15, 292–313, 333, 384–387
Photodynamische erscheinung, 292
Photoelectric, 12, 13, 96, 97
Photofrin I & II, 298–300
Photomultiplier, 279, 280, 441, 467, 571
Photon, 23, 28, 28, 51, 62, 85, 97, 122
  Hitting density, 183–186
  Power, 121, 141
  Two, 228, 439, 442–447, 555
Photooxidation, 295, 296
Photophone, 233
Photophysical
  Diagnostic, 425
  Mechanical, 219, 264
  Therapeutic, 333, 345
Photoradiation therapy, 292
Photo-reagent, 298
Photosensitivity, 299, 300, 312
Photosensitizer, 291, 312, 379, 384
Photosynthesis, 115, 290, 336, 337
Phototherapy, 292
Photothermal, 220, 228, 317, 321
Phototoxicity, 15, 292, 296
Photovoltaic dye, 571
Phycoerythrin, 553
Piezoelectric, 241, 450, 453, 454, 455, 542
Piezoelectric transducer (PZT), 116, 207, 472
Pigment, 14, 290, 329, 336, 384, 531, 569
Pigmented lesion, 61, 382, 385, 403
Pixel, 185, 186, 449, 455, 548
Planck, Maxwell Karl Ernst Ludwig, 12
Planck's constant, 28, 86, 122, 272, 324
Planck's law of radiation, 9, 13, 28, 29, 50, 510
Plaque, 62, 267, 270, 350, 353, 354
Plasma, 220, 242, 243, 264
p-n junction, 54
Pneuma, see Optic nerve
Pneumotharax, 465, 466
Pohl, D.W., 448
Poisson ratio, 236, 238
Poisson, Simeon, 10
Polarization, 10, 80, 101, 219, 250, 253
  Angle
  Direction, 29, 80, 123, 255, 259
  Sensitive OCT, 498
  Material
Polyetheyleneglycol, 299
Polyhydramnios, 404
Polymer, 241, 354
  Luminescent, 454
  Microsphere, 545
  Solder, 334
Polynomial, LaGuerre, 155
Polynomials, Legendre, 155
Polyp, nasal, 281, 415
Polypeptide, 337, 542
Population inversion, 47, 52, 53
Porcine, 265
Porcine aorta, 265
Porphyrin, 293, 298, 300, 380, 562
Port-wine stain, 15, 327, 333, 346, 404, 408
Post infarction, 391, 394
Posterior, 554
Potential energy, 86, 102, 106
Potter, U, 298
Power density, 125, 129, 141, 224, 242
Power density
Power, resolving, 438
Poynting vector, 36, 138, 147, 276, 277, 364, 365, 366
Premature baby, 18, 465, 466
Principle axis, 245
Principle, quantum, 12, 85, 510
Probability, 138
  Event, 126, 133, 139, 183
  Scattering, 95, 101, 143, 147, 160, 182
  Stimulation, 49
Proliferation, cell, 313, 320, 354
Proliferation, fibroblast, 313, 319
Propagation, 30, 38, 524
  Direction, 277
  Law of rectilinear, 71
  Light, 139, 142, 145
  Simulation, light, 187
  Sound, 237, 242, 524
  Speed, 26, 38
  Wave, 11, 24, 39, 123
Properties,
  Chemical, 545
  Mechanical, 39
  Optical, 95, 133, 134, 145, 146, 329
  Quantum dot, 337

Thermal, 135, 136, 523
Wave, 13, 73
Propidium iodine, 536
Prostate, 417, 420
Prostatic hyperplasia, benign, 417, 420
Protease, 295
Protein, 15, 62, 255, 322–326, 447, 461, 562
Proximal, 136
Psoriasis, 382
Public health, 561
Pulmonary silicosis, 253, 254, 463, 464
Pulse, 132, 222–227, 349
Length, 132, 140, 350, 379
Oxymeter, 560
Rate, 224, 330, 357, 526
Pulsed, 20, 58, 222, 223
Pulse-width, 132, 358, 445, 516, 526
Pump, 48, 52
Energy, 52
Source, 53–58
Purpurin, 299
Pustule, 382

Q
Q-switched, 55, 225, 238, 279
QTL, see Quencher-Tether-Ligand
Quanta, 86, 533
Quantization, 28, 118, 510
Quantum, 229
Dot, 337, 572
Principle, 13, 28
Theory, 85, 324, 510
Well, 460
Yield, 298, 305, 533
Quartz, 73, 241, 418, 452
Quench, 304, 450, 536
quencher-tether-ligand, 459
Quinine, 14, 545

R
Raab, Oscar, 14, 291, 292
Rabi, Isidor Isaac, 14, 291, 292
Radachlorin, 298
Radiant energy, 27, 148, 224
Density, 27, 148
Fluence rate, 27, 147, 148, 301, 469
Flux, 27, 147, 148, 321
Radiation, 30, 45, 85, 86
Black-body, 11, 28, 45, 50, 51
First law of, 27
Second law of, 28
Third law of, 11, 28, 510
Radiationless, 293, 303, 532
Radiative transfer, 137, 140, 155, 166, 182, 513, 469
radiative transport, 155, 166, 177, 181
Raman assisted histology, 554, 555
Raman line, 273, 274, 276
Raman scattering, 96, 101, 109, 274
Raman scattering spectroscopy, 114, 115, 271, 279
Raman spectroscopy, 114, 275, 541
Raman, Chadrasekara, 101, 102
Random walk, 187–189
Rate constant, 213, 230, 303, 324
Rate, bleaching, 302
Ratio fluorescence microscopy, 537, 541
Ray optics, 78
Rayleigh criterion, 77, 438
Rayleigh scattering, 96, 99, 274, 387
Rayleigh, John William Strutt Lord, 12
Reactant, 273, 279, 324
Reaction equation, 213
Real-time, 203, 476
Rectilinear propagation, law of, 71, 79
Rectoscope, 281
Reduced
Scattering coefficient, 139, 145, 202, 210, 301
Total attenuation coefficient, 139, 205
Reflect, 4, 12, 147, 348
Reflectance, 53, 147, 171, 479, 494
Spectroscopy, 556, 557
Time-rsolved, 208
Reflection, 10, 81, 82, 130, 131, 147
Black-backing, 171
Coefficient, 170, 194, 371
Coefficient, bulk, 347, 348
Diffuse, 145, 147, 198, 206
Law of, 71, 79, 241
Specular, 148, 203, 402
Reflectometry, 206, 478, 506
Refraction, 72, 79, 83, 243, 370
Index, 29, 39, 41, 73, 134, 147, 244, 250, 526
Law of, 71, 79, 241
Regeneration, 55, 270, 306, 312
Remittance, 207, 466
Removal, tattoo, 379, 381, 403

Repetition, 26, 132, 222
Repetition rate, 132, 222, 224, 359, 517
Resolution, 248, 429, 437, 481, 486
  Spatial, 249, 448, 456, 476
  Temporal
Resolving power, 438
Resonance, 50, 453
  Energy transfer (RET), 536, 540
  Frequency, 40, 238, 274, 518, 524
Resonant cavity, 47
Respiration, 290, 318, 465
Restenosis, 306
Resurfacing, skin, 61, 62, 379
Retardation time, 486, 491
Retina, 6, 327, 336, 363, 477
Retine, 336
Reversible, 321, 322, 324, 327, 463
Rh-237, 572
Rhazes, 5
Rhinophyma, 382
Rhodamine, 56, 338, 443, 535, 540
Rhodopsin, 336, 337
Roemer, Ole, 7
Ronchi ruling, 248, 249, 250
Röntgen, 43, 460
Rontgen, Wilhelm Conrad, 17, 43
Rosacea, 7, 15, 382, 384, 385, 408
Rosencwaig, 234
Ruby, 9, 60, 62, 359, 381, 385, 408
Rufus, Quintus Curtius, 5
Ruling, Ronchi, 248–250
Russian roulette, 177, 181, 194

S

Sacrococcygeal teratoma (SCT), 407, 409
Sakata, I, 298
Salt, 73, 536, 543, 547
Sapphire, 20, 58, 254, 265, 411, 507
Saturation, 270, 304, 398, 514
  Detector, 203
  Oxygen, 137, 499, 567
Scanning confocal microscopy, 18, 439
Scanning, lifetime, 446
Scar, 306, 313, 319, 327, 345, 390
Scattering, 97, 99–101, 133, 146, 155, 189, 193, 492
  Angle distribution, 133, 189
  Anisotropy factor, 142, 144, 147, 162, 187
  Coefficient, 148, 172, 200
  Coefficient, reduced, 145, 202, 210, 301
  Cross-section, 159, 160, 172, 368
  Dominated, 97, 99–101, 133, 146, 155, 189, 193, 492
  Isotropic, 143, 158
  Phase function, 143, 147
  Raman, 101, 109, 114, 275, 281
  Spectroscopy, 269, 271, 459
SCD, see sudden cardiac death
Schawlow, Arthur, 9
Schieren imaging, 244–250
Schlieren, 17, 41
Schuster, Arthur, 13, 145
Schwarzschild, Karl, 13
Scrotal sac, 465
Scrotum, 418, 465
Seminaphthorhodamine-1 carboxylate, 537
Semiconductor laser, 54, 56, 459
Sensor, luminescent, 454
Septum, 394, 409
SERDS, see Shifted Excitation Raman Difference Spectroscopy
Series (development), 117, 123, 155–157, 163, 164
  Binomial, 432, 433, 438, 513
  Fourier, 222, 480, 516
  Lagrangian, 471
  Recursive, 176, 348, 438
  Taylor, 123, 164, 226, 232, 329
Shadow imaging, 18, 465, 466, 503
Shear-force, 452–455
Shift
  Doppler, 116, 117, 472, 500
  Frequency, 101, 115, 273, 458, 498
  Phase, 106, 454, 475
Shifted Excitation Raman Difference Spectroscopy (SERDS), 114, 115
Sight, 4, 6
Signal, 91, 210, 241, 472–476
  Fluorescence, 305, 546, 570, 571
  Interference, 479, 482, 486, 496, 500, 565
Signaling, 313, 447
Signal-to-noise ratio, 268, 459, 474, 478, 482
Silicon, 259, 264, 299, 543–548
Silicosis, pulmonary, 253, 254, 463, 464
SilverChloride, 463

Simulation, 177, 181, 187, 196, 472
Single scattering, albedo, 146
Single-mode fiber, 207, 482, 559
singlet Oxygen, 292, 302, 304, 379, 384
Singlet state, 293–295, 304, 532, 533
Skin resurfacing, 61, 62, 378
Slab, 166, 176, 185, 504, 572
SLD, see superluminescent diode
Slit, 74–77, 115, 248
Smallpox, 3, 14
Smooth muscle, 266, 306, 354
SN ratio, see Signal-to-noise ratio
SNARF, see seminaphthorhodamine-1 carboxylate
Snell, Willibrord von Roijen, 6, 29
Snell's law, 29, 72, 73, 81, 82, 130, 244, 257, 363
Snyder, E, 298
Soft tissue, 211, 351, 380, 402
Sognnaes, 359
Solar, 10, 11, 312
Solar eclipse, 4, 5, 15
Solder, 333–335, 378
Solid state laser, 57, 60, 499
Soret band, 380
Source, 543, 544, 547
  Broad band light, 45
  Function, 158, 160, 163, 242, 469
  Heat, 211, 244, 399, 405, 514, 515
  Laser, 4, 6, 61, 121
  Light, 11, 15, 18, 19, 23, 44, 53, 129, 310, 479
Spatial frequency resolution, 90, 91, 449
Spatial resolution, 249, 250, 449, 476, 499, 521, 562
Specific heat, 212, 223, 235, 242, 393, 523
Spectra, 86, 100, 102
Spectra, Raman, 281, 458, 542
Spectral
  Absorbance, 459, 506, 556, 567
  Bleed-through, 540
  Content
  Density function, 488, 489
  Domain OCT, 501
  Line, 11, 28, 276, 459
  Range, 457, 477, 556
Spectroscope
Spectroscopy, 11, 269–271, 457–459, 554
  Absorption, 280, 458, 459
  Coherent anti-Stokes Raman, 279, 280
  Intracavity laser absorption, 459, 260
  Light scattering, 268, 269, 459
  Raman, 278, 280, 281, 457, 541
  Raman scattering, 114, 271
  Time resolved, 269, 271
  Time-resolved Raman, 279
  Ultrafast, 269, 271
Spectrum, 9, 23, 42, 80, 481, 493, 495, 510
  Continuous
  Electromagnetic
  Visible
Specular reflection, 70, 131, 148
Speed of light, 7, 8, 12, 35, 514
Spherical, 27, 101, 307
  Coordinates, 86, 101, 187, 240, 434
  Harmonics, 155–157
  Wave, 70, 74, 75, 429, 437
Spontaneous emission, 47, 49, 51, 52, 97, 271
Spotsize, 232
Spring, 98, 102–104, 106, 107, 112, 276
Spring constant, 102, 237
Squamous cell carcinoma, 307
Square wave, 562
Standard deviation, 126, 548
State
  Damage, 214
  Excited, 47–49, 56, 229, 271, 294, 303
  Ground, 47–49, 98, 271, 274, 303
  Metastabe, 47, 50, 532
  Singlet, 293–295, 298, 304, 532
  Triplet, 293–295, 298, 302, 304
Steady state, 140, 301, 303, 305, 331
Steady state, quasi, 232, 303, 353
Stefan-Boltzmann, 509, 510
Stellar, 4, 7, 13, 14
Stenosis, 355, 391, 414, 415
Step, 33, 103, 187, 191, 321
Sterilization, 220, 310, 345, 411, 416
Stern, 359
Stimulated emission, 9, 49–51, 481
Stimulation, 49, 318, 320
Stoichiometry, 281
Stokes, 272–275, 279
Stomach, 282, 283, 497, 564
Strain, 236, 237, 239, 247, 517, 519
Stress, 234, 236, 237, 239, 519
Striation, 254, 463
Subwavelength, 449–452, 456

Sudden cardiac death, 390
Sugar, 256, 545
Sunburn, 404, 555
Sunge, Edward Hutchinson, 448
Super Gaussian, 127
Super Lorentzian, 128
Superluminescent diode, 116, 474, 481
Superpulsed, 220
Surface, 27, 31, 36, 71, 79, 83, 126, 130, 430
Surface contour, 135
Surgery, cosmetic, 62, 378, 379
Surgery, eyelid, 378
Surrey, 318
Swammerdam, Jan, 426
Sweet, 543, 545, 547
Synchronization, 280
Synchronize, 280
Syndrome, twin-to-twin transfusion, 407

T

Tachycardia, ventricular, see Ventricular tachycardia
Tadpole, 563, 564
Takagi, 282
Tapered, 450–452, 559
Tapping mode, 452, 454, 455
Taste, 543–545, 547, 548
Taste transduction, 544
Tattoo, 61, 378, 379, 381
Tattoo removal, 379, 381
Taylor expansion, 155, 156, 158, 160, 471, 517
T-cell, 206
Tear-ducts, 5
Teeth, 359, 361, 362, 521, 554, 555
Telangiectasia, 403, 404, 407, 408, 410
Television, 24, 42, 44
Temperature, 12, 50, 134, 135, 211–214
  Average, 15, 235
  Coagulation, 226, 341
  Gradient, 211, 244, 245, 247, 406, 514
  Tissue, 135, 188, 212, 223
Temporal
  Change, 302
  Coherence, 36, 482
  Information, 267, 562
  Profile, 221, 223, 467, 505, 515
  Resolution, 221, 467
Tension, 7, 229, 237, 239
Terahertz, 526–528
Terahertz imaging, 18, 526–528
Testes, 465
Tetramethylrhodamineisothiocyanate, 553
Thermal
  Conductivity, 136, 211, 235, 236, 350, 358, 393, 404, 514, 523
  Damage, 130, 214, 253
  Denaturation, 15, 265, 326, 327
  Diffusion, 234, 235, 406, 517
  Diffusion coefficient, 235, 352
  Diffusion length
  Diffusion time, 379, 404, 515, 517
  Diffusivity, 136, 235, 264, 404, 514, 517
  Dispersion, 248
  Energy, 15, 102, 132, 239, 520
  Equilibrium, 50–52, 515
  Expansion, 233
  Injury, 264, 265, 327, 332
  Motion, 110, 272
  Properties, 362
  Relaxation time, 132, 264, 350, 404, 513, 515–517
Thermal vaporization, 264
Thermally confined, 261, 264, 353, 404, 405
Thermionic, 58
Thermodynamic, 235, 324, 350, 351, 394
  First law of
  Properties, 135, 333
  Second law of
  Third law of
Thermography, 509, 512
Thin-lens, 84, 88
Three dimensional, 30, 144, 353, 442, 564
  Gaussian, 372
  Geometry, 27, 101, 182
  Graphics, 182
  Grit, 463
  Image, 250, 441, 476, 494
  Rendering, 439
  Rotation matrix, 189
  Space, 30, 79
  Spectroscopy, 500
  Tissue block, 195, 396, 439, 446
  Vector, 35, 470

Three-flux, 165, 172, 173, 347
Threshold, 229
  ablation, 221, 223, 225, 230–323, 357
  condition, 52
Thrombosis, 296, 560
Time
  Delay, 486
  Dependent, 138, 208, 211, 229, 303, 330, 513
  Domain OCT, see time-domain Optical Coherence Tomography
  Domain Optical Coherence Tomography, 500, 502, 563
  Of-flight, 223, 467, 562
  Of-flight, optical, 561, 562
  Resolved, 208, 467
  Resolved spectroscopy, 271
  Resolved spectroscopy, 271, 279
  Retardation, 486, 491
Ti:Sapphire, see Titanium Sapphire
Tissue
  Properties
  Properties, thermal, 135
  Soft, 211, 351, 353, 380, 402, 349–359
  Welding, 333, 334, 377
Titanium Sapphire, 20, 58, 60, 521, 523
TMLR, see transmyocarial revascularization
Toepler imaging
Toepler, August, 17
Tomography, diffuse optical, 468, 469
Tomography, transillumination, 468, 471
Tongue, 543
Tonsil, 281, 415
Tooth, 241, 359, 360, 362, 521–524, 555
Topography, 117, 448, 456, 463, 554
Total attenuation coefficient, 139, 145, 200, 205, 211
Townes, Charles, 9
Trabeculoplasty, 363, 413
Transdermal, 62
Transfer, heat, 14, 211, 352, 515
Transfusion syndrome, twin-to-twin, 407
Transillumination, 18, 244, 464, 465, 467, 568
Transillumination tomography, 425, 469
Transmission, 81, 82, 148, 198–201, 207, 268, 371, 504
Transmittance, 145, 148, 171, 198–200,
Transmyocarial revascularization
Transparent, 62, 365, 357, 389
Transport coefficient, 162, 205, 209
Transport cross-section, 162
Transport theory, 137, 138, 155, 158, 166, 513
Transurethral resection of the prostate, 417–420
Trebes, James, 462
Triplet state, 293, 298, 302, 304, 533
TRITC, see Tetramethylrhodamineisothiocyanate
Tryptophan, 291
Tumor, 15, 269, 292, 294, 296, 305, 307, 308
Tumor, embryonic, 407, 409
Tuning fork, 452–455
Turbid media, 13, 85, 121, 140, 167, 504
TURP, see transurethral resection of the prostate
Tweezer, optical, 364, 367, 373
Twin-to-twin transfusion syndrome, 407
Two-dimensional, 144, 247, 439, 491
  Attenuation, 142
  Diffusion equation, 140, 190
  Fourier transform, 91, 491
  Geometry, 13, 434
  Graphics, 182
  Image, 449, 478
  Phase distortion, 208
  Plane, 182
  Surface, 79
  Tissue block, 195
  Wave equation, 90
  Wave, 91
Two-flux, 166, 199, 201
Two-photon, 228, 442, 555
Tyndall, John, 12

U

Ultrafast spectroscopy, 271
Ultrasound, 238, 241, 476, 518
Ultraviolet, 43, 46, 57, 100, 228, 380, 531
Universal gas constant, 213, 323, 350
Urethra, 283, 417, 418, 420

Urology, 283, 416–418
Uvea, 5
Uvula, 281, 414, 415
Uvulopalaplasty, 414, 415

V

Valve debridement, 354, 355
Van Leeuwenhoek, Antonie, 7, 426
Vaporization,220, 221, 223, 264, 350, 357, 418, 514
Vaporized water volume, 223
Vascular, 62, 305, 327, 377, 403, 511
Vascular welding, 334, 354, 377
Vascularized, 61
Vasculature, 62, 296
Vasodilation,
VECSEL, see Vertical external cavity surface emitting semiconductor laser
Vector, 29, 30, 36, 79, 87, 91, 111, 122, 184, 370, 431, 470
Vector, unit, 36, 79, 80, 130, 145, 190, 251, 431
Vein, 355, 564
  Varicose, 62, 407
Velocimetry, 559
Velocity
  Ablation, 223, 224, 226, 351–353
  Angular, 25
  Displacement, 242, 247, 515, 519
  Electron, 61
  Flow, 116, 498, 499, 564
  Group, 38, 41, 500
  Linear, 25
  Mirror, 475
  Observer, 116
  Particle, 240, 360
  Phase, 24, 38, 41, 272, 500, 502
  Potential, 240
  Profile, 116
  Rotational, 352
  Source, 116
  Vaporization, 350
Venetian, 6
Ventricular tachycardia (VT), 390, 391, 394
Venus welding, 377
Vericose vein, 62, 407
Vertical external cavity surface emitting semiconductor laser, 459, 460
Vessel, 116, 305, 405, 406
vibration, two-mass, 104, 106, 107, 112, 209, 272
Virtual, 83, 426, 526, 527
Visibility, 245, 252
  Function, 483, 484, 490–494
  Fringe, 482, 489, 490
Visible spectrum, 42, 43, 69, 85, 100, 400
Vision, 4–7, 12, 336, 345, 412
  Color, 43
  Polarized, 251, 252
Vitreous humor, 5, 500
Voltage sensitive dye, 570, 572
Volume, 130, 139, 158, 213, 224, 239, 366, 519, 565
Von Fraunhofer, Joseph, 11, 76
Von Helmholtz, Ludwig Ferdinand , 8
Von Tappeiner, Herman, 291, 292
Vorosmarthy, Daniel, 15
VT, see Ventricular tachycardia

W

Waist, 124, 148, 516, 521
Waist length, 124, 125
Wart, 382, 283, 416
Wave, 9, 24–26, 30
  Acoustic, 233, 241, 512, 519
  Square, 562
Wavelength, 25, 26, 28, 525, 526
Wavenumber, 25, 38, 122, 148, 273
Wavicle, 12, 13
Webb, 319
Webb, Robert, 363, 413
Webb, Watt W., 442
Weight, 134
  Photon, 177, 182, 190, 193
  Radiation, 50
Weishaupt, Doughery K., 298
Welding, tissue, 333, 334, 337, 418
welding, vascular, 334, 377
welding, venus, 377
Wentzel-Kramers-Brillouin, 302
Wien, Wilhelm, 28
Wien's displacement law
Wiggler, 38, 58–61, 511
WKB, see Wentzel-Kramers-Brillouin
Wood's lamp, 531, 532

Wound, 220, 345
  Healing, 289, 290, 312, 313, 319, 333
  Sterilization

X

XCl, see Xenon Chloride
Xenon Chloride, 57
Xenon light source, 115, 310
X-ray, 17, 43, 400, 401, 504
X-ray radiography, 466, 476

Y

Young, 10
Young, Thomas, 10
Young's modulus, 236–239, 517, 518, 524

Z

Zinc(II)-naphtoalocyanine, 299
Zinc-Sulfide, 259, 264, 508
ZnS, see Zinc-Sulfide

www.ingramcontent.com/pod-product-compliance
Ingram Content Group UK Ltd.
Pitfield, Milton Keynes, MK11 3LW, UK
UKHW051941210425
457613UK00026BA/82

9780750309387